Analytical Chemistry of
Polycyclic Aromatic Compounds

Analytical Chemistry of Polycyclic Aromatic Compounds

MILTON L. LEE
Department of Chemistry
Brigham Young University
Provo, Utah

MILOS V. NOVOTNY
Department of Chemistry
Indiana University
Bloomington, Indiana

KEITH D. BARTLE
Department of Physical Chemistry
University of Leeds
Leeds, United Kingdom

1-3
55-60
62-73
98-102
123-186
242-286
329-349

1981

ACADEMIC PRESS
A Subsidiary of Harcourt Brace Jovanovich, Publishers
New York   London   Toronto   Sydney   San Francisco

ACADEMIC PRESS, INC.
111 Fifth Avenue, New York, New York 10003

United Kingdom Edition published by
ACADEMIC PRESS, INC. (LONDON) LTD.
24/28 Oval Road, London NW1 7DX

Library of Congress Cataloging in Publication Data

Lee, Milton L.
  Analytical chemistry of polycyclic aromatic compounds.

  Includes bibliographical references and index.
  1. Polycyclic compounds--Analysis. 2. Aromatic
compounds--Analysis. I. Novotny, Milos, Date.
II. Bartle, Keith D. III. Title.
QD335.L48   547'.5   80-68559
ISBN 0-12-440840-0

PRINTED IN THE UNITED STATES OF AMERICA

81 82 83 84   9 8 7 6 5 4 3 2 1

Contents

8   Mass Spectrometry

9   Ultraviolet Absorption and Luminescence Spectroscopy

10   Nuclear Magnetic Resonance and Infrared Spectroscopy

11   Approaches to Problem Solving in PAC Analysis

Preface

THE RAPID GROWTH in industry over the last century and particularly within the last several decades has required more critical evaluations of environmental health hazards associated with combustion products and effluents from these industries. The recent increase in the utilization of coal and other synthetic fuels to meet changing world energy demands places even greater emphasis on these evaluations. Fortunately, developments in analytical methodology and instrumentation have largely paralleled industrial growth, and many of the chemical factors relating to the environment have been identified. Among these are the polycyclic aromatic compounds, the largest class of chemical carcinogens known today.

The carcinogenic and mutagenic properties of numerous polycyclic compounds have been documented, and many others are presently under investigation. The modern age of chemical instrumentation has already had a significant impact in areas related to this class of compounds: environmental toxicology, experimental carcinogenesis, environmental chemistry, chemical and fuels engineering, etc. The known health hazards associated with the increasing emission of polycyclic aromatic compounds into our environment, along with the developing societal environmental awareness dictate the need for both further structural identification and more accurate and precise quantitative measurement of these substances.

Although many excellent books have been written on various aspects of carcinogenesis, and although both biological and chemical properties of polycyclic aromatic compounds have been discussed at many international symposia, it is our belief that such important subjects as the chemical separation, the structural identification, and the quantitative measurement of these compounds had not been drawn together and treated in a comprehensive manner.

This book is mainly devoted to the discussion and critical evaluation of various chromatographic (Chapters 5–7) and spectroscopic (Chapters 8–10) methods. It is shown how gas chromatography and high-performance liquid chromatography can be both competitive and complementary analytical methods. Ancillary techniques of both are emphasized for the structural elucidation of individual polycyclic aromatic compounds in complex mixtures. The merits of spectroscopic methods in both structural work and quantitation are compared. New directions for these analytical techniques are also discussed.

Chapters 1–3 are descriptive in nature and provide the reader with background information concerning the chemistry, occurrence, and toxicology of polycyclic aromatic compounds. These chapters were included to give the reader basic information pertinent to the understanding and appreciation of the analytical chemistry of these compounds.

The isolation of polycyclic aromatic compounds from a wide variety of materials and matrices necessitates a number of different approaches; and therefore sample collection, extraction, separation, and purification are discussed in Chapter 4. Numerous methods and examples which cover most applications are discussed.

An important point stressed in this book is that no one analytical technique is sufficient to solve all analytical problems associated with this class of compounds; and that therefore, various multi-technique approaches are required. This is demonstrated in Chapter 11 with two practical examples representative of the widely differing problems that may be encountered. This book does not pretend to have the universal analytical solution to all studies involving polycyclic aromatic compounds. Instead, the different analytical techniques and approaches are treated in some detail in separate sections and their advantages and disadvantages explained. It is hoped that sufficient information is provided concerning the chemical and physical properties of the polycyclic aromatic compounds as well as the principles behind the various analytical techniques so that the reader can approach any particular analytical problem with sufficient understanding. The correct nomenclature of the polycyclic aromatic compounds is contained in the appendices, and should provide a valuable resource for those new to this class of compounds.

This book is recommended to all scientists involved with the study of polycyclic aromatic compounds, to analysts who need to acquire routine data, as well as to individuals charged with formulating environmental policies and drafting regulations. In addition, this book may also be appreciated by engineers concerned with emission-control and energy-related industries.

The timely completion of this book was aided by fellow scientists who gave permission to reproduce their data and who often provided as yet unpublished manuscripts and valuable criticisms. We would also like to thank Mary Fencl, Paul Peaden, Dan Vassilaros, Cherylyn Willey, and Bob Wright, for valuable help in assembling and proofreading the text, compiling the information in the appendices, and preparing the figures for reproduction. The untiring assistance of Ms. Peggy Gore is gratefully acknowledged for typing and retyping the text, and for invaluable assistance in the preparation of the manuscript. Finally, we are grateful to our families who gave the necessary encouragement and provided an atmosphere conducive for us to initiate and complete this book.

Milton L. Lee
Milos Novotny
Keith D. Bartle

1

Physical and Chemical Properties

I. NOMENCLATURE

Polycyclic aromatic compounds (PAC) have been studied for well over a century, and during this time many compounds have been named unsystematically. Some names reflect the initial isolation of compounds from coal tar (naphthalene, pyrene, etc.); some reflect their color (fluoranthene and chrysene—the latter erroneously, because of contamination with naphthacene which is orange); and some reflect the shape of their molecules (coronene, ovalene). Such names passed into general use, and it proved impractical to change them when systematic nomenclature was introduced. Thus, many important PAC systems are named nonsystematically.

Recently, IUPAC (International Union of Pure and Applied Chemistry) attempted to systematize PAC nomenclature, prefixing to the name of a parent ring system the names of other component parts (*1*). Appendices 1 through 4 list the names and structures of many of the hydrocarbons and their heterocyclic analogs discussed in the remainder of the book. An exhaustive list of PAC is contained in "The Ring Index" (*2*). All the ring systems used in organic chemistry are classified according to the number and identity of the atoms in each ring and are named and numbered.

The numbering and names of the compounds in Appendices 1–4 are based on the following rules:

1. Rings are drawn with two sides vertical wherever possible.
2. Irrespective of their size, as many rings as possible are drawn in a horizontal line.
3. As much as possible of the rest of the structure is arranged in the top right quadrant and as little as possible in the bottom left quadrant (the middle of the first row is taken as the center of the circle).

4. Starting with the first carbon atom not engaged in ring fusion in the right-hand ring of the top row, numbering proceeds clockwise around the molecule (anthracene and phenanthrene are exceptions).

5. Atoms engaged in ring fusion are given the letters *a*, *b*, *c*, etc., after the number of the preceding atom, e.g., triphenylene (**1**):

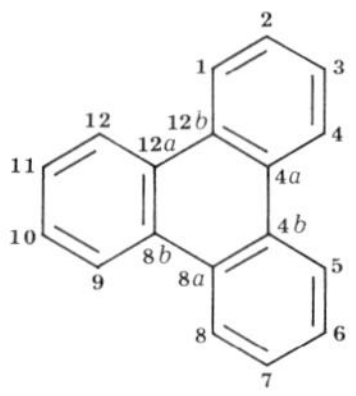

(I)

6. Certain trivial names are retained. Otherwise the name of a fused-ring system is made up of a prefix of the fixed part (benzo, cyclopenta, or a group of rings such as indeno—see abbreviations listed in Appendix 1) followed by an italic letter or letters denoting the bond or bonds of the base (which has as many rings as possible) at which fusion occurs. *a* refers to the 1,2-bond, and all bonds are then lettered sequentially whether or not they carry hydrogen atoms; the name of the parent compound follows. Examples are benz[*a*]anthracene (**2**) and 4*H*-cyclopenta[*def*]phenanthrene (**3**).

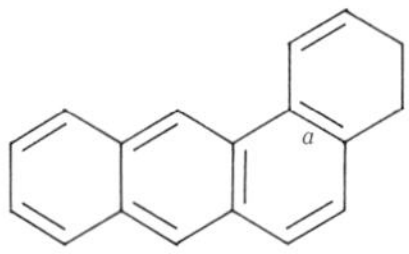

(2) 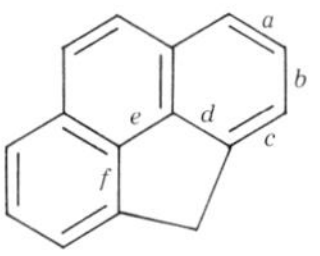   (3)

If more than one ring is fused, the italic letters are separated by a comma, e.g., dibenz[*a,c*]anthracene (**4**).

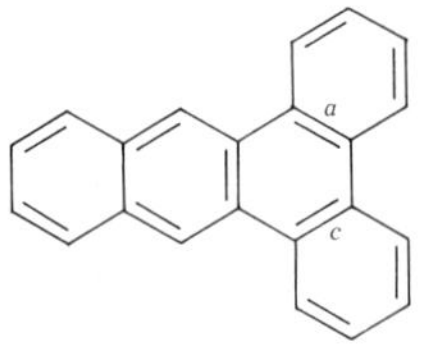

(4)

The fusion position on the first ring system is shown, if necessary, by the appropriate numeral which precedes the italic letters, e.g., indeno[1,2,3-*cd*]-pyrene (**5**).

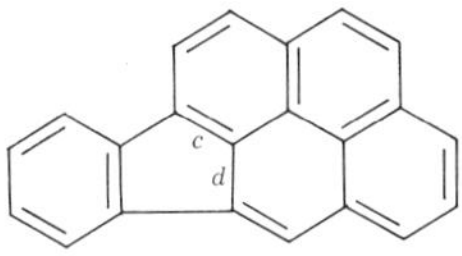

(5)

7. A component CH_2 of a ring is indicated by the italic *H* preceded by the appropriate numeral, except where its position is assumed, as for example in the compounds trivially named indene and fluorene (Appendix 1), e.g., 1*H*-benz[*de*]anthracene (**6**). Where there is a choice, the carbon atom carrying an indicated hydrogen atom is numbered as low as possible.

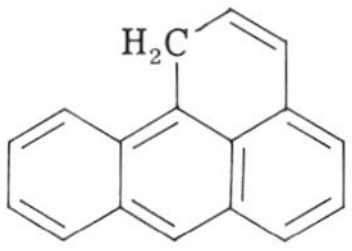

(6)

8. Hydrogenation is denoted by prefixes such as dihydro, etc., followed by the name of the corresponding unreduced hydrocarbon, e.g., 1,4-dihydronaphthalene (**7**):

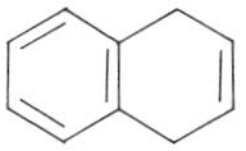

(7)

An even greater variety of structure is presented by the heterocyclic analogs and derivatives of polycyclic aromatic hydrocarbons (PAH), even if only compounds containing a single heteroatom are considered (Appendices 2, 3, and 4). Similar rules of nomenclature apply as outlined above, since, in general, the name is composed of a hydrocarbon ring system followed by a trivial heterocycle name. Numbering also follows the rules for hydrocarbons; where there is a choice of orientation, low numbers are assigned to heteroatoms. In an alternative IUPAC-approved scheme, heterocyclic systems may be named by prefixing aza (N), oxa (O), or thio (S) to the name of the corresponding hydrocarbon: thus benz[*g*]isoquinoline is named 2-azaanthracene.

II. PHYSICAL PROPERTIES

The physical and spectroscopic properties of PAC are dominated by the conjugated π-electron systems which also account for their chemical stability.

   All PAC, with the exception of a few hydrogenated derivatives, are solids at ambient temperatures and are the least volatile of the hydrocarbons. PAC occur in air, mainly adsorbed on particles (Chapter 2). The boiling points of PAC are markedly higher than those of the *n*-alkanes of the same carbon number. Nonetheless, losses from environmental PAC samples are probable without stringent precautions (*3*). Particulates from urban air stored in an open container, but in the dark, showed the following losses (*4*): pyrene 88%, benzo[*a*]pyrene 32%, benzo[*ghi*]perylene 10%, but coronene only 1%.

   The availability of high-energy π-bonding orbitals and of relatively low-energy π*-antibonding orbitals in PAC leads to the absorption of visible or ultraviolet (UV) radiation by the transition of an electron from the the π- to π*-orbital which gives characteristic absorption and fluorescence spectra. The processes occurring when UV or visible light is absorbed by PAC are illustrated in Fig. 1-1. Excitation E takes the molecule from the ground-state singlet S_0 to the first excited state S_1. In condensed systems, any excess vibrational energy is lost within a picosecond by transfer to neighboring ground-state molecules. From the lowest vibrational level of S_1 the molecule may either: (a) return to the ground-state S_0 by radiationless internal conversion (IC); (b) return to S_0 by fluorescent emission (F); or (c) be converted to the excited vibrational levels of the lowest triplet state T_1, and then by very rapid (less than 1 ps) internal conversion (intersystem

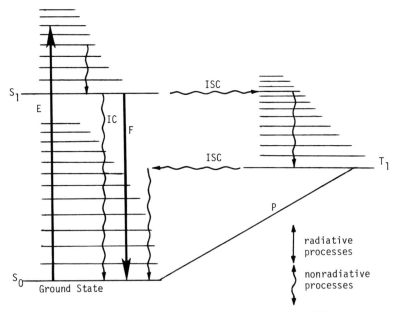

Fig. 1-1.   Electronic energy levels and transitions for PAC.

crossing—ISC) go to the lowest vibrational level of T_1. The lifetime of S_1 is short (10–100 ns for PAH), but process (a) above is generally unimportant: for anthracene in ethanol at 20°C, about 30% of the excited singlet disappears by the $S_1 \rightarrow S_0$ fluorescence process and 70% by the $S_1 \rightarrow T_1$ process.

Once in the triplet state, the radiative process $T_1 \rightarrow S_0$, called phosphorescence (P), is slow because the transition involves a change in spin multiplicity and is "forbidden." Weak phosphorescence is often observed, however, and the variation with time of the intensity of the phosphorescence emission may be used to measure the lifetime of the triplet state.

The energy differences between S_0 and S_1, and therefore the wavelengths of exciting radiation, depend on the separations between the various molecular orbitals. Certain features of their UV/visible spectra thus are common to all PAC and can be interpreted as follows. The strong absorption bands designated (5, 6) α, ρ, and β, which occur in that order at decreasing wavelengths, have extinction coefficients (E_{max}) of generally $\sim 10^2$, 10^4, and $10^5\ M^{-1}\ cm^{-1}$, respectively. The ρ-band is assigned (Fig. 1-2) to transitions from the highest occupied molecular orbital to the lowest unoccupied molecular orbital, and the α- and β-bands to transitions from the next highest occupied to lowest unoccupied molecular orbital, and from the highest occupied to the next higher unoccupied molecular orbital.

The phenomenon named annellation by Clar may be used (5, 6) to explain why the α-, ρ-, and β-bands retain their characteristic features while shifting towards the red with ring number. Briefly, in fused-ring PAH, some rings give

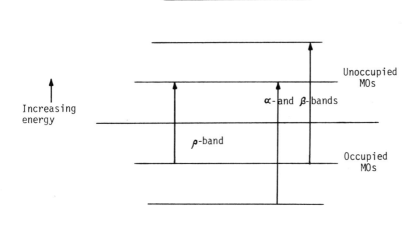

Fig. 1-2. Electron transitions corresponding to α-, β-, and ρ-bands in UV/visible spectra of PAH.

up part of their aromaticity to adjacent rings, and the physical and chemical properties of a system then depend on the number of aromatic sextets, denoted by a full circle. In this way, one of the sextet structures of phenanthrene (**8**) emphasizes the reactivity of the 9,10-bond, while conversion of anthracene to the 9,10-endoperoxide (Fig. 1-3) yields a system with two aromatic sextets.

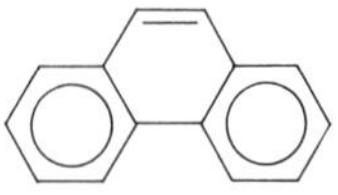

(8)

Annellation can also be invoked to explain similarities between the electronic spectra of benzologs of certain PAH, especially in the pyrene series: dibenzo[de,qr]naphthacene, dibenzo[fg,st]pentacene, and dibenzo[hi,uv]-hexacene all have similar UV spectra (Fig. 1-4). In this way, the parent ring system of an unknown PAH may be recognized.

The characteristic low-field chemical shift of resonances of the protons of PAC in their ^{1}H-NMR spectra (Chapter 10) may also be discussed in terms of the annellation principle (5). However, the more usual approach is via the ring-current concept (7). A magnetic field induces a circulation of π-electrons in delocalized molecular orbitals extending over the rings. This induced ring current has associated with it a magnetic field, which in the region of the peripheral aromatic hydrogens of PAC and (less strongly) of methyl protons of methyl derivatives reinforces the main (applied) field and brings about a deshielding (Fig. 1-5). The magnitude of this chemical shift depends on the intensity of the ring currents and on the proximity of the given proton to the current loops. Detailed geometrical procedures for calculating ring current chemical shifts in PAC have been published (8), but in qualitative terms there is a sequence of increasing downfield shifts, which is also followed by the protons of substituent methyl groups (9).

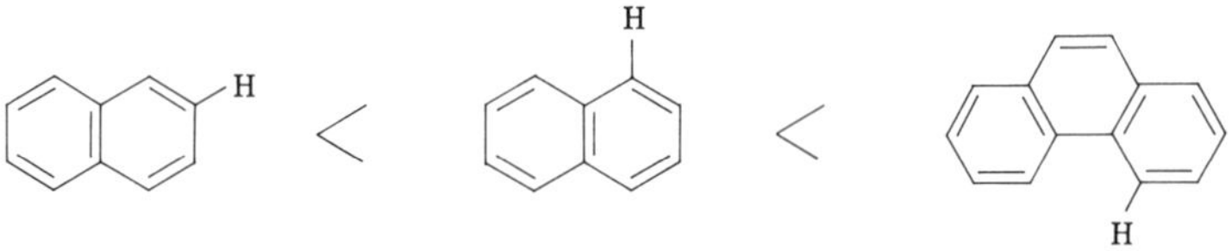

Spin–spin coupling between such protons and the aromatic protons of PAC (benzylic coupling) (10) and between the aromatic protons themselves (11) depends on the intervening molecular orbital π-bond order.

The removal of π-electrons from PAC by the impact of comparatively low-energy electrons and the stability of the ions so produced lead to the characteristic mass spectra of these compounds (Chapter 8). Also, because the π-electrons of PAC are in high-energy orbitals, they are available for

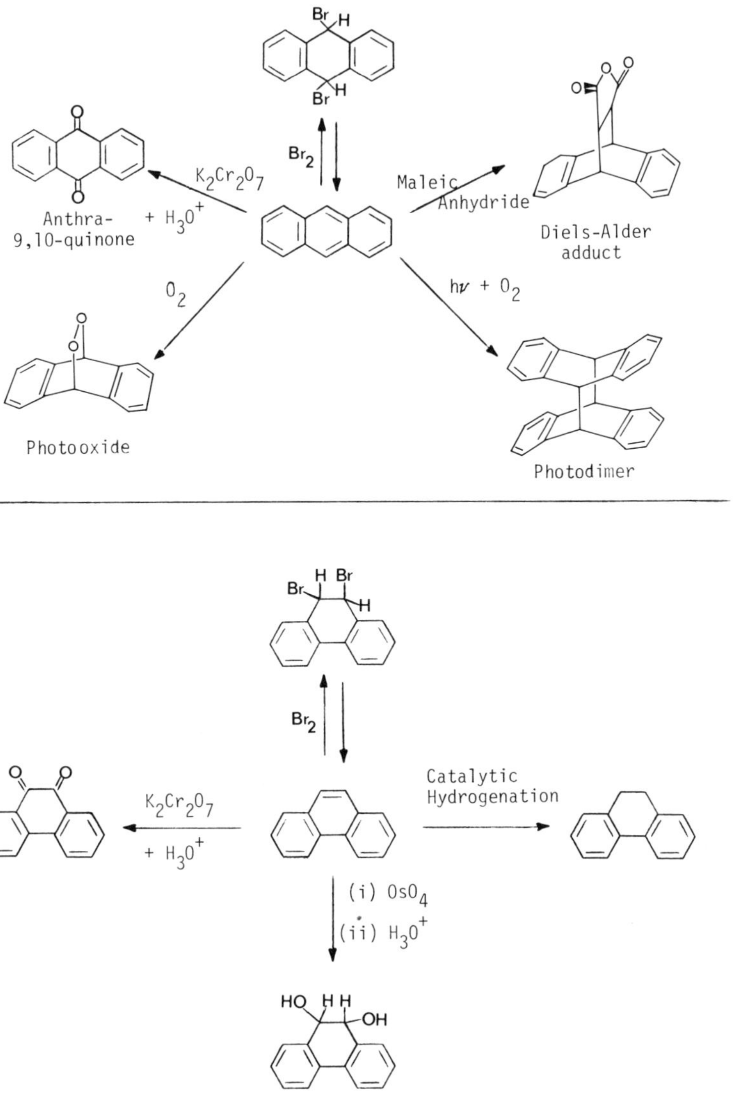

Fig. 1-3. Addition and redox reactions of anthracene and phenanthrene.

sharing. PAC act as donors in charge-transfer molecular π-complexes with a variety of acceptors with low-energy vacant orbitals (*12, 13*): polynitroaromatics, nitriles, quinones (with electron withdrawing groups, etc.), anhydrides, tetracyanoethylene, etc. The best characterized PAC complexes, usually with 1:1 mole ratio, but sometimes 1:2 or 2:1 stoichiometry, are

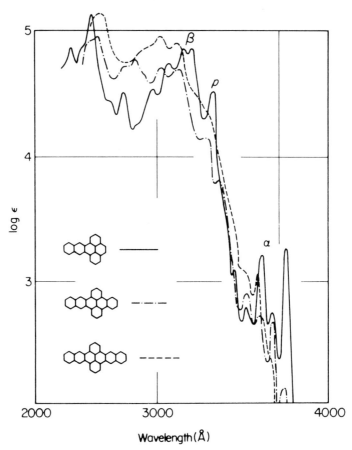

Fig. 1-4. UV spectra of dibenzo[*de,qr*]naphthacene, dibenzo[*fg,st*]pentacene, and dibenzo-[*hi,uv*]hexacene. (Reproduced with permission from E. Clar, *The Aromatic Sextet*, Wiley, London, 1972, p. 74. Copyright, John Wiley and Sons.)

those with picric acid and 2,4,7-trinitrofluoren-9-one, which are often stable enough to be characterized by melting point. The formation of these compounds involves complete transfer of an electron, but other acceptors may interact more weakly with less complete orbital overlap. The complex is too unstable to be isolated in the pure state and exists only in solution in equilibrium with its components; detection is then usually by difference in a physical property, such as the UV/visible absorption spectrum (*12, 13*).

The use of donor–acceptor complexes in the separation of PAC is mentioned in Chapters 4, 5, and 6, but the retention of PAC by column and thin-layer chromatographic materials such as silica gel and alumina, and by

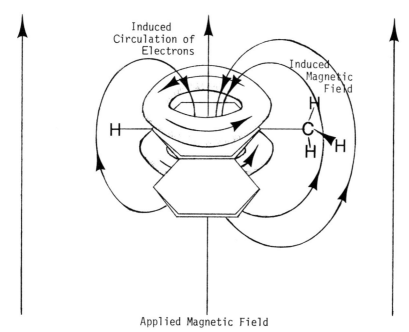

Fig. 1-5. The ring-current effect for 1-methylnaphthalene.

polar gas and liquid chromatographic packings is a manifestation of weaker associations originating in interactions with the polarizable π-electron systems. In the same way, the preferential solubilities of PAC in polar solvents such as nitromethane and dimethyl sulfoxide are results of interactions of the latter kind (Chapter 4).

On the other hand, the high molecular weights of PAH and the absence of polar substituent groups make these compounds very insoluble in water. There is a rough correlation between increasing molecular weight and decreasing solubility (Table 1-1). Linearly fused PAH are less soluble than their angular isomers. However, micelle formation by detergents and other soluble organic compounds such as caffeine at levels of only 10–50×10^{-3} g/liter can increase the solubilities of PAH by up to a factor of 10 (*14*, *15*). Nonionic detergents are better solubilizing agents than ionic detergents (*16*). In general, concentrations of detergent in industrial wastes are normally insufficient to bring about micelle formations, but the presence of organic solvents in polluted water can markedly increase the solubilities of PAC, particularly if emulsions are formed. Moreover, sorption of PAC on suspended particles such as mineral matter, sediments, and soil in natural waters may also lead to higher apparent concentrations (*17*, *18*).

TABLE 1-1

Solubilities of PAH in Water[a]

Compound	Solubility (μg/liter)			
	(19)[b]	(20)[b]	(21)[b]	Other
Naphthalene	31,700	30,200		
1-Methylnaphthalene	28,500	30,000		
2-Methylnaphthalene	25,400			
1,3-Dimethylnaphthalene	8,000			
1,4-Dimethylnaphthalene	11,400			
1,5-Dimethylnaphthalene	3,380			
2,3-Dimethylnaphthalene	3,000			
2,6-Dimethylnaphthalene	2,000			
1-Ethylnaphthalene	10,700	9,980		
1,4,5-Trimethylnaphthalene	2,100			
Biphenyl	7,000			
Acenaphthene	3,930			
Fluorene	1,980			
Phenanthrene	1,290	1,150	1,002	
1-Methylphenanthrene			269	
Anthracene	73	40	45	
2-Methylanthracene	39		21	
9-Methylanthracene	261			
9,10-Dimethylanthracene	56			
Pyrene	135	129	132	
Fluoranthene	260		206	
Benzo[a]fluorene	45			
Benzo[b]fluorene	2.0			
Chrysene	2.0		1.8	
Triphenylene	43		6.6	
Naphthacene	0.6			
Benz[a]anthracene	14		9.4	
7,12-Dimethylbenz[a]anthracene	61			
Perylene	0.4			
Benzo[a]pyrene				0.01 (22), 0.05 (23)[c] 0.4 (24)
Benzo[e]pyrene	3.8	5.5		
Benzo[ghi]perylene	0.3			
Coronene	0.1			

[a] The italic numbers in parentheses are reference numbers.
[b] At 25°C.
[c] At 20°C.

III. CHEMICAL PROPERTIES

PAC are classed chemically as rather inert compounds (25). Thus we may cite the widespread distribution in the geosphere, discussed in Chapter 2, of perylene and various alkyl phenanthrenes, stable end-products of the degradation of biochemicals (26). However, when PAC do react, they tend to retain their conjugated ring systems by forming derivatives by electrophilic substitution rather than by addition; PAH chemistry has been reviewed at length most recently by Clar (6). Earlier summaries also exist (27, 28), but an authoritative sequel to one of these only extends to the chemistry of naphthalene (29).

Two series of addition reactions are of special interest in a consideration of the environmental chemistry of PAH. These are respectively the reactions undergone by the 9,10-positions of anthracene and phenanthrene (Fig. 1-3). Thus anthracene reacts to give Diels–Alder adducts at these positions and in the presence of light reacts with oxygen to give an endoperoxide; when irradiated in the absence of oxygen, it dimerizes. By contrast, phenanthrene is inert to maleic anhydride under normal conditions and does not form a photodimer; however, phenanthrene does undergo certain addition reactions readily at the 9,10-positions, and, like anthracene, is oxidized to the 9,10-quinone (Fig. 1-3). These marked differences between the linearly and angularly fused tricyclic PAH are maintained with the increasing ring number of benzologs. Moreover, down the series of linear PAH to the seven-ring heptacene there is an increasing tendency to undergo addition and redox reactions; the angular PAH, while retaining the properties typified by the 9,10-bond of phenanthrene, show little tendency to become more reactive.

The oxidation of PAH to endoperoxides is particularly relevant to possible losses during analysis. There have been many reports of photochemical transformation when adsorbed on a variety of support materials such as filters, carbon particles, and chromatographic media (4). For example, the photooxidation of anthracene adsorbed on alumina and silica gel by UV light is well-known (30); a number of nitrogen-containing compounds have also been detected in the product anthra-9,10-quinone (31). Inscoe made a detailed study of the stability of 15 PAH adsorbed on thin-layer chromatographic supports to UV irradiation in air and found that the angular phenanthrene, chrysene, picene, and triphenylene were stable, but that compounds with anthracenic structure—anthracene, benz[a]anthracene, naphthacene, and dibenz[a,c]- and -[a,h]anthracenes—showed changes which also were more marked for spots of the more condensed hydrocarbons: pyrene, benzo[a]- and -[e]pyrenes, perylene, benzo[ghi]perylene, and coronene. The spots also lost their fluorescence (32). The changes were less evident on cellulose and acetylated cellulose than on alumina and silica gel, and the chief products of reaction were confirmed as quinones from the isolation of pyrenediones from pyrene (*32*). Similar changes are brought about by sunlight, although more slowly (*4*). More recently, Hellmann (*33*) has investigated the change in fluorescence intensity with time of PAH adsorbed on TLC layers and has confirmed the susceptibility of PAH to oxidation on silica gel (Fig. 1-6). Oxidation is much reduced on mixed plates of alumina and acetylcellulose. Benzo[*a*]pyrene is particularly prone to photooxidation on silica gel (*34*) (Fig. 1-6).

Several quinones, including benzo[*a*]pyrene-1,6-, -3,6-, and -6,12-diones have been identified in urban air particulates (*35*), and such compounds may have properties which may account, in part, for the extra carcinogenicity of air particulates in comparison with that of the identified PAH (*36*). Katz *et al.* (*37*) showed that photolysis of a variety of PAH on cellulose under simulated atmospheric conditions followed first-order kinetics with half-lives of the order of hours (Table 1-2), and that the above benzo[*a*]pyrene diones could be isolated from the products of photolysis of the hydrocarbon. When dispersed into atmospheric particulate matter, anthracene is photooxidized by air and sunlight to a variety of products (Fig. 1-7), including the 9,10-endoperoxide and various quinones (*38*). On the other hand, the photooxidation of PAH, other than those with benzylic carbon atoms, is markedly inhibited if they are adsorbed on coal fly-ash particles (*39*).

Based on model experiments (*40*), the formation of PAH quinones in air has been attributed to the action of ozone, but in experiments in which HPLC

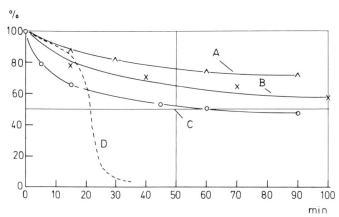

Fig. 1-6. Decay of the fluorescence intensity (365/445 nm) for PAH on various adsorbents. (A) 20 ng fluoranthene on silica gel (Kieselgel F-60); (B) 500 ng environmental PAH mixture on paper; (C) 20 ng benzo[*ghi*]perylene on Kieselgel F-60; and (D) 30 ng benzo[*a*]pyrene on Kieselgel F-60. (A), (B), and (C) reproduced with permission from M. Hellmann, *Z. Anal. Chem.* **295**, 24 (1979). Copyright, Springer-Verlag. (D) reproduced with permission from B. Seifert, *J. Chromatogr.* **131**, 417 (1977). Copyright, Elsevier Scientific Publishing Co.]

TABLE 1-2

Half-Lives (in hr) of Polycyclic Aromatic Hydrocarbons under Simulated Atmospheric Conditions[a]

	Simulated sunlight	Simulated sunlight + ozone (0.2 ppm)	Dark reaction + ozone (0.2 ppm)
Anthracene	0.20	0.15	1.23
Benz[a]anthracene	4.20	1.35	2.88
Dibenz[a,h]anthracene	9.60	4.80	2.71
Dibenz[a,c]anthracene	9.20	4.60	3.82
Pyrene	4.20	2.75	15.72
Benzo[a]pyrene	5.30	0.58	0.62
Benzo[e]pyrene	21.10	5.38	7.60
Benzo[b]fluoranthene	8.70	4.20	52.70
Benzo[k]fluoranthene	14.10	3.90	34.90

[a] From Katz et al. (37).

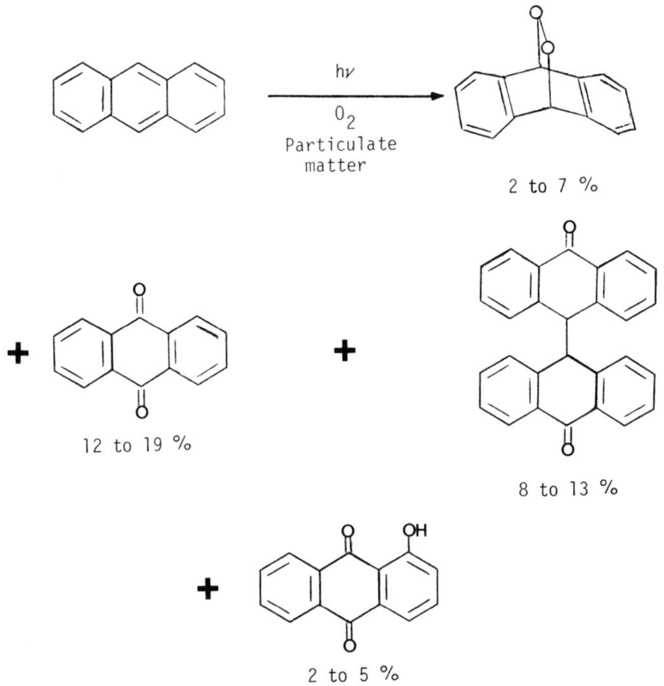

Fig. 1-7. Products of photooxidation of anthracene on atmospheric particulate matter. Percentage yields are based on consumed anthracene. [According to M. A. Fox and S. Olive, *Science* **205**, 582 (1979).]

rather than TLC was used in the separation step, no quinones were detected in the products of reaction of benzo[a]pyrene and ozone (*41*). Photooxidation of PAC may also take place through the action of active oxygen $O(^3P)$ (*42*), which can react with a double bond of an aromatic compound, but it is generally believed that singlet oxygen ($^1O_2$, excited molecular oxygen) is the key intermediate (*38, 43*), especially for derivatives of anthracene (*44*). An alternative route to PAH quinones in the atmosphere is via nitro derivatives. The formation of the latter, which are themselves strong mutagens, has been demonstrated (*41*) by exposure of PAH to ppm levels of NO_2, one of the gaseous components of photochemical smog. PAH quinones are also constituents of cigarette smoke condensate—especially anthracene-9,10-diones and 2,3,6-trimethylnaphthalene-1,4-dione (*45*).

Many PAC are transformed in water under the influence of sunlight, but there are conflicting accounts of the importance of singlet oxygen in these reactions (*37, 46*). Products of ozonation reactions in aqueous solutions may also be of particular interest (*47*) because of the use of ozone in the purification of chemical waste waters. Relatively short contact times with ozone may remove a significant proportion of the PAH present in an aqueous solution (*14*).

The most potent carcinogens among the PAC usually undergo addition and redox reactions at the equivalent of the phenanthrene 9,10-bond. Such reactions led to the proposal of the "K region" as reaction center in carcinogenesis (*48*); thus the K region of benzo[a]pyrene is the 4,5-bond. More

recent studies have established that many of the carcinogenic properties of PAC are associated with metabolites (Chapter 3) such as arene oxides, hydroxy and dihydroxy derivatives, dihydrodiols, and quinones (*49*). Numerous K-region derivatives of this kind may be prepared (*50*) by first reacting osmium tetroxide with PAC to yield the osmate ester (e.g., Fig. 1-3).

REFERENCES

1. International Union of Pure and Applied Chemistry, "Nomenclature of Organic Chemistry," Section A, 3rd ed. (1971). Butterworths, London, 1977.
2. A. M. Patterson and L. T. Capell, *in* "The Ring Index," 2nd ed., Am. Chem. Soc., Washington, 1960, with subsequent supplements.
3. R. Tomingas, *Fresenius Z. Anal. Chem.* **297**, 97 (1979).
4. E. Sawicki, *Crit. Rev. Environ. Control* **1**, 280 (1970).

5. E. Clar, "The Aromatic Sextet." Wiley, New York, 1972.
6. E. Clar, "Polycyclic Hydrocarbons." Academic Press, New York, 1964.
7. J. D. Memory, "Quantum Theory of Magnetic Resonance Parameters," p. 127. McGraw-Hill, New York, 1968.
8. C. W. Haigh and R. B. Mallion, *Org. Magn. Reson.* **6**, 203 (1972).
9. K. D. Bartle and D. W. Jones, *J. Phys. Chem.* **73**, 293 (1969).
10. K. D. Bartle, D. W. Jones, and R. S. Matthews, *Rev. Pure Appl. Chem.* **19**, 191 (1969); *Tetrahedron* **25**, 2701 (1969).
11. M. A. Cooper and S. L. Manatt, *J. Am. Chem. Soc.* **91**, 6325 (1969).
12. L. J. Andrews and R. M. Keefer, *in* "Molecular Complexes in Organic Chemistry." Holden-Day, San Francisco, 1964.
13. S. Brieglib, *in* "Molekulverbingdungen und Koordinationverbingdungen in Eizeldarstellungen: Electron-Donator-Acceptor-Komplexe." Springer-Verlag, Berlin and New York, 1961.
14. R. M. Harrison, R. Perry, and R. A. Wellings, *Water Res.* **9**, 331 (1975).
15. J. B. Andelman and J. E. Snodgrass, *Crit. Rev. Environ. Control* **4**, 69 (1974).
16. P. Ekwall, K. Setala, and L. Sjöblom, *Acta Chem. Scand.* **5**, 175 (1951).
17. S. W. Karickhoff, D. S. Brown, and T. A. Scott, *Water Res.* **13**, 241 (1979).
18. J. C. Means, J. J. Hassett, S. G. Wood, and W. L. Banwart, *in* "Polynuclear Aromatic Hydrocarbons" (P. W. Jones and P. Leber, eds.), p. 327. Ann Arbor Science Publ., Ann Arbor, Michigan, 1979.
19. D. Mackay and W. Y. Shiu, *J. Chem. Eng. Data* **22**, 399 (1977).
20. F. P. Schwarz, *J. Chem. Eng. Data* **22**, 273 (1977).
21. W. E. May, J. M. Brown, S. N. Chesler, F. Guenther, L. R. Hilpert, M. S. Hertz, and S. A. Wise, *in* "Polynuclear Aromatic Hydrocarbons" (P. W. Jones and P. Leber, eds.), p. 411. Ann Arbor Science Publ., Ann Arbor, Michigan, 1979.
22. A. P. Il'nitskii and K. P. Ershova, *Vopr. Onkol.* **16**, 86 (1970).
23. D. C. Locke, *J. Chromatogr. Sci.* **12**, 433 (1974).
24. J. Eisenbrand, *Dtsche. Lebensm. Rundsch.* **67**, 435 (1971).
25. L. Fishbein, *in* "Chemical Mutagens" (A. Hollaender, ed.), Vol. 1, p. 219. Plenum, New York, 1976.
26. R. E. Laflamme and R. A. Hites, *Geochim. Cosmochim. Acta* **42**, 289 (1978).
27. G. M. Badger, "The Structure and Reactions of the Aromatic Compounds." Cambridge Univ. Press, London and New York, 1956.
28. E. H. Rodd, ed., "Chemistry of Carbon Compounds," Vol. III, Part B, Chapters 20–22. Elsevier, Amsterdam, 1956.
29. S. Coffey, ed., "Rodd's Chemistry of Carbon Compounds," 2nd ed., Vol. III, Part G, Chapter 27. Elsevier, Amsterdam, 1978.
30. G. Körtum and W. Braun, *Ann.* **632**, 104 (1960).
31. E. Voyatzakis, *C. R. Hebd. Seances Acad. Sci.* **251**, 2696 (1960).
32. M. N. Inscoe, *Anal. Chem.* **36**, 2505 (1964).
33. M. Hellmann, *Fresenius Z. Anal. Chem.* **295**, 24 (1979).
34. B. Seifert, *J. Chromatogr.* **131**, 417 (1977).
35. R. C. Pierce and M. Katz, *Environ. Sci. Technol.* **10**, 45 (1976).
36. J. N. Pitts, Jr., D. Grosjean, T. M. Mischke, V. F. Simmon, and D. Poole, *Toxicol. Lett.* **1**, 65 (1977).
37. M. Katz, C. Chan, H. Tosine, and T. Sakuma, *in* "Polynuclear Aromatic Hydrocarbons" (P. W. Jones and P. Leber, eds.), p. 171. Ann Arbor Sci. Publ., Ann Arbor, Michigan, 1979.
38. M. A. Fox and S. Olive, *Science* **205**, 582 (1979).
39. W. A. Korfmacher, D. F. S. Natusch, D. R. Taylor, E. L. Wehry, and G. Mamantov, *in*

"Polynuclear Aromatic Hydrocarbons" (P. W. Jones and P. Leber, eds.), p. 165. Ann Arbor Sci. Publ., Ann Arbor, Michigan, 1979.

40. M. Katz, T. Sakuma, and A. Ho, *Environ. Sci. Technol.* **12,** 909 (1978).
41. J. N. Pitts, Jr., K. A. Van Cauwenberghe, D. Grosjean, J. P. Schmid, D. R. Fitz, W. L. Belsen, Jr., E. B. Knudson, and P. M. Hynds, *Science* **202,** 519 (1978).
42. V. Saravanja-Bozanic, S. Gäb, K. Hustert, and F. Korte, *Chemosphere* **6,** 21 (1977).
43. B. J. Dowty, N. E. Brightwell, J. L. Laseter, and G. W. Griffin, *Biochem. Biophys. Res. Commun.* **57,** 452 (1974).
44. J. Saito and T. Matsuura, *in* "Singlet Oxygen" (H. H. Wasserman and R. W. Murray, eds.), p. 511. Academic Press, New York, 1979.
45. I. Schmeltz, J. Tosk, G. Jacobs, and D. Hoffmann, *Anal. Chem.* **49,** 1924 (1977).
46. R. G. Zepp and P. F. Schlotzauer, *in* "Polynuclear Aromatic Hydrocarbons" (P. W. Jones and P. Leber, eds.), p. 141. Ann Arbor Sci. Publ., Ann Arbor, Michigan, 1979.
47. P. N. Chen, G. A. Junk, and H. J. Svec, *Environ. Sci. Technol.* **13,** 451 (1979).
48. R. Robinson, *Brit. Med. J.* **1,** 965 (1946).
49. D. M. Jerina, R. E. Lehr, H. Yagi, O. Hernandez, P. M. Dansette, P. G. Wislochi, A. W. Wood, W. Levin, and A. M. Connery, *in* "In Vitro Metabolic Activation in Mutagenesis Testing" (F. J. deSerres, J. R. Fouts, J. R. Bend, and R. M. Philpot, eds.), p. 159. Elsevier, Amsterdam, 1976.
50. D. M. McCaustland, D. L. Fischer, K. C. Kolwyck, W. P. Duncan, J. C. Wiley, Jr., C. S. Menon, J. F. Engel, J. K. Selkirk, and P. P. Roller, *in* "Carcinogenesis: a Comprehensive Survey" (R. I. Freudenthal and P. W. Jones, eds.), Vol. 1, p. 349. Raven, New York, 1976, and references therein.

2

Occurrence

I. FORMATION

Polycyclic aromatic compounds (PAC) can be formed from both natural and anthropogenic sources, although the latter are by far the major contributors of environmentally hazardous compounds of this class (*1*). Natural sources may include biosynthesis, natural combustion, and long-term degradation followed by synthesis from biological material. The importance of each of these processes has been discussed in a recent paper (*1*).

A. Natural

The biosyntheses of various PAH by algae (*2*), plants (*3, 4*), or bacteria (*5–11*) have been reported. Hites *et al.* (*12*) found no evidence of bacterial biosynthesis of any PAH when a mixed anaerobic bacteria culture isolated from the Charles River was studied. Criticism was made of incomplete experimental descriptions, especially the way in which blank measurements were conducted, and the lack of consideration of the bioaccumulation effect in these previous studies. Grimmer and Duevel (*13*), also in contrast to earlier work (*3, 4*), showed that when strict care was taken to exclude all outside contamination, no PAH could be found in plants.

Forest and prairie fires are the major contributors to production of PAC by natural combustion. Blumer and Youngblood (*14*) attributed the widespread distribution of PAH in recent sediments to forest fires. Another source has recently been emphasized by Ilnitsky *et al.* (*15, 16*). From their studies, they estimated that the total quantity of benzo[*a*]pyrene released in the atmosphere from all world volcanoes is 12–14 tons per year.

The *in situ* synthesis of PAC from degraded biological material is probably the most interesting and least understood natural method for production of PAC. Fossil fuels (petroleum, coal, shale oil, etc.) contain extremely complex mixtures of these compounds. In addition, various sediments (*17–23*) and fossils (*24, 25*) have been shown to contain PAH that most likely result from biological material.

For instance, alkylated naphthalenes are common constituents of coal (*26, 27*), petroleum (*28*), silicified conifers (*25*), and shales (*29, 30*). It has been postulated that they are dehydrogenation products of sesquiterpenoids (*30*). Alternatively, they could be the products of the reaction of β-carotene (*31*) [Eq. (1)].

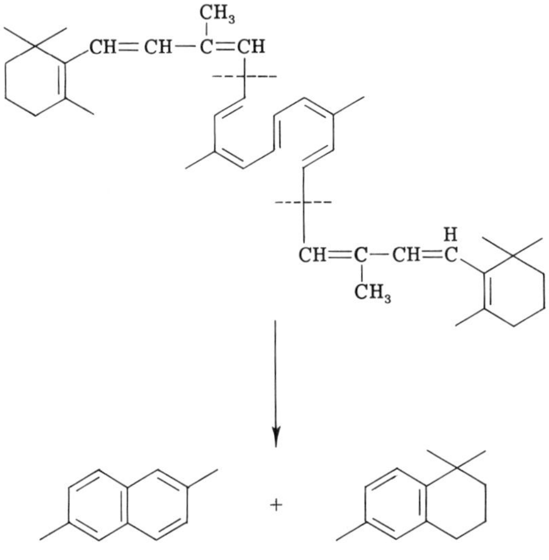

$$\text{(1)}$$

2,6-Dimethyl-   +   α-Ionene
naphthalene

Douglas and Mair (*32*) have shown that naphthalenes, other aromatic hydrocarbons, and organosulfur compounds can be produced by the reaction of elemental sulfur with terpenoids and steroids.

Retene (1-methyl-7-isopropylphenanthrene) occurs in fossilized pine trunks and peat beds (*33*), soils from pine forests (*1*), and lignite extracts (*27, 34*). It is believed to be the result of dehydrogenation of abietic acid, which is a major component of pine rosin (*35*). Laflamme and Hites (*1*) identified a number of possible intermediates and proposed the hypothetical conversion of abietic acid to retene as shown in Fig. 2-1.

Pentacyclic and tetracyclic aromatic hydrocarbons have been reported in shales (*36–40*), crude oil (*41*), lignite (*42*), and sediments (*43, 44*). The skeletal structures of these compounds suggest triterpenoid precursors of the hopane and β-amyrin types. Several hypothetical aromatization schemes have been suggested (*23, 39*).

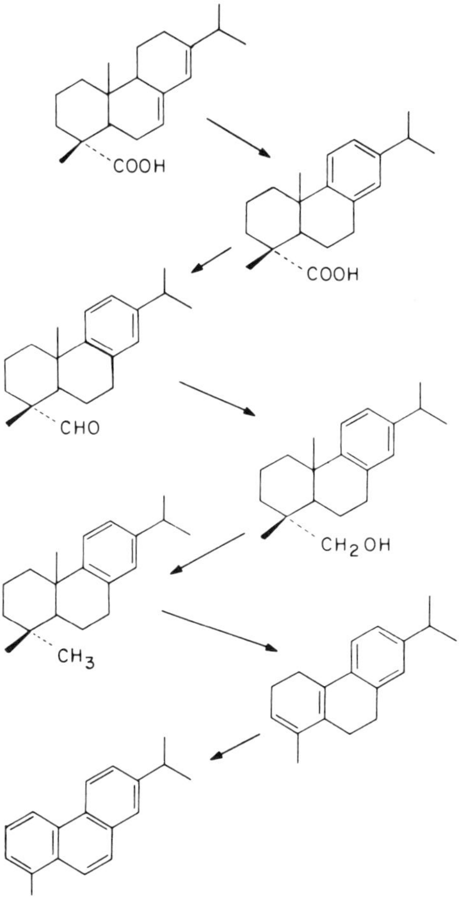

Fig. 2-1. Hypothetical degradation scheme of abietic acid to retene.

A number of authors have found perylene in sediments in substantially high concentrations (*1, 17–21*). Grady (*17*) and Aizenshtat (*19*) suggested that the reduction of pigments such as erythroaphin pigments and 4,9-dihydroxyperylene-3,10-quinone (see Fig. 2-2) would produce perylene. These extended quinone pigments have been found in insects (*44*) and fungi (*45, 46*). Aizenshtat (*19*) attributes the perylene in the Vema Fracture Zone to the deposition of sediment from the Amazon River 1200 km away. The finding of perylene in the Amazon River system by Laflamme and Hites (*1*) supports this hypothesis.

Several rare polycyclic aromatic hydrocarbon minerals have been found and characterized. Pendletonite (*47, 48*) is nearly pure coronene; it contains

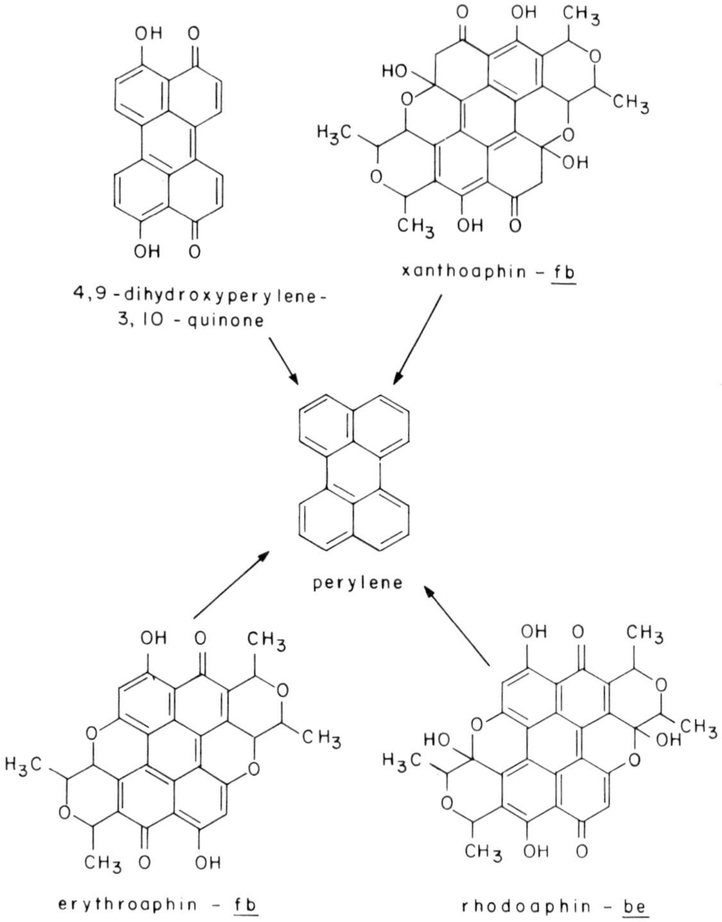

Fig. 2-2. Structures of some extended quinone pigments which may be possible precursors of perylene in sediments. Stereochemistry is not given. [Reproduced with permission from R. E. Laflamme and R. A. Hites, *Geochim. Cosmochim. Acta* **42**, 289 (1978). Copyright, Pergammon Press, Inc.]

only a small amount of methyl derivatives (1.4%) and possible traces of C_2–C_4 homologs. Curtisite (*48, 49*) contains dibenzofluorene, picene, chrysene, and a number of their alkyl homologs. The most abundant PAH series of idrialite (*48*) is thought to be the tribenzofluorenes and their homologs. The second most abundant series is probably naphthenologs of picene, containing one five-membered ring with two alicyclic carbon atoms. Both curtisite and idrialite are thought to contain at least 100 different chemical compounds.

It has been suggested (*48*) that the first stage in the formation of the curtisite–idrialite minerals involves pyrolysis of organic matter at depth, at temperatures that are far higher than those involved in the generation of petroleum, but not high enough to produce a liquid silicate phase. The next stage involves hydrothermal transport with crystallization and recrystallization occurring along the migration path, just as is the case in the formation of many inorganic minerals. The deposits formed along the migration pathway vary in their purity, average molecular weight, and molecular weight range, and the composition at any particular location reflects the formation temperature of the crystallized mineral itself, rather than of its chemical contributors.

Although nearly pure coronene, pendletonite probably had a similar origin to that of curtisite and idrialite. It has a very high (440°C) melting point and may have crystallized close to its source.

B. Anthropogenic

The mechanism(s) of formation of PAC during incomplete combustion of organic material is far from being completely understood. It is believed that two distinct reaction steps are involved, pyrolysis and pyrosynthesis. At high temperatures, organic compounds are partially cracked to smaller, unstable molecules (pyrolysis). These fragments, mostly radicals, recombine to yield larger, relatively stable aromatic hydrocarbons (pyrosynthesis). The pyrosynthesis of PAC in combustion systems has been discussed by Crittenden and Long (*50*) and Schmeltz and Hoffmann (*51*). The latter paper is mainly centered on tobacco combustion.

The hypotheses that have been proposed for the formation of PAC have generally been founded upon the results of pyrolysis experiments. In 1958, Badger *et al.* (*52*) proposed the classical stepwise synthesis of benzo[*a*]pyrene, as shown in Fig. 2-3, in which species (**1**) is acetylene, species (**2**) is a four-

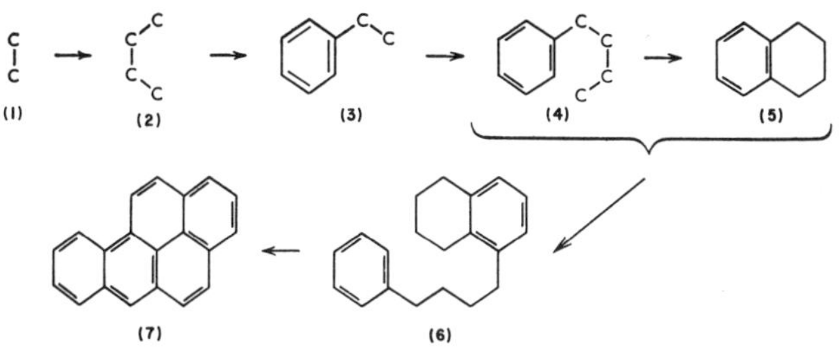

Fig. 2-3. Proposed stepwise synthesis of benzo[*a*]pyrene. [Reproduced with permission from G. M. Badger, R. G. Buttery, R. W. L. Kimber, G. E. Lewis, A. G. Moritz, and I. M. Napier, *J. Chem. Soc.* 2449 (1958). Copyright, Richard Clay and Company, Ltd.]

carbon unit such as vinylacetylene or 1,3-butadiene, and species (3) is styrene or ethylbenzene. Other studies have emphasized the presence of intermediates such as acetylene (53–55), ethylene (56, 57), 1,3-butadiene (52, 56, 58), and their associated free radicals in PAH pyrosynthesis.

In a recent study (59) on the combustion of methane at atmospheric pressure, it was found that ethylene was formed before acetylene and that its concentration decreased while, at the same time, the concentration of acetylene increased.

Crittenden and Long (50) identified phenylacetylene, styrene, and indene in an oxyacetylene flame, and styrene and indene in oxyacetylene and oxyethylene flames, and added these species to the list of key intermediates in the formation of higher molecular weight PAC. Davies and Scully (60) also found that alkyl-substituted aromatic species (including indene) promoted soot formation in a rich town gas/air flame. They proposed that indene underwent bond cleavage in the flame according to Eq. (2).

$$\text{(structures)} \qquad (2)$$

The formation of PAH such as benz[a]anthracene and chrysene could easily be accomplished by combination of such diradicals.

Crittenden and Long (50) have also suggested that compound (5) in Badger's reaction scheme (Fig. 2-3) might be dihydronaphthalene rather than tetralin.

Once formed, simple PAC might undergo further pyrosynthetic reactions. Lang et al. (61) found that pyrolysis of PAH themselves yielded higher condensed ring structures. With nonsubstituted PAH, polyaryls are formed followed by ring closure to highly condensed hydrocarbons. For instance, in the pyrolysis of naphthalene, the isomeric binaphthyls 7, 8, and 9 (Fig. 2-4) were formed. Isomer 9 predominated since it is unable to enter into ring-closure reactions. Perylene (10) and the benzofluoranthenes (11 and 12) were also found in the reaction products and are probably formed by cyclodehydrogenation of the binaphthyls (7 and 8). In addition, terrylene (14) and the naphthoperylenes (15 and 16) were probably formed similarly from precursors containing three naphthalene nuclei.

With alkyl-substituted PAH, direct nuclear condensation takes place with the primary bonding of the hydrocarbon occurring preferentially across the alkyl group (61). Dinaphthylethane (17) and dinaphthylethylene (18) (Fig. 2-4) were formed on pyrolysis of 2-methylnaphthalene. Pentaphene

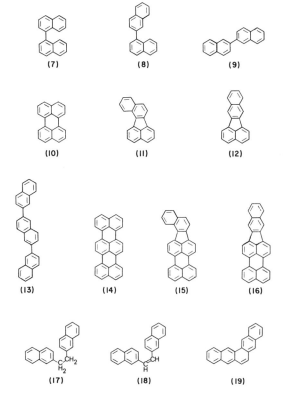

Fig. 2-4. Pyrolysis products of naphthalene.

(19) was also found, and was probably formed from dinaphthylethylene **(18)** by cyclization.

As just discussed, complex hydrocarbons need not necessarily break down into small fragments before pyrosynthesis. Compounds with several rings already present in the molecule can undergo partial cracking followed by dehydrogenation of the expected primary radicals. Common tobacco-leaf constituents such as phytosterols, isoprenoid alcohols (phytol and solanesol), and the amyrins have been shown to be highly efficient precursors of PAH (*51, 62*). Figure 2-5 gives a suggested mechanism for the pyrosynthesis of chrysene from stigmasterol (*63*). Pyrolysis of other leaf constituents such as nicotine (*64*), peptides (*64, 65*), pigments (*66, 67*), and certain tobacco additives (*68*), also result in the formation of PAH. Studies of other pre-

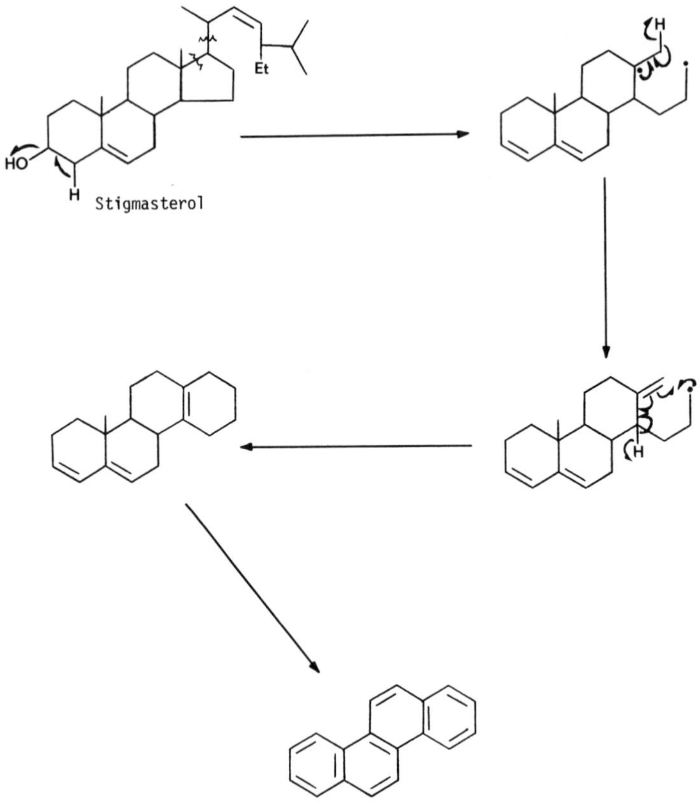

Fig. 2-5. Proposed pyrosynthesis of chrysene from stigmasterol. [Reproduced with permission from G. M. Badger, J. K. Donnelly, and T. M. Spotswood, *Aust. J. Chem.* **18,** 1249 (1965). Copyright, CSIRO.]

cursors from materials other than tobacco include lipids (*69*), carbohydrates (*70*), lignins, terpenes (*71*), and Δ^9-tetrahydrocannabinol (*72*).

In general, all organic compounds containing carbon and hydrogen may serve as precursors of PAH. However, pyrolysis of a substance displaying chain branching or unsaturation results in an overall increase in the production of PAH. Furthermore, PAH are formed more easily from pyrolysis of compounds that already contain cyclic structures. Where oxygen, nitrogen, or sulfur is present in the starting material, oxygen-, nitrogen-, or sulfur-containing analogs of the PAH can be expected.

The results of the pyrolysis of butylbenzene (*73*) over a range of temperatures from 300–900°C indicate that optimum temperatures for the formation of PAH are in the range of 660–740°C. In particular, the optimum temperature for the formation of benzo[a]pyrene was found to be 710°C. The maximum temperature of the burning zone of a cigarette has been found to be 880°C (74).

The yields of PAH and the distributions of alkyl-substituted PAH can be quite different depending on conditions such as temperature, residence time in the hot zone, and starting material. Low temperatures (such as 880°C in a cigarette) will yield an aromatic fraction with a high concentration of substituted PAH (75), whereas very high temperatures (such as 1900°C, as in a carbon black furnace) yield a fraction devoid of alkyl PAH (76). Extended periods of time (10^6 yr) at low temperatures (200°C) will yield PAH mixtures in which the unsubstituted species is not at all abundant— for example, in petroleum (77).

II. SOURCES

Recent estimates of total benzo[a]pyrene (BaP) emissions from major sources in the United States are given in Table 2-1 (78). The major data base for these numbers was reported by Hangebrauck et al. in 1968 (79). More recent studies were used to update this information. Minimum, maximum, and intermediate estimates of recent emissions and the intermediate estimate of 1985 emissions are given. Major sources are considered to be those with national annual emissions of one metric ton per year or greater.

As seen from Table 2-1, greatest emissions are from coal refuse burning and coke production. Estimates on burning coal refuse banks and forest fires cannot be accurately made because of their uncertain extent and the great variability in fuels and burning conditions. Residential sources, such as fireplaces and furnaces, are estimated to emit significant quantities of BaP, and will remain constant or increase slightly in the future. BaP emissions from oil-fired commercial/institutional and industrial boilers and coal-fired power plants are expected to decrease as fuel consumption changes, while emissions from coke production and gasoline consumption in automobiles should decrease as improved control techniques are developed. Most other emissions from major sources are expected to remain nearly constant. Emissions from coal-fired power plants are expected to be in the neighborhood of one metric ton per year by 1985.

Several reviews are available that discuss in much greater detail the sources of PAC (78–81). One review (81) deals mainly with energy-related sources.

TABLE 2-1

Estimates of Total BaP Emissions from Major Sources[a]

Source type	Year	Estimates of current BaP emissions (Mg/yr)[b]			Intermediate estimate of 1985 BaP emissions[b]
		Minimum	Maximum	Intermediate	
Burning coal refuse banks	1972	200.0	310.0	unknown[c]	unknown[c]
Coke production	1975	0.050	300.0	110.0	21.0
Residential fireplaces	1975	52.0	110.0	73.0	77.0
Forest fires	1972	9.5	127.0	unknown[d]	unknown[d]
Coal-fired residential furnaces	1973	0.85	740.0	26.0	14.0
Rubber tire wear	1977	0.0	11.0	unknown[e]	unknown[e]
Motorcycles	1976			5.6	5.6
Automobiles (gasoline)	1975	1.6	3.8	2.7	0.21
Commercial incinerators	1972	0.90	4.7	2.1	2.1
Oil-fired commercial/ institutional boilers	1973			2.0	1.3

[a] Major sources are considered to be those for which the intermediate estimate of current or projected emissions is greater than 1 Mg/yr.

[b] Emission estimates are reported in megagrams (Mg) per year. A megagram is equivalent to one million grams or one metric ton.

[c] Because coal refuse banks can ignite spontaneously and may flare up, smolder, or go out naturally, no reliable estimates can be made of the number of burning banks.

[d] The variation in the burning process and fuel types in different geographical, climatological, and seasonal conditions for prescribed or wild fires is extremely great. The current understanding and quantitative knowledge of those variations does not allow an intermediate estimate to be made.

[e] The quantity of emissions from rubber tire wear has not been adequately determined. Preliminary experimental work detected no BaP, while a roughly estimated emission factor proportional to population was based on particulate emission and some analytical work. Using that factor, 1985 BaP emissions would be projected to be 12 Mg/yr.

III. DISTRIBUTION

The sources of environmental PAC, as discussed earlier, are due to a variety of processes—both natural and man-controlled. High-temperature combustion processes have become the major contributors in most urban areas. Until the beginning of this century there existed a natural balance between the production and natural degradation of PAC which kept the background concentration low and fixed (*82*). However, with the increasing industrial development throughout the world, the natural balance has been disturbed and the production and accumulation rates of PAC are constantly rising.

In the following sections, the distributions of PAC in air, water, sediments, food, tobacco smoke, and fossil fuels will be discussed. It is virtually impossible to cite all papers that have been published on these subjects. A bibliography of 1055 references dealing with the occurrence and analysis of PAC has recently been prepared (*83*). In each section, then, reference will be made to major papers and published reviews, and emphasis will be placed on more recent studies involving advanced analytical techniques, such as gas chromatography/mass spectrometry.

Detailed lists of PAC found in air particulate matter and tobacco-smoke condensate are given. It is felt that these examples are representative of what might be expected in other samples, except for the fossil fuels. In this case, the mixtures are too complex to list all components found. Instead, a discussion on group-type analysis is given.

A. Air

The distribution of PAC in the atmosphere is dependent on a number of variables. Airborne PAC are always associated with various types of aerosols, both natural and man-made (*80, 84–87*). Sources of natural aerosols include dust-rise by wind, sea spray, forest fires, volcanic dust, meteoritic dust, and some vegetation. Again, of much greater significance is the production of aerosols by combustion and industry. It is believed that most airborne PAC are primarily associated with carbonaceous aerosols produced by incomplete combustion of fossil fuels. Thus, the distribution of PAC in the atmosphere is governed by aerosol transport from place to place primarily by wind currents, fallout through wet and dry precipitation, and chemical degradation as well as by source production. Effects of atmospheric physics, chemical degradation, and meteorological conditions have been reviewed in a report by the United States National Academy of Sciences (*80*).

During the last 30 years, numerous studies throughout the world have been undertaken to characterize the PAC content in airborne particulate matter with respect to geographical location, point sources, seasonal variation, and meteorological conditions (*88–107*). In most cases, 1 to 20 compounds were identified and quantified as indicative of the total PAC content of the air. In two studies, very detailed and comprehensive identifications were made by combined gas chromatography/mass spectrometry (*98, 104*). Table 2-2 lists the PAC that have been identified in air particulate matter to date.

TABLE 2-2

Identified Airborne PAH

Compound	Molecular weight	Reference
Biphenyl	154	*98*
Fluorene	166	*98*
Benzindenes	166	*98*
Methylbiphenyls	168	*98*
Dihydrofluorenes	168	*98*
Phenanthrene	178	*98, 104*
Anthracene	178	*98, 104*
Dihydrophenanthrene	180	*98*
Dihydroanthracene	180	*98*
1-Methylfluorene	180	*98*
2-Methylfluorene	180	*98*
9-Methylfluorene	180	*98*
Octahydrophenanthrene and/or octahydroanthracene	186	*98*
4*H*-Cyclopenta[*def*]phenanthrene	190	*104*
Methylphenanthrene	192	*98, 104*
1-Methylphenanthrene	192	*104*
2-Methylphenanthrene	192	*104*
3-Methylphenanthrene	192	*104*
9-Methylphenanthrene	192	*104*
Methylanthracene	192	*98, 104*
1-Methylanthracene	192	*104*
2-Methylanthracene	192	*104*
Fluoranthene	202	*98, 104*
Pyrene	202	*98, 104*
Benzacenaphthylene	202	*104*
Methyl-4*H*-cyclopenta[*def*]phenanthrene	204	*104*
Dihydrofluoranthene	204	*98*
Dihydropyrene	204	*98*
Dimethylphenanthrenes[a] and/or dimethylanthracenes[a]	206	*98, 104*
Octahydrofluoranthene	210	*98*
Octahydropyrene	210	*98*
Methylfluoranthene	216	*98, 104*
1-Methylfluoranthene	216	*104*
2-Methylfluoranthene	216	*104*
3-Methylfluoranthene	216	*104*
7-Methylfluoranthene	216	*104*
8-Methylfluoranthene	216	*104*
Methylpyrene	216	*98, 104*
1-Methylpyrene	216	*104*
2-Methylpyrene	216	*104*
4-Methylpyrene	216	*104*

(Continued)

TABLE 2-2 (*Continued*)

Compound	Molecular weight	Reference
Benzo[*a*]fluorene	216	*98, 104*
Benzo[*b*]fluorene	216	*98, 104*
Benzo[*c*]fluorene	216	*98*
Dimethyl-4*H*-cyclopenta[*def*]phenanthrene[a]	218	*104*
Dihydrobenzo[*a*]fluorene and/or dihydrobenzo[*b*]fluorene	218	*98*
Dihydrobenzo[*c*]fluorene	218	*98*
Trimethylphenanthrenes[b] and/or trimethylanthracenes[b]	220	*104*
Benzo[*ghi*]fluoranthene	226	*98, 104*
Benzo[*c*]phenanthrene	228	*98, 104*
Benz[*a*]anthracene	228	*98, 104*
Chrysene	228	*98, 104*
Triphenylene	228	*98*
Dimethylfluoranthenes[a] and/or dimethylpyrenes[a]	230	*104*
Dihydrobenzo[*c*]phenanthrene	230	*98*
Dihydrobenz[*a*]anthracene and/or dihydrochrysene and/or dihydrotriphenylene	230	*98*
Trimethyl-4*H*-cyclopenta[*def*]phenanthrenes[b]	232	*104*
Hexahydrochrysene	234	*98*
Methylbenzo[*ghi*]fluoranthene	240	*104*
Dihydromethylbenzo[*ghi*]fluoranthene	242	*98*
Methylbenzo[*c*]phenanthrenes	242	*98*
Methylchrysene	242	*98, 104*
1-Methylchrysene	242	*104*
2-Methylchrysene	242	*104*
3-Methylchrysene	242	*104*
6-Methylchrysene	242	*104*
Methylbenz[*a*]anthracenes	242	*98, 104*
Methyltriphenylene	242	*98*
Trimethylfluoranthenes[b] and/or trimethylpyrenes[b]	244	*98, 104*
Tetrahydromethyltriphenylene and/or tetrahydromethylbenz[*a*]anthracene and/or tetrahydromethylchrysene	246	*98*
Benzo[*j*]fluoranthene	252	*98, 104*
Benzo[*k*]fluoranthene	252	*98, 104*
Benzo[*b*]fluoranthene	252	*98*
Benzo[*e*]pyrene	252	*98, 104*
Benzo[*a*]pyrene	252	*98, 104*
Perylene	252	*98, 104*
Binaphthyls	254	*98, 104*
2,2′-Binaphthyl	254	*98, 104*

(*Continued*)

TABLE 2-2　*(Continued)*

Compound	Molecular weight	Reference
Dimethylchrysenes[a] and/or dimethylbenz[a]anthracenes[a]	256	*98, 104*
Dimethyltriphenylene[b]	256	*98*
Methylbenzopyrenes and/or methylbenzofluoranthenes	266	*98, 104*
Methylbinaphthyls	268	*98, 104*
Methyl-2,2′-binaphthyl	268	*98*
3-Methylcholanthrene (3-methylbenz[j]aceanthrylene)	268	*98*
Benzo[ghi]perylene	276	*98, 104*
Anthanthrene	276	*98, 104*
Indeno[1,2,3-cd]fluoranthene	276	*98*
Indeno[1,2,3-cd]pyrene	276	*98*
Dibenzanthracenes	278	*98, 104*
Dibenz[a,c]anthracene	278	*98*
Pentacene	278	*98*
Benzo[b]chrysene	278	*98*
Picene	278	*98*
Dimethylbenzofluoranthenes[a]	280	*98*
Dimethylbenzopyrenes[a]	280	*98*
Methylindeno[1,2,3-cd]fluoranthene	290	*98*
Methylindeno[1,2-cd]pyrene	290	*98*
Methylbenzo[ghi]perylene and/or methylanthanthrene	290	*98*
Methyldibenzanthracenes	292	*98, 104*
Methylbenzo[b]chrysene and/or methylbenzo[c]tetraphene	292	*98*
Methylpicene	292	*98*
Coronene	300	*98, 104*
Dibenzopyrenes	302	*98, 104*
Diphenylacenaphthalene	304	*104*
Quaterphenyl	306	*104*

[a] Could be ethyl.
[b] Could be ethylmethyl or propyl.

Results of the previous studies can be summarized as follows:

1. The range of benzo[a]pyrene (BaP) in unpolluted nonurban air varied between 0.1 and 0.5 ng/m^3, while the concentration in polluted air reached 74 ng/m^3 (approximately 150 times higher) (*80*).

2. There was considerable variation in the concentrations of PAC in air particulate matter collected at different sampling sites which were representative of different point sources.

3. Inconsistent variations in the concentrations of PAC in air particulate matter sampled for short periods at the same sampling site were dependent on variations in meteorological conditions (rainfall, wind direction change, etc.).

4. Air at commercial sites was more polluted during the day than at night, corresponding to automobile traffic and industrial emissions as major sources.

5. Pollution was more pronounced in the autumn and winter than in spring or summer, indicating the higher use of combustion sources in the winter, presumably for heating purposes, or seasonal meteorological variations such as lower than average wind-speed and more frequent thermal inversions during the winter months.

The direct comparison and eventual correlation of airborne PAC with point-source effluents oftentimes is quite difficult. For instance, it has been estimated that aerosols smaller than 1 μm in diameter will stay in the lower atmosphere without precipitation for a few days to several weeks, but will persist only a few days to less than a day if 1–10 μm in diameter (*80*). It has recently been reported that there are different distributions of PAH (see Chapter 4) for differently sized aerosols (*87, 108, 109*). Furthermore, more reactive PAH photodecompose very rapidly in the atmosphere by reaction with ozone, nitrogen oxides, sulfur oxides, and various other oxidants (see Chapter 1), while the more stable PAH can exist for much longer periods of time (*80*).

The determination of sulfur- and nitrogen-containing PAC in air particulate matter has been pursued with less intensity than the identification of the hydrocarbons. Nevertheless, a number of nitrogen-containing PAC ranging from two to five rings have been identified (*92, 110–112*) and are listed in Table 2-3. Only one study (*104*) to date reports the identification of sulfur-containing PAC in airborne particulates. These compounds are listed in Table 2-4.

TABLE 2-3

Identified Airborne *N*-Heterocycles

Compound	Molecular weight	Reference
Quinoline	129	*111, 112*
Isoquinoline	129	*111*
Methylquinolines and/or methylisoquinolines	143	*111*
Dimethylquinolines[a] and/or dimethylisoquinolines[a]	157	*111*

(*Continued*)

TABLE 2-3   (*Continued*)

Compound	Molecular weight	Reference
4-Azafluorene	167	*112*
Trimethylquinolines[b] and/or trimethylisoquinolines[b]	171	*111*
Benzoquinolines and/or benzisoquinolines	179	*98, 110–112*
Acridine	179	*98, 110–112*
Benzo[f]quinoline	179	*111, 112*
Benzo[h]quinoline	179	*111, 112*
Phenanthridine	179	*110–112*
Tetramethylquinolines[c] and/or tetramethylisoquinolines[c]	185	*111*
Methylacridines and/or methylphenanthridines and/or methylbenzoquinolines and/or methylbenzisoquinolines	193	*111*
Azafluoranthenes and/or Azapyrenes	203	*111, 112*
7-Azafluoranthene	203	*112*
1-Azafluoranthene	203	*110*
4-Azapyrene	203	*112*
Dimethylacridines[a] and/or dimethylphenanthridines[a] and/or dimethylbenzoquinolines[a] and/or dimethylbenzisoquinolines	207	*111*
Azabenzofluorenes and/or methylazapyrenes and/or methylazafluoranthenes	217	*111*
11H-indeno[1,2-b]quinoline	217	*110*
Azabenz[a]anthracenes and/or azachrysenes and/or dibenzo[f,h]quinoline and/or dibenz[f,h]isoquinoline	229	*110, 111*
Benz[a]acridine	229	*110*
Benz[c]acridine	229	*110*
Methylbenzacridines and/or methylbenzophenanthridines and/or methyldibenzo[f,h]quinoline and/or methyldibenz[f,h]isoquinoline	243	*111*
Azabenzopyrenes and/or azabenzofluoranthenes	253	*111*
Dibenz[a,h]acridine	279	*110, 111*
Dibenz[a,j]acridine	279	*110, 111*

[a] Could be ethyl.
[b] Could be ethylmethyl or propyl.
[c] Could be methylpropyl, diethyl, or dimethylethyl.

TABLE 2-4

Identified Airborne S-Heterocycles

Compound	Molecular weight
Dibenzothiophene	184
Methyldibenzoth'ophenes	198
Phenanthro[4,5-bcd]thiophene	208
Naphthobenzothiophenes	234
Methylnaphthobenzothiophenes	248

B. Water

The presence of PAC in polluted air is a problem that has been extensively studied while, on the other hand, the water environment has received much less attention. Nevertheless, evidence is mounting up rapidly to support the fact that PAC are ubiquitous water pollutants and must be controlled similarly to air particulate PAC. Several extensive reviews of the occurrence of PAC in the aqueous environment have recently been published (82, 113–115).

The distribution of PAC in different bodies of water is dependent on their point sources. Spillage of crude oil or refined products from a tanker, oil drilling area, or storage area; terrestrial run-off of rain water which scrubs PAC from the air and leaches them from soil, pavements, slag dumps, and coal storage piles; and effluents from industrial plants contribute to the natural production of PAC from plants and organisms. Groundwater that remains uncontaminated by human activities has a concentration range of 0.001–0.10 µg/liter of carcinogenic PAC, while uncontaminated freshwater lakes have approximately a 10 times higher concentration (0.010–0.025 µg/liter (82, 114). It is assumed that the groundwater PAC were leached from upper soil layers, while the higher concentration in freshwater lakes originated from aquatic biota in sediments and soils. In comparison, contaminated river water contained 0.05 to over 1 µg/liter of carcinogenic PAH, the corresponding BaP concentration range being 0.001–0.05 µg/liter (82, 114, 116). The solubilities of a number of PAH in water have been determined (114, 117–119) and are listed in Table 1-1 (see Chapter 1). The high molecular weight and low polarity of PAH render them fairly insoluble, and therefore it must be assumed that a large proportion of the PAH content of polluted water is adsorbed on suspended solids.

Problems with point-source identification are also present in the aqueous environment. Selected PAC may evaporate into the atmosphere, disperse in the water column, precipitate and settle to the bottom sediments, concentrate in aquatic biota, be chemically oxidized, or be biodegraded. The photodecomposition of PAC depends on water depth, daily and seasonal fluctuation of solar radiation, ambient temperature, and dissolved oxygen. All of these processes have been extensively covered in recent reviews (*82, 114, 115, 120, 121*).

C. Recent Sediments

PAC found in bottom sediments originate from two processes. As discussed earlier, PAC are quite insoluble in water and tend to precipitate to the bottom sediment or adsorb on suspended solids that are eventually incorporated in the sediment. In a recent study (*122*) of a New England river basin, the more water-soluble PAH, the naphthalenes, were found in the water only; those of intermediate solubility, the anthracenes and phenanthrenes, were found both in water and sediment; and those of very low solubility, molecular weights above 228, were found only in the sediment. Secondly, as discussed in an earlier section, some build-up of PAC in the sediment could be due to synthesis by microorganisms and plants, although there is some question as to the validity of these processes (*12, 14*).

It has only recently been discovered that PAC mixtures in sediments are extremely complex (*14, 122–129*). A large number of alkyl and cycloalkyl derivatives have been detected. Only two sulfur-containing PAC, dibenzothiophene and naphthothiophene, have been identified in sediment (*123*).

The natural level of PAC concentration in sediments is much lower than in polluted areas. Dried material taken from a depth of 30 m off the remote shores of Greenland has shown a concentration of 5.5 μg/kg of dry substances (*82*). In comparison, the concentration of BaP in the suspended solids and river bottom sediments may reach many hundreds to many thousands of μg/kg of dry sample, concentrations in one river (*113*) being as high as 15 mg/kg.

The stability of PAC in the sediment is greater than in air or water (see Chapter 1). The chemical half-life of PAC under intense sunlight may be only hours or days (*80, 82, 120, 121, 130*). However, PAC incorporated in the bottom sediments will have a much smaller rate of degradation. On the other hand, the bioaccumulation of PAC by various organisms is well known (*12, 131–135*), and preferential bioaccumulation or biodegradation can change the PAC distribution in sediments with time.

D. Soils and Ancient Sediments

It was determined as early as 1947 that PAC were minor constituents in the soil (*136*). Since that time, it has been found that the compositional complexity of PAC mixtures in the earth's crust ranges from one and two components to many components. Simple mixtures include those in asbestos (*137–139*), mineral ores (*47–49*), and recent clays and carbonate muds (*10, 17, 19*). More complex mixtures include those in ancient sedimentary rocks (*140*) and fossils (*24, 25*), while there are literally thousands of homologous and isomeric PAC in many ancient sediments and fossil fuels (*124*).

Precipitation and fallout from air particulate matter, whether from pollution or forest fires, is the most likely source of soil PAC. In more remote areas, they could originate from low-temperature pyrolysis of wood and plant material or are products of biosynthesis. BaP has been found to range from 40 to 1,300 μg/kg in U.S. soil (*141*) and up to 21 mg/kg in a Swiss soil sample (*136*).

E. Food

The sources of PAC contamination of foodstuffs are numerous and varied. Curing smokes, contaminated soils, polluted air and water, food additives, food processing, modes of cooking, and endogenous sources all contribute to PAC concentrations in different foods. Several reviews that cover the occurrence and determination of PAC in foods have been published (*142–146*). In addition to these reviews, recent reports of PAC in smoked foods (*147–151*); liquid smoke flavors (*152*); total diet composites (*153*); high-protein foods, oils, and fats (*154, 155*); plants and vegetables (*156–158*); and seafood (*135, 159, 160*) are available.

Investigations have shown that frying, grilling, roasting, and smoking of fish and meat products produce the highest levels of BaP found in food products. Concentrations range from a few to several hundred μg/kg of dry material, with an average concentration of approximately 50 μg/kg (*82*).

F. Tobacco Smoke

A number of variables influence the yields of PAC in tobacco smokes. The frequency and duration of puffs, the type of tobacco, the moisture content of the tobacco, and the type and permeability of cigarette paper and filter all contribute to the distribution of PAC. A summary of the effects of these variables is available (*161*). Although many individual PAC had been identified in tobacco smoke by 1972 (*162–164*), two recent studies report detailed identifications of over 150 different PAH (*75, 165–168*). Compounds identified to date are listed in Table 2-5. The numerous alkylated PAC presumably originate in the cooler regions of the burning zone. The nitrogen-containing PAC which have been identified in tobacco-smoke condensate to date are listed in Table 2-6 (*169–171*).

The range of BaP produced in the mainstream of cigarette smoke is from

TABLE 2-5

Tobacco Smoke PAH

Compound	Molecular weight	Reference
Indene	116	*162, 167*
Naphthalene	128	*162, 167*
Azulene	128	*162*
Methylindenes	130	*167*
1-Methylindene	130	*167*
3-Methylindene	130	*167*
Methylnaphthalenes	142	*162, 167*
1-Methylnaphthalene	142	*162, 167*
2-Methylnaphthalene	142	*162, 167*
Dimethylindenes[a]	144	*167*
Acenaphthylene	152	*162, 167*
Acenaphthene	154	*162, 167*
Biphenyl	154	*162, 167*
1-Vinylnaphthalene	154	*167*
2-Vinylnaphthalene	154	*167*
Dimethylnaphthalenes[a]	156	*162, 167*
1,6-Dimethylnaphthalene	156	*162, 167*
1,8-Dimethylnaphthalene	156	*162, 167*
2,6-Dimethylnaphthalene	156	*162, 167*
2,7-Dimethylnaphthalene	156	*162, 167*
2,5-Dimethylnaphthalene	156	*162*
1,3-Dimethylnaphthalene	156	*167*
2,3-Dimethylnaphthalene	156	*167*
1,4-Dimethylnaphthalene	156	*167*
1,5-Dimethylnaphthalene	156	*167*
1,7-Dimethylnaphthalene	156	*167*
1-Ethylnaphthalene	156	*167*
2-Ethylnaphthalene	156	*167*
Trimethylindenes[b]	158	*167*
Methylacenaphthylenes	166	*75, 167*
1-Methylacenaphthylene	166	*167*
Fluorene	166	*162, 167*
Benz[f]indene	166	*167*
Methylacenaphthenes	168	*167*
3-Methylbiphenyl	168	*167*
4-Methylbiphenyl	168	*167*
Trimethylnaphthalenes[b]	170	*167*
1,3,6-Trimethylnaphthalene	170	*162*
Phenanthrene	178	*162, 75, 167*
Anthracene	178	*162, 75, 167*
1-Methylfluorene	180	*162, 164, 75, 167*
2-Methylfluorene	180	*164, 75, 167*

(*Continued*)

TABLE 2-5 (*Continued*)

Compound	Molecular weight	Reference
3-Methylfluorene	180	*164, 167*
4-Methylfluorene	180	*164, 167*
9-Methylfluorene	180	*162, 164, 167*
9,10-Dihydroanthracene	180	*162*
Dimethylacenaphthylene[a]	180	*167*
Methylbenz[*f*]indenes	180	*167*
Dimethylacenaphthenes[a]	182	*167*
Tetramethylnaphthalenes[c]	184	*167*
4*H*-Cyclopenta[*def*]phenanthrene	190	*75, 168*
1-Methylanthracene	192	*75, 167*
2-Methylanthracene	192	*162, 75, 167*
9-Methylanthracene	192	*162, 167*
1-Methylphenanthrene	192	*162, 75, 167*
2-Methylphenanthrene	192	*75, 167*
3-Methylphenanthrene	192	*75, 167*
4-Methylphenanthrene	192	*167*
9-Methylphenanthrene	192	*162, 75, 167*
Dimethylfluorenes[a]	194	*164, 167*
Dimethylbenz[*f*]indene[a]	194	*167*
Trimethylacenaphthylenes[b]	194	*167*
Trimethylacenaphthenes[b]	196	*167*
Fluoranthene	202	*162, 75, 167*
Pyrene	202	*162, 75, 167*
Benzacenaphthylene	202	*75, 167*
Methyl-4*H*-cyclopenta[*def*]phenanthrene	204	*75*
Dimethylphenanthrenes[a] and/or dimethylanthracenes[a]	206	*75, 167*
2,5-Dimethylphenanthrene	206	*162*
Trimethylfluorenes[b]	208	*167*
Tetramethylacenaphthylenes[c]	208	*167*
Tetramethylacenaphthenes[c]	210	*167*
1-Methylpyrene	216	*162, 75, 167*
2-Methylpyrene	216	*162, 75, 167*
4-Methylpyrene	216	*75, 167, 162*
1-Methylfluoranthene	216	*75, 167*
2-Methylfluoranthene	216	*75, 167*
3-Methylfluoranthene	216	*75*
7-Methylfluoranthene	216	*75*
8-Methylfluoranthene	216	*162, 75, 167*
Benzo[*a*]fluorene	216	*75, 167*
Benzo[*b*]fluorene	216	*162, 75, 167*
Benzo[*c*]fluorene	216	*75, 167*
5*H*-Benzo[*a*]fluorene	216	*162*

(*Continued*)

TABLE 2-5   (*Continued*)

Compound	Molecular weight	Reference
11*H*-Benzo[*a*]fluorene	216	*162*
7*H*-Benzo[*c*]fluorene	216	*162*
11*H*-Benzo[*b*]fluorene	216	*162*
Dimethyl-4*H*-cyclopenta[*def*]phenanthrenes[a]	218	*75*
Trimethylphenanthrenes[b] and/or trimethylanthracenes[b]	220	*75, 167*
Benzo[*ghi*]fluoranthene	226	*162, 75, 167*
Cyclopenta[*cd*]fluoranthene	226	*75*
Cyclopenta[*cd*]pyrene	226	*75, 167*
Benz[*a*]anthracene	228	*162, 75, 167*
Benzo[*c*]phenanthrene	228	*162, 167*
Chrysene	228	*162, 75, 167*
Naphthacene	228	*162*
Triphenylene	228	*167*
Dimethylfluoranthenes[a] and/or dimethylpyrenes[a] and/or methylbenzofluorenes	230	*162, 75, 167*
Methylbenzo[*a*]fluorene	230	*167*
Methylbenzo[*b*]fluorene	230	*167*
Methylbenzo[*c*]fluorene	230	*167*
11-Methyl-11*H*-benzo[*a*]fluorene	230	*162*
Methylbenzo[*ghi*]fluoranthene	240	*167*
4*H*-Cyclopenta[*def*]triphenylene	240	*168*
4*H*-Cyclopenta[*def*]chrysene	240	*168*
1-Methylchrysene	242	*75, 167*
2-Methylchrysene	242	*75, 167*
3-Methylchrysene	242	*75, 167*
4-Methylchrysene	242	*167*
5-Methylchrysene	242	*75*
6-Methylchrysene	242	*75, 167*
2-Methylbenz[*a*]anthracene	242	*75*
3-Methylbenz[*a*]anthracene	242	*162, 75*
4-Methylbenz[*a*]anthracene	242	*75*
5-Methylbenz[*a*]anthracene	242	*162, 75*
6-Methylbenz[*a*]anthracene	242	*75*
8-Methylbenz[*a*]anthracene	242	*75*
9-Methylbenz[*a*]anthracene	242	*75*
10-Methylbenz[*a*]anthracene	242	*75*
Methyltriphenylenes and/or methylbenz[*a*]anthracenes and/or methylchrysenes	242	*75, 167*
Trimethylfluoranthenes[b] and/or trimethylpyrenes[b]	244	*75, 167*

(*Continued*)

TABLE 2-5 (*Continued*)

Compound	Molecular weight	Reference
Dimethylbenzofluorenes[a]	244	*167*
Benzofluoranthenes	252	*75, 167*
Benzo[a]fluoranthene	252	*167*
Benzo[b]fluoranthene	252	*162, 167*
Benzo[j]fluoranthene	252	*162, 75, 167*
Benzo[k]fluoranthene	252	*162, 75, 167*
Benzo[e]pyrene	252	*162, 75, 167*
Benzo[a]pyrene	252	*162, 75, 167*
Perylene	252	*162, 75, 167*
3,4-Dihydrobenzo[a]pyrene	254	*162*
Binaphthyls	254	*75*
Dimethylchrysenes[a] and/or dimethylbenz[a]anthracenes[a] and/or dimethyltriphenylenes[a]	256	*162, 75, 167*
9,10-Dimethylbenz[a]anthracene	256	*162*
Tetramethylpyrene[c]	258	*167*
Methylbenzopyrenes and/or methylbenzofluoranthenes	266	*162, 75, 167*
Dibenzo[a,i]fluorene	266	*162*
11H-Naphtho[2,1-a]fluorene	266	*162*
Methylbinaphthyls	268	*75*
5,6-Dihydro-8H-benzo[a]cyclopent[h]anthracene	268	*162*
10,11-Dihydro-9H-benzo[a]cyclopent[i]anthracene	268	*162*
16,17-Dihydro-15H-benzocyclopenta[a]phenanthrene	268	*162*
Trimethylchrysene[b] and/or trimethylbenz[a]anthracene[b] and/or trimethyltriphenylene[b]	270	*75, 167*
Benzo[ghi]perylene	276	*162, 75*
Anthanthrene	276	*162, 75, 168*
Indeno[1,2,3-cd]fluoranthene	276	*162, 168*
Indeno[1,2,3-cd]pyrene	276	*162, 167, 168*
Dibenz[a,h]anthracene	278	*162, 75, 168*
Dibenz[a,c]anthracene	278	*75, 168*
Dibenz[a,j]anthracene	278	*168*
Dibenz[a,i]anthracene	278	*162, 75*
Dibenzo[b,h]phenanthrene	278	*162*
Dimethylbenzopyrenes[a] and/or dimethylbenzofluoranthenes[a]	280	*75, 167, 168*
Dimethylbinaphthyls[a]	282	*75*
Methylbenzo[ghi]perylenes and/or methylanthanthrenes and/or methylindeno[1,2,3-cd]pyrenes	290	*75, 168*
Methyldibenzanthracenes	292	*168*

(*Continued*)

TABLE 2-5 (*Continued*)

Compound	Molecular weight	Reference
Trimethylbenzopyrenes[b]	294	*168*
Coronene	300	*162, 168*
Dibenzofluoranthenes	302	*168*
Dibenzo[b,j]fluoranthene	302	*168*
Dibenzo[a,e]fluoranthene	302	*168*
Dibenzopyrenes	302	*162, 75, 168*
Dibenzo[a,e]pyrene	302	*168*
Dibenzo[a,i]pyrene	302	*162, 168*
Dibenzo[a,h]pyrene	302	*162, 168*
Dibenzo[a,l]pyrene	302	*162, 168*
Dibenzo[e,l]pyrene	302	*168*
Benzo[b]perylene	302	*168*
Naphtho[2,3-a]pyrene	302	*162*
Diphenylacenaphthylene	304	*75*
Dimethylindeno[1,2,3-cd]pyrenes[a] and/or dimethylbenzo[ghi]perylenes[a] and/or dimethylanthanthrenes[a]	304	*168*
Quaterphenyls	306	*75*
Methylcoronene	314	*168*
Trimethylindeno[1,2,3-cd]pyrenes[b] and/or trimethylbenzo[ghi]perylenes[b] and/or trimethylanthanthrenes[b]	318	*168*
Dibenzo[a,c]naphthacene	328	*162*
Dibenzo[a,j]naphthacene	328	*162*
Tribenz[a,c,h]anthracene	328	*162*
Dimethylcoronene[a]	328	*168*

[a] Could be ethyl.
[b] Could be ethylmethyl or propyl.
[c] Could be diethyl, methylpropyl, or dimethylethyl.

8 ng/cigarette to 122.5 ng/cigarette, with the average value being approximately 30 ng/cigarette (*161*). The concentration of BaP in the sidestream of cigarette smoke is higher ($\sim$ 130 ng/cigarette) than in the mainstream.

G. Fossil Fuels

PAC fractions from fossil fuels are generally much more complex than fractions obtained from combustion sources. They are extremely rich in alkylated and hydroaromatic species. Figure 2-6 illustrates the structural range of petroleum hydrocarbons as a three-dimensional, continuous array (*172*). Generally, each of the compound types implied occurs in a given

TABLE 2-6

Tobacco Smoke *N*-Heterocycles

Compound	Molecular weight	Reference
Indole	117	*169, 170*
Quinoline	129	*169, 171*
Isoquinoline	129	*169, 171*
Methylindole(s)	131	*169*
1-Methylindole	131	*169*
2-Methylindole	131	*169*
3-Methylindole	131	*169, 170*
5-Methylindole	131	*169*
7-Methylindole	131	*169*
Methylquinoline(s)	143	*169, 171*
2-Methylquinoline	143	*171*
3-Methylquinoline	143	*171*
4-Methylquinoline	143	*169, 171*
5-Methylquinoline	143	*171*
6-Methylquinoline	143	*171*
7-Methylquinoline	143	*171*
8-Methylquinoline	143	*171*
Methylisoquinoline	143	*171*
Dimethylindole(s)	145	*169, 170*
1,2-Dimethylindole	145	*169*
1,3-Dimethylindole	145	*169*
1,5-Dimethylindole	145	*169*
1,6-Dimethylindole	145	*169*
1,7-Dimethylindole	145	*169*
2,3-Dimethylindole	145	*169*
1-Ethylindole	145	*169*
3-Ethylindole	145	*169, 170*
Dimethylquinoline(s)	157	*169, 171*
2,6-Dimethylquinoline	157	*169*
C_2-Isoquinolines	157	*171*
Trimethylindole(s)	159	*169, 170*
1,2,3-Trimethylindole	159	*169*
Ethylmethylindole(s)	159	*170*
Propylindole	159	*170*
3-Propylindole	159	*169*
Carbazole	167	*169, 170*
4-Azafluorene	167	*169, 171*
Azacarbazole	168	*169*
C_3-Quinolines	171	*171*
Tetramethylindole	173	*170*
Ethyldimethylindole(s)	173	*170*
Propylmethylindole	173	*170*

(Continued)

TABLE 2-6 (*Continued*)

Compound	Molecular weight	Reference
Benzo[*h*]quinoline	179	*171*
Acridine	179	*171*
Phenanthridine	179	*171*
Benzo[*f*]quinoline	179	*171*
1-Methylcarbazole	181	*169, 170*
2-Methylcarbazole	181	*169, 170*
3-Methylcarbazole	181	*169, 170*
4-Methylcarbazole	181	*169, 170*
9-Methylcarbazole	181	*169*
Propyldimethylindole	187	*170*
Propylethylindole	187	*170*
3-Phenylindole	193	*170*
Dimethylcarbazole(s)	195	*169, 170*
1,9-Dimethylcarbazole	195	*169*
2,9-Dimethylcarbazole	195	*169*
3,9-Dimethylcarbazole	195	*169*
4,9-Dimethylcarbazole	195	*169*
Ethylcarbazole	195	*170*
1-Azafluoranthene	203	*171*
4-Azapyrene	203	*171*
Methylphenylindole(s)	207	*170*
Trimethylcarbazole(s)	209	*170*
Benzo[*a*]carbazole	217	*170*
Benzo[*b*]carbazole	217	*170*
Benzo[*c*]carbazole	217	*170*
Tetramethylcarbazole(s)	223	*170*
Methylbenzocarbazole(s)	231	*170*
Dimethylbenzocarbazole(s)	245	*170*
7*H*-Dibenzo[*c,g*]carbazole	267	*169*
Dibenz[*a,h*]acridine	279	*169*
Dibenz[*a,j*]acridine	279	*169*

petroleum as a continuous homologous series. Figure 2-7 diagrams the sulfur-containing PAC composition found in petroleum. Similar diagrams could be constructed to represent the oxygen and nitrogen heterocyclic fractions. In general, coal- and shale-derived oils may be expected to contain larger quantities of alkylated PAC, heteroatomic constituents, and hydroaromatics than does petroleum. In short, these fractions are extremely complex, and even with the high resolving power of modern analytical instrumentation a detailed compositional analysis is far from being possible.

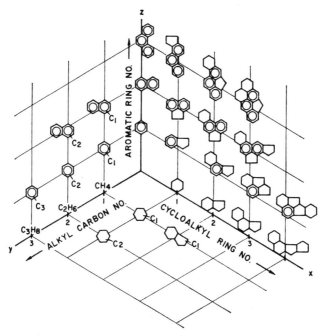

Fig. 2-6. Petroleum hydrocarbons: their structural range. [Reproduced with permission from L. R. Snyder, *Acc. Chem. Res.* **3**, 290 (1970). Copyright, American Chemical Society.]

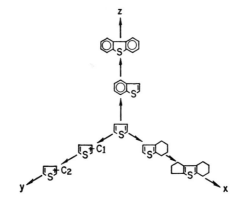

Fig. 2-7. Structural range of petroleum sulfur heterocycles. [Reproduced with permission from L. R. Snyder, *Acc. Chem. Res.* **3**, 290 (1970). Copyright, American Chemical Society.]

The approach to characterizing these fractions has centered mainly on mass spectrometric group-type analysis, and the reader is referred to Chapter 8 for a discussion on this topic. An excellent review of the distribution of PAH in products and emissions from fossil fuel processes has recently been published (*81*).

Studies described in the previous sections have demonstrated that PAC are ubiquitous in nature. They are usually present as complex mixtures that vary greatly in the relative contribution of their individual components. The goals of all analytical studies of environmental PAC are to define the PAC sources, to elucidate the pathways of hydrocarbon dispersal, and eventually to perform complete quantitative analyses. The latter goal hinges on the development of analytical methods with greater resolving power.

In the past, most comparisons between different PAC samples were done by selecting one to four representative PAC in each sample and quantitatively determining their levels. Benzo[*a*]pyrene, the most notorious carcinogen in the PAC class, has usually been chosen for these comparative studies. For example, the levels of BaP concentration in air, water, sediments, soils, food, and tobacco were quoted in each of the previous sections.

It is obvious that BaP does not reflect the true carcinogenic potential of each PAH mixture. Many other compounds are carcinogenic, and the quantitative determination of their levels is certainly important. As more powerful analytical techniques become available and more routinely usable, more accurate data on the composition of materials discussed in this chapter will also be obtained.

REFERENCES

1. R. E. Laflamme and R. A. Hites, *Geochim. Cosmochim. Acta* **42,** 289 (1978).

2. J. Borneff, F. Selenka, H. Kunte, and A. Mazimos, *Environ. Res.* **2,** 22 (1968).

3. W. Graef and H. Diehl, *Arch. Hyg. Bakteriol.* **150,** 49 (1966).

4. J. L. Hancock, H. G. Applegate and J. D. Dodd, *Atmos. Environ.* **4,** 363 (1970).

5. J. Brisou, *C. R. Seances Soc. Biol. Ses Fil.* **163,** 772 (1969).

6. M. Knorr and D. Schenk, *Arch. Hyg. Bakteriol.* **152,** 282 (1968).

7. C. DeLima-Zanghi, *Cah. Oceanogr.* **20,** 203 (1968).

8. L. Mallet and M. Tissier, *C. R. Seances Soc. Biol. Ses Fil.* **163,** 63 (1969).

9. P. Niaussat, L. Mallet, and J. Ottenwaelder, *C. R. Hebd. Seances Acad. Sci. Ser. D* **268,** 1109 (1969).

10. P. Niaussat, C. Auger, and L. Mallet, *C. R. Hebd. Seances Acad. Sci. Ser. D.* **270,** 1042 (1970).

11. C. E. ZoBell, *N.Z. Oceanogr. Inst. Rec.* **3,** 39 (1959).

12. A. Hase and R. A. Hites, *Geochim. Cosmochim. Acta* **40,** 1141 (1976).

13. G. Grimmer and D. Duevel, *Z. Naturforsch. Teil B* **25,** 1171 (1970).

14. M. Blumer and W. W. Youngblood, *Science* **188,** 53 (1975).

15. A. P. Ilnitsky, G. A. Belitsky, and L. M. Shabad, *Cancer Lett.* (*Amsterdam*) **1**, 291 (1976).

16. A. P. Ilnitsky, V. S. Mischenko, and L. M. Shabad, *Cancer Lett.* (*Amsterdam*) **3**, 227 (1977).

17. W. L. Orr and J. R. Grady, *Geochim. Cosmochim. Acta* **31**, 1201 (1967).

18. F. S. Brown, M. J. Baedecker, A. Nissenbaum, and I. R. Kaplan, *Geochim. Cosmochim. Acta* **36**, 1185 (1972).

19. Z. Aizenshtat, *Geochim. Cosmochim. Acta* **37**, 559 (1973).

20. R. Ishiwatari and T. Hanya, *Proc. Jpn. Acad.* **51**, 436 (1975).

21. S. G. Wakeham, *Environ. Sci. Technol.* **11**, 272 (1977).

22. B. R. T. Simoneit, *Geochim. Cosmochim. Acta* **41**, 463 (1977).

23. R. E. Laflamme and R. A. Hites, *Geochim. Cosmochim. Acta* **42**, 289 (1978).

24. M. Blumer, *Science* **149**, 722 (1965).

25. A. C. Sigleo, *Geochim. Cosmochim. Acta* **42**, 1397 (1978).

26. V. L. Berkofer and W. Pauly, *Brennst. Chem.* **50**, 376 (1969).

27. C. M. White and M. L. Lee, *Geochim. Cosmochim. Acta* **44**, 1825 (1980).

28. B. J. Mair and A. G. Douglas, *Geochim. Cosmochim. Acta* **28**, 1303 (1964).

29. D. E. Anders, F. G. Doolittle, and W. E. Robinson, *Geochim. Cosmochim. Acta* **39**, 1423 (1975).

30. G. Eglinton and M. T. J. Murphy, eds., "Organic Geochemistry." Springer-Verlag, Berlin and New York, 1969.

31. R. Ikan, Z. Aizenshtat, M. J. Baedecker, and I. R. Kaplan, *Geochim. Cosmochim. Acta* **39**, 173 (1975).

32. A. G. Douglas and B. J. Mair, *Science* **147**, 499 (1965).

33. E. C. Sterling and M. T. Bogert, *J. Org. Chem.* **4**, 20 (1939).

34. R. Hayatsu, R. E. Winans, R. G. Scott, L. P. Moore, and M. H. Studier, *Nature* (*London*) **275**, 116 (1978).

35. W. D. Stonecipher and R. W. Turner, *Encycl. Polym. Sci. Technol.* **12**, 139 (1970).

36. A. Van Dorsselaer, A. Ensminger, C. Spyckerelle, M. Dastillong, O. Sieskind, P. Arpino, P. Albrecht, G. Ourisson, P. W. Brooks, S. J. Gaskell, B. J. Kimble, R. P. Philip, J. R. Maxwell, and G. Eglinton, *Tetrahedron Lett.* **14**, 1349 (1974).

37. B. J. Kimble, J. R. Maxwell, R. P. Philip, G. Eglinton, P. Albrecht, A. Ensminger, P. Arpino, and G. Ourisson, *Geochim. Cosmochim. Acta* **38**, 1165 (1974).

38. A. C. Greiner, C. Spyckerelle, and P. Albrecht, *Tetrahedron* **32**, 257 (1976).

39. A. C. Greiner, C. Spyckerelle, P. Albrecht, and G. Ourisson, *J. Chem. Res.* (S), 334 (1977).

40. C. Spyckerelle, A. C. Greiner, P. Albrecht, and G. Ourisson, *J. Chem. Res.* (S), 332 (1977).

41. W. Carruthers and D. A. M. Watkins, *Chem. Ind.* 1433 (1963).

42. V. Jarolim, K. Hejno, F. Hemmert, and F. Sorm, *Collect. Czech. Chem. Commun.* **30**, 873 (1965).

43. C. Spyckerelle, A. C. Greiner, P. Albrecht, and G. Ourisson, *J. Chem. Res.* (S), 330 (1977).

44. D. W. Cameron, R. I. T. Cromartie, Y. K. Hamied, P. M. Scott, and L. Todd, *J. Chem. Soc.*, 62 (1964).

45. R. H. Thompson, "Naturally Occurring Quinones," 2nd ed., Academic Press, New York, 1971.

46. D. C. Allport and J. D. Bu'lock, *J. Chem. Soc.*, 654 (1960).

47. J. Murdoch and T. A. Geissman, *Am. Mineral.* **52**, 611 (1967).

48. M. Blumer, *Chem. Geol.* **16**, 245 (1975).

49. T. A. Geissman, K. Y. Sim, and J. Murdoch, *Experientia* **23**, 793 (1967).

50. B. D. Crittenden and R. Long, *in* "Carcinogenesis—A Comprehensive Survey" (R. I. Freudenthal and P. W. Jones, eds.), Vol. 1, p. 209. Raven, New York, 1976.

51. I. Schmeltz and D. Hoffman, *in* "Carcinogenesis—A Comprehensive Survey" (R. I. Freudenthal and P. W. Jones, eds.), Vol. 1, p. 225. Raven, New York, 1976.

52. G. M. Badger, R. G. Buttery, R. W. L. Kimber, G. E. Lewis, A. G. Moritz, and I. M. Napier, *J. Chem. Soc.*, 2449 (1958).
53. H. P. A. Groll, *Ind. Eng. Chem.* **25**, 784 (1933).
54. B. B. Chakraborty and R. Long, *Combust. Flame* **12**, 469 (1968).
55. G. M. Badger, "The Chemical Basis of Carcinogenic Activity." Charles C. Thomas, Springfield, Illinois, 1962.
56. E. N. Hague and R. V. Wheeler, *Fuel* **8**, 560 (1929).
57. R. E. Kinney and D. J. Crowley, *Ind. Eng. Chem.* **46**, 258 (1954).
58. T. Kunugi, T. Sakai, K. Soma, and Y. Sasaki, *Ind. Eng. Chem. Fundam.* **8**, 374 (1969).
59. A. D'Allessio, A. Di Lorenzo, A. F. Sarofim, F. Beretta, S. Masi, and C. Venitozzi, *Int. Symp. Combust.*, *15th* 1975, p. 1427.
60. R. A. Davies and D. B. Scully, *Combust. Flame* **10**, 165 (1966).
61. K. F. Lang, H. Buffleb, and M. Zander, *Erdoel. Kohle* **16**, 944 (1963).
62. W. S. Schlotzhauer and I. Schmeltz, *Beitr. Tabakforsch.* **5**, 5 (1969).
63. G. M. Badger, J. K. Donnelly, and T. M. Spotswood, *Aust. J. Chem.* **18**, 1249 (1965).
64. I. Schmeltz, W. S. Schlotzhauer, and E. B. Higman, *Beitr. Tabakforsch.* **6**, 134 (1972).
65. E. B. Higman, I. Schmeltz, and W. S. Schlotzhauer, *J. Agric. Food Chem.* **18**, 636 (1970).
66. O. T. Chortyk, W. S. Schlotzhauer, and R. L. Stedman, *Beitr. Tabakforsch.* **3**, 421 (1966).
67. W. S. Schlotzhauer and I. Schmeltz, *Tob. Sci.* **11**, 89 (1967).
68. I. Schmeltz and W. S. Schlotzhauer, *Nature* (*London*) **219**, 370 (1968).
69. G. A. Halaby and I. S. Fagerson, *Proc. Int. Congr. Food Sci. Technol. 3rd, 1970*, p. 820 (1971).
70. Y. Masuda, K. Mori and M. Kuratsune, *Gann* **58**, 69 (1967).
71. A. A. Liverovskii, E. I. Shmulevskaya, L. S. Romanovskaya, E. I. Pankina, V. N. Kun, P. P. Dikun, and L. D. Kostenko, *Izv. Vyssh. Uchebn. Zaved. Lesn. Zh.* **15**, 99 (1972).
72. M. Novotny, M. L. Lee, and K. D. Bartle, *Experientia* **32**, 280 (1976).
73. G. M. Badger, R. W. L. Kimber, and J. Novotny, *Aust. J. Chem.* **17**, 778 (1964).
74. G. P. Touey and R. C. Mumpower, *Tobacco* **144**, 17 (1957).
75. M. L. Lee, M. Novotny, and K. D. Bartle, *Anal. Chem.* **48**, 405 (1976).
76. M. L. Lee and R. A. Hites, *Anal. Chem.* **48**, 1890 (1976).
77. B. J. Mair, *Geochim. Cosmochim. Acta* **28**, 1303 (1964).
78. P. C. Siebert, C. A. Craig, and E. B. Coffey, Preliminary Assessment of the Sources, Control and Population Exposure to Airborne Polycyclic Organic Matter (POM) As Indicated by Benzo[*a*]pyrene (BaP), Final Rep. EPA Contr. No. 68-02-2836, Environ. Protect. Agency, 1978.
79. R. P. Hangebrauck, D. J. von Lehmden, and J. E. Meeker, Sources of Polynuclear Hydrocarbons in the Atmosphere, U.S. Dep. HEW, Public Health Service, AP-33, PB 174-706, Washington, D.C. 1967.
80. Biologic Effects of Atmospheric Pollutants: Particulate Polycyclic Organic Matter, Natl. Acad. Sci., Washington, D.C., 1972.
81. M. R. Guerin, *in* "Polycyclic Hydrocarbons and Cancer" (H. V. Gelboin and P. O. P. Ts'o, eds.), Vol. 1, p. 3. Academic Press, New York, 1978.
82. M. J. Suess, *Sci. Total Environ.* **6**, 239 (1976).
83. A. Martin and M. Blumer, "Polycyclic Aromatic Hydrocarbons: Occurrence and Analysis—A Partial Bibliography," Rep. WHOI-75-22, Woods Hole Oceanographic Inst., Woods Hole, Massachusetts, 1975.
84. J. F. Thomas, M. Mukai, and B. D. Tebbens, *Environ. Sci. Technol.* **2**, 33 (1968).
85. B. D. Tebbens, J. F. Thomas, and M. Mukai, *Am. Ind. Hyg. Assoc. J.* **32**, 365 (1971).
86. S. G. Chang and T. Novakov, *Atmos. Environ.* **9**, 495 (1975).
87. M. Katz and R. C. Pierce, *in* "Carcinogenesis—A Comprehensive Survey" Vol. 1: Poly-

nuclear Aromatic Hydrocarbons, (R. Freudenthal and P. W. Jones, eds.), Raven, New York, 1976, p. 413.

88. A. Liberti, G. P. Cartoni, and V. Cantuti, *J. Chromatogr.* **15,** 141 (1964).
89. G. R. Clemo, *Tetrahedron* **23,** 2389 (1967).
90. R. E. Waller and B. T. Commins, *Environ. Res.* **1,** 295 (1967).
91. E. Sawicki, *Proc. Arch. Environ. Health* **14,** 46 (1967).
92. D. Hoffmann and E. L. Wynder, *in* "Air Pollution" (A. C. Stern, ed.), 2nd ed., p. 187. Academic Press, New York, 1968.
93. G. Chatot, M. Castegnaro, J. L. Roche, R. Fontanges, and P. Obaton, *Chromatographia* **3,** 507 (1970).
94. L. E. Stromberg and G. Widmark, *J. Chromatogr.* **47,** 27 (1970).
95. J. M. Colucci and C. R. Begeman, *Environ. Sci. Technol.* **5,** 145 (1971).
96. G. Grimmer and H. Bohnke, *Fresenius Z. Anal. Chem.* **261,** 310 (1972).
97. R. J. Gordon and R. J. Bryan, *Environ. Sci. Technol.* **7,** 1050 (1973).
98. R. C. Lao, R. S. Thomas, H. Oja, and L. Dubois, *Anal. Chem.* **45,** 908 (1973).
99. K. D. Bartle, M. L. Lee, and M. Novotny, *Int. J. Environ. Anal. Chem.* **3,** 349 (1974).
100. R. C. Pierce and M. Katz, *Anal. Chem.* **47,** 1743 (1975).
101. G. Ketseridis and J. Hahn, *Fresenius Z. Anal. Chem.* **273,** 257 (1975).
102. R. J. Gordon, *Environ. Sci. Technol.* **10,** 370 (1976).
103. G. Ketseridis, J. Hahn, R. Jaenicke, and C. Junge, *Atmos. Environ.* **10,** 603 (1976).
104. M. L. Lee, M. Novotny, and K. D. Bartle, *Anal. Chem.* **48,** 1566 (1976).
105. M. L. Lee, G. P. Prado, J. B. Howard, and R. A. Hites, *Biomed. Mass Spectrom.* **4,** 182 (1977).
106. M. Katz, T. Sakuma, and A. Ho, *Environ. Sci. Technol.* **12,** 909 (1978).
107. E. Sawicki, *Crit. Rev. Anal. Chem.* **1,** 275 (1970).
108. R. C. Pierce and M. Katz, *Environ. Sci. Technol.* **9,** 347 (1975).
109. J. D. Butler and P. Crossley, *Sci. Total Environ.* **11,** 53 (1979).
110. E. Sawicki, S. P. McPherson, T. W. Stanley, J. Meeker, and W. C. Elber, *Int. J. Air Water Pollut.* **9,** 515 (1965).
111. W. Cautreels and K. Van Cauwenberghe, *Atmos. Environ.* **10,** 447 (1976).
112. M. Dong and D. C. Locke, *J. Chromatogr. Sci.* **15,** 32 (1977).
113. J. B. Andelman and M. J. Suess, *Bull. W. H. O.* **43,** 479 (1970).
114. R. M. Harrison, R. Perry, and R. A. Wellings, *Water Res.* **9,** 331 (1975).
115. J. Borneff, *Adv. Environ. Sci. Technol.* **8,** 393 (1977).
116. R. A. Hites and K. Biemann, *Science* **178,** 158 (1972).
117. F. P. Schwarz and S. P. Wasik, *Anal. Chem.* **48,** 524 (1976).
118. D. MacKay and W. Shiu, *J. Chem. Eng. Data* **22,** 4 (1977).
119. W. E. May, J. M. Brown, S. N. Chesler, F. Guenther, L. R. Hilpert, H. S. Hertz, and S. A. Wise, *in* Trace Organic Analysis: A New Frontier in Analytical Chemistry, NBS Spec. Publ. 519, p. 219. U.S. Gov. Printing Office, Washington, D.C., 1979.
120. M. J. Suess, *Zentralbl. Bakteriol. Parasitenkd. Infektionskr. Hyg. Abt. Orig. Reihe B.* **155,** 541 (1972).
121. J. P. Andelman and M. J. Suess, *in* "Organic Compounds in Aquatic Environments" (S. J. Faust and J. V. Hunter, eds.), p. 439. Marcel Dekker, New York, 1971.
122. R. A. Hites and W. G. Biemann, *Adv. Chem. Ser.* **147,** 188 (1975).
123. W. Giger and M. Blumer, *Anal. Chem.* **46,** 1663 (1974).
124. W. W. Youngblood and M. Blumer, *Geochim. Cosmochim. Acta* **39,** 1303 (1975).
125. H. Hellman, *Fresenius Z. Anal. Chem.* **275,** 109 (1975).
126. M. Blumer, T. Dorsey, and J. Sass, *Science* **195,** 283 (1977).
127. R. A. Hites, R. E. Laflamme, and J. W. Farrington, *Science* **198,** 829 (1977).

128. R. H. Bieri, M. K. Cueman, C. L. Smith, and C.-W. Su, *Int. J. Environ. Anal. Chem.* **5,** 293 (1978).

129. W. Giger and C. Schaffner, *Anal. Chem.* **50,** 243 (1978).

130. R. F. Lee, W. S. Gardner, J. W. Anderson, J. W. Blaylock, and J. Barwell-Clark, *Environ. Sci. Technol.* **12,** 832 (1978).

131. M. Blumer, G. Souza, and J. Sass, *Mar. Biol.* **5,** 195 (1970).

132. R. F. Lee, R. Sauerheber, and G. H. Dobbs, *Mar. Biol.* **17,** 201 (1972).

133. G. Grimmer and D. Duevel, *Z. Naturforsch. Teil B* **25,** 1171 (1970).

134. H. O. Hettche, *Staub* **31,** 72 (1971).

135. F. I. Onuska, A. W. Wolkoff, M. E. Comba, R. H. Larose, M. Novotny, and M. L. Lee, *Anal. Lett.* **9,** 451 (1976).

136. W. Kern, *Chim. Acta* **30,** 1595 (1947).

137. J. S. Harrington and B. T. Commins, *Chem. Ind.* 1427 (1964).

138. J. S. Harrington, *Nature (London)* **193,** 43 (1962).

139. H. L. Boiteau, M. Robin, and S. Gelot, *Arch. Mal. Prof. Med. Trav. Secur. Soc.* **33,** 261 (1972).

140. G. W. Hodgson, B. Hitchon, K. Taguchi, B. L. Baker, and E. Peake, *Geochim. Cosmochim. Acta* **32,** 737 (1968).

141. M. Blumer, *Science* **134,** 474 (1961).

142. J. W. Howard and T. Fazio, *J. Agric. Food Chem.* **17,** 527 (1969).

143. J. W. Howard and T. Fazio, *Ind. Med. Surg.* **39,** 435 (1970).

144. T. Saito, *Kaguku To Seibutsu* **8,** 178 (1970).

145. M.-T. Lo and E. Sandi, *Residue Rev.* **69,** 35 (1978).

146. R. Hamm, *Pure Appl. Chem.* **49,** 1655 (1977).

147. A. J. Malanoski, E. L. Greenfield, J. M. Worthington, and F. L. Joe, *J. Assoc. Off. Anal. Chem.* **51,** 114 (1968).

148. G. Grimmer, *Dtsch. Apoth.-Ztg.* **108,** 529 (1968).

149. D. J. Tilgner and H. Daun, *Residue Rev.* **27,** 19 (1969).

150. K. Fretheim, *J. Agric. Food Chem.* **24,** 976 (1976).

151. K. S. Rhee and L. J. Bratzler, *J. Food Sci.* **35,** 146 (1970).

152. R. H. White, J. W. Howard, and C. J. Barnes, *J. Agric. Food Chem.* **19,** 143 (1971).

153. J. W. Howard, T. Fazio, R. H. White, and B. A. Klimeck, *J. Assoc. Off. Anal. Chem.* **51,** 122 (1968).

154. G. Grimmer and A. Hildebrandt, *J. Assoc. Off. Anal. Chem.* **55,** 631 (1972).

155. G. Grimmer and H. Bohnke, *J. Assoc. Off. Anal. Chem.* **58,** 725 (1975).

156. I. A. Kalinina, *Vopr. Onkol.* **18,** 112 (1972).

157. J. Borneff, F. Selenka, H. Kunte, and A. Maximus, *Environ. Res.* **2,** 22 (1968).

158. G. Grimmer and D. Duevel, *Z. Naturforsch Teil B* **25,** 1171 (1970).

159. M. Giaccio, *Quad. Merceol.* **10,** 21 (1971).

160. J. S. Warner, *Anal. Chem.* **48,** 578 (1976).

161. E. L. Wynder and D. Hoffmann, "Tobacco and Tobacco Smoke, Studies in Experimental Carcinogenesis." Academic Press, New York, 1967.

162. R. L. Stedman, *Chem. Rev.* **68,** 153 (1968).

163. D. Hoffmann, G. Rathkamp, and S. Nesnow, *Anal. Chem.* **41,** 1256 (1969).

164. D. Hoffmann and G. Rathkamp, *Anal. Chem.* **44,** 899 (1972).

165. R. F. Severson, M. E. Snook, R. F. Arrendale, and O. T. Chortyk, *Anal. Chem.* **48,** 1866 (1976).

166. R. F. Severson, M. E. Snook, H. C. Higman, O. T. Chortyk, and F. J. Akin, *in* "Carcinogenesis—A Comprehensive Survey," Vol. 1: Polynuclear Aromatic Hydrocarbons, (R. I. Freudenthal and P. W. Jones, eds.), p. 253. Raven, New York, 1976.

167. M. E. Snook, R. F. Severson, H. C. Higman, R. F. Arrendale, and O. T. Chortyk, *Beitr. Tabakforsch.* **8,** 250 (1976).

168. M. E. Snook, R. F. Severson, R. F. Arrendale, H. C. Higman, and O. T. Chortyk, *Beitr. Tabakforsch.* **9,** 79 (1977).

169. I. Schmeltz and D. Hoffmann, *Chem. Rev.* **77,** 295 (1977).

170. M. E. Snook, R. F. Arrendale, H. C. Higman, and O. T. Chortyk, *Anal. Chem.* **50,** 88 (1978).

171. M. Dong, I. Schmeltz, E. LaVoie, and D. Hoffmann, *in* "Carcinogenesis—A Comprehensive Survey," Vol. 3, (P. W. Jones and R. I. Freudenthal, eds.), p. 97. Raven, New York, 1978.

172. L. R. Snyder, *Acc. Chem. Res.* **3,** 290 (1970).

3

Toxicology and Metabolism

I. INTRODUCTION

Occupational hazards resulting from exposure to combustion products were indicated for the first time more than two centuries ago (*1*), and several convincing proofs of the relationship between exposure to environmental tars and cancer date back to the early part of this century. Pioneering studies on the chemical composition of combustion materials were conducted several decades ago. Even at that time, many PAH were isolated and their structures determined. Subsequent biological testing of the isolated and synthetic compounds has provided substantial evidence that PAC are carcinogenic. For this reason, scientific interest in this class of compounds has rapidly increased.

Modern analytical chemistry has assumed a very important role in answering the highly complex questions arising from PAC carcinogenesis. This role concerns not only the detection and quantitation of PAC in complex environmental mixtures, but also the products formed during metabolism.

Thus, reliable detection, identification, and quantitation of PAC metabolites, their metabolic conjugates, various products of the interactions of PAC-related molecules with cellular components and biological molecules, etc., are among the many tasks of modern analytical chemistry in this area. Clearly, a multidisciplinary approach is needed; the analytical chemist has become indispensable in research on problems of carcinogenesis.

In spite of the rapidly advancing knowledge of the chemistry, biochemistry, and toxicology of PAC, many important questions remain unanswered. Although it is obvious that some solutions may not be easily found in the near future, much useful information has been gathered in recent years. In fact, literature data on the subject of PAC carcinogenesis are now becoming so numerous that orientation in this field will soon be extremely difficult.

The present chapter is thus written with the objective of providing a brief survey of the most important facts in the field of PAC carcinogenesis for workers who are primarily interested in analytical aspects. This is deemed particularly desirable for stimulating further interest among analytical chemists in the complexity of problems in this area at both cellular and biochemical levels. It is indeed likely that systematic improvements in the measurement techniques currently utilized, as well as significant break-throughs in analytical methodology, may be essential before further progress can be made in this field. However, this chapter is not intended to provide an exhaustive survey of the field of chemical carcinogenesis. Readers interested in a more detailed treatment of toxicological and metabolic aspects are referred to one of the many books and reviews currently available on these subjects (2–5).

II. TOXICITY

Under the expression "cancer," the medical profession now recognizes a variety of neoplastic diseases with different symptoms and, most likely, of different origins. Their common characteristic is uncontrolled cell growth. Wynder and Mabuchi (6) estimate that some 70–90% of human and animal cancer is caused by environmental factors. Among the environmental chemicals concerned, PAC comprise the largest group of carcinogens (7).

Polycyclic compounds are widespread in our environment, primarily due to their escape from combustion processes. Air pollutants (and, most notably, PAH adsorbed on airborne particulates) are believed to be major contributors to the higher death rate from lung cancer in urban areas as compared to rural areas (8).

Occupational skin cancer is another matter of concern. Prolonged periods of exposure of industrial workers' skin to soot, petroleum fractions, asphalts, coal products, etc., present a serious hazard. It seems that development of simple devices that will monitor PAC levels directly at industrial sites will soon become a necessity.

Many adverse effects of tobacco smoking on human health are now generally known. Lung cancer has long been attributed to the use of tobacco products. According to our present knowledge of smoke effects, PAC are likely to act as major tumor initiators, but the presence of other chemicals in smoke is necessary to account fully for the overall carcinogenicity. Research in this area has been hindered by the extreme chemical complexity of tobacco-smoke condensates. Production of "safe" cigarettes through elimination of hazardous chemicals is a desirable future goal. Although many aspects of tobacco carcinogenesis are discussed in a monograph by Wynder and

Hoffmann (9), it is less widely known that cannabis smoking may also be related to cancer (10–12). Cannabis smoke produces greater cellular abnormalities in lung tissue than tobacco smoke.

Carcinogenic properties of PAC have been tested with both individual substances and mixtures. Whereas the most common compounds that are under study in numerous laboratories are the potent carcinogens such as benzo[a]pyrene, 3-methylbenz[j]aceanthrylene (3-methylcholanthrene), and 7,12-dimethylbenz[a]anthracene, many other PAC have now been extensively tested for their carcinogenicity and overall toxicity. Some toxicity data have been compiled (13–17), but continuous updating of such findings must take place. In particular, some of the older biological investigations are often found to be inaccurate when newer, more standardized techniques are used. Appendix 5 of this book provides a guide for various compounds that have been tested for carcinogenicity. Current extensive research is likely to extend the list of tested compounds.

The amount of a given chemical carcinogen which is needed to produce cancerous changes in the animal is a question of increasing importance. Although epidemiological studies are ultimately most valuable, the possible hazard to humans must often be extrapolated from animal studies. Such studies are initially performed with relatively high doses of chemical carcinogens. Carcinogens exhibit dose–response relationships similar to those of drugs. Furthermore, differences in target-tissue sensitivity can often be misleading. For example, 7H-dibenzo[c,g]carbazole is a significantly more potent carcinogen in lung tissue than on skin (7).

The complexity of events leading to tumor growth subsequent to the application of chemical carcinogens is now generally recognized. A particularly crucial event is the promotion of tumor growth following the initiation step. It is accepted that many PAC act as tumor initiators, and the process can be further aided by either a PAC of different structure or by an entirely different chemical. Thus, for example, phorbol esters, or certain phenols contained in tobacco smoke (18), will significantly promote the tumor-growth process initiated by PAC. Even more interestingly, some PAC which exhibit little or no tumorigenic effects alone (for example, dibenz[a,c]anthracene, chrysene, benz[a]anthracene, pyrene, and fluoranthene) can act as cocarcinogens (18, 19).

The toxic effects of PAC must always be considered in relation to both external and internal factors. Thus, in addition to the above-mentioned cocarcinogenicity and tumor-promoting activities of both man-made and naturally occurring chemicals, the combined effects of carcinogens and viruses must also be taken into consideration (20). In addition, hormones (and endocrine factors in general) appear heavily implicated in the process of carcinogenesis.

There is also little doubt concerning the importance of genetic factors in natural defense against cancer. Since certain animals are known to be less suitable for carcinogenesis investigations (and, thus, presumably more resistant to tumor growth), different humans are also likely to have metabolic differences translating into different susceptibilities to cancer. Recent research appears to indicate that such phenomena are likely to be related to different metabolic rates and routes by which a given carcinogen can be altered in living cells. Further studies in this area are highly desirable from the point of view of cancer prevention.

III. BIOLOGICAL TESTING

Various experimental animals have been used to demonstrate tumor development, most typically mice of different strains, rats, hamsters, Guinea pigs, dogs, hens, and ducks. Since it is now known that a number of relationships such as genetic factors, species, environmental effects, age, diet, route of carcinogen introduction, etc., can all be crucial to the development of cancer, many of the earlier studies must be evaluated with caution. The species correlation is most obvious, since it can be demonstrated that a certain PAC carcinogen may be active in mice, but not, for example, in Guinea pigs. Thus, the primary consideration in selection of an experimental animal is the convenience for reasonably quantitative bioassay. Experimental mice have been inbred for years to produce highly genetically homogeneous strains. Furthermore, they can be easily bred in numbers that allow good statistical evaluations of chemical carcinogens. At least for initial carcinogenicity screening studies, there seems to be no particular advantage in using animals other than mice. However, the consideration of strain is important. In order to achieve a high degree of standardization, most researchers in the field now agree on using mouse strains available from certain specialized animal colonies.

The most commonly performed test is based on inducing epidermoid cancer or benign papillomas following the repeated application of the material under test to the shaved backs of mice. Subsequent statistical evaluation is important. For example, the results are typically judged as significant if at least 12 out of 50 tested animals show any sign of tumors. If tumors develop in up to 33% of the cases, the carcinogenic activity is considered "slight"; 33–66% incidence is viewed as "moderate"; and above 66% as "high" (21).

Statistics is an important tool in these and related biological assays. Some earlier evaluations of carcinogens were sometimes based on insufficient numbers of tested animals, and repeated tests were needed to rectify this

(22). When the initial carcinogenicity screenings show positive results, additional evaluation is needed, including dose–response studies, different routes of carcinogen administration, checks for possible formation of multiple tumors, tumor potency, etc.

Tests by the subcutaneous route are performed to establish whether the carcinogen under study is also effective in other organs. Sarcomas are then sought in various organs after the injection. Experimental animals can develop tumors up to several months after the subcutaneous introduction of a carcinogen. During the time of tumor growth, it is also informative to evaluate certain physiological criteria such as weight loss, general behavior, changes in the composition of physiological fluids, etc. The selected vital organs of dead animals are searched for evidence of cancer, and tumors are eventually examined histologically (23).

The occurrence of multiple tumors is not uncommon with very potent carcinogens. The induction of pulmonary adenomas and cellular abnormalities of the respiratory system of mice (24) led to more frequent studies involving exposure of experimental animals to carcinogenic smoke aerosols (25, 26).

Progress in chemical carcinogenesis investigations has been largely dependent on the selection of suitable animal models for extensive studies. A good example is the use of the Syrian hamster as test animal for respiratory cancer. Since the first work of this kind was reported in 1958 (27), this species has been widely used in tobacco carcinogenesis studies.

Two rules appear to be particularly important for animal studies in chemical carcinogenesis: (a) initial screening studies should be carried out with the same animal species, preferably those used by the majority of researchers; and (b) experimental conditions should be strictly specified because of the biological variations mentioned above.

Tumor initiation and growth are associated with significant morphological and biochemical alterations in the affected mammalian cells. When observed under standard conditions, such changes can be widely useful in evaluating the toxicity and other properties of a given chemical carcinogen. Thus, the correlation between carcinogenicity and DNA damage, mutagenicity, mitosis, fragmentation of chromosomes, or even more closely defined biochemical changes can be useful in making these evaluations.

Whereas carcinogenicity investigations in animals and epidemiological studies of human cancer ultimately provide far better assessment of potential hazards associated with a given chemical, experiments with tissue cultures and cell homogenates can also provide much useful information. In fact, the importance of *in vitro* experiments for screening carcinogens (or mutagens) has been rapidly increasing. The rapidity and simplicity of such tests are the main advantages.

The effects of cigarette smoke on certain cellular preparations were demonstrated many years ago (9); alterations were produced in mitosis, and significant cell damage and strong basal-cell hyperplasia in fetal lung cultures occurred. Since many carcinogens are mutagens, sensitive and relatively simple tests can now be designed for screening. It is assumed that the molecules of interest interact with DNA base pairs, and certain sites are metabolized to form reactive groups; cancer is caused by somatic mutations. Ames *et al.* (28) have shown that PAC metabolites are very powerful mutagens.

A mutagenicity test described by Ames *et al.* (29) in a different publication is based on the conversion of PAC by human autopsy or rat-liver homogenates to oxygenated metabolites which are subsequently assayed for mutagenic activity by *Salmonella* histidine mutants. For conversion of PAC to their corresponding mutagens, the oxygenating liver enzymes must first be induced. Tests are relatively simple to perform, yet impressive sensitivities, down to a few nanograms were demonstrated for several known carcinogens. Various modifications of this procedure as well as extensions to mammalian mutagenicity models are currently under development.

IV. STRUCTURE–ACTIVITY RELATIONSHIPS

Complex mixtures of potentially carcinogenic materials should first be assessed by animal experiments and mutagenicity tests. Even though many components of the mixture may exhibit synergistic effects during the primary action of PAC carcinogens, detailed studies of the tumor-causing properties of individual PAC are likely to yield fundamental information. Systematic structure–activity studies on the separate PAC undoubtedly provide an important data base for a better understanding of the biochemical causes and mechanisms of tumor induction.

Thus, investigations with model PAC at both molecular and cellular levels are needed. The experiments performed hitherto gave very clear evidence that among the many types of molecules that may be encountered in complex PAC mixtures, enormous differences exist in terms of biological activity. The fact that minor structural differences between PAC molecules often produce drastic changes in carcinogenicity has puzzled many scientists (see Appendix 5).

Studies with model PAC have been somewhat limited because of the difficulties encountered in isolating pure compounds from complex PAC mixtures, the unavailability of synthetic standard PAC, and the tedium and high cost involved in biological testing with experimental animals.

Although systematic studies aimed at recognition of relationships between the structures of PAC and their carcinogenicity date as far back as the 1930s, the problem is still far from being solved. Nevertheless, many observations have been made over the years and certain approximate, but useful, correlations are now available. Exceptions to these empirical rules are understandable, since according to the current theories of cancer, there are several possible routes leading to malignancy (21). Even before experimental evidence for metabolic products of PAH was obtained, it had been felt that certain selective reactions of such compounds determine their degree of biological activity.

Even though PAC are considered to be quite stable compounds, a certain degree of chemical reactivity results from the π-electron delocalization within the carcinogen molecules. Electron distributions partly predictable from quantum-mechanical calculations appear to correlate with both reactivity and carcinogenicity (30). The size of PAC molecules (31) and their "molecular thickness" (32) are criteria also involved in such correlations.

As defined by Pullman (33) long ago, this high electron density is located in the so-called "K region" of a PAC molecule (for example, in benz[a]anthracene).

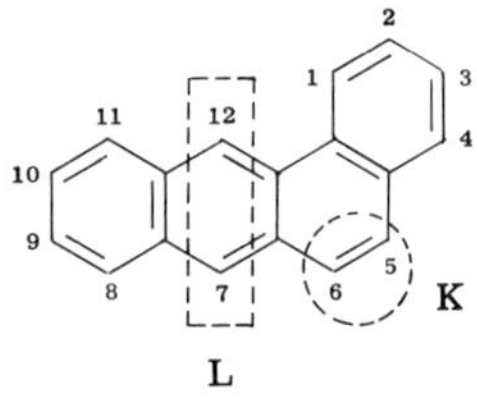

L

However, according to Pullman and Pullman (30), the existence of a so-called "L region" will make the molecule relatively noncarcinogenic. Thus, the regions of distinct electron density are the positions 5,6 and 7,12 (K region and L region, respectively). Similarly, the K region of phenanthrene is at the 9,10-position:

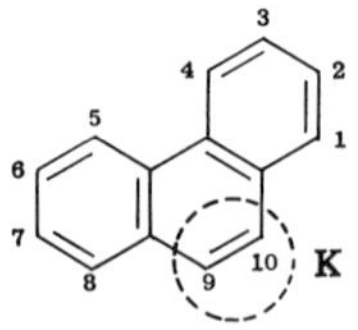

Both the K region and L region of a given molecule are reactive, but in different ways. According to the present knowledge of metabolism, a reactive carcinogen can be formed via epoxidation at the K region, while reaction at the L region primarily leads to a product that is easily detoxified. Although there are exceptions to the rule [e.g., benzofluoranthenes, as discussed by Dipple (*21*)], the requirement that a K region is present in a carcinogenic PAC is one of the useful suggested rules.

Additional criteria advocated by others include affinities of PAC for thermal energy electrons (*34*), photodynamic action (*35*), and ionization potential (*21*). From the point of view of chemical reactivity, the experimentally determined nucleophilicity for silver ion (*36*) and the ease of osmium tetroxide addition at the K region (see also Chapter 1) are also quite informative for the estimation of carcinogenicity. Similarly, biochemical criteria such as the rate of enzymatic oxidation at the K region and binding to DNA and other biomolecules may also be useful. While using pattern-recognition computational techniques, carcinogenicity can be predicted by analysis of both theoretical and certain measurable variables (*37, 38*).

Discoveries in the area of carcinogen metabolism have recently provided additional insights into structure–activity relationships. Structural similarities between the active metabolites of different PAC have brought about the proposed "bay-region" theory (*39*). The bay-region theory will be discussed later, in relation to PAC metabolism.

A. Parent PAH

Carcinogenicity of PAH has mainly been observed for tri-, tetra-, penta-, and hexacyclic compounds. As already discussed, the carcinogenic activity of a particular compound is dependent on various structural features of the molecule. Shape, size, and steric factors all seem to be important. Isomeric parent PAH may differ markedly in their activities; thus, benzo[*a*]pyrene has significantly greater activity than benzo[*e*]pyrene; activity varies within the series of dibenzanthracenes; and benzo[*g*]chrysene appears to be more active than benzo[*c*]chrysene (*40*).

Among the PAH molecules that have more than six rings, only a few have been found to exhibit any appreciable activity (*21*). Coronene, which is quite commonly encountered in many mixtures, is capable of producing only minor skin papillomas (*41*). Reasons for the limited activity of the larger PAH are not clear, although limited solubility may play a part; but it should be noted that many high-molecular-weight compounds remain to be isolated and/or synthesized and tested. Thus, any generalizations in this regard would be somewhat premature.

B. Alkylated PAH

Alkyl substitution at various positions on the parent PAH sometimes has an activating effect, but decreases in activity have also been observed (*42*).

For example, benzo[rst]pentaphene is a very potent carcinogen, while 5,8-dimethylbenzo[rst]pentaphene is inactive.

The location of an alkyl group within the PAH molecule may play a crucial role in its activity, since such substitution may alter the electron distribution within the molecule. Certain relationships between the structures of methylated PAH and their activities have been extensively studied. Selected cases are now discussed below. Within the series of methylated benz[a]anthracenes, the isomers have very different biological activities. Biological tests in mice (skin painting or subcutaneous injections) with monomethylbenz[a]anthracenes showed marked differences (43) in both carcinogenicity and chemical reactivity of various isomers.

From the synthesis and biological testing of x-fluoro-7-methyl-benz[a]-anthracenes (a total of 11 isomers), Newman hypothesized (43) that benz[a]-anthracenes may be metabolized in two ways: (a) metabolism at position 7 results in a series of reactions that lead to detoxification; whereas (b) metabolism at position 5 is associated with cancer activity. Blocking of position 5 by fluorination cancels the carcinogenicity of the compound. 7,12-Dimethylbenz[a]anthracene was found to be considerably more active than the 7-methyl derivative. Although the 12-methyl isomer is only weakly carcinogenic, the presence of a second methyl group in the 7-position results in more steric crowding. In general, steric factors may determine whether cancer-producing metabolism is favored relative to detoxification.

The preparation of trimethylbenz[a]anthracenes (44) yielded isomers of varying toxicity. Interpretation of the results of tests within this series of compounds is difficult until more studies become available. Again, steric

TABLE 3-1

Comparative Carcinogenic Activity of Benz[a]anthracenes

Compound	Number of rats	% with tumors	Number of tumors	Mean latent period ± probable error in days
7-Methyl	60	93	97	201 ± 2.7
6-Methyl	58	71	58	235 ± 4.1
8-Methyl	59	61	41	265 ± 8.7
12-Methyl	61	52	39	292 ± 10.7
10-Methyl	57	5	3	396
9-Methyl	59	5	3	413

factors appear important. Examples of the activities of various methylated benz[a]anthracenes are given in Table 3-1 (45).

Within the series of methylchrysenes, a strong dependence of carcinogenicity on the position of the methyl group has also been found (46). While chrysene itself has only slight carcinogenicity, substitution of a methyl group at position 5 converts it to a strong carcinogen (46).

C. Heterocyclic PAC

Numerous nitrogen-containing PAC have also been tested for their tumorigenicity, since such compounds are known to be present in the environment. Following earlier observations that some dibenzocarbazoles and dibenzacridines are active, detailed studies on the nitrogen PAC and their biological activities were carried out by Lacassagne et al. (47). These authors claim that the carcinogenicity of certain of these compounds may be greater than that of 3-methylbenz[j]aceanthrylene, a very potent carcinogen. Within the series of benzacridines, high activity was encountered with the angular

TABLE 3-2

Relation between pK_a and Carcinogenic Activity

Series	pK_a	P^a	I^b
Benz[c]acridine			
7,9-Dimethyl	4.26	81	41
7,10-Dimethyl	3.99	56	64
7,8,9,11-Tetramethyl	3.98	50	20
10,11-Dimethyl	3.74	0	0
10-Methyl	3.68		0
8-Methyl	3.67		0
5,7-Dimethyl	3.63	0	0
Unsubstituted	3.24		
Benz[a]acridine			
9,12-Dimethyl	5.13	11	0
12-Methyl	4.60		
8,9,12-Trimethyl	4.59	0	0
9-Methyl	4.52		
10-Methyl	4.22		
Unsubstituted	3.95		

[a] P = carcinogenic index by painting.
[b] I = carcinogenic index by injection.

types (benz[*a*]acridine and benz[*c*]acridine) as opposed to the linear benz[*b*]-acridine. Among the large number of alkylated benzacridines tested for carcinogenicity, the benz[*c*]acridines possess greater toxicity than the benz[*a*]acridines. It has been reported that there is a correlation between pK_b values of such compounds and their carcinogenic activities. Table 3-2 clearly suggests this (*46*). Studies of the dibenzacridines produce fewer correlations of carcinogenic activity with structure.

A great number of other heterocyclic PAC have been tested more recently, including compounds with more than one nitrogen atom per molecule, or containing a nitrogen with other heteroatoms such as sulfur or oxygen (*48–57*).

V. BINDING OF PAC TO CELLULAR AND SUBCELLULAR UNITS

The mechanism of cancer induction after a carcinogen has entered the body still remains relatively unknown, and clarification of this problem at the morphological level using scanning electron microscopy (*58*), and at the cellular and biochemical levels, have been sought for years by numerous investigators.

Chemical interactions between the constituents of the cells involved and the tumorigenic molecules have also been studied. Thus, *in vivo* studies demonstrated that PAH will bind to both DNA (*59*) and protein (*60, 61*) of mouse skin. Initial difficulties in proving chemically the existence of car-cinogen–biomacromolecule interactions in direct experiments with animals led to numerous *in vitro* studies. The binding of labeled PAC to important cellular components such as DNA, RNA, histones, etc., is well established, but the relative importance of carcinogen binding to different types of macromolecules cannot be fully assessed at this stage (*62*).

Testing the mitotic activities of condensates derived from tobacco smoke by applying them to growing root tips of *Allium cepa* led Venema (*63*) to postulate that a decrease of the mitotic frequency with an increasing amount of the carcinogen could be caused by either (a) hindered DNA synthesis; (b) possible interference with protein synthesis; (c) prevention of chromosome spiralization; or (d) a combination of all three factors. Although results suggest that disturbances in RNA rather than DNA synthesis occurred, the exact cause of mitotic irregularities could not be deduced.

Although the physical binding of PAH to various biological molecules (*64, 65*) has been proposed, it is now commonly accepted that covalent binding takes place. For example, tritiated benzo[*a*]pyrene bound to calf thymus DNA can only be released upon hydrolysis (*62*). In a different study

(66), a labeled carcinogen bound to DNA could not be dissociated from it by heat denaturation, solvent extraction, or gel chromatography.

It is quite evident that PAC as such cannot enter into chemical reactions with polar biopolymers, and it is now established that conversion to more polar and reactive metabolites is a necessary step that must precede the binding process. The chemistry related to such binding will be discussed in more detail later, but it should be pointed out that the various hypotheses relating to the nature of active PAC metabolites existed prior to the experimental verification of their chemical nature. In particular, it has been shown that the enzyme complex (67) isolated from microsomes of numerous tissues is essential in facilitating formation of the carcinogen–DNA complex.

Much research will be needed in the future to provide quantitative data on the binding of carcinogens. It will be important to consider the type of biopolymer and the activity of the oxidative enzyme complex in any given cell. In addition, the results of Grover and Sims (68) indicate that various PAC bind differently to protein and DNA (see Table 3-3). It is interesting to note that covalent binding of PAH carcinogens to mouse DNA has a strong genetic dependence (69, 70).

It may be significant that the binding of benzo[a]pyrene to calf thymus DNA takes place in the ratio of one carcinogen molecule per 50,000 nucleotides, and 1:500,000 for 7,12-dimethylbenz[a]anthracene; when RNA is substituted for DNA, a similar degree of binding occurs (62, 71).

TABLE 3-3

Enzyme-Catalyzed Reaction of Tritiated Polycyclic Hydrocarbons with Protein and DNA in Vitro

Hydrocarbon	Reaction with protein (mmoles/mole[a])	Reaction with DNA (μmoles/g atom of DNA P[b])
Benzo[a]pyrene	0.78	1.41
3-Methylcholanthrene (3-methylbenz[j]aceanthrylene)	0.73	0.78
7,12-Dimethylbenz[a]anthracene	0.78	0.64
Dibenz[a,h]anthracene	0.95	0.44
Dibenz[a,c]anthracene	1.07	0.56
Benz[a]anthracene	0.76	0.70
Pyrene	1.15	0.31
Phenanthrene	0.72	0.05

[a] MW of bovine plasma albumin taken as 64,000.

[b] Calculated on a basis of 8% phosphorus.

VI. METABOLISM

Although it is well established that certain carcinogenic PAC bind to various cellular structures and to macromolecules contained in living cells, there is still much to be explained as to which biochemical events take place in the processes of tumor initiation and growth. Progress in this field was recently reviewed (5), and it is beyond the scope of this chapter to discuss the many relevant communications. Instead, some of the most important findings of recent years will be summarized. It should be pointed out that the present state of knowledge in the area of PAC metabolism could not have been achieved without the recent significant advances in separation and identification techniques.

A. The "Ultimate Carcinogen"

It has long been believed that oxygenated and reactive metabolites, rather than the original PAC, are the active carcinogens. Many metabolites of PAC have subsequently been identified. Although the reactions of these compounds with biomacromolecules *in vitro* and *in vivo* as well as their mutagenic and carcinogenic properties were eventually proven, there are still many controversies as to which metabolic processes are most important. For example, is binding of PAC metabolites to proteins more important than similar reactions with nucleic acids? It is most likely that no single mechanism of carcinogenicity can be proposed, and that the associations of carcinogens with different biologically important molecules can all be important. Thus, there is little doubt that alterations of the structures of RNA and DNA through their reactions with oxygenated PAC derivatives are likely to affect their biological functions, cause mutations, cause chromosomal damage, etc. On the other hand, the importance of their reactions with proteins is supported by the correlation of carcinogenicity with the ability of a soluble protein of mouse skin to bind with PAC (72), and the association of carcinogens with a steroid-binding protein (65).

The active PAC may cause a number of events while in contact with biological material. Certain microsomal enzymes appear to be important in such reactions. Naturally, various PAC are likely to yield different oxygenated metabolites at different rates. These steps appear to be important, since they may determine which of the two alternative paths will predominate (see Fig. 3-1). Specifically, will a PAC molecule give rise to the "ultimate carcinogen," or will it be detoxified? Although many metabolites of the common carcinogens have been detected after treatment with cellular enzyme preparations, until recently (73), no *in vivo*-formed metabolite had been found to be more carcinogenic than the original PAC.

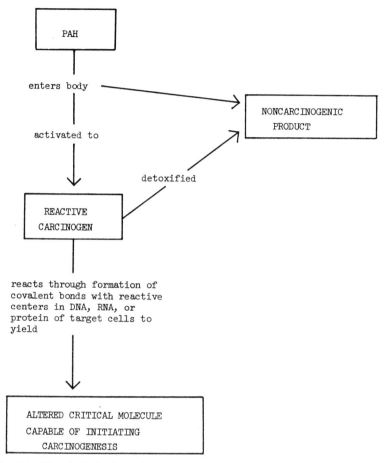

Fig. 3-1. Hypothetical scheme for the conversion of PAH to cancer-inducing molecular species. (M. L. Lee, Ph.D. Thesis, Indiana University, Bloomington, 1975.)

Which oxygenated products can be derived from carcinogenic PAC through the action of living cells? Although *in vitro* studies with various enzyme preparations provide some insights, it has been argued that the actual processes occurring *in vivo* (e.g., in experimental animals treated with a carcinogen) may be significantly different (*21, 74*).

Since, according to Fig. 3-1, the detoxification route is competing with that leading to the formation of a carcinogen–biomacromolecule complex, it is very likely that inhibition of specific enzyme(s) is the determining factor. The deactivation process may be "solubilization" of the PAH molecule through the usual process of biological conjugation in the liver, and elimination of conjugates through the kidneys. Thus, the formation of glucuronides, sulfates, and mercapturic acids (glutathione conjugates) is not surprising.

Since the formation of diols or phenols must precede such conjugation, it was suggested as early as 1950 by Boyland (75) that epoxides (also now commonly referred to in the literature as "arene oxides") are the most probable precursors of diols. It was also suggested that such reactive species could be the actual carcinogens. Subsequent synthesis of arene oxides and studies of their metabolic conversions have provided strong evidence for the precursor hypothesis (21). However, the carcinogenicity of the arene oxides is generally less than that of the original PAC.

Numerous studies on the formation of oxygenated metabolites have involved naphthalene as the substrate. This is a rather simple model, since all metabolites of naphthalene seem to arise from metabolic attack at the 1,2-bond. However, a survey of results from other PAC compounds indicates that the vulnerability of the K regions of various compounds to metabolic attack is quite different, and that other sites can also be oxygenated.

Interesting suggestions have been raised concerning the role of glutathione conjugation in the process of carcinogenesis (72, 76, 77). Metabolic studies suggest (74) that carcinogenic activity is associated with the absence of K-region phenols or dihydrodiols as metabolites (supposedly, the products of detoxification), whereas the presence of glutathione derivatives may have some relationship to cancer. Clearly, a rigorous quantitative evaluation of the degree of conjugation and the role of enzymes involved in the conjugation (78, 79) should be carried out.

The metabolism of several potent carcinogens such as benzo[a]pyrene, 3-methylbenz[j]aceanthrylene, dibenz[a,h]anthracene, and 7,12-dimethylbenz[a]anthracene has been studied in some detail (21). Various oxygenated metabolites were found, but K-region phenols and dihydrodiols were absent.

Some attention has also been paid to the keto forms of K-region phenols as possible candidates for the "ultimate carcinogen." Ketone intermediates can be formed during the rearrangement of certain arene oxides to phenols. The corresponding keto form of the oxygenated benzo[a]pyrene has been isolated (80). A similar case of tautomerism has been proven for 7,12-dimethylbenz[a]anthracene by Newman and Olson (81), who propose that the keto form may be a crucially important carcinogenic agent.

$$\text{(1)}$$

The formation of free radicals and their possible role in carcinogenesis have also been suggested (*82, 83*).

Methyl groups of PAH can also be metabolized, yielding hydroxymethyl derivatives and carboxylic acids (*84, 85*). Reactive esters of oxygenated methyl PAC have been suggested (*86*) as possible "ultimate carcinogens" for these potent tumorigenic agents. Subsequently, precursors of the reactive esters were extensively studied (*87, 88*) to determine their toxicity and binding to mouse skin and nucleic acids. A recent study by Rogan *et al.* (*89*) also suggests that benzylic esters might be the ultimate carcinogens from alkylated arenes; they bind to DNA by an ATP-mediated process.

B. Interactions of PAC Metabolites with Nucleic Acids

Much research activity in recent years has been concentrated on the mechanism of binding of PAC metabolites to DNA and nucleotides. The early findings (*90*) that the water-solubility of PAC increases in the presence of purines can now be regarded as the precursor of such studies. However, it has also long been known that the binding of PAC or their metabolites to nucleic acids can occur in different ways under different circumstances; the occurrence of more than one metabolite in such binding studies has been a major complicating factor (*74*).

It is now accepted that a covalent bond is required for binding. However, both a carcinogenic PAH or its metabolites are known to bind in various ways with or without enzymatic catalysis (*21*). Thus, it is not surprising that *in vitro* and *in vivo* investigations produce different results.

The reactions of potent carcinogens such as benzo[*a*]pyrene and 7,12-dimethylbenz[*a*]anthracene were investigated with various natural and synthetic polynucleotides; different reactivities with poly(A), poly(G), poly(I), poly(C), poly(U), etc., were found in both microsomal (*91*) and nonenzymatic systems (*92*). Although many PAC metabolites can react with nucleic acids, they may not necessarily be carcinogenic. Different reactivities of arene oxides with various nucleotides were observed by Grover and Sims (*93*).

An understanding of the mechanism involving binding of benzo[*a*]pyrene and benz[*a*]anthracene to nucleic acids seems to be at hand; this involves the initial formation of a non-K-region diol, and a subsequent enzymatic conversion of the diol to a diol epoxide. Thus, metabolic conversion of benz[*a*]-anthracene-8,9-dihydrodiol to 8,9-dihydro-8,9-dihydroxybenz[*a*]anthracene 10,11-oxide was accomplished (*94*); very strong binding of this compound to DNA was observed. Similar conclusions were also reached for benzo[*a*]-pyrene (*95*).

The diol epoxides can exist in two racemic forms:

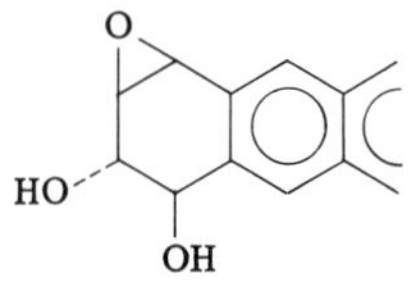

(1) 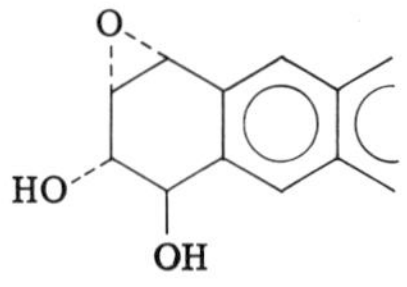   (2)

Hulbert predicted (96) that structure **1** should have a much greater reactivity than structure **2**. Indeed, an *in vivo*-formed adduct of the following structure has now been positively identified by a combination of structural methods (97):

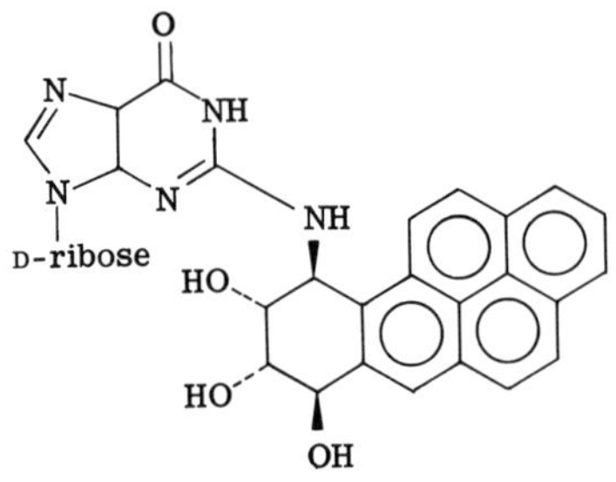

Moreover, benzo[a]pyrene-7,8-dihydrodiol has been shown to be more carcinogenic than benzo[a]pyrene (93). The elucidation of PAC metabolic routes has resulted in the proposal of the so-called "bay-region" theory of carcinogenicity that is briefly outlined below.

C. The Bay-Region Theory

Apparent similarities between the diol epoxide structures of benzo[a]-pyrene and substituted benz[a]anthracenes led Jerina *et al.* (39) to propose that the position of the reactive epoxide ring within the molecule is highly related to its biological activity. The activity is high if the epoxide ring becomes part of the "bay region" of the original PAC. The positions of bay regions are illustrated below and indicated by the arrows on the structures of phenanthrene, benz[a]anthracene, chrysene, and benzo[a]pyrene (39).

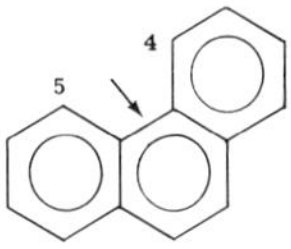

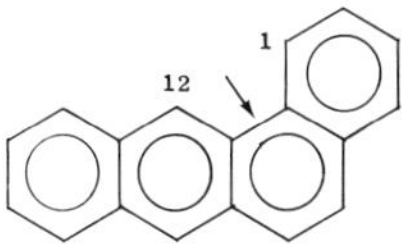

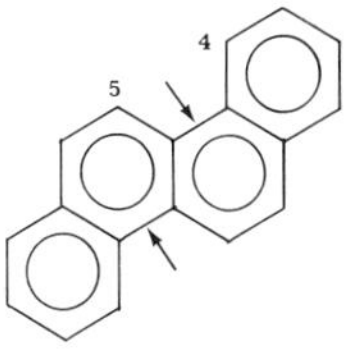

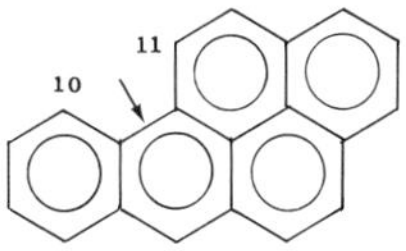

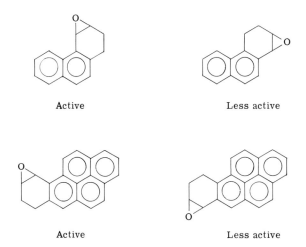

Although other diol epoxide structures may be produced by the metabolic conversion of a PAC, only those metabolites where the epoxide is part of the "bay" are highly active. For example, the following structures can be compared:

Active   Less active

Active   Less active

The bay-region theory is supported by the available carcinogenicity data (98) on substituted PAC. It also holds for 3-methylbenz[j]aceanthrylene (3-methylcholanthrene) and benzo[rst]pentaphene. This is supported by the ease of carbonium ion formation at benzylic positions of saturated terminal rings of PAC as indicated by molecular orbital calculations. A similar approach can be made in predicting the reactivities of diol epoxides (99).

It should be pointed out that the bay-region theory appears to correlate the observed carcinogenicity/mutagenicity effects with structural features of PAC more accurately than the previously popular K-region theory (*30, 33*). However, some overlap between the two theories exists (*39*), as a number of PAH with K regions also possess bay regions.

D. Enzymatic Systems Involved in PAC Metabolism

Unlike many other carcinogens that are polar and can bind directly to biological molecules, PAC need metabolic activation. As described above, formation of oxygenated metabolites is required prior to covalent binding with nucleic acids. Many cells contain the microsome-bound enzyme complex that provides this metabolic activation. Therefore, some cells have greater sensitivity than others toward the toxic effects of a PAC carcinogen.

Because of the membrane-bound nature of enzymes and their unavailability in a purified state, microsomal fractions of tissue homogenates that contain the enzyme complex are commonly used in carcinogenesis experiments. Rat-liver nuclei have been shown to contain the enzymes (*100*). Mason (*101*) refers to these enzymes as "mixed-function oxygenases" because it is proposed that they convert aromatic molecules to phenols, dihydrodiols, and epoxides. Mixed-function oxidases use one molecule of oxygen per molecule of substrate.

$$AH + 2\,e^- + O_2 \rightarrow AOH + O^{2-} \qquad [A = \text{aromatic moiety}]$$

The enzymatic hydroxylating system of the liver requires molecular oxygen and a source of electrons for its function. Thus, for example, the reduced form of nicotinamide adenine dinucleotide phosphate (NADP) is necessary for metabolic activation of PAC molecules. The electron transport processes are associated with the role that cytochrome P-450 apparently plays in such enzymatic systems. Its role in the metabolism of PAC and carcinogenesis has now been extensively investigated (*102–104*).

Hydroxylation is the common metabolic fate of aromatic compounds that are foreign to normal metabolism. Whereas these oxidative enzymes appear to have a broad spectrum of possible substrates, stereospecificity is important in PAC metabolism (*105, 106*).

Dawidow and Radomski (*107*) suggested that epoxides are key intermediates in the formation of phenols. The metabolism of PAC can be envisioned in the following way, with all steps carried out enzymatically.

$$\text{PAH} \xrightarrow{\text{oxidation}} \underset{\text{(arene oxide)}}{\text{epoxide}} \xrightarrow{\text{hydrolysis}} \text{dihydrodiol} \quad\Big\downarrow{\scriptstyle\text{dehydration}} \tag{2}$$

$$\text{phenol}$$

However, according to Mason (*101*), it may not be necessary to suggest individual precursors since "the nature of enzyme–oxygen–substrate interaction may determine whether phenol, epoxide, or dihydrodiol is formed, and in what amounts." Nevertheless, in connection with PAC metabolism research, the enzymes specifically mentioned are monoxygenase (PAC → arene oxides), epoxide hydrase (arene oxides → dihydrodiols), and arylhydrocarbon hydroxylase (which is responsible for the formation of a phenol from PAC). Clearly, the enzymatic reactions involved in the metabolism of PAC are complex.

Arylhydrocarbon hydroxylase and its biological role and functions have attracted much interest. This enzyme was also called "benzopyrene hydroxylase" before it became apparent that it is not specific for benzo[*a*]pyrene (*108*).

Gelboin *et al.* (*109*) demonstrated that arylhydrocarbon hydroxylase activity varies for different cells and that the cell sensitivity to the toxic effects of benzo[*a*]pyrene correlates well with the levels of the enzyme. Cells that are sensitive to the cytotoxic effects of the hydrocarbon contain the enzyme and NADPH, and are able to metabolize carcinogens readily. Enzyme activity is markedly reduced in cytotoxically resistant cells. Numerous studies also agree that the enzyme activity can be easily induced by pretreatment with one of the common PAH carcinogens. It has been shown that arylhydrocarbon hydroxylase is encountered in a great many cells of different origin, and the presence of the enzyme was shown histochemically in a variety of tissues (*110*), different animal species, and cell cultures (*111*). However, it is also known that different tissues and cell cultures yield somewhat different reaction products. Further, various tissues may exhibit different induction of enzyme activity after treatment with carcinogens (*112*). Thus, the processes of activation as well as detoxification of carcinogens may be species- and tissue-variable.

Characterization of arylhydrocarbon hydroxylase has been thus far insufficient, apparently owing to the fact that it is a membrane-bound enzymatic system. Enzyme kinetics were studied by Roberfroid *et al.* (*113*) and Hunt *et al.* (*114*), who found that Lineweaver–Burk plots of activity were nonlinear, presumably due to the use of homogenates.

The carcinogenic process and binding of PAC to DNA, RNA, and protein are inhibited by benzoflavones (*115–117*). Such compounds apparently reduce arylhydrocarbon hydroxylase activity.

Recent interest in arylhydrocarbon hydroxylase centers on using enzyme levels to identify individuals with a high probability of developing cancer. Trell *et al.* (*118*) measured the enzyme activity in lymphocytes of patients with invasive laryngeal carcinoma and found significantly high activities of the enzyme. There is now some evidence that the level of response of arylhydrocarbon hydroxylase is genetically determined (*119, 120*). Other interesting studies involve the activity of the enzyme in human skin (*121, 122*).

The functions of several other enzyme systems have also received some attention. For example, UDP glucuronosyl transferase (123) or β-glucuronidase (124) were observed to cause increased binding of PAC metabolites to DNA. Recently, the role of prostaglandin synthetase in the activation of PAC metabolism has been the subject of several studies (125–129). Prostaglandin-related systems may be implicated in the carcinogenicity of PAC.

VII. ANALYTICAL METHODS

With the currently increasing interest in PAC metabolism, future research efforts are likely to depend on analytical methods for the determination of oxygenated and conjugated PAC. Since the reactions of model PAC with microsomal enzymes and tissue cultures generally result in the formation of several metabolites, greater use will be made of chromatographic methods. Separations of various metabolites are needed for both structural studies and precise quantitation. Furthermore, very sensitive detection techniques are required for *in vivo* metabolic studies.

Unlike PAC which can be successfully analyzed by gas chromatography and combined GC/MS, the analysis of polar PAC metabolites and their

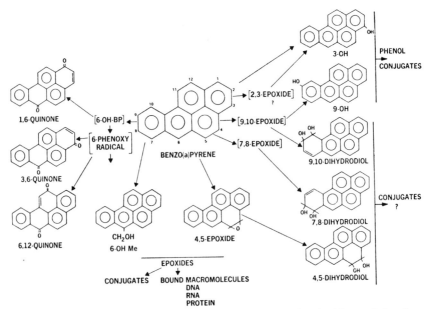

Fig. 3-2. Pathways of benzo[a]pyrene metabolism. [Reproduced with permission from J. K. Selkirk, *Adv. Chromatogr.* **16**, 1 (1978). Copyright, Marcel Dekker, Inc.]

conjugates will need to rely to a large extent on liquid chromatographic methods. However, there are indications that polar metabolites, although difficult to chromatograph in the gas phase, can be converted into more volatile derivatives by the well-established techniques of biomedical analysis. For example, silylation of the polar metabolites of naphthalene (*130*) and benzo[*a*]pyrene (*131*) prior to gas chromatography has been reported.

Although classical column chromatography or thin-layer chromatography is occasionally useful, the speed, efficiency, and convenience of high-pressure liquid chromatography are more advantageous. High-performance liquid chromatography has been used for the majority of studies of PAC metabolites. High column efficiencies are needed to cope with the complex products of metabolic pathways, as is exemplified (*132*) by the case of benzo[*a*]pyrene activation (Fig. 3-2). If greater resolution is still needed, recycle chromatography may be applicable (*132*).

PAC metabolites are strong UV absorbers, and the conventional detectors (UV absorption and fluorescence) of liquid chromatography are adequate for most purposes.

The separation capability of HPLC in this context is illustrated in Fig. 3-3, which shows the resolution of the metabolic (phenolic) products of benzo[*a*]pyrene (*133*).

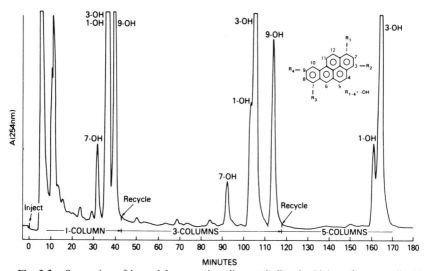

Fig. 3-3. Separation of benzo[*a*]pyrene phenolic metabolites by high-performance liquid chromatography with recycling. [Reproduced with permission from R. G. Croy, J. K. Selkirk, R. G. Harvey, J. F. Engel, and H. V. Gelboin, *Biochem. Pharmacol.* **25**, 227 (1976). Copyright, Pergamon Press.]

Although the question of the formation of *cis*- and *trans*-dihydrodiol derivatives appears settled, the knowledge of absolute stereochemistry is yet another essential step in understanding interactions of PAC metabolites with biologically important molecules. For example, nucleic acids themselves are known to be asymmetric. A recent application of high-performance liquid chromatography in the resolution of diastereoisomeric PAC dihydrodiol derivatives (*134*) demonstrates the feasibility of separation and quantitative analyses at the levels expected for *in vivo* experiments.

More recently, high-performance liquid chromatography was employed as a highly efficient and sensitive method for studies of PAC nucleoside conjugate formation (*135*). Previous studies on the binding of PAC to nucleic acids were complicated by the number of metabolites produced and the very low level of nucleic acid modification. Figure 3-4 shows an example of a reverse-phase chromatographic separation of the conjugates. Prior to this separation, the nucleic acids were modified with 7,12-dimethylbenz[*a*]-anthracene 5,6-oxide and hydrolyzed with alkali to mononucleotides.

Structural studies and detailed characterization of the metabolites and their conjugates can be performed after trapping the corresponding fractions at the exit of a liquid chromatograph. Depending on the amount of material available, different spectroscopic techniques can be utilized. During recent years, the possibilities of studying minute quantities of samples have been significantly extended by modern techniques; examples may be quoted of

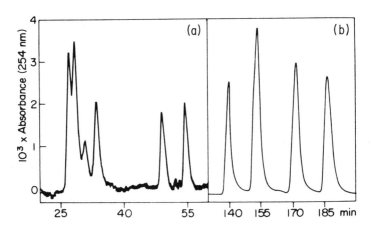

Fig. 3-4. Separation by reverse-phase chromatography of the nucleotides conjugated with metabolic products of 7,12-dimethylbenz[*a*]anthracene 5,6-oxide: (*a*) conjugation with tRNA; (*b*) conjugation with poly(G). [Reproduced with permission from A. Jeffrey, S. H. Blobstein, B. Weinstein, and R. G. Harvey, *Anal. Biochem.* **73**, 378 (1976). Copyright, Academic Press.]

the characterization of a carcinogen–DNA conjugate by low-temperature fluorescence (*136*) and the total characterization of the *in vitro*-formed benzo[*a*]pyrenetetrahydrodiol–epoxide–guanosine adduct by the combination of NMR, circular dichroism, and mass spectroscopic techniques (*97*).

REFERENCES

1. P. Pott, "Chirurgical Observations," 1775, article reproduced in *Natl. Cancer Inst. Monogr.* **10,** 7 (1963).

2. C. E. Searle, ed., "Chemical Carcinogens." ACS Monogr. 173, Am. Chem. Soc., Washington, D.C., 1976.

3. R. I. Freudenthal and P. W. Jones, eds., "Polynuclear Aromatic Hydrocarbons: Chemistry, Metabolism and Carcinogenesis," Vol. 1. Raven, New York, 1976.

4. P. W. Jones and R. I. Freudenthal, eds., Carcinogenesis—A Comprehensive Survey, Vol. 3: Polynuclear Aromatic Hydrocarbons" Raven, New York, 1978.

5. H. V. Gelboin and P. O. P. Ts'o, eds., "Polycyclic Hydrocarbons and Cancer." Academic Press, New York, 1978.

6. E. L. Wynder and K. Mabuchi, *Prev. Med.* **1,** 300 (1972).

7. P. Shubik, *Proc. Natl. Acad. Sci. U.S.A.* **69,** 1052 (1972).

8. L. B. Lave and E. P. Seskin, *Science* **169,** 723 (1970).

9. E. L. Wynder and D. Hoffmann, "Tobacco and Tobacco Smoke." Academic Press, New York, 1967.

10. C. Leuchtenberger, R. Leuchtenberger, and A. Schneider, *Nature* (*London*) **241,** 137 (1973).

11. C. Leuchtenberger, R. Leuchtenberger, V. Ritter, and N. Inui, *Nature* (*London*) **242,** 403 (1973).

12. J. Marcotte, F. S. Skelton, M. G. Côté, and H. Witschi, *Toxicol. Appl. Pharmacol.* **33,** 231 (1975).

13. J. L. Hartwell, "Survey of Compounds Which Have Been Tested for Carcinogenic Activity." U.S. Public Health Services Publ. 149, Washington, D.C., 1951.

14. P. Shubik and J. L. Hartwell, "Survey of Compounds Which Have Been Tested for Carcinogenic Activity," Suppl. 1. U.S. Public Health Service Publ. 149, Washington, D.C., 1957.

15. P. Shubik and J. L. Hartwell, "Survey of Compounds Which Have Been Tested for Carcinogenic Activity," Suppl. 2. U.S. Public Health Service Publ. 149, Washington, D.C., 1969.

16. J. I. Thompson and Co., "Survey of Compounds Which Have Been Tested for Carcinogenic Activity," 1968–1969. U.S. Public Health Service, Washington, D.C.

17. A. Dipple, *in* "Chemical Carcinogens" (C. E. Searle, ed.), pp. 258–271. ACS Monogr. 173. Am. Chem. Soc., Washington, D.C., 1976.

18. B. J. Van Duuren, *in* "Chemical Carcinogens" (C. E. Searle, ed.), p. 24. ACS Monogr. 173, Am. Chem. Soc., Washington, D.C., 1976.

19. D. Hoffmann and E. L. Wynder, *J. Air Pollut. Contr. Assoc.* **13,** 322 (1963).

20. J. S. Rhim, W. Vass, H. Y. Cho, and R. J. Huebner, *Int. J. Cancer* **7,** 65 (1971).

21. A. Dipple, *in* "Chemical Carcinogens" (C. E. Searle, ed.), p. 245. ACS Monogr. 173, Am. Chem. Soc., Washington, D. C., 1976.

22. J. D. Scribner, *J. Natl. Cancer Inst.* **50,** 1717 (1973).

23. E. A. Graham, A. B. Croninger, and E. L. Wynder, *Cancer Res.* **17,** 1058 (1957).
24. J. M. Essenberg, *Science* **116,** 561 (1952).
25. Reference 9, p. 210.
26. Reference 9, p. 224.
27. C. Leuchtenberger, R. Leuchtenberger, P. F. Doolin, and P. Shaffer, *Cancer* **11,** 490 (1958).
28. B. N. Ames, P. Sims, and P. L. Grover, *Science* **176,** 47 (1972).
29. B. N. Ames, W. E. Durston, E. Yamasaki, and F. D. Lee, *Proc. Natl. Acad. Sci. U.S.A.* **70,** 2281 (1973).
30. A. Pullman and B. Pullman, *Adv. Cancer Res.* **3,** 117 (1955).
31. J. C. Arcos and M. F. Argus, *Adv. Cancer Res.* **11,** 305 (1968).
32. J. Pataki and R. Malick, *J. Med. Chem.* **15,** 905 (1972).
33. A. Pullman, *C.R. Hebd. Seances Acad. Sci.* **221,** 140 (1945).
34. J. E. Lovelock, A. Zlatkis, and R. S. Becker, *Nature (London)* **193,** 540 (1962).
35. S. S. Epstein, M. Small, H. L. Falk, and N. Mantel, *Cancer Res.* **24,** 855 (1964).
36. R. E. Kojahl and H. J. Lucas, *J. Am. Chem. Soc.* **76,** 3931 (1954).
37. G. M. Badger, *Cancer Res.* **2,** 73 (1954).
38. B. Norden, U. Eglund, and S. Wold, *Acta Chem. Scand. B.* **32,** 602 (1978).
39. D. M. Jerina, H. Yagi, R. E. Lehr, D. R. Thakker, M. Schaefer-Ridder, J. M. Karle, W. Levin, A. W. Wood, R. L. Chang, and A. H. Cohney, *in* "Polycyclic Hydrocarbons and Cancer" (H. V. Gelboin and P. O. P. Ts'o, eds.), p. 173. Academic Press, New York, 1978.
40. W. C. Hueper and W. D. Conway, "Chemical Carcinogenesis and Cancers." C. C. Thomas, Springfield, Illinois, 1964.
41. B. L. Van Buuren, A. Sivak, L. Langseth, B. M. Goldschmidt, and A. Segal, *Natl. Cancer Inst. Monogr.* **28,** 173 (1968).
42. R. Schoental, *in* "Polycyclic Hydrocarbons" (E. Clar, ed.), p. 133. Academic Press, New York, 1964.
43. M. S. Newman, *in* "Polynuclear Aromatic Hydrocarbons: Chemistry, Metabolism, and Carcinogenesis" (R. I. Freudenthal and P. W. Jones, eds.), Vol. 1, p. 203. Raven, New York, 1976.
44. M. S. Newman and W. M. Huang, *J. Med. Chem.* **20,** 179 (1977).
45. W. F. Dunning and M. R. Curtis, *J. Natl. Cancer Inst.* **25,** 387 (1960).
46. S. S. Hecht, M. Loy, R. Mazzarese, and D. Hoffmann, *in* "Polycyclic Hydrocarbons and Cancer" (M. V. Gelboin and P. O. P. Ts'o, eds.), p. 119. Academic Press, New York, 1978.
47. A. Lacassagne, N. P. Buu-Hoi, R. Daudel, and F. Zajdela, *Adv. Cancer Res.* **9,** 316 (1965).
48. A. Lacassagne, N. P. Buu-Hoi, F. Zajdela, F. Perin, and P. Jacquignon, *Nature (London)* **191,** 1005 (1961).
49. A. Lacassagne, N. P. Buu-Hoi, F. Zajdela, P. Jacquignon, and F. Perin, *C.R. Hebd. Seances Acad. Sci.* **257,** 818 (1963).
50. A. Lacassagne, N. P. Buu-Hoi, F. Zajdela, and P. Magille, *C.R. Hebd. Seances Acad. Sci.* **258,** 3387 (1964).
51. A. Lacassagne, N. P. Buu-Hoi, F. Zajdela, O. Perrin-Roussel, P. Jacquignon, F. Perin, and J.-P. Hoeffinger, *C.R. Hebd. Seances Acad. Sci.* **271,** 1474 (1970).
52. A. Lacassagne, N. P. Buu-Hoi, F. Zajdela, C. Stora, M. Mangane, and P. Jacquignon, *C.R. Hebd. Seances Acad. Sci.* **272,** 3102 (1971).
53. A. Lacassagne, N. P. Buu-Hoi, F. Zajdela, P. Jacquignon, and M. Mangane, *Science* **158,** 387 (1967).
54. F. Zajdela, N. P. Buu-Hoi, P. Jacquignon, and M. Dufour, *Br. J. Cancer* **26,** 262 (1972).
55. A. Sellakumar and P. Shubik, *J. Natl. Cancer Inst.* **48,** 1641 (1972).
56. M. R. Guerin, J. L. Epler, W. H. Griest, B. R. Clark, and T. K. Rao, *in* "Carcinogenesis—

A Comprehensive Survey, Vol. 3: Polynuclear Aromatic Hydrocarbons" (P. W. Jones and R. I. Freudenthal, eds.), p. 21. Raven, New York, 1978.

57. H. Schwind, F. Oesch, F. Zajdela, P. Jacquignon, A. Croisy, F. Perin, and H. Glatt, *Experientia* **34**, 918 (1978).

58. A. E. Williams, J. Beattie, J. M. Allen, and J. F. Murphy, *Br. J. Cancer* **34**, 311 (1976).

59. P. Brooks and P. D. Lawley, *Nature (London)* **202**, 781 (1964).

60. E. C. Miller, *Cancer Res.* **11**, 100 (1951).

61. C. Heidelberger and M. G. Moldenhauer, *Cancer Res.* **16**, 442 (1956).

62. H. V. Gelboin, *Cancer Res.* **29**, 1272 (1969).

63. G. Venema, *Chromosoma* **10**, 679 (1959).

64. D. J. Williams and B. R. Rabin, *Nature (London)* **232**, 102 (1971).

65. G. Litwack, B. Ketterer, and I. M. Arias, *Nature (London)* **234**, 466 (1971).

66. M. M. Coombs, A. M. Kissonerghis, and J. A. Allen, *Cancer Res.* **36**, 4387 (1976).

67. A. H. Conney, E. C. Miller, and J. A. Miller, *J. Biol. Chem.* **228**, 753 (1957).

68. P. L. Grover and P. Sims, *Biochem. J.* **110**, 159 (1968).

69. D. H. Phillips, P. L. Grover, and P. Sims, *Int. J. Cancer* **22**, 487 (1978).

70. A. R. Boobis, R. E. Kouri, and D. W. Herbert, *Cancer Detect. Prev.* **2**, 83 (1979).

71. T. Meehan and K. Straub, *Nature (London)* **277**, 410 (1979).

72. C. W. Abel and C. Heidelberger, *Cancer Res.* **22**, 921 (1962).

73. J. Kapitulnik, W. Levin, A. H. Conney, H. Yagi, and D. M. Jerina, *Nature (London)* **266**, 378 (1977).

74. P. Sims, *Biochem. Pharmacol.* **19**, 795 (1970).

75. E. Boyland, *Biochem. Soc. Symp.* **5**, 40 (1950).

76. P. L. Grover and P. Sims, *Biochem. Pharmacol.* **19**, 2251 (1970).

77. T. Kuroki and C. Heidelberger, *Biochemistry* **11**, 2116 (1972).

78. J. R. Bend, Z. Ben-Zui, J. Van Anda, P. M. Dansette, and D. M. Jerina, *in* "Polynuclear Aromatic Hydrocarbons: Chemistry, Metabolism and Carcinogenesis" (R. I. Freudenthal and P. W. Jones, eds.), Vol. 1, p. 63. Raven, New York, 1976.

79. M. S. Moron, J. W. Pierre, and B. Mannervik, *Biochim. Biophys. Acta* **582**, 67 (1979).

80. C. R. Raha, *Bull. Soc. Chim. Biol.* **52**, 105 (1970).

81. M. S. Newman and D. R. Olson, *J. Am. Chem. Soc.* **96**, 6207 (1974).

82. G. Nagata, M. Kodama, and Y. Tagashira, *Gann* **58**, 493 (1967).

83. G. Nagata, M. Kodama, and Y. Ioki, *in* "Polycyclic Hydrocarbons and Cancer" (H. V. Gelboin and P. O. P. Ts'o, eds.), p. 247. Academic Press, New York, 1978.

84. E. Boyland and P. Sims, *Biochem. J.* **95**, 780 (1965).

85. E. Boyland and P. Sims, *Biochem. J.* **104**, 394 (1967).

86. E. C. Miller and J. A. Miller, *Proc. Soc. Exp. Biol. Med.* **124**, 915 (1967).

87. A. Dipple and T. A. Slade, *Eur. J. Cancer* **6**, 417 (1970).

88. A. Dipple and T. A. Slade, *Eur. J. Cancer* **7**, 473 (1971).

89. E. G. Rogan, R. W. Roth, and E. L. Cavalieri, *Proc. Am. Assoc. Cancer Res.* **20**, 54 (1979).

90. H. Weil-Malherbe, *Biochem. J.* **40**, 351 (1946).

91. C. Pietropaolo and I. B. Weinstein, *Cancer Res.* **35**, 2191 (1975).

92. S. H. Blobstein, I. B. Weinstein, D. Grunberger, J. Weisgras, and R. G. Harvey, *Biochemistry* **14**, 3451 (1975).

93. P. L. Grover and P. Sims, *Biochem. J.* **129**, 41 (1972).

94. A. J. Swaisland, A. Hewer, L. Pal, G. R. Keysell, J. Booth, P. L. Grover, and P. Sims, *FEBS Lett.* **47**, 34 (1974).

95. P. Sims, P. L. Grover, A. Swaisland, K. Pal, and A. Hewer, *Nature (London)* **252**, 326 (1974).

96. P. B. Hulbert, *Nature (London)* **256**, 146 (1975).

97. A. M. Jeffrey, K. W. Zennette, S. H. Blobstein, I. B. Weinstein, F. A. Beland, R. G. Harvey, H. Kasai, I. Miura, and K. Nakanishi, *J. Am. Chem. Soc.* **98,** 5714 (1976).

98. D. M. Jerina and J. W. Daly, *in* "Drug Metabolism" (D. V. Parke and R. L. Smith, eds.), p. 13. Taylor and Francis, London, 1977.

99. D. M. Jerina and R. E. Lehr, *in* "Microsomes and Drug Oxidations" (J. R. Gillette, A. H. Conney, G. J. Cosmides, R. W. Estabrook, J. R. Fouts, and G. J. Mannering, eds.), p. 709. Pergamon, Oxford, 1977.

100. A. S. Kwandwala and C. B. Kasper, *Biochem. Biophys. Res. Commun.* **54,** 1241 (1973).

101. H. S. Mason, *Adv. Enzymol.* **19,** 79 (1957).

102. R. W. Estabrook, J. Werringloer, J. Capdevila, and R. A. Prough, *in* "Polycyclic Hydrocarbons and Cancer" (H. V. Gelboin and P. O. P. Ts'o, eds.), p. 285. Academic Press, New York, 1978.

103. M. J. Coon and K. P. Vatsis, *in* "Polycyclic Hydrocarbons and Cancer" (H. V. Gelboin and P. O. P. Ts'o, eds.), p. 336. Academic Press, New York, 1978.

104. J. Lampe and G. Butschak, *Pharmazie* **33,** 407 (1978).

105. P. L. Grover, A. Hewer, and P. Sims, *Biochem. Pharmacol.* **21,** 2713 (1972).

106. D. M. Jerina, H. Selander, H. Yagi, M. C. Wells, J. F. Davey, V. Mahaderan, and D. T. Gibson, *J. Am. Chem. Soc.* **98,** 5988 (1976).

107. B. Dawidow and J. L. Radomski, *J. Pharmacol. Exp. Ther.* **107,** 259 (1953).

108. L. J. Alfred and H. V. Gelboin, *Science* **157,** 75 (1967).

109. H. V. Gelboin, E. Huberman, and L. Sachs, *Proc. Natl. Acad. Sci. U.S.A.* **64,** 1188 (1969).

110 L. W. Wattenberg and J. L. Leong, *J. Histochem. Cytochem.* **10,** 412 (1962).

111. D. W. Nebert and H. V. Gelboin, *J. Biol. Chem.* **243,** 6250 (1968).

112. M. D. Burke and R. A. Prough, *Biochem. Pharmacol.* **25,** 2187 (1976).

113. M. Roberfroid, J. Cumps, and C. Razzouk, *Arch. Int. Physiol. Biochim.* **83,** 396 (1975).

114. W. G. Hunt, S. E. Knight, and L. F. Soyka, *Int. Symp. Detect. Prev. Cancer, 3rd,* p. 288, 1976.

115. L. Diamond, R. McFall, J. Miller, and H. V. Gelboin, *Cancer Res.* **32,** 731 (1972).

116. L. W. Wattenberg and J. L. Leong, *Cancer Res.* **30,** 1922 (1970).

117. L. Diamond, R. McFall, J. Miller, and H. V. Gelboin, *Cancer Res.* **32,** 731 (1972).

118. B. Trell, R. Korsgaard, B. Hood, P. Kitzing, G. Norden, and B. G. Simonsson, *Lancet* **2,** 140 (1976).

119. T. D. Shultz, C. E. Smith, and C. D. Stewart, *Int. Symp. Detect. Prev. Cancer, 3rd,* p. 311, 1976.

120. S. A. Atlas, E. S. Vesell, and D. W. Nebert, *Cancer Res.* **36,** 4619 (1976).

121. P. H. Chapman, M. D. Rawlins, and S. Shuster, *Br. J. Clin. Pharmacol.* **7,** 499 (1979).

122. P. H. Chapman, M. D. Rawlins, and S. Shuster, *Lancet* **1,** 297 (1979).

123. W. E. Fahl, A. L. Schen, and C. R. Jefcoate, *Biochem. Biophys. Res. Commun.* **85,** 891 (1978).

124. N. Kinoshita and H. V. Gelboin, *Science* **199,** 307 (1978).

125. A. Lupulescu, *Nature (London)* **272,** 634 (1978).

126. A. Lupulescu, *Experientia* **34,** 785 (1978).

127. L. J. Marnett, G. A. Reed, and D. J. Dennison, *Biochem. Biophys. Res. Commun.* **82,** 210 (1978).

128. K. Sivarajah, M. W. Anderson, and T. E. Eling, *Fed. Proc. Fed. Am. Soc. Exp. Biol.* **37,** 607 (1978).

129. L. J. Marnett, J. T. Johnson, and G. A. Reed, *Proc. Am. Assoc. Cancer Res.* **20,** 60 (1979).

130. W. G. Stillwell, M. G. Horning, R. M. Hill, and G. W. Griffin, *Fed. Proc. Fed. Am. Soc. Exp. Biol.* **3,** 585 (1979).

131. G. Takahashi, K. Kinoshita, K. Hashimoto, and K. Yasuhira, *Cancer Res.* **39,** 1814 (1979).

132. J. K. Selkirk, *Adv. Chromatogr.* **16,** 1 (1978).

133. R. G. Croy, J. K. Selkirk, R. G. Harvey, J. F. Engel, and H. V. Gelboin, *Biochem. Pharmacol.* **25,** 227 (1976).

134. R. G. Harvey and H. Cho, *Anal. Biochem.* **80,** 540 (1977).

135. A. Jeffrey, S. H. Blobstein, B. Weinstein, and R. G. Harvey, *Anal. Biochem.* **73,** 378 (1976).

136. V. Ivanovic, N. E. Geacintov, and I. B. Weinstein, *Biochem. Biophys. Res. Commun.* **70,** 1172 (1976).

4

Collection, Extraction, and Fractionation

I. INTRODUCTION

An environmental sample originating from a source rich in PAC, such as coke-oven effluents (1–4) or coal-tar volatiles (5), may be analyzed directly if a selective separation/detection method such as GC/MS or HPLC/fluorimetry is used. However, most materials from the environment contain only traces of PAC. For example, benzo[a]pyrene, the most studied hydrocarbon in cigarette smoke condensate, is present only at levels of between 1 and 2 ppm (6); while individual PAC are present in water (7, 8), foods (9), soils (10), and sediments (10) at the ppb level or below (see Chapter 2). Efficient extraction, and then separation from other extractable compounds is hence necessary so that the concentration of PAC is adequate for the various analyses discussed in this monograph.

Before clean-up methods are applied, however, the need to establish more closely the parameters affecting environmental pollution means that sampling objectives and methods must be carefully defined and standardized. For example, in sampling the PAC associated with airborne particulates, the range of particle sizes on which PAC are adsorbed must be known, and methods devised for the collection of these and also the aerosols in which PAC are also present. In assessing the PAC originating from vehicle exhausts, attention must be paid to the fuel, engine performance, load, etc.

In this chapter, particular attention is therefore paid to the *efficiencies* of the collection, extraction, and clean-up stages during the analysis of PAC from air, water, soil, and food. Only with these considerations can the proper significance of analytical data be assessed.

II. COLLECTION AND EXTRACTION

A. Air and Combustion Effluents

PAC are generally considered not to exist as pure substances in the atmosphere, but to be adsorbed on suspended particulate matter with average diameter less than 10 μm (*11–16*). A detailed study by Katz and Pierce (*13, 14*) of the quantitative distribution of PAH in relation to particle size suggested that 70–90% of these compounds are associated with aerosol particles in the respirable size range (*14*) (below 5 μm) (*17*), with a large proportion of the benzo[*a*]pyrene-containing particle mass having a diameter less than 1.1 μm (*13*). Similar work by DeWiest confirmed that 90% of the total mass of benzo[*a*]pyrene is present in particles with aerodynamic diameter less than 2 μm, and as much as 50% may be associated with particles less than 0.7 μm (*18*) in diameter. Very similar results were obtained in independent studies by Bjorseth (*12*) and by Broddin *et al.* (*19*) for the size distribution of PAH-containing particles from coke-oven emission sources. In Broddin's work, maximum concentrations of the 12 major compounds present were found for aerodynamic diameters between 0.85 and 1.8 μm (Fig. 4-1), with an average of 94% of all PAH on particles below 2.9 μm (*19*). For 20 PAH, Bjorseth (*12*) found the largest amounts adsorbed on particles in the diameter ranges 0.9–3 μm, with less than 1% on nonrespirable particles. Particles in air with diameters larger than 10 μm tend to settle out as dust. Particles in the range of 1–10 μm have mean residence times of the order of 10–100 hr, while for submicron particles residence times are of the order of 100–1000 hr (*20*). Since the particles which penetrate into the respiratory tract are just those associated with most of the PAC present in the air, the efficient collection of suspended matter is a primary concern in environmental analysis. Methods of taking samples of particulate matter in the air have been reviewed by Hrudey (*21*), by Cheremisinoff and Morresi (*22*), and by Warner (*23*). Collection is by impingement, precipitation (thermal or electrostatic), cyclone, cascade impactor, or most commonly by filtering a known volume of air through a porous filter with a lower size limit of 0.1–1 μm at a regulated flow-rate.

In the high-volume (Hi-Vol) sampling method (*24*), up to 100 m^3/hr air is drawn through 20 × 25-cm glass-fiber filters, so that up to 1 g of particulate matter may be collected in 24 hr. Air entering the sampler is caused to change direction (Fig. 4-2) by at least 90° before entering a horizontal filter, so that the definition of a truly suspended dust (*23*) is met. A rotameter is used to calibrate (23) the flow rate (0.85–1.7 m^3/min); Hrudey has reviewed (*21*) the factors leading to irreproducibility of sampling by the Hi-Vol

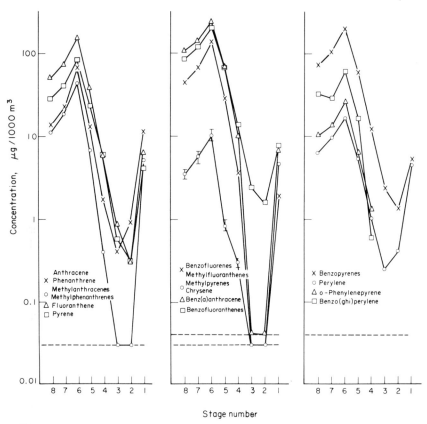

Fig. 4-1. Particle size distribution of polycyclic aromatic hydrocarbons in coke-oven emission sample. Standard deviation is shown on the methylpyrene curve. Detection limits are indicated by spaced bar lines. Stage numbers correspond to aerodynamic size diameters in μm: 1, $r > 10$; 2, $6.2 > r > 10$; 3, $4.0 > r > 6.2$; 4, $2.9 > r > 4.0$; 5, $1.8 > r > 2.9$; 6, $0.85 > r > 1.8$; 7, $0.52 > r > 0.85$; 8, $0.35 > r > 0.52$. [Reproduced with permission from G. Broddin, L. Van Vaeck, and K. Van Cauwenberghe, *Atmos. Environ.* **11,** 1061 (1977). Copyright, Pergamon Press, Inc.]

method, although the average deviation over 239 pairs of particulate mass concentration results has been determined as only 4.5% (25).

Glass-fiber filters used in high-volume sampling should have collection efficiencies of at least 99% (specified often by manufacturers as 99.9%) for particles of 0.3 μm or larger in diameter. These filters are also generally used in "nonstandard" sampling of air for analysis of PAC. Cellulose filters offer more resistance to air flow and are more useful in sampling at lower flow rates, e.g., 2 m^3/day, in the equipment used by the National Survey of Air Pollution in the United Kingdom (26); the blackness of these filters is

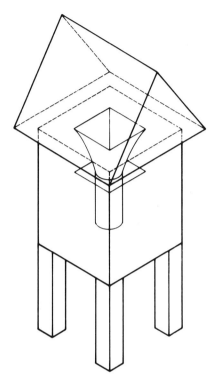

Fig. 4-2. High-volume air sampler in shelter. [Reproduced with permission from S. E. Hrudey, *in* "Handbook of Air Pollution Analyses" (R. Perry and R. J. Young, eds.), Chapman Hall, London, 1977, Chap. 1.]

used to determine total particulates (27), but may also correlate (28) with the PAC content. Relative efficiencies of glass fiber, cellulose, and other filters have been compared (27, 29), but silver membranes with 0.8 μm porosity are recommended by both the National Institute of Occupational Safety and Health (NIOSH) and the Occupational Safety and Health Administration (OSHA) in filters for personal monitoring for occupational health purposes; a glass fiber prefilter is used ahead of the silver membrane to avoid blockage (30, 31).

The very fine particles associated with PAC in air, extending down to molecular clusters of the order of 5 nm in diameters (12), may be collected on polyurethane plugs located in the airstream behind the glass-fiber filter (32). Such plugs minimize, by adsorption, the inevitable losses from filters arising from the saturated or equilibrium vapor concentrations of PAC (33). Without these precautions, the collection efficiency of PAC with four rings or less is always open to question (33), although for benzopyrenes and benzofluoranthenes, losses may be minimized by adsorption effects, and for high-molecular-weight PAC such as benzo[*ghi*]perylene and coronene, losses are thought to be insignificant. Krstulovic *et al.* (*32*) have shown how the ratio of weight of PAH collected by a glass-fiber filter to the weight collected by a polyurethane-foam adsorbent depends on both the compound concerned and on the location. For low PAH levels, more is collected on the polyurethane than on the filter. More recently, Cautreels and Van Cauwenberghe (*34*) used a tube packed with Tenax GC to trap compounds not retained on glass-fiber filters in an investigation of the distribution of organic pollutants between airborne particulate matter and the corresponding gas phase. A distribution factor, defined as concentration of PAH in sampled air adsorbed on particulates divided by concentration in the gas phase, varied from 0.03 (anthracene and phenanthrene) to 11.5 (benzofluoranthene). A sampling system using only Tenax GC as adsorbent has been described (*35*).

The above problems associated with collection efficiency are accentuated during the sampling of PAH from combustion effluents such as stack gases and vehicle exhausts. Filters alone are rendered ineffective by the marked temperature dependence of the relation between the vapor and particulate phases. The ASTM standard procedure for in-stack sampling of particulates requires filtration through glass-wool filter tubes (up to 480°C) or alundum thimbles (up to 550°C) with a condenser arrangement for materials in the gas phase (*36*). The British Standard equipment is similar but lacks a condenser; only particles over 1 μm in diameter are collected with greater than 98% efficiency (*37*). However, the high-volume sampling equipment most commonly used in Britain is probably the internal cyclone and filter sampling apparatus developed by the British Coal Utilization Research Association (*38*).

The United States Environmental Protection Agency Method 5 procedure (*39*) for in-stack sampling employs a heated filter followed by a series of impingers (Fig. 4-3) and has been modified (*40*) to sample a larger volume. A newer, more efficient version (*40*) developed by the Battelle, Columbus, Laboratories incorporates an adsorbent sampler (Fig. 4-4) between the filter and impingers, and has been accepted by the EPA (*41*). The adsorbent [Tenax GC (*40*) or Amberlite XAD resin (*42*)] is maintained at a temperature low enough to retain efficiently PAH not held on filterable particulates, but high enough to preclude condensation of water. In parallel studies, more than twice the weight of PAH was obtained with the adsorbent sampler system than with the original EPA method. XAD-2 is in fact preferred as the adsorbent since it has a higher capacity than Tenax GC, especially for lower boiling compounds (*43*). On the other hand, high collection efficiencies have been found for the vapors of naphthalene, 1-methylnaphthalene, and

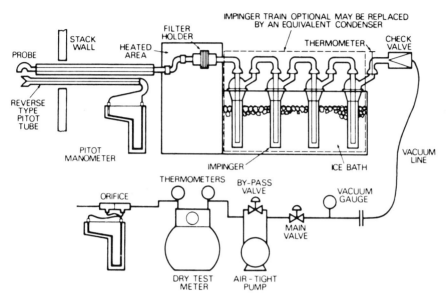

Fig. 4-3. Particulate sampling train for in-stack combustion effluents. (Reproduced with permission from *Fed. Regist.*, **36**, No. 247, 24876.)

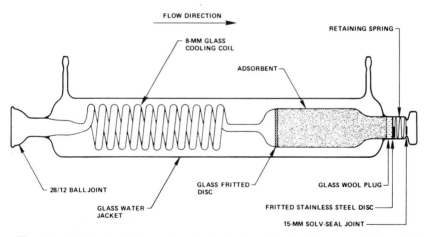

Fig. 4-4. Adsorbent sampling system for combustion effluents. [Reproduced with permission from P. E. Strup, R. D. Giammar, T. B. Stanford, and P. W. Jones, *in* "Carcinogenesis: A Comprehensive Survey: (R. I. Freudenthal and P. W. Jones, eds.), Vol. 1, p. 241. Raven, New York, 1976.]

indene on Tenax GC (*44*). Moreover, decomposition of Tenax GC in contact
with hot exhaust gases may result in sample contamination with diphenyl
quinones (*42*).

Particulate sampling from engine exhausts also presents difficult problems,
and has been reviewed by Hrudey (*21*). Among systems specifically designed
for sampling exhausts for PAH determination, an apparatus designed by
del Vecchio *et al.* (*45*) (Fig. 4-5) is typical. The exhaust is sampled via a probe
fitted with a container for condensed water, and is then drawn through a
condenser and a series of cold traps before passing through a glass-fiber and
two cellulose-triacetate filters. A similar approach was used by Grimmer
(*46*); the vehicle exhausts were cooled and then passed through glass fiber and
(double) silica gel filters. More recently, Spindt (*47*) and Lee *et al.* (*48*) used
air dilution to improve the efficiency of collection of PAH from exhaust
gases; the hot exhaust aerosol is mixed with cold air before filtration and
final adsorption on a Chromosorb-102 (*47*) or Amberlite XAD-2 resin trap.
None of these procedures has been subjected to full tests for reproducibility
and efficiency, however, and there is need for standardization in the collection
of samples from exhausts.

PAH are soluble in many organic solvents and a variety have been recommended (*48–50*) for the Soxhlet extraction of solid environmental samples,

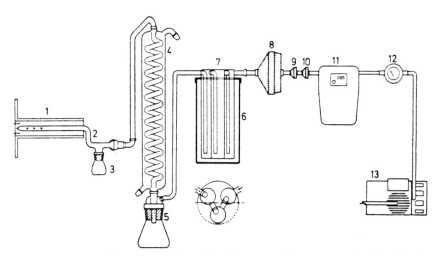

Fig. 4-5. Schematic drawing of the engine exhaust sampling apparatus for determination
of polycyclic hydrocarbons. Key: 1, exhaust pipe; 2, probe; 3, condensed water flask; 4, condenser; 5, condensed water flask; 6, Dewar flask; 7, traps in dry ice; 8, glass-fiber filter; 9,
cellulose-triacetate filters; 10, cellulose-triacetate filters; 11, dry gas meter; 12, vacuum gauge;
13, vacuum pump. [Reproduced with permission from V. del Vecchio, P. Valori, C. Melchiori,
and A. Grella, *Pure Appl. Chem.* **24**, 739 (1970). Copyright, Butterworths, London.]

particularly air and combustion-effluent particulates collected on filters (e.g., acetone, benzene, cyclohexane, chloroform, methanol and other alcohols, acetic acid, benzene–methanol, petroleum ether, dichloromethane, and tetrahydrofuran). Among these, the first three have all been shown (49) to be nearly 100% efficient in Soxhlet extraction of benzo[a]pyrene from filters. Pierce and Katz (51) have determined extraction curves (Fig. 4-6) for five PAH from glass-fiber filters (Hi-Vol samples); extraction with benzene was essentially complete after 6 hr. Extraction yields of major PAH by benzene were determined by Broddin et al. (19) who found more than 99% extraction

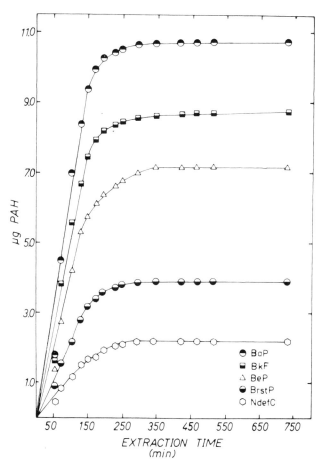

Fig. 4-6. Extraction curves of benzo[a]pyrene (BaP), benzo[k]fluoranthene (BkF), benzo[e]-pyrene (BeP), benzo[rst]pentaphene (BrstP), and naphtho[1,2,3,4-def]chrysene (NdefC). [Reproduced with permission from R. C. Pierce and M. Katz, *Anal. Chem.* **47**, 1743 (1975). Copyright, American Chemical Society.]

after only 2 hr, or 20 Soxhlet cycles. The same group recommended this solvent in preference to a number of others investigated, finding high efficiencies for various PAH (Table 4-1). However, since cyclohexane extracts fewer extraneous materials (3, 52–54), such as asphaltic tars and other uncharacterized materials, and is less hazardous (55) than benzene, its use was advocated by Moore and Monkman (54, 56–58); this suggestion was endorsed by the World Health Organization (WHO) (59) and the Intersociety Committee (60). All principal UV absorbers have been shown (4) to be extracted by cyclohexane from coke-oven dust; benzene brought out no more PAH of interest. A variety of extraction times have been recommended between 2 hr (61) and 20–30 hr (45), with 6–8 hr (62) a commonly used compromise between the long periods necessary for extraction from certain particulate matter on which PAH are strongly adsorbed [such as carbon black (63)] and the short times necessary to avoid losses during Soxhlet extraction (64). Low ($<30\%$) recoveries of [^{14}C]benzo[a]pyrene from spiked fly ash, which were markedly greater for ^{14}C-labeled naphthalene and phenanthrene, are indicative of strong adsorption of high-molecular-weight PAH on this material (64). Ultrasonication allows high recoveries of two- and three-ring PAH from fly ash, but recoveries of four-ring and larger PAH may be incomplete (64).

Methanol has also been proposed as a more efficient solvent than cyclohexane for the extraction of organic matter from glass-fiber filters. Both Grosjean (65) and Gordon (66) found that methanol extracted more PAH from air particulate matter than both benzene (ratios 1.35 and 2.73) and cyclohexane (ratios 1.01 and 4.40). Hill et al. (67) found that more inorganic matter was extracted by methanol, but also confirmed Grosjean's figures by gas chromatographic measurements on the organic portion of the extract; only 2 hr were found to be sufficient for complete extraction. Cautreels and van Cauwenberghe (68) have also recommended extraction of air-particulate filters with methanol. Although methanol is clearly a more efficient solvent for organic acids, salts, and esters of organic acids (2, 68), it is not clear whether extra PAH are extracted.

For extraction of PAH from polyurethane plugs, tetrahydrofuran has been shown to be especially suitable because of its high volatility and increased solvent power (32); virtually complete recoveries have been illustrated (Table 4-1). The Tenax GC of the Battelle adsorbent sampler (40–42) is stripped of PAH by continuous extraction with pentane; near 100% extraction efficiencies have been shown (44) (Table 4-1). Dilution of PAH with a volatile solvent may lead to losses during concentration; removal of n-pentane from PAH solutions using a Kuderna–Danish evaporation concentration leads, however, to minimal losses of PAH with vapor pressures greater than that of naphthalene (44).

An interesting alternative to the Soxhlet extraction of atmospheric dust (with its possible attendant losses) is the use of ultrasonic vibration at room temperature; extraction of benzo[a]pyrene and "total PAH" was found to be complete within 30 min by Chalot et al. (50). Golden and Sawicki (69) refined the procedure by suspending the glass-fiber filter sample in cyclohexane and sonicating in the presence of silica powder to adsorb polar extractives. Recovery of PAH between 95 and 98% was concluded in a full evaluation (70); both extraction efficiency and reproducibility were superior in comparison with the Soxhlet procedure, and this method has been adopted by NIOSH for the determination of coal-tar pitch volatiles (71).

Other less commonly used extraction methods include dissolving the glass-fiber filters in hydrofluoric acid (72), and thermally stripping PAH from the collection device directly into the analytical system. The latter method has been applied to both high-volume filters (72) and Tenax GC cartridges (35); in both studies, a stream of nitrogen was passed through the heated samples and PAH condensed on the (cool) early portion of a GC column. The comparable technique of vacuum sublimation of PAH from particulates is claimed to be a more rapid procedure than Soxhlet extraction (73); sublimation is complete after 40 min at 300°C and 10^{-2} torr.

B. Water

PAH may be extracted from water by liquid–liquid partition, reverse osmosis, or the use of lipophilic adsorbents (7, 74–76); but the importance of checks for contamination during sampling, especially of sea water, must be emphasized (77). Extraction by a solvent involves the highest risk of further contamination (76), but the method is commonly used. The chosen solvent should combine high solubility for PAH, low miscibility with water, and high volatility. Numerous materials have been recommended (7, 77) (e.g., C_5–C_8 alkanes, benzene, cyclohexane, chloroform, carbon tetrachloride, and dichloromethane), generally with shaking in a separatory funnel, or with violent stirring (78), as in a mixer–homogenizer (79). A method of analysis recommended by WHO (80) and based on definitive work by Borneff (81) uses benzene as a solvent, but this may be replaced by cyclohexane (82).

Acheson et al. (74) systematically investigated the factors affecting the efficiency of solvent extraction of PAH from environmental water samples: initial concentration of PAH, presence of suspended solids, and prolonged storage of the sample. Extraction in a mixer–homogenizer with dichloromethane was found to be most efficient. Both prolonged mixing before sampling and the presence of suspended Fuller's earth markedly reduced the efficiency of extraction of pyrene and benzo[ghi]perylene, probably due to adsorption on the solids. Sample losses may also occur through adsorption

TABLE 4-1

Percentage Recovery of PAH from Air Particulates, Combustion Effluents, and Water

Compound	Extraction[a] from air particulates with benzene	Stripping[b] from polyurethane plug air sampler with THF	Stripping[c] from Tenax GC combustion effluent sampler with n-pentane	Extraction[d] from water by sorption on Tenax GC, acetone stripping	Extraction[g] from water by sorption on polyurethane, acetone/cyclohexane stripping	Extraction[j] from water by sorption on HPLC column
Naphthalene		97.8	103[k]			
Biphenyl		99.9				
Anthracene			101[k]	96.6[e]		
Phenanthrene	96	99.9				
Fluoranthene	99	99.9		96.5[f]	93.5[h], 118.9[i]	101
Pyrene	96	97.9	91, 98, 104 ± 4	98.5[f]		
Benzo[b]fluorene	101					
Benz[a]anthracene	102					
Chrysene		95.9	90, 92, 106 ± 5			

Compound						
Benzo[j]fluoranthene					98.1[h], 121.1[i]	
Benzo[k]fluoranthene					97.9[h], 84.1[i]	
Benzo[a]pyrene	75	97.6		96.6[f]	92.3[h], 80.1[i]	94
Perylene		95.6		92.0[e]		98
Indeno[1,2,3-cd]pyrene			91, 105, 102 ± 5			
Benzo[ghi]perylene	39	98.0		86.0[e]	93.7[h], 71.8[i]	
Dibenz[a,c]anthracene			101, 106, 103 ± 10		87.6[h], 118.4[i]	96
Dibenz[a,h]anthracene	43					
Coronene		99.8	80, 92, 100 ± 10			

[a] According to Cautreels and Van Cauwenberghe (68a).
[b] According to Krstulovic et al. (32a). Analysis by HPLC.
[c] According to Strup et al. (68b). Analysis by GC/MS.
[d] According to Leoni et al. (86). Analysis by spectrofluorimetry.
[e] From 30 liters water; PAH at 0.13-ppb level.
[f] From 25 liters water; PAH at 0.08-ppb level.
[g] According to Dasu and Saxena (92). Analysis by TLC/fluorimetry.
[h] From 60 liters finished water; PAH at 0.1–0.5-ppb level.
[i] From 50 liters raw water; PAH at 0.1–0.5-ppb level.
[j] According to Ogan et al. (93). Analysis by HPLC.
[k] According to White et al. (44). Analysis by UV; 96–98% recoveries also found for a mixture of anthracene, phenanthrene and pyrene.

of PAH on glass containers, and in-field sampling may be necessary (75).

Preconcentration methods involve passing water samples through plugs of active carbon (76, 83), Amberlite XAD macroreticular cross-linked polystyrene resins (75, 76, 84, 85), Tenax GC (86, 87), ion-exchange resins (88), and open-pore polyurethane (88). The PAH may then be removed by stripping with a small volume of solvent (84), or by direct desorption into a GC/MS system (1, 89). The criteria for the selection of the appropriate sorbent are: selectivity, breakthrough capacity, particle size (which limits flow rate), and solvent compatibility. For instance, humic and fulvic acids, which are commonly present in raw and potable waters, and which may interfere with subsequent determinations by UV, are not preconcentrated on polyurethane (88). The collection efficiency of polyurethane was not as great as that of Amberlite XAD for naphthalene (90), but better for pyrene on polyurethane prepared in situ (88). The breakthrough capacity (weight of test substance that can be adsorbed by the polymer before the collection efficiency falls below 90%) of polyurethane is correspondingly greater for this adsorbent than for Amberlite XAD-2, or Bio-Rad AG MP-50; naphthalene, phenanthrene, and fluoranthene all had good recovery from polyurethane when stripped with methanol or ethanol if the capacity (~ 3 μg for 5-cm columns) was not exceeded (88).

Saxena has investigated the concentration on polyurethane foam from water of radiolabeled benzo[a]pyrene (91) and of mixtures of the representative PAH (92) selected by the WHO. Heating the water to $62 \pm 2°C$ maximized retention, presumably by aiding desorption from suspended particles. The retention efficiencies of the individual PAH of the foams (Table 4-1) were not less than 88% from finished water, and 72% from raw water. As little as 0.1 ng/liter could be detected by this procedure.

Polyurethane may thus replace Amberlite XAD-2 in the concentration of PAH, although the latter material was used with great success in the analysis of PAH in drinking water in Ames, Iowa (85). A variety of two-ring aromatics at concentrations between 0.4 and 19.3 ppb were determined after passing 150 liters of water through a 1.5×7.0-cm column of Amberlite XAD-2 (100–150 mesh) at 3 liter/hr. Stripping was achieved with 15 ml of diethyl ether (85).

Because of its high thermal stability, Tenax GC may be the appropriate sorbent for the preconcentration of PAH if direct thermal desorption is used (1, 88); slight solubility in polar organic solvents may render it less suitable with solvent stripping (40). However, Leoni et al. have successfully used acetone extraction of this material to recover PAH from water at the 0.1 ppb level (Table 4-1) (86).

An elegant method for concentrating PAH at levels as low as 0.2 ng/liter from water onto a HPLC precolumn packed with 40 μm C_{18} bonded-phase pellicular material has been described (*93*). Recovery is quantitative (Table 4-1) when the extraction column is backflushed with LC mobile phase (38% acetonitrile, 15% methanol in water) onto an analytical HPLC column.

C. Soils and Sediments

The PAH of ancient sediments may be Soxhlet extracted (*94, 95*) with 1:1 benzene–methanol as described by Blumer and Rodrum (*96*); UV fluorescence is used to test for completion. The extraction of organic matter from the aromatic hydrocarbon minerals curtisite, pendletonite, and idrialite was achieved by stirring with hot benzene followed by brief ultrasonic agitation (*97*).

Sonication and ball-milling have also been applied to soils and recent sediments (*77*), and a method due to Brown *et al.* (*98*) and applied by Aizenshtat (*99*) involves triple extraction of the wet sediment with 0.1 *M* HCl in 70:30 benzene–methanol using a homogenizer. However, Soxhlet extraction of wet sediments is also widely applied, usually with a benzene–methanol mixture; in a modification (*100*) of a method recommended by Farrington and Trip (*101, 102*), partially thawed samples are transferred to Soxhlet thimbles for extraction with the above solvents in 2:3 ratio. In another widely applied procedure (*94, 95, 103*), Giger and Blumer (*10*) proposed first extracting with methanol alone followed by the addition of benzene to the refluxing solvent after the water has been removed from the sample. The refluxing solvent then changes to the methanol–benzene azeotrope. Extracts of soils and sediments may contain elemental sulfur which can be removed by percolation over precipitated copper (*104, 105*).

D. Food and Biota

Extraction from oils and fats is straightforward if the materials are soluble in a nonpolar solvent such as cyclohexane (*106–108*); PAH concentrates may be obtained directly by one of the partition methods described in Section III. Recoveries near 100% for PAH added at the 10-ppb level to sunflower oil were shown by Grimmer (*107*) (Table 4-2), while comparable efficiencies (71–100%) were obtained by Howard for additions at the 2-ppb level (*106*).

Further, PAH may be obtained from fruit (*109*) and vegetable matter (*110*) by Soxhlet extraction with a hydrocarbon solvent, or methanol followed by cyclohexane. If the vegetable matter is shredded in a blender, care must be taken not to produce too fine particles which may slow extraction (*111*). Stirring with 1:1 hexane–benzene proved an adequate method for extracting the PAH from plant leaf surfaces (*112*).

TABLE 4-2

Recovery of PAH from Food

Compound	Percentage recovery from				
	Sausage[a]	Cheese[a]	Fish[a]	Meat[b]	Sunflower oil[b]
Anthracene					107, 102
Phenanthrene					106, 105
3,6-Dimethylphenanthrene[c]					100, 100
Pyrene					104, 101
Benz[a]anthracene	83	79	88	101, 99, 99	
Chrysene					93, 95
Benzo[b]fluoranthene				99, 102, 98	104, 107
Benzo[ghi]fluoranthene				102, 103, 100	
Benzo[a]pyrene	87–100	73–76	90–100	100, 102, 99	98, 95
Benzo[e]pyrene				100, 102, 99	102, 105
Perylene				101, 97, 97	96, 92
Dibenz[a,h]anthracene	70	80	70	101, 100, 98	
Dibenz[a,j]anthracene				105, 102, 100	96, 103
Benzo[b]chrysene[c]				100, 100, 100	100, 100
Indeno[1,2,3-cd]pyrene					97, 102
Benzo[ghi]perylene	80	75	80	101, 96, 97	97, 108
Coronene				99, 98, 97	

[a] Added at 2-ppb level to 500 g food. According to Howard *et al.* (*113*). Analysis by spectrophotometry.

[b] Added at 10 ppb to 200 g food. According to Grimmer and Bohnke (*107*).

[c] Internal standard.

However, for insoluble fats, protein-rich foods (meat, fish, cheese, yeast, etc.) and biotic samples, extraction with saturated hydrocarbons is incomplete unless saponification is carried out. Digestion of the sample in alcoholic KOH disrupts the cells and makes extraction of PAH more efficient when the mixture is partitioned with a nonpolar solvent (*77*). Care must be taken to avoid difficult-to-break emulsions.

A procedure due to Howard (*113*) has been widely applied (*114, 115*), particularly to smoked foods. Samples are extracted with ethanol and saponified with KOH in a Soxhlet apparatus. After concentration, the extracts are diluted with water and partitioned with isooctane; 70–100% recoveries have been shown (Table 4-2) at the 2-ppb level.

In the method of Grimmer (*107, 116*), the samples are dissolved homogenously by saponification in 2 M methanolic KOH. PAH are then extracted into cyclohexane. Only about 30% of the PAH is extractable from smoked herring by methanol alone (*107*), whereas further alkaline hydrolysis in-

TABLE 4-3

Extraction and Saponification of the Extraction Residue of a Sample of Smoked Herring[a]

Compound	ppb extracted		
	Methanol[a]	KOH in methanol[b]	Total
Benz[a]anthracene + chrysene	1.29	2.90	4.19
Benzofluoranthene	0.69	1.58	2.27
Benzo[a]pyrene	0.36	0.90	1.26
Benzo[e]pyrene	0.33	0.85	1.18
Perylene	0.07	0.18	0.25
Dibenzo[a,j]anthracene	0.19	0.46	0.65
Indeno[1,2,3-cd]pyrene + dibenz[a,h]anthracene	0.26	0.64	0.90
Benzo[ghi]perylene	0.38	0.72	1.10
Anthanthrene	0.08	0.17	0.25

[a] According to Grimmer and Bohnke (107).

[b] Smoked herring (450 g) was boiled with 700 ml methanol for 3 hr. The methanol extract was decanted off and the insoluble residue washed with 300 ml methanol.

[c] The insoluble methanol residue (see b) was saponified for 3 hr with 750 ml 2 M KOH.

creases the extraction yield by approximately 60% (Table 4-3). This procedure has been especially widely applied (117) to smoked (118) and protein-rich foods (119), yeast (120), coffee (121), and tea (121). Excellent recoveries (Table 4-2) have been obtained for numerous PAH added to foods at the 10-ppb level. Propylene carbonate has also been claimed as an efficient solvent for extracting PAH from meat (122).

III. CONCENTRATION AND CLEAN-UP

The extracts of environmental samples obtained, as outlined above, in solution in cyclohexane or other solvents inevitably contain substantial quantities of materials other than PAH which may interfere with subsequent analysis. If the "final" analytical method is selective and the concentration high enough, PAH may be determined without further separation; but, in general, concentration and clean-up procedures are usually applied for the separation of PAH from extraneous compounds—in which either solvent partition or column chromatography of various kinds, or both, are used.

A. Solvent Partition

PAH may be enriched by partitioning the original solution with a second, immiscible, solvent. If the partition coefficients for PAH differ from those of the other materials present, they will be preferentially concentrated in one or the other layer. For example, hydrophilic and polar compounds may be removed from cyclohexane solutions of PAH by partitioning with methanol–water (*123*) or acetone–water (*124*). Partition coefficients for PAH in these extractions strongly favor the cyclohexane layer (Table 4-4), and the loss is small. If substantial quantities of bases and acids are present (see Section IV), for example in tobacco-smoke condensate, washing with dilute

TABLE 4-4

Compound	Cyclohexane/ methanol– H_2O^a $(C_{C_6H_{12}}/ C_{MeOH-H_2O})$	Nitromethane/alkane	
		$(C_{CH_3NO_2}/ C_{cyclohexane})^a$	$(C_{CH_3NO_2}/ C_{n-heptane})^b$
Naphthalene			
Fluorene			
Anthracene			
9-Methylanthracene			
9,10-Dimethylanthracene			
Phenanthrene			
Fluoranthene	140	1.30	
Pyrene	150	4.40	
1-Methylpyrene			
Benz[a]anthracene			
7-Methylbenz[a]anthracene			
7,12-Dimethylbenz[a]anthracene			
Chrysene	14.6	1.65	
Triphenylene			
Benzo[b]fluoranthene	14.7	1.94	
Benzo[j]fluoranthene	14.8	1.75	
Benzo[k]fluoranthene	14.9	1.76	
Benzo[a]pyrene	14.9	1.53	1.7
Benzo[e]pyrene	14.5	1.68	
Perylene	14.9	1.69	
Dibenz[a,h]anthracene	14.9	1.80	1.9
Benzo[ghi]perylene	14.6	1.77	
Coronene			

[a] According to Hoffmann and Wynder (*123*).
[b] According to Haenni et al. (*135*).
[c] According to Natusch and Tomkins (*139*).

acid and alkali follows the partitioning of the cyclohexane solution with methanol–water (125, 126).

Hoffmann and Wynder proposed (123) that after methanol–water/cyclohexane partition, PAH might be further concentrated by extraction from cyclohexane into nitromethane. Five partitions are necessary to achieve >99% extraction, because many of the partition coefficients (Table 4-4) are only between 1.5 and 2.0. This method has been incorporated into several schemes [e.g., Schmeltz et al. (127), modified by Lee, Novotny, and Bartle (128–132), Acheson et al. (74), and Laflamme and Hites (133)]. The nitromethane extract is evaporated down to a small volume or to dryness under reduced pressure at a relatively low temperature (e.g., $\sim 50°C$), but Acheson

Partition Coefficients for PAH Between Different Solvent Pairs

Dimethyl sulfoxide/alkane			Dimethylformamide/alkane			
$(C_{DMSO}/C_{n\text{-pentane}})^c$	$(C_{DMSO}/C_{n\text{-heptane}})$	$(C_{DMSO}/C_{isooctane})$	$(C_{DMF}/C_{n\text{-heptane}})^b$	$(C_{DMF}/C_{isooctane})^d$	$(C_{DMF\text{-}H_2O}/C_{cyclohexane})$	$C_{cyclohexane}$
2.3	3.0^b					
3.0						
4.9	3.7^b	6.2^e			2.2^e	2.4^f
4.7						
3.0						
4.4		4.3^e			2.5^e	2.7^f
9.0	5.1^c	8.7^c			3.4^e	3.3^f
					2.2^e	
		20^e			5.0^e	4.3^f
					2.6^e	
		6.0^e			1.7^e	
19.7	8.8^b					
28.5						
	9.3^b	25^d–22^e	9.9	24	5.5^e	6.9^f
	18^b	55	18		12.0^e	8.7^f
					8.4^e	7.4^f
36.7		27^d		29	10.0^e	9.3^f

[d] According to Jentoft and Gouw (175).
[e] According to Robbins (143).
[f] According to Grimmer and Bohnke (176).

et al. (*74*) have shown that PAH were lost, presumably by volatilization and thermal degradation, during the extraction of these compounds from water with dichloromethane and clean-up and concentration via partition between methanol–water and cyclohexane and thence into nitromethane (Table 4-5). Recoveries varied between < 10% and 83%.

An alternative solvent for the extraction of PAH from alkane solvents is dimethyl sulfoxide (DMSO), originally suggested by Haenni (*134–136*). Partition coefficients are more favorable than for nitromethane (Table 4-4), and are governed by the extent of interaction between the S atom and the aromatic π-system (*137*). Isolation is carried out by dilution with water and back extraction into cyclohexane, which may be evaporated at a lower temperature than nitromethane. Moreover, PAH are stable in solution in dimethyl sulfoxide for at least 90 days at ambient temperature (*137*). Greatly improved efficiencies of PAH recovery (84–110%, Table 4-5) are then obtained (*74*). Extraction with dimethyl sulfoxide has been used in the clean-up of PAH from foods by Howard (*106, 113, 114*) and from tobacco-smoke condensate by Stedman (*126*) and Snook (*138*).

Natusch and Tomkins (*139*) have made a thorough study of the use of DMSO in liquid–liquid extraction procedures for isolating PAH from complex mixtures. Partition coefficients for pyrene distributed between DMSO and a number of alkane solvents showed that, while *n*-hexane provided the largest value, virtually complete removal was obtained for all the solvents after three equivolume partitions. *n*-Pentane is preferred, however, because of its high volatility; partition coefficients for various PAH between DMSO and *n*-pentane are listed in Table 4-4. The effect of the water : DMSO ratio in the stripping step was also investigated; at 2 : 1 by volume, quantitative removal of pyrene was possible in a single equivolume partition with *n*-pentane. Organic bases and phthalates were found to behave similarly to PAH, but compounds strongly associating with DMSO such as phenols and acids cannot be removed by adding water. Based on these results, a simple separation scheme was proposed (*139*), and validated for an airborne particulate sample, which allows three groups of compounds to be separated in a few minutes. Three successive extractions from the original *n*-pentane solution into DMSO, followed by three successive back extractions from DMSO/water into *n*-pentane, yield: the aliphatic hydrocarbons in the original pentane; the alcohols, acids, and phenols in the DMSO/water; and the PAH, phthalates, and bases in the final pentane solution.

A general impediment in fractionation schemes involving dimethyl sulfoxide is the unfavorable partition coefficients reported for methyl derivatives (*136, 139*) (Table 4-4); these are thought to arise because the methyl group blocks a region of high electron density (*139*). Moreover, overall loss of PAH may be significant; Stedman found that [^{14}C]benzo[*a*]-

TABLE 4-5

Efficiencies of Recovery of PAH from Water by Solvent Extraction Followed by Alternative Partition Schemes[a]

	Low-level addition			High-level addition		
	Initial weight (μg)	Percentage recovery		Initial weight (μg)	Percentage recovery	
		Nitromethane procedure[b]	Dimethyl sulfoxide modification		Nitromethane procedure[b]	Dimethyl sulfoxide modification
Fluoranthene	0.52	48	92	10.4	83	96
Pyrene	1.21	42	94	24.2	38	96
Benz[a]anthracene + chrysene	1.56	50	90	30.8	64	92
Benzo[k]fluoranthene	0.60	54	91	12.0	59	94
Benzo[a]pyrene + benzo[e]pyrene	1.01	36	95	20.2	33	99
Perylene	0.95	24	84	19.0	<10	90
Indeno[1,2,3-cd]pyrene	0.38	49	97	7.6	41	100
Benzo[ghi]perylene	0.79	45	91	15.8	39	90
Coronene	0.50	44	93	10.0	82	110

[a] According to Acheson et al. (74).
[b] According to Hoffmann and Wynder (123).

pyrene suffered significant loss in DMSO partition (*126, 140*), while Natusch and Tompkins recovered only 91% and 87% respectively of naphthalene and anthracene (*139*). On the other hand, Acheson *et al.* (*74*) found generally higher recoveries for this technique (Table 4-5).

Partition coefficients (Table 4-4) for the extraction of PAH into dimethylformamide from alkanes (*135*) are similar to those for DMSO; Grimmer (*107, 108, 124*) has thus preferred the use of partition between cyclohexane and 9:1 dimethylformamide–water (Table 4-4) in the analysis of foods, air pollutants, and exhaust gases, and his procedure has been applied in studies of the PAH of long-range transported aerosols (*141*). Addition of further water again precipitates the PAH and allows extraction into cyclohexane. Wallcave *et al.* (*142*) also used dimethylformamide–water in the clean-up of PAH from asphalts and coal-tar pitches (*142*).

Robbins examined (*143*) the general extraction characteristics of PAH by determining the partition coefficients in a number of extraction systems, mainly with the solvents dimethyl sulfoxide, dimethylformamide, and *N*-methylpyrrolidone. Systematic responses of PAH to changes in extraction conditions were observed, and it was established that the logarithm of the partition coefficient for a polar/nonpolar solvent pair decreases linearly with (a) the volume percentage of diluent added to the extractant; (b) the reciprocal of absolute temperature; (c) the volume percentage of aromatic solvent present; and (d) the degree of alkyl substitution; and it increases with molecular weight. The selectivity of extraction conditions may thus be tailored to the separation problem.

B. Column Chromatography

An early chromatographic procedure for the separation of the neutral fraction of environmental samples on activated silica gel, or latterly silicic acid (see Chapter 5), was devised by Rosen (*83*). Elution is with an alkane solvent (isooctane, hexane, pentane, or petroleum ether) to yield the aliphatic fraction, and benzene, or benzene–petroleum ether followed by benzene, to yield the aromatics. Tests have shown that 97% of benzo[*a*]pyrene in tobacco smoke condensate is concentrated in the benzene–petroleum ether fraction (*126, 140*). The Rosen separation has been modified by Thomas and Monkman for air-pollutant PAH (*144*), and Stedman *et al.* (*126, 140*) for tobacco-smoke condensate, and is still in wide use (*138, 145*).

The application of lipophilic gels has, however, now largely replaced alumina chromatography (*143, 146*) previously recommended (*53*) as the method of choice for obtaining PAH fractions for analysis after preliminary clean-up. Classical adsorbents such as alumina and silica gel suffer the disadvantages (*129*) of (a) high adsorptivity which often results both in losses of trace constituents and in marked peak tailing with consequent less sharp fractionation, particularly for higher molecular weight PAH, and (b) poor reproducibility of fractionation because of modification of the adsorbent with traces of water. Enrichment of the PAH fractions of environmental samples has been achieved by means of gels working in both steric exclusion and adsorption modes (see Chapter 5). Thus, Snook's group used the Bio-Beads series to concentrate tobacco-smoke PAH (*147, 148*). With benzene as eluent, Bio-Beads SX-12 (a neutral, porous, styrene divinylbenzene copolymer, exclusion 400) yields fractions containing compounds in sequence of increasing ring number (Fig. 4-7) and may also be used to separate multialkylated PAH from parent plus monoalkylated PAH. Interfering material is eluted immediately after the void volume.

Giger and Blumer (*10*) removed lipids and pigments from extracts of soils and sediments prior to Rosen separation by use of Sephadex LH-20. These materials are eluted near the column void by benzene–methanol. Klimisch ingeniously made use of Sephadex LH-20 in two ways in the clean-up and concentration of PAH from cigarette smoke (*149*): lipophilic and hydrophilic substances were separated by gel permeation chromatography (elution with 85% methanol–hexane); paraffinic substances were then removed by chromatography with isopropanol to yield a PAH concentrate

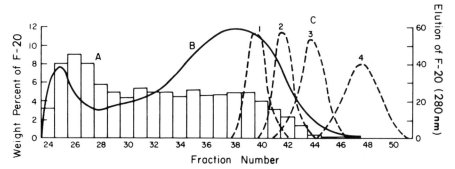

Fig. 4-7. Gel filtration chromatography of PAH portion of cigarette-smoke condensate. Key: A, percent weight distribution; B, 280-nm absorbance curve for the elution (100 = 1.28 absorbance units); C, elution curves for standard PAH (1 = 3.6-dimethylphenanthrene; 2 = phenanthrene; 3 = pyrene; 4 = benzo[*a*]pyrene). [Reproduced with permission from M. E. Snook, R. F. Severson, H. D. Higman, R. F. Arrendale, and O. T. Chortyk, *Beitr. Tabakforsch.* **8,** 250 (1976). Copyright, Wissenschaftliche Forschungsstelle im Verband der Cigarettenindustrie.]

comprising only 0.95% of the condensate, compared with the 7% obtained by a partition procedure. The aliphatics elute first from the column, in order of decreasing molecular weight, while the aromatics are retained. The aromatics are then eluted in order of increasing ring number (see also Chapter 5).

Jones *et al.* also made full use of the peculiar nature of LH-20 in various solvents as they fractionated shale oil and coal oil several times (*150*). As in the work of Klimisch, their fractionation used a hexane eluent to separate the hydrophilic and lipophilic components of the oils. The lipophilic subfraction was chromatographed on LH-20 swollen with THF. The nonpolar constituents were separated from the polar components by the exclusion principle, and the polar compounds were retained by H-bonding to the gel matrix. The nonpolar subfraction was then subjected to further separation on LH-20 with isopropanol. After the two-ring systems had eluted, THF was again introduced into the column (in place of the isopropanol), and the remainder of the PAH (three rings and up) were washed rapidly out of the column. The changeover from isopropanol to THF does not change the volume of the gel bed, since the gel swells equally in both solvents.

Lee, Novotny, and Bartle applied Sephadex LH-20/isopropanol chro-

TABLE 4-6

Bulking Scheme for Fractions Obtained through Sephadex LH-20 Chromatography of PAH from Tobacco- and Marijuana-Smoke Condensates[a]

Fraction number	Standard	Retention volume[b] (cm^3)	Bulking volume[b] (cm^3)
I			0–258
II	Naphthalene	300	259–336
III	Anthracene	372	337–402
IV	Fluoranthene	432	403–480
V	Triphenylene	534	481–558
VI	Benzo[a]pyrene	588	559–660
VII	Dibenz[a,c]anthracene	732	661–810
VIII			811–1500

[a] According to Lee *et al.* (*131*).

[b] For 115-cm column (1.5 cm i.d.) with isopropanol as mobile phase.

matography to PAH fractions obtained from air particulates and tobacco-
and marijuana-smoke condensates (*129–132*). One-hour fractions from a
115-cm column (1.5 cm i.d.) were bulked according to the predetermined
retention volumes of standard PAH (Table 4-6 and Fig. 4-8). The optimum
flow rate of 6 ml/hr was determined from experiments with pure compounds
(e.g., Fig. 4-9); the graph of HETP against flow rate resembled that for a
lipophilic gel in exclusion mode (*151*); the efficiency corresponded to 3000
plates for anthracene and 2500 plates for fluoranthene. Eight final fractions
were collected, corresponding to those in Table 4-6. Fraction weights are

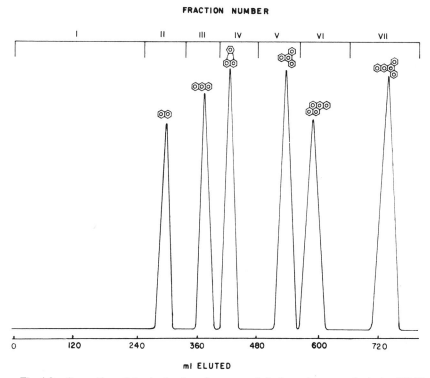

Fig. 4-8. Separation of standard polynuclear aromatic hydrocarbons on a Sephadex LH-20
column. Conditions: column, 115 cm × 1.5 cm i.d.; mobile phase, isopropanol; flow rate,
6 ml/hr. Fractions were collected at 1-hr intervals with a fraction collector (Buchler Instruments,
Fort Lee, N. J.), and their UV absorption was measured. [Reproduced with permission from
M. L. Lee, M. Novotny, and K. D. Bartle, *Anal. Chem.* **48**, 1566 (1976). Copyright, American
Chemical Society.]

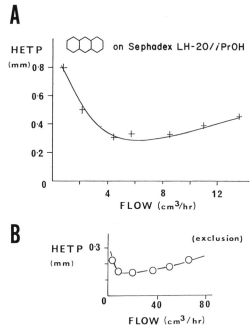

Fig. 4-9. Graphs of HETP against flow rate for: A, Sephadex LH-20 with isopropanol eluent; and B, Sephadex in exclusion mode. [Reproduced with permission from J. Sjovall, E. Nystrom, and E. Haahti, *Adv. Chromatogr.* **6**, 119 (1967). Copyright, Marcel Dekker, Inc.]

compared in the histograms in Fig. 4-10. The degree of separation is illustrated in the HPLC analyses in Fig. 4-11.

A number of other groups have made similar use of the Sephadex LH-20/isopropanol system (*107, 124, 152, 153*). Thus Grimmer's widely applied procedures allow separation of PAH into two fractions: PAH with two and three rings, and PAH with four to seven rings (*107, 124*). Hoffmann and Rathkamp (*152*) made use of the similarity of relative retention times for parent and methyl derivatives in separating a methylfluorene fraction of cigarette smoke condensate.

C. Other Concentration and Clean-Up Methods

A TLC version (*154*) of the Rosen procedure involving separating PAH from other groups (saturated compounds, heterocyclics) on silica gel with benzene–cyclohexane (1.5:1) has been applied to air particulates (*52*). The PAH were located by viewing under long-wavelength UV light, and scraped from the plate. Clean-up of atmospheric PAH by adsorption TLC on

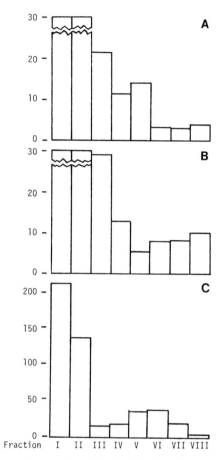

Fig. 4-10. Histograms showing the weight distribution of fractions obtained by Sephadex LH-20 gel chromatography of the air-pollution nitromethane extracts of: A, tobacco-smoke condensate; B, marijuana-smoke condensate; and C, air particulates. [Reproduced with permission from M. L. Lee, M. Novotny, and K. D. Bartle, *Anal. Chem.* **48**, 405 (1976). Copyright, American Chemical Society.]

neutral alumina with hexane–ether (19:1) was used by Pierce and Katz (*51*) to isolate three isomeric groups of arenes from each other and from nonpolar fluorescent organic compounds of different molecular weight.

Separation of PAH from soil from the long-chain methyl ketones which may contaminate them can be effected by adduction with 2,4,7-trinitro-fluoren-9-one (*10*). Complexes with picric acid (*155*), 1,3,5-trinitrobenzene (*155*, *156*), and 2,4,7-trinitrofluoren-9-one (*155*) have also been used in other work to separate PAH. Aromatic hydrocarbons also form complexes with

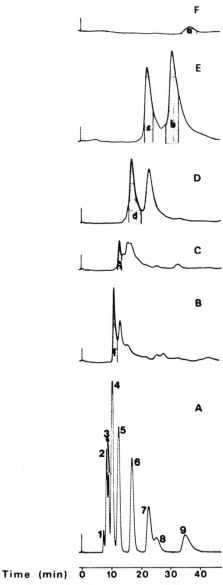

Fig. 4-11. High-resolution liquid chromatograms of fractions obtained by Sephadex LH-20 fractionation of air-particulate PAH. Chromatograms B, C, D, E, and F represent fractions III, IV, V, VI, and VII of Table 4-6. Chromatogram A represents chromatography of standard compounds. Conditions: column 4.25 m × 20 mm i.d., packed with oxypropio-nitrile/Porasil C; mobile phase, *n*-hexane; flow rate, 2 ml/min. Key: 1, benzene; 2, biphenyl; 3, fluorene; 4, anthracene; 5, benzo[*a*]fluorene; 6, triphenylene; 7, benzo[*a*]pyrene; 8, perylene; 9, dibenz[*a,c*]anthracene. [Reproduced with permission from M. L. Lee, M. Novotny, and K. D. Bartle, *Anal. Chem.* **48,** 1566 (1976). Copyright, American Chemical Society.]

1,3,7,9-tetramethyluric acid, which may be concentrated by partitioning between cyclohexane and 90% methanol (*157*). Rothwell and Whitehead (*158*) separated PAH from complex mixtures by high-voltage electrophoresis of complexes (*157*) with caffeine; 98.6% of added PAH were contained in the concentrate (*158*).

IV. SELECTED ANALYTICAL SCHEMES

In Figs. 4-12 through 4-17, a number of proven schemes for the clean-up and concentration of PAH from a variety of environmental sources are summarized. All employ chromatographic procedures, generally coupled with solvent partition.

Figure 4-12 illustrates the simple Rosen procedure—chromatography on silica gel—still useful in separations of air particulates when analyses for relatively few compounds are required (*73*, *144*). Grimmer's procedure (Fig. 4-13) represents a more vigorous approach for these materials (*124*) and broadly resembles the scheme found equally applicable to air particulates and tobacco- and marijuana-smoke condensates by Lee, Novotny, and Bartle (*128–132*) (Fig. 4-18), except that dimethylformamide partition is used in the former, not nitromethane. Figure 4-14 outlines Giger and Schaffner's method (*159*) [a refinement of that of Giger and Blumer (*10*)] for the clean-up of PAH from environmental samples such as soils and sediments. The Grimmer method can also be used for the analysis of foods

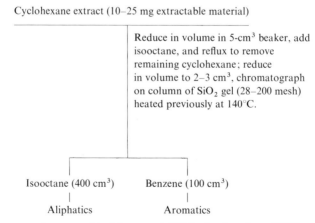

Cyclohexane extract (10–25 mg extractable material)

Reduce in volume in 5-cm³ beaker, add isooctane, and reflux to remove remaining cyclohexane; reduce in volume to 2–3 cm³, chromatograph on column of SiO₂ gel (28–200 mesh) heated previously at 140°C.

Isooctane (400 cm³)   Benzene (100 cm³)
Aliphatics   Aromatics

Fig. 4-12. Rosen procedure as used by Moore *et al.* (*57*) for fractionation of PAH particularly from air particulates. [Reproduced with permission from G. E. Moore, R. S. Thomas, and J. L. Monkman, *J. Chromatogr.* **26,** 456 (1976). Copyright, Elsevier Scientific Publishing Co.]

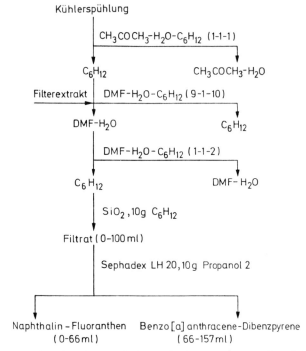

Fig. 4-13. Fractionation of air-particulate and exhaust-gas extracts for PAH. [Reproduced with permission from G. Grimmer and H. Bohnke, *Fresenius Z. Anal. Chem.* **261**, 310 (1972). Copyright, Springer-Verlag.]

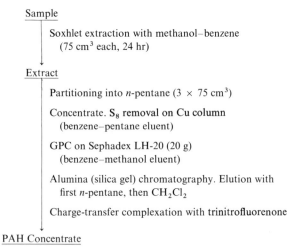

Fig. 4-14. Extraction and clean-up of PAH in soils and sediments. [Reproduced with permission from W. Giger and M. Blumer, *Anal. Chem.* **46**, 1663 (1976). Copyright, American Chemical Society.]

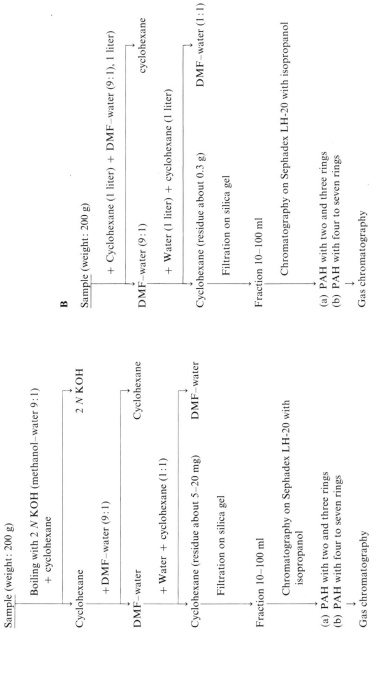

Fig. 4-15. Scheme for extraction and concentration of PAH from: A, meat, fish and yeast; and B, oils and fats. [Reproduced with permission from G. Grimmer and H. Bohnke, *J. Assoc. Offic. Anal. Chem.* **58**, 725 (1975). Copyright, AOAC.]

A

Sample (weight: 200 g)

Boiling with 2 *N* KOH (methanol–water 9:1)
+ cyclohexane

→ 2 *N* KOH

Cyclohexane

+DMF–water (9:1)

→ Cyclohexane

DMF–water

+ Water + cyclohexane (1:1)

→ DMF–water

Cyclohexane (residue about 5–20 mg)

Filtration on silica gel

→

Fraction 10–100 ml

Chromatography on Sephadex LH-20 with
isopropanol

→ (a) PAH with two and three rings
(b) PAH with four to seven rings

Gas chromatography

B

Sample (weight: 200 g)

+ Cyclohexane (1 liter) + DMF–water (9:1), 1 liter)

DMF–water (9:1)

→ cyclohexane

+ Water (1 liter) + cyclohexane (1 liter)

→ DMF–water (1:1)

Cyclohexane (residue about 0.3 g)

Filtration on silica gel

→

Fraction 10–100 ml

Chromatography on Sephadex LH-20 with isopropanol

→ (a) PAH with two and three rings
(b) PAH with four to seven rings

Gas chromatography

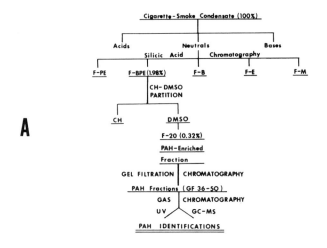

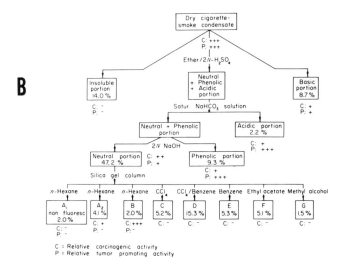

Fig. 4-16. Large-scale fractionation of cigarette-smoke condensate. [A reproduced with permission from M. E. Snook, R. F. Severson, N. C. Higman, R. F. Arrendale, and O. T. Chortyk, *Beitr. Tabakforsch* **8**, 250 (1976). Copyright, Wissenschaftliche Forschungsstelle im Verband der Cigarettenindustrie. B reproduced with permission from E. L. Wynder and D. Hoffmann, "Tobacco and Tobacco Smoke," p. 730. Academic Press, New York, 1967.]

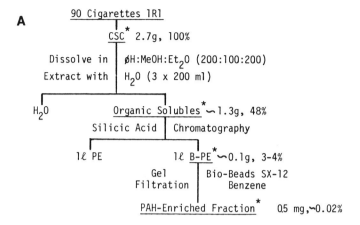

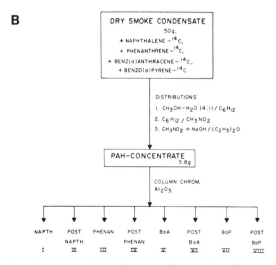

Fig. 4-17. Separation of PAH from small quantities of cigarette-smoke condensate. [A reproduced with permission from R. F. Severson, M. E. Snook, H. C. Higman, O. T. Chortyk, and F. J. Akin, *in* "Carcinogenesis—A Comprehensive Survey" (R. Freudenthal and P. E. Jones, eds.), Vol. 1, p. 253. Raven Press, New York, 1976. B reproduced with permission from D. Hoffmann, G. Rathkamp, K. D. Brunnemann, and E. L. Wynder, *Sci. Total Environ.* **2,** 151 (1973). Copyright, Elsevier Scientific Publishing Co.]

(*107*)—both with (Fig. 4-15A) and without (Fig. 4-15B) a saponification step. Large-scale fractionation of cigarette-smoke condensate is shown in Fig. 4-16. The first version is that used by a U.S. Department of Agriculture group (*138*), and the second (*160*) is that favored by the group of Wynder and Hoffmann. Revised versions of these schemes, applicable to small quantities of cigarette-smoke (*161, 162*) and marijuana-smoke (*163*) condensates, are outlined in Fig. 4-17.

The outstanding feature of many of these methods is the use of chromatography on lipophilic gels—particularly in adsorption/partition mode: Sephadex LH-20/isopropanol in Fig. 4-13, 4-15, and 4-18, and Bio-Beads SX-12/benzene in Figs. 4-16 and 4-17A.

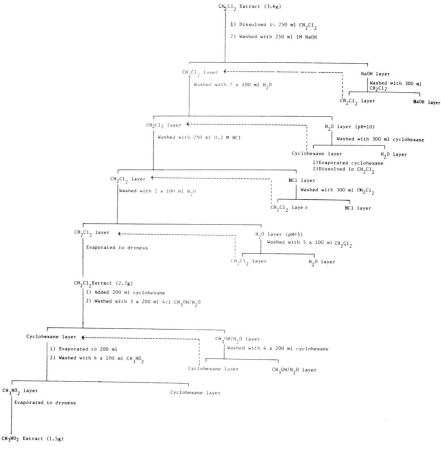

Fig. 4-18. Fractionation scheme for concentration of PAH from air particulates, or tobacco- or marijuana-smoke condensates. [Reproduced with permission from M. Novotny, M. L. Lee, and K. D. Bartle, *J. Chromatogr. Sci.* **12**, 606 (1974). Copyright, Preston Tech. Abstr. Company.]

V. ISOLATION OF THE NITROGEN HETEROCYCLE FRACTION

As early as 1965, Sawicki *et al.* (*164, 165*) isolated a nitrogen-rich fraction from air particulate matter and identified 19 nitrogen heterocycles. Particulate matter was extracted in a Soxhlet with chloroform for 4 hr; triethylamine was then added to ensure the extraction of all the soluble basic material, and the extraction was continued for another 4 hr. The chloroform was evaporated, and the residue was dissolved in ether. The ether solution was extracted eight times with equal volumes of 10% aqueous sulfuric acid. The cold acid solution was neutralized with sodium carbonate, extracted with chloroform, and chromatographed on alumina. Cautreels and Van Cauwenberghe (*68*) followed the same general scheme outlined by Sawicki for fractionation of air particulate matter.

Dong *et al.* (*166, 167*) extracted soiled air filters with benzene–methanol

TABLE 4-7

Distribution of Certain Petroleum Compound Types among the Standard Fractions
from Alumina Chromatography[a]

Compound type	$\varepsilon_1{}^b$	\% of compound type in indicated fraction					
		A_0	A_1	A_2	A_3	A_4	A_5
Benzofurans	0.03–0.04	100					
Dibenzofurans	0.10–0.13	38	49	13			
Dibenzothiophenes	0.15	51	49				
N-Alkylindoles	0.16–0.17		42	58			
Triaromatic hydrocarbons[c]	0.16–0.19		37	63			
Naphthobenzothiophenes	0.20–0.22			100			
Naphthobenzofurans	0.22–0.23			98	2		
Tetraaromatic hydrocarbons[d]	0.20–0.26			76	24		
Indole[e]	0.28–0.33			67	33		
Sulfoxides	0.34–0.38			4	2	22	72
Carbazoles[e]	0.34–0.41				37	63	
Benzocarbazoles[e]	0.41					100	
N-Alkylquinolines	0.44–0.58					100	
Phenols	0.45–0.62				1	15	84
Quinolines[e]	0.6 –0.8					6	94
ε_1 range		<0.12	0.12–0.17	0.17–0.24	0.24–0.30	0.30–0.45	0.45+

[a] Data from the separation of a 700–850°F crude distillate.

[b] Compound type adsorptivity.

[c] Fluorenes and phenanthrenes; principally C_nH_{2n-16} through C_nH_{2n-20}.

[d] Benzofluorenes, chrysenes, etc.; C_nH_{2n-22} through C_nH_{2n-23}.

[e] N-H derivatives (not alkyl-substituted on the nitrogen atom).

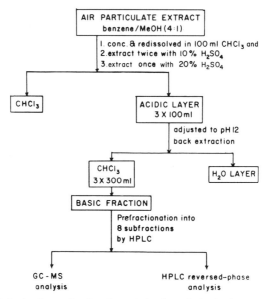

Fig. 4-19. Analytical scheme for the characterization of the basic organic fraction of suspended particulate matter. [Reproduced with permission from M. W. Dong, D. C. Locke, and D. Hoffmann, *Environ. Sci. Technol.* **11**, 612 (1977). Copyright, American Chemical Society.]

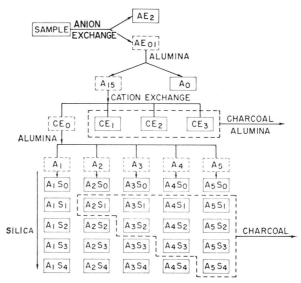

Fig. 4-20. Summary of the integrated separation scheme for the analysis of petroleum heterocompounds. [Reproduced with permission from L. R. Snyder and B. E. Buell, *Anal. Chem.* **40**, 1295 (1968). Copyright, American Chemical Society.]

(4:1) because it gave the highest amount of extractable organic matter. Furthermore, since azaarenes are weak bases, and salt formation with inorganic and organic acids on the filter is a possibility, a polar solvent must be used. The extract was evaporated to near dryness, redissolved in $CHCl_3$, and extracted with 10% and 20% sulfuric acid. The acid extract was neutralized with NaOH and then back-extracted in $CHCl_3$. The final prefractionation was accomplished by HPLC on Lichrosorb SI-60 (30 μm). The total analytical scheme is given in Fig. 4-19.

Figure 4-20 represents an integrated, general-purpose scheme for the separation and subsequent analysis of petroleum heterocycles (*168*). Final fractions are indicated by solid boxes, intermediate fractions by dashed

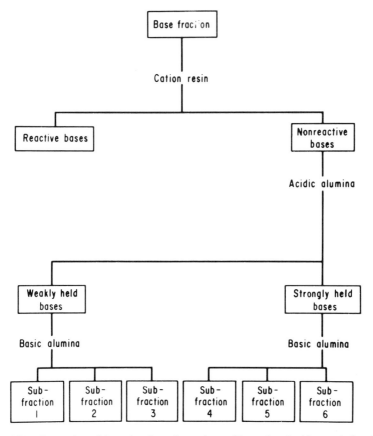

Fig. 4-21. Separation of base fraction of petroleum. [Reproduced with permission from J. F. McKay, J. H. Weber, and D. R. Latham, *Anal. Chem.* **48**, 891 (1976). Copyright, American Chemical Society.]

lines. The sample is separated initially by anion exchange to remove carboxylic acids into the AE_2 fraction. The sample is next separated on alumina. The A_{15} fraction is separated by cation exchange into four fractions. The CE_0 fraction is further separated on alumina, then on silica, into a total of 25 fractions. A number of these fractions were then separated by charcoal chromatography into final fractions for analysis. Table 4-7 (*168*) (see p. 111) gives the distribution of compound types among the fractions from alumina chromatography.

More recently (*169, 170*), the base fractions from petroleum were obtained by removing the acid fraction by anion exchange chromatography, followed by chromatography on Amberlyst 15 cation-exchange resin. The materials retained by the resin were defined as the base fraction. The bases were removed from the resin using benzene–methanol–isopropylamine (55:37:8). Further separation of the bases followed the scheme illustrated in Fig. 4-21. Various groups of nitrogen heterocycles were found in the subfractions.

In the separation of nitrogen heterocycles from sediments, Blumer *et al.* (*103*) Soxhlet extracted with methanol and then benzene, followed by partitioning into *n*-pentane. The residue was chromatographed on Sephadex LH-20 followed by chromatography on acidic silica gel. A series of partitioning and adsorption chromatography steps followed. One step involved

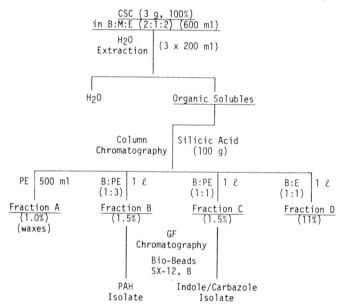

Fig. 4-22. PAH and N-PAH isolation procedure. [Reproduced with permission from M. E. Snook, R. F. Arrendale, H. C. Higman, and O. T. Chortyk, *Anal. Chem.* **50,** (1978). Copyright, American Chemical Society.]

precipitation with picric acid. The final fraction was collected in the pentane–methylene chloride eluate from an alumina column.

Snook et al. (171, 172) followed the scheme illustrated in Fig. 4-22 for the separation of a nitrogen-rich fraction from tobacco-smoke condensate. The smoke condensate in benzene–methanol–ether (2:1:2) was washed with water and then chromatographed on silicic acid. Table 4-8 shows the distribution of compounds in the various fractions, which required an additional purification step, accomplished by gel filtration on Bio-Beads SX-12.

TABLE 4-8

Silicic Acid Chromatography of Heteroatom PAH

Compound	Heteroatom	PE	B–PE (1:3)	B–PE (1:1)	B	B–E (1:1)	E
			% Distribution in solvents[a]				
Naphthalene–^{14}C	—	—	100	—	—	—	—
Anthracene–^{14}C	—	—	100	—	—	—	—
Benzo[a]pyrene–^{14}C	—	—	100	—	—	—	—
Dibenzofuran	O	—	100	—	—	—	—
Benzonaphthofuran	O	—	100	—	—	—	—
Dibenzothiophene	S	—	98	1.5	—	—	—
Indole	N	—	—	98[b]	2[b]	—	—
Indole–^{14}C	N	—	0.08[c]	68.9[c]	2.1[c]	15.3[c]	8.6[c]
3-Methylindole (Skatole)	N	—	—	100	—	—	—
5-Methylindole	N	—	—	100	—	—	—
5-Ethylindole	N	—	—	99	1	—	—
2,3-Dimethylindole	N	—	—	100	—	—	—
Carbazole–^{14}C	N	—	—	99[c]	1[c]	—	—
2-Methylcarbazole	N	—	—	100	—	—	—
Benzo[c]carbazole	N	—	10	90	—	—	—
1-Methylindole[d]	N	—	48	52	—	—	—
1,2-Dimethylindole[d]	N	—	—	100	—	—	—
9-Ethylcarbazole[d]	N	—	95	5	—	—	—
Dibenzo[c,g]carbazole	N	—	—	9	91	—	—
Quinoline	N	—	—	—	—	79	21
6-Methylquinoline	N	—	—	—	—	83	17
Acridine	N	—	—	—	—	100	—
Benz[c]acridine	N	—	—	—	69	31	—
Phenanthridine	N	—	—	—	—	100	—
1-Azapyrene	N	—	—	—	—	100	—

[a] PE, petroleum ether; B, benzene; E, ethyl ether.

[b] Percent distribution of indole as determined by GC.

[c] Percent distribution of radioactivity.

[d] N-Alkylated derivative.

VI. ISOLATION OF THE SULFUR HETEROCYCLE FRACTION

The high concentration of sulfur in fossil fuels is leading to considerable interest in the polycyclic sulfur heterocycles. A procedure originally devised to isolate the sulfur-rich fraction from petroleum (*173*) has been applied to coal-derived products (*174*) by Lee *et al.* (Fig. 4-23). The PAC concentrate from silicic acid chromatography (Section III) was taken up in 50 ml benzene and refluxed with an equal volume of glacial acetic acid with the addition of 20 ml of 30% H_2O_2 over a period of 1 hr, and the refluxing was continued for 16 hr. This procedure quantitatively oxidizes the sulfur heterocycles to sulfones, which were separated from unoxidized material (mainly PAH) by chromatography on silicic acid, where elution with benzene–methanol elutes the oxidized fraction. The sulfones were then reduced back to the sulfides by refluxing with a suspension of $LiAlH_4$ in dry ether. Further column chromatography, on silica gel, was necessary to separate the product from hydroquinones formed by reduction of oxidized PAH.

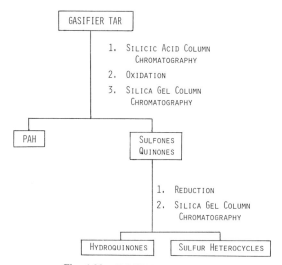

Fig. 4-23. S-PAH isolation scheme.

REFERENCES

1. R. C. Lao, R. S. Thomas, and J. L. Monkman, *J. Chromatogr.* **112**, 681 (1975).

2. F. W. Karasek, D. W. Denney, K. W. Chan, and R. E. Clement, *Anal. Chem.* **50**, 82 (1978).

3. B. S. Das and G. H. Thomas, *Anal. Chem.* **50**, 967 (1978).

4. T. D. Searl, F. J. Cassidy, W. H. King, and R. A. Brown, *Anal. Chem.* **42**, 956 (1970).

5. H. Boden, *J. Chromatogr. Sci.* **14,** 392 (1976).
6. R. L. Stedman, *Chem. Rev.* **68,** 153 (1968).
7. R. M. Harrison, R. Perry, and R. A. Wellings, *Water Res.* **9,** 331 (1975).
8. J. B. Andelman and J. E. Snodgrass, *Crit. Rev. Environ. Control* **4,** 69 (1974).
9. E. O. Haenni, *Residue Rev.* **24,** 41 (1968).
10. W. Giger and M. Blumer, *Anal. Chem.* **46,** 1663 (1974).
11. S. G. Chang and T. Novakov, *Atmos. Environ.* **9,** 495 (1975).
12. A. Bjorseth, *in* "Polynuclear Aromatic Hydrocarbons" (P. W. Jones and P. Leber, eds.), p. 371. Ann Arbor Sci. Publ., Ann Arbor, Michigan, 1979.
13. M. Katz and R. C. Pierce, *in* "Carcinogenesis—A Comprehensive Survey" (R. Freudenthal and P. W. Jones, eds.), Vol. 1, p. 413. Raven, New York, 1976.
14. R. C. Pierce and M. Katz, *Environ. Sci. Technol.* **9,** 367 (1975).
15. J. F. Thomas, M. Mukai, and B. D. Tebbens, *Environ. Sci. Technol.* **2,** 33 (1968).
16. B. D. Tebbens, J. F. Thomas, and M. Mukai, *Am. Ind. Hyg. Assoc. J.* **32,** 365 (1971).
17. M. Lippman, *Amer. Ind. Hyg. Assoc. J.* **31,** 138 (1970).
18. F. DeWiest, *Atmos. Environ.* **12,** 1705 (1978).
19. G. Broddin, L. Van Vaeck, and K. Van Cauwenberghe, *Atmos. Environ.* **11,** 1061 (1977).
20. N. A. Esmen and M. Corn, *Atmos. Environ.* **5,** 571 (1971).
21. S. E. Hrudy, *in* "Handbook of Air Pollution Analysis" (R. Perry and R. J. Young, eds.), Chap. 1. Chapman and Hall, London, 1977.
22. P. N. Cheremisinoff and A. C. Morresi, "Air Pollution Sampling and Analysis Deskbook." Ann Arbor Sci. Publ., Ann Arbor, Michigan, 1978.
23. P. O. Warner, "Analysis of Air Pollutants," Chap. 5. Wiley (Interscience), New York, 1976.
24. R. S. Sholtes, *Health Lab. Sci.* **7,** 279 (1970); Methods of Air Sampling and Analysis Intersociety Method Nos. 11104-01-69T and 11106-02-69T, Am. Public Health Assoc., Washington, D.C., 1972; Environmental Protection Agency Reference Method for the Determination of Suspended Particulates in the Atmosphere (High Volume Method) *Fed. Regist.* **36,** No. 84 (1971).
25. H. A. Clements, *J. Air Pollut. Control Assoc.* **22,** 955 (1972).
26. Warren Spring Laboratory, National Survey of Smoke and SO_2. Instructional Manual, Ministry of Technology (1966); British Standard Method for the Measurement of Air Pollution, Part II, Determination of Concentration of Suspended Matter, B.S. 1747, Part II, 1969.
27. R. M. Harrison, *in* "Handbook of Air Pollution Analysis" (R. Perry and R. J. Young, eds.), Chap. 2. Chapman and Hall, London, 1977.
28. L. Dubois, C. J. Baker, A. Zdrojewski, and J. L. Monkman, *Pure Appl. Chem.* **24,** 695 (1970).
29. M. Katz, *in* "Measurement of Air Pollutants, Guide to Selection of Methods." WHO, Geneva, 1969.
30. K. A. Schulte, D. L. Larsen, R. W. Horning, and J. V. Crable, Analytical Methods Used in Coke-Oven Effluent Study. HEW Publ. NIOSH 74-105, Natl. Inst. Occupational Safety and Health, Washington, D.C., 1974.
31. Exposure to coke oven emissions. Occupational Safety and Health Admin., Dept. Labor: *Fed. Regist.* **41** No. 206 46741 (1976).
32. A. M. Krstulovic, D. M. Rosie, and P. R. Brown, *Anal. Chem.* **48,** 1384 (1976).
32a. A. M. Krstulovic, D. M. Rosie, and P. R. Brown, *Int. Lab.* **5,** 11 (1977).
33. C. Pupp, R. C. Lao, J. J. Murray, and R. F. Pottie, *Atmos. Environ.* **8,** 915 (1974).
34. W. Cautreels and K. Van Cauwenberghe, *Atmos. Environ.* **12,** 1133 (1978).
35. W. Bertsch, R. C. Chang, and A. Zlatkis, *J. Chromatogr. Sci.* **12,** 175 (1974).

36. ASTM Annual Book of Standards: Standard Methods of Test for Sampling Stacks for Particulate Matter, D2928-71, Part 23 (1973). Am. Soc. Testing Mat., 1973.

37. British Standard Simplified Methods for Measurement of Grit and Dust Emissions (Metric Units), B.S. 3405: 1971.

38. D. D. B. H. Munns, *Filtr. Sep.* **14,** 656 (1977).

39. *Fed. Regist.* **36,** No. 247, 24878 (1971).

40. P. W. Jones, R. D. Giammar, P. E. Strup, and T. B. Stanford, *Environ. Sci. Technol.* **10,** 806 (1976).

41. P. W. Jones, A. P. Graffeo, R. Detrick, P. A. Clarke, and R. J. Jakobsen, Tech. Manual for Analysis of Organic Material in Process Streams, NTIS Publ. EPA-600/2-76-072 (1976).

42. P. W. Jones, *Proc. Anal. Div. Chem. Soc.* **15,** 158 (1978).

43. J. Adams, K. Menzies, and P. Levins, NTIS Rep. No. PB-268559, April 1977.

44. C. M. White, A. G. Sharkey, Jr., M. L. Lee, and D. L. Vassilaros, *in* "Polynuclear Aromatic Hydrocarbons" (P. W. Jones and P. Leber, eds.), p. 261. Ann Arbor Sci. Publ., Ann Arbor, Michigan, 1979.

45. V. del Vecchio, P. Valori, C. Melchiori, and A. Girella, *Pure Appl. Chem.* **24,** 739 (1970).

46. V. G. Grimmer, A. Hildebrandt, and M. Bohnke, *Erdoel, Kohle, Erdgas, Petrochem.* **25,** 442, 531 (1972).

47. R. S. Spindt, Polynuclear Aromatic Content of Heavy Duty Diesel Engine Exhaust Gases, NTIS Rep. No. PB-267774 (1977).

48. F. S. C. Lee, T. J. Prater, and F. Ferris, *in* "Polynuclear Aromatic Hydrocarbons" (P. W. Jones and P. Leber, eds.), p. 83. Ann Arbor Sci. Publ., Ann Arbor, Michigan, 1979.

49. T. W. Stanley, J. E. Meeker, and M. J. Morgan, *Environ. Sci. Technol.* **1,** 927 (1967).

50. G. Chalot, M. Castegnaro, J. L. Roche, R. Fontagnons, and P. Obatan, *Anal. Chim. Acta* **53,** 259 (1971).

51. R. C. Pierce and M. Katz, *Anal. Chem.* **47,** 1743 (1975).

52. M. Dong, D. C. Locke, and E. Ferrand, *Anal. Chem.* **48,** 368 (1976).

53. E. Sawicki, *Crit. Rev. Anal. Chem.* **1,** 275 (1970).

54. J. L. Monkman, G. E. Moore, and M. Katz, *Am. Ind. Hyg. Assoc. J.* **23,** 487 (1962).

55. Emergency Temporary Standard for Occupational Exposure to Benzene, *Fed. Regist.* **42,** No. 85 (1977).

56. G. E. Moore, M. Katz, and W. B. Drowley, *J. Air. Pollut. Contr. Assoc.* **16,** 492 (1966).

57. G. E. Moore, R. S. Thomas, and J. L. Monkman, *J. Chromatogr.* **26,** 456 (1967).

58. L. Dubois, A. Zdrojewski, and J. L. Monkman, *Mikrochim. Acta* **170** (1967).

59. Working Group on Air Standardization of Sampling and Analytical Procedure for Estimation of Polynuclear Aromatic Hydrocarbons in the Environment, WHO, Geneva, December, 1969.

60. E. Sawicki, R. C. Carey, A. E. Dooley, J. B. Giseland, J. L. Monkman, R. E. Neligan, and L. A. Ripperton, *Health Lab. Sci.* **7,** 45 (1970).

61. A. J. Lindsay and J. R. Stanbury, *Int. J. Air Water Pollut.* **6,** 327 (1962).

62. E. Sawicki, R. C. Carey, A. E. Dooley, J. B. Giseland, J. L. Monkman, R. E. Neligan, and L. A. Ripperton, *Health Lab. Sci.* **7,** 60 (1970).

63. H. L. Falk and P. E. Steiner, *Cancer Res.* **12,** 30, 60 (1952).

64. W. H. Griest, J. E. Caton, M. R. Guerin, L. B. Yeatts, Jr., and C. E. Higgins, *Anal. Chem.* **52,** 199 (1980).

65. D. Grosjean, *Anal. Chem.* **47,** 797 (1975).

66. R. J. Gordon, *Atmos. Environ.* **8,** 189 (1974).

67. H. H. Hill, Jr., K. W. Chan, and R. W. Karasek, *J. Chromatogr.* **131,** 245 (1977).

68. W. Cautreels and K. Van Cauwenberghe, *Atmos. Environ.* **10,** 447 (1976); *J. Chromatogr.* **131,** 253 (1977).

68a. W. Cautreels and K. Van Cauwenberghe, *Water Air Soil Pollut.* **6,** 103 (1976).

68b. P. E. Strup, R. D. Giammar, T. R. Stanford, and P. W. Jones, *in* "Polynuclear Aromatic Hydrocarbons: Chemistry, Metabolism, and Carcinogenesis" (R. I. Freudenthal and P. W. Jones, eds.), p. 241. Raven, New York, 1976.

69. C. Golden and E. Sawicki, *Int. J. Environ. Anal. Chem.* **4,** 9 (1975).

70. E. Sawicki, T. Belsky, R. A. Friedel, D. L. Hyde, J. L. Monkman, R. A. Rasmussen, L. A. Ripperton, and L. D. White, *Health Lab. Sci.* **12,** 407 (1975).

71. "NIOSH Manual of Analytical Methods." HEW Publ. NIOSH 77-157 Natl. Inst. Occupational Safety and Health, Washington, D.C., 1977.

72. H. P. Burchfield, E. E. Green, R. J. Wheeler, and S. M. Billedeau, *J. Chromatogr.* **99,** 697 (1974).

73. M. Matsushita, K. Arashidani, and H. Hayashi, *Bunseki Kagaku* **25,** 412 (1976).

74. M. A. Acheson, R. M. Harrison, R. Perry, and R. A. Wellings, *Water Res.* **10,** 207 (1976).

75. P. E. Strup, J. E. Wilkinson, and P. W. Jones *in* "Carcinogenesis—A Comprehensive Survey, Vol. 3: Polynuclear Aromatic Hydrocarbons" (P. W. Jones and R. I. Frueudenthal, eds.), p. 131. Raven, New York, 1978.

76. R. A. Hites, *Adv. Chromatogr.* **15,** 69 (1977).

77. J. W. Farrington and P. A. Meyer, *in* "Environmental Chemistry," Vol. 1, Chap. 5. Chem. Soc. Spec. Per. Rep., Chemical Society, London, 1975.

78. L. Scholz and H. J. Altmann, *Fresenius Z. Anal. Chem.* **240,** 81 (1968).

79. J. Borneff and H. Kunte, *Arch. Hyg. Bakteriol.* **148,** 585 (1964).

80. International Standards for Drinking Water, 3rd Ed., WHO, Geneva, 1971.

81. J. Borneff and H. Kunte, *Arch. Hyg. Bakteriol.* **153,** 220 (1969).

82. M. Fielding and B. Crathorne, *Proc. Anal. Div. Chem. Soc.* **15,** 155 (1978).

83. A. A. Rosen and F. M. Middleton, *Anal. Chem.* **27,** 790 (1955).

84. G. A. Junk, J. J. Richard, M. D. Grieser, D. Witiak, J. L. Witiak, M. D. Arguello, R. Vick, H. J. Svec, J. S. Fritz, and G. V. Calder, *J. Chromatogr.* **99,** 745 (1974).

85. A. K. Burnham, G. V. Calder, J. S. Fritz, G. A. Junk, M. J. Svec, and R. Willis, *Anal. Chem.* **46,** 139 (1972).

86. V. Leoni, G. Puccetti, and A. Girella, *J. Chromatogr.* **106,** 119 (1975).

87. W. E. May, S. N. Chesler, S. P. Cram, B. H. Gump, H. S. Hertz, D. P. Enagonio, S. M. Dyszel, *J. Chromatogr. Sci.* **13,** 535 (1975).

88 J. D. Navratil, R. E. Sievers, and H. F. Walton, *Anal. Chem.* **49,** 2260 (1977).

89. W. Bertsch, E. Anderson, and G. Holzer, *J. Chromatogr.* **112,** 701 (1975).

90. R. G. Webb, Isolating Organic Water Pollutants—XAD Resins, Urethane Foams, Solvent Extraction EPA-66014-75-003, Environmental Protection Agency, 1975.

91. J. Saxena, J. Kozuchowski, and D. K. Basu, *Environ. Sci. Technol.* **11,** 683 (1977).

92. D. K. Dasu and J. Saxena, *Environ. Sci. Technol.* **12,** 791 (1978).

93. K. Ogan, E. Katz, and W. Slavin, *J. Chromatogr. Sci.* **16,** 517 (1978).

94. M. Blumer and W. W. Youngblood, *Science* **188,** 53 (1975).

95. W. W. Youngblood and M. Blumer, *Geochim. Cosmochim. Acta* **39,** 1303 (1975).

96. M. Blumer and M. Rodrum, *J. Inst. Petrol.* **58,** 99 (1970).

97. M. Blumer, *Chem. Geol.* **16,** 245 (1975).

98. F. S. Brown, M. J. Baedecker, A. Nissenbaum, and I. R. Kaplan, *Geochim. Cosmochim. Acta* **36,** 1185 (1972).

99. Z. Aizenshtat, *Geochim. Cosmochim. Acta* **37,** 559 (1973).

100. P. D. Keizer, J. Dale, and D. C. Gordon, *Geochim. Cosmochim. Acta* **42,** 185 (1978).

101. J. W. Farrington and B. W. Tripp, *in* "Marine Chemistry in the Coastal Environment" (T. M. Church, ed.) ACS Symp. Ser. **18,** p. 267 Am. Chem. Soc., Washington, D.C., 1975.

102. J. W. Farrington and B. W. Tripp, *Geochim. Cosmochim. Acta* **41,** 1627 (1977).

103. M. Blumer, T. Dorsey, and J. Sass, *Science* **195,** 283 (1977).

104. M. Blumer, *Anal. Chem.* **29,** 1039 (1957).

105. D. W. Farrington, *in* "Marine Pollution Control Monitoring" (E. D. Goldberg, ed.), NOAA, U.S. Dep. of Commerce, Washington, 1972.

106. J. W. Howard, E. W. Turicchi, R. H. White, and T. Fazio, *J. Assoc. Offic. Anal. Chem.* **49,** 1237 (1966).

107. G. Grimmer and H. Bohnke, *J. Assoc. Offic. Anal. Chem.* **58,** 725 (1975).

108. G. Grimmer and A. Hildebrandt, *Chem. Ind.* (*London*), 2000 (1967).

109. G. C. Croft and S. Norman, *J. Assoc. Offic. Anal. Chem.* **49,** 625 (1966).

110. G. Grimmer and D. Duevel, *Z. Naturforsch.* **25B,** 1171 (1970).

111. J. J. Schmidt-Collerius, F. Bonomo, K. Gala, and L. Leffler, *in* "Science and Technology of Oil Shale" (T. F. Yen, ed.), p. 115. Ann Arbor Sci. Publ. Ann Arbor, 1976.

112. J. L. Hancock, H. G. Applegate, and J. D. Dodd, *Atmos. Environ.* **4,** 362 (1970).

113. J. W. Howard, R. T. Teague, R. H. White, and B. E. Fry, Jr., *J. Assoc. Offic. Anal. Chem.* **49,** 595 (1966).

114. J. W. Howard, R. H. White, B. E. Fry, Jr., and E. W. Turicchi, *J. Assoc. Offic. Anal. Chem.* **49,** 611 (1966).

115. A. J. Malonski, C. L. Greenfield, C. J. Barnes, J. M. Worthington, and F. L. Joe, *J. Assoc. Offic. Anal. Chem.* **51,** 114 (1968).

116. G. Grimmer and A. Hildebrandt, *J. Assoc. Offic. Anal. Chem.* **55,** 631 (1972).

117. G. Grimmer, *Dtsch. Apoth. Ztg.* **108,** 529 (1968).

118. G. Grimmer, *Z. Krebsforsch.* **69,** 223 (1967).

119. G. Grimmer, A. Hildebrandt, and H. Bohnke, *Dtsch. Lebensm. Rundsch.* **71,** 93 (1975).

120. G. Grimmer, *Dtsch. Lebensm. Rundsch.* **65,** 229 (1969).

121. G. Grimmer, *Dtsch. Lebensm. Rundsch.* **62,** 19 (1966).

122. K. Potthast and G. Eigner, *J. Chromatogr.* **103,** 173 (1975).

123. D. Hoffmann and E. Wynder, *Anal. Chem.* **32,** 295 (1960).

124. G. Grimmer and H. Bohnke, *Fresenius Z. Anal. Chem.* **261,** 310 (1972).

125. P. O. Warner, "Analysis of Air Pollutants" p. 65. Wiley (Interscience), New York, 1976.

126. R. L. Stedman, R. L. Miller, L. Lakpitz, and W. J. Chamberlain, *Chem. Ind.* (*London*), 394 (1966).

127. I. Schmeltz, C. J. Dooley, R. L. Stedman, and W. J. Chamberlain, *Phytochemistry* **6,** 33 (1967).

128. K. D. Bartle, M. L. Lee, and M. Novotny, *Int. J. Environ. Anal. Chem.* **3,** 349 (1974).

129. M. Novotny, M. L. Lee, and K. D. Bartle, *J. Chromatogr. Sci.* **12,** 606 (1974).

130. M. L. Lee, K. D. Bartle, and M. V. Novotny, *Anal. Chem.* **47,** 540 (1975).

131. M. L. Lee, M. Novotny, and K. D. Bartle, *Anal. Chem.* **48,** 405 (1976).

132. M. L. Lee, M. Novotny, and K. D. Bartle, *Anal. Chem.* **48,** 1566 (1976).

133. R. E. Laflamme and R. A. Hites, *Geochim. Cosmochim. Acta* **42,** 289 (1978).

134. E. O. Haenni, F. L. Joe, J. W. Howard, and R. L. Leibel, *J. Assoc. Offic. Agric. Chem.* **45,** 59 (1962).

135. E. O. Haenni, J. W. Howard, and F. L. Joe, *J. Assoc. Offic. Agric. Chem.* **45,** 67 (1962).

136. J. W. Howard and E. O. Haenni, *J. Assoc. Offic. Agric. Chem.* **46,** 933 (1963).

137. J. E. Wilkinson, P. E. Strup, and P. W. Jones, *in* "Polynuclear Aromatic Hydrocarbons" (P. W. Jones and P. Leber, eds.), p. 217. Ann Arbor Sci. Publ., Ann Arbor, Michigan, 1979.

138. M. E. Snook, R. F. Severson, M. C. Higman, R. F. Arrendale, and O. T. Chortyk, *Beitr. Tabakforsch.* **8,** 250 (1976).

139. D. F. S. Natusch and B. A. Tomkins, *Anal. Chem.* **50,** 1429 (1978).

140. A. P. Swain, J. E. Cooper, R. L. Stedman, and F. E. Bock, *Beitr. Tabakforsch.* **1,** 97 (1969).

141. G. Lunde and A. Bjorseth, *Nature (London)* **268,** 518 (1977).

142. L. Wallcave, H. Garcia, R. Feldman, W. Lijinsky, and P. Shubik, *Toxicol. Appl. Pharmacol.* **18,** 41 (1971).

143. W. K. Robbins, *in* "Polynuclear Aromatic Hydrocarbons: Chemistry and Biological Effects" (A. Bjørseth and A. J. Dennis, eds.), p. 841. Battelle Press, Columbus, Ohio, 1980.

144. R. S. Thomas and J. L. Monkman, CD Rep. No. 215, Chem. Div. Tech. Dev. Branch, Air Pollution Control Directorate, Department of the Environment, Ottawa, Ontario, Canada 1972.

145. R. C. Lao, R. S. Thomas, M. Oja, and L. Dubois, *Anal. Chem.* **45,** 908 (1973).

146. G. J. Cleary, *J. Chromatogr.* **9,** 204 (1962), and references therein.

147. R. F. Severson, M. E. Snook, R. F. Arrendale, and O. T. Chortyk, *Anal. Chem.* **48,** 1866 (1976).

148. M. E. Snook, *Anal. Chim. Acta* **81,** 423 (1976).

149. H. J. Klimisch, *Fresenius Z. Anal. Chem.* **264,** 275 (1973).

150. A. R. Jones, M. R. Guerin, and B. R. Clark, *Anal. Chem.* **49,** 1766 (1977).

151. J. Sjovall, E. Nystrom, and E. Haahti, *Adv. Chromatogr.* **6,** 119 (1967).

152. D. Hoffmann and E. Rathkamp, *Anal. Chem.* **44,** 899 (1972).

153. R. Tomingas, G. Voltmer, and R. Bednarik, *Sci. Total Environ.* **7,** 261 (1977).

154. D. Brocco, V. Cantuti, and G. P. Cartoni, *J. Chromatogr.* **49,** 66 (1969).

155. B. L. van Duuren, *J. Natl. Cancer Inst.* **21,** 1 (1958).

156. R. Tye and Z. Bell, *Anal. Chem.* **36,** 1612 (1964).

157. D. D. Mold, T. B. Walker, and L. E. Veasey, *Anal. Chem.* **35,** 2071 (1963).

158. K. Rothwell and J. K. Whitehead, *Nature (London)* **213,** 797 (1967).

159. W. Giger and C. Schaffner, *Anal. Chem.* **50,** 243 (1978).

160. E. L. Wynder and D. Hoffmann, "Tobacco and Tobacco Smoke," p. 730. Academic Press, New York, 1967.

161. R. F. Severson, M. E. Snook, H. C. Higman, O. T. Chortyk, and F. J. Akin, *in* "Carcinogenesis—a comprehensive Survey" (R. I. Freudenthal and P. W. Jones, eds.), p. 253. Raven, New York, 1976.

162. D. Hoffmann, G. Rathkamp, K. D. Brunnemann, and E. L. Wynder, *Sci. Total Environ.* **2,** 151 (1973).

163. D. Hoffmann, K. D. Brunnemann, G. B. Gori, and E. L. Wynder, *Recent Adv. Phytochem.* **9,** 63 (1975).

164. E. Sawicki, J. E. Meeker, and M. J. Morgan, *Int. J. Air Water Pollut.* **9,** 291 (1965).

165. E. Sawicki, S. P. McPherson, T. W. Stanley, J. Meeker, and W. C. Elbert, *Int. J. Air Water Pollut.* **9,** 515 (1965).

166. M. W. Dong, D. C. Locke, and D. Hoffmann, *Environ. Sci. Technol.* **11,** 612 (1977).

167. M. Dong, I. Schmeltz, E. LaVoie, and D. Hoffmann, *in* "Carcinogenesis—A Comprehensive Survey, Vol. 3: Polynuclear Aromatic Hydrocarbons" (P. W. Jones and R. I. Freudenthal, eds.), p. 97. Raven, New York, 1978.

168. L. R. Snyder and B. E. Buell, *Anal. Chem.* **40,** 1295 (1968).

169. D. M. Jewell, J. H. Weber, J. W. Bunger, H. Plancher, and D. R. Latham, *Anal. Chem.* **44,** 1391 (1972).

170. J. F. McKay, J. H. Weber, and D. R. Latham, *Anal. Chem.* **48,** 891 (1976).

171. M. E. Snook, *in* "Carcinogenesis—A Comprehensive Survey, Vol. 3; Polynuclear Aromatic Hydrocarbons" (P. W. Jones and R. I. Freudenthal, eds.), p. 203. Raven, New York, 1978.

172. M. E. Snook, R. F. Arrendale, H. C. Higman, and O. T. Chortyk, *Anal. Chem.* **50,** 88 (1978).

173. H. V. Drushel and A. L. Sommers, *Anal. Chem.* **39,** 1819 (1967).

174. M. L. Lee, C. Willey, R. N. Castle, and C. M. White, *in* "Polynuclear Aromatic Hydrocarbons: Chemistry and Biological Effects" (A. Bjørseth and A. J. Dennis, eds.), p. 59. Battelle Press, Columbus, Ohio, 1980.

175. R. E. Jentoft and T. H. Gouw, *Anal. Chem.* **40,** 1787 (1968).

176. G. Grimmer and H. Bohnke, *Chromatographia* **9,** 30 (1976).

5

Column, Paper, and Thin-Layer Chromatography

I. INTRODUCTION

The classical techniques of column adsorption, paper, and thin-layer chromatography have been widely used in the separation of complex PAC mixtures into simpler fractions prior to further separation by higher resolution techniques and/or identification of mixture constituents. Before the development of high-performance liquid chromatography (HPLC) and GC/MS, these techniques coupled with UV absorption or fluorescence spectroscopy provided the most powerful means of analysis. Extensive reviews of these methods have been published (1, 2). In the early 1970s, the superior performance of HPLC techniques almost completely replaced the use of classical column adsorption chromatography for the analysis of PAC except for the separation and purification of the total PAC fraction from other classes of compounds, as was discussed in Chapter 4.

In this chapter, a review and discussion of the classical analytical methods of column adsorption, paper, and thin-layer chromatography are given. New techniques of high-performance TLC are described. Recent developments in column chromatography using lipophilic gels are also included.

II. COLUMN ADSORPTION CHROMATOGRAPHY

The earliest and most widely used method for simplifying complex PAH mixtures for identification purposes is fractionation by column adsorption chromatography. The mechanism of separation depends on the relative affinities of the solutes for the adsorbent and on their solubilities in the eluting solvent. It is a form of liquid–solid chromatography (LSC). Solvent molecules in the moving carrier liquid compete with sample molecules for adsorptive sites on the surface of the solid adsorbents.

Most classical techniques of column LSC make use of relatively wide diameter columns (10 to 20 mm) and large ($> 100 \mu$m), totally porous packings, particularly silica gel and alumina (3). Generally speaking, columns with diameter/length ratios of at least 1:25 and packed with adsorbents of uniform particle size (60–80, 80–100, 100–120 mesh) are used. Separation of a wide range of compounds is accomplished by gradually increasing the polarity of the eluting solvents. The variables which affect chromatographic performance include the percentage of water in the sorbent, the percentage of polar solvent in the solvent mixture, and the weight of the organic fraction. Although adsorbents such as alumina and silica gel have been most extensively used, applications of Florisil and cellulose acetate have also been reported.

A. Alumina

The separation of many PAC is more readily achieved using alumina than when using silica adsorbents (3). Alumina is normally prepared by low-temperature ($< 700°$C) dehydration of alumina trihydrate and is a mixture of γ-alumina and a small percentage of alumina monohydrate ($Al_2O_3 \cdot H_2O$). If the alumina is heated above $900°$C, complete dehydration results and the less active α-alumina is produced. γ-Alumina is highly active when freshly prepared but rapidly loses its reactivity by adsorption of water when exposed to a damp atmosphere (4). The reactivity or extent of hydration of the material can be determined easily by grading according to the Brockmann test (5).

Various pretreatments, such as heating, grinding, and treatment with acids, alkalis, and salts, change the properties of natural alumina. The natural alumina lattice which consists of hexagonal close-packed γ-alumina will be damaged, especially near the surface, by grinding and abrasion. This damage to the surface will expose positively charged aluminum atoms and negatively charged oxygen atoms. From the surrounding water, hydrogen and hydroxyl ions will be adsorbed on the O^- and Al^+ sites, respectively, to form surface hydroxyl groups. Extensively damaged parts are generally hydrated

$$Al_2O_3 + 3 H_2O \rightarrow 2 Al(OH)_3 \tag{1}$$

whereas less damaged parts are not completely hydrated:

$$Al(OH)_3 \rightleftharpoons Al(OH)_2^+ + OH^- \tag{2}$$

$$>\!Al \cdot O \cdot OH \rightleftharpoons Al \cdot O \cdot O^- + H^+ \tag{3}$$

Hydrochloric acid-treated alumina powder has a positive surface containing both covalently bound and ionized chloride.

$$\text{>AlOH} + \text{HCl} \rightleftharpoons \text{>Al—Cl}(+\text{H}_2\text{O}) \rightleftharpoons \text{>Al}^+ + \text{Cl}^- \tag{4}$$

All organic solutes, except possibly aliphatic hydrocarbons, are adsorbed by the polar alumina surface groups to a greater or lesser extent. PAH, which are nonpolar, are adsorbed probably by π-electron bonding with aluminum atoms exposed by damage to the surface (4).

Alumina is commercially available in a number of grades, sizes, pH, and activity. The sample/adsorbent ratio can vary between $1:100$ and $1:1000$. The alumina can be deactivated by adding $0\text{--}15\%$ water prior to use. Columns and solvent systems used by different workers are listed in Table 5-1. The standardization of alumina for chromatography of PAH has been described (13).

Figure 5-1 shows the typical elution profiles of PAH standard compounds on alumina using cyclohexane as the mobile phase (24).

The major disadvantages in using alumina column chromatography are that it is very time consuming, sensitive compounds may decompose on the adsorbent, and reproducibility is difficult to achieve.

B. Silica

Silica can be obtained from purified ground quartz and sand, or it can be formed by acid precipitation from silicate salts and by hydrolysis of silicon compounds such as $SiCl_4$ in the liquid or vapor phase. Silica gel is the form

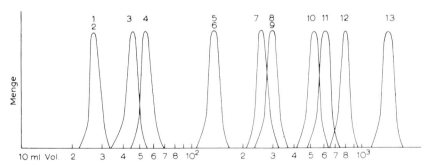

Fig. 5-1. Elution profile for column adsorption chromatography of 13 PAH on alumina with cyclohexane as the mobile phase: 1 = anthracene; 2 = phenanthrene; 3 = pyrene; 4 = fluoranthene; 5 = benz[a]anthracene; 6 = chrysene; 7 = benzo[a]pyrene; 8 = benzo[e]pyrene; 9 = perylene; 10 = anthanthrene; 11 = benzo[ghi]perylene; 12 = dibenz[a,h]anthracene; 13 = coronene. [Reproduced with permission from G. Grimmer and A. Hildebrandt, J. Chromatogr. **20**, 89 (1965). Copyright, Elsevier Scientific Publishing Co.]

TABLE 5-1

Alumina Column Chromatography of PAH

Adsorbent	Column dimensions	Weight of adsorbent	% H_2O	Solvent system	Sample type	Ref.
Alumina			"Activated"	Petroleum ether → petroleum ether + ether	Carbon blacks	(6)
Alumina	1.5 to 10 × 1 cm		"Moderate activity"	Cyclohexane	Industrial effluents	(7)
Alumina	30 × 0.8 cm	2 g	"Activated"	Petroleum ether + ether	Auto exhaust	(8)
Alumina 20 mesh	45 × 1.7 cm		"Activated"	Petroleum ether → petroleum ether + ethyl ether	Air particulates	(9)
Spence type H 100 to 200 mesh			"Suitably activated"	Cyclohexane	Town air	(10)
Alumina	10 × 2.8 cm		Bottom half = "low activity," top half = "high activity	Cyclohexane → benzene	Tobacco smoke	(11)
Alumina 200 mesh (Takeda Chem. Co.)	15 × 0.8 cm 10 × 1.2 cm 15 × 1.5 cm		"Activated"	Petroleum ether → petroleum ether + benzene	Various combustion products	(12)
Spence type H 100 to 200 mesh	1 cm, i.d.	5.5 g	"Brought into equilibrium with water vapor over 75% sulfuric acid"	Cyclohexane	Air particulates, cigarette smoke	(13)

Adsorbent	Dimensions	Amount	Activity	Eluent	Application	Ref.
Al-0109P (Harshaw Chemical Co.)	260 × 9 mm to 60 × 6 mm		3–6%	Cyclohexane, pentane, or isooctane with increasing portions of benzene or acetone	Oysters	(14)
Spence 100 to 200 mesh				Petroleum ether → petroleum ether + acetone	Tobacco smoke	(15)
Woelm neutral grade	28 mm, i.d.	115 g	activity 1	Light petroleum → light petroleum + benzene	Tobacco smoke	(16)
Spence type H 100 to 200 mesh	7.5 × 1.4 cm		4% w/w	Cyclohexane	Air particulates around diesel-bus garages	(17)
Alcoa grade F-20	30–80 cm	150–250 g		Petroleum ether → petroleum ether + benzene	Standard compounds	(18)
Alumina	10 × 1.5 cm	18 g	"Deactivated"	Cyclohexane → cyclohexane + benzene	Air particulates, diesel exhausts	(19)
Merck acid-washed aluminum oxide	15 × 0.5 in	9 g + 0.5 g silica gel	13.7%	Pentane → pentane + ether	Air particulates	(20)
Spence 100 to 200 mesh	3 × 0.5 in		5% w/w	Cyclohexane	Air particulates	(21)
Alcoa F-20	Various		0.5–3.6%	n-pentane	Petroleum	(22)
B.D.H. 100 to 200 mesh	23 × 0.5 in		Brought to equilibrium over 50% sulfuric acid (13–13.5%)	Cyclohexane → cyclohexane + ether	Air particulates	(23)

(Continued)

TABLE 5-1 (*Continued*)

Absorbent	Column dimensions	Weight of adsorbent	% H$_2$O	Solvent system	Sample type	Ref.
Merck (0.08–0.09 mm)	200 × 8 mm	15 g	4%	Cyclohexane	Environmental samples	(24)
Spence type H 100 to 200 mesh	12 × 1.0 cm		1.8%	Cyclohexane → cyclohexane + ether	Air particulates	(25)
Spence 100 to 200 mesh			13%	Cyclohexane → cyclohexane + ether	Soot	(26)
Spence type H 100 to 200 mesh	1 cm, i.d.		1.8%	Cyclohexane → cyclohexane + ether	Air particulates	(27)
Merck acid-washed aluminum oxide, 100 to 200 mesh	13 mm, i.d.	9 g	13.7%	Pentane → pentane + ether	Air particulates	(28)
Alumina	40 × 1.2 cm		2%	Cyclohexane	Auto exhaust	(29)
Alumina			2%	Cyclohexane → cyclohexane + ether	Air particulates	(30)

of silica most widely used, and, like alumina, it is white so that bands of material can be seen clearly.

The surface of the silica is covered with strongly bonded hydroxyl groups. Physisorbed water is removed from the surface by heating in air at 120°C, and all water is removed at 900°C. Upon dehydration, a strained siloxane bridge is formed which may be represented as:

$$\underset{-Si-Si-}{\overset{OH\ OH}{|\ \ |}} \rightarrow \underset{-Si-Si-}{\overset{O}{/\backslash}} \tag{5}$$

Depending upon the arrangement and number of the hydroxyl and siloxane groups, the silica is reactive. Again, the interaction of the adsorbate with adsorbent is π-electron bonding interactions with the polar surface (4).

Since there are many commercially available silica gel adsorbents which differ widely in their characteristics, and the adsorbent which gives best separation of the components is desired, a method of choosing silica gel adsorbents has been given by Scott (31). The method is based on resolution, scope, load, and speed. Silica gel adsorbents tend to have a smaller pore size than the alumina adsorbents. The sample/adsorbent ratio varies, usually between 1:50 and 1:500.

Table 5-2 lists the silica gel systems used by various workers. Recoveries from both alumina and silica gel columns usually range from 50 to 100% for different PAC. Silica gel chromatography is rarely used as the sole separation process. It has been used most often as a preseparation and clean-up step before further chromatography on alumina. Hirsch et al. (35) reported the use of a dual-packed (silica gel–alumina) adsorption column to separate samples of high-boiling petroleum distillates.

Column chromatography with silica gel suffers from the same disadvantages as encountered with alumina. In addition, flow-rates are usually slower through silica gel because of the greater porosity.

C. Other Adsorbents

Other adsorbents have not been used nearly as extensively as alumina or silica gel. Florisil, a synthetic magnesium silicate, has been used for isolation of PAH as a group from other compounds (36, 37), as well as for PAH fractionations (38). The main advantage in using Florisil is the rather good PAH separations despite short elution times.

The use of cellulose acetate as an adsorbent for column chromatography has been somewhat limited, even though it has been successfully used in planar adsorption techniques. Spotswood (39) described the use of columns packed with partially acetylated cellulose powder. This technique was found

TABLE 5-2

Silica Gel Column Chromatography of PAH

Adsorbent	Column dimensions	Weight of adsorbent	% H$_2$O	Solvent system	Sample type	Ref.
Davison chromatographic grade	Various		1%	n-Pentane	Petroleum	(22)
Silica gel sieve fraction 0.125–0.160 mm		20 g	9–10%	Cyclohexane	Environmental samples	(24)
Silica gel	30 × 0.25 cm		5%	Hexane → hexane + benzene	Air particulates	(27)
Davison grade 922	6 mm, i.d.	2.4 g	0 to 24%	Cyclohexane → cyclohexane + benzene	Benzo[a]pyrene	(32)
Silica gel		30 g	"Deactivated"	Hexane → hexane + benzene	Tobacco smoke	(33)
Mallinckrodt analytical reagent silicic acid 100 mesh	22 mm, o.d.	50 g	"Highly activated"	Hexane → hexane + benzene	Tobacco smoke	(34)

to be of particular value in the separation of larger polycyclic aromatic hydrocarbons, which are eluted almost simultaneously from an alumina column. Cellulose acetate has recently been used as an adsorbent for the purification of PAH fractions (37).

The formation of complexes of PAH with π-acids has been used for many years as a means of purification and identification by organic chemists. Columns containing an adsorbent which has been impregnated with 1,3,5-trinitrobenzene (40), picric acid (41), or tetrachlorophthalic anhydride (42) have been used with varying degrees of success. Tye and Bell (43) described the use of 1,3,5-trinitrobenzene in polyethylene glycol as a selective stationary phase in column chromatography of PAH.

D. Structure–Retention Correlations

The correlation between structure and retention for PAH in adsorption chromatography, whether it be column or thin-layer, is not completely understood. However (44) it is a well-known fact that the retention increases with a greater number of double bonds. Therefore, larger PAH have more rings, more double bonds, and hence, longer retention.

Further insight into structural relationships can be obtained by comparing the retention characteristics of anthracene to phenanthrene. According to Fries' rule (45), the most stable form of a polycyclic hydrocarbon exists when the highest possible number of rings has the normal benzoic conformation of three double bonds. These configurations are shown in Fig. 5-2. It can be seen that, while anthracene has only two rings in the "benzoic conformation," phenanthrene has three. While this approach is interesting in view of the fact that phenanthrene is more strongly retained than anthracene, it is not consistent with quantum-mechanical calculations of ring currents; it is more likely that the high bond order of the phenanthrene 9,10-bond

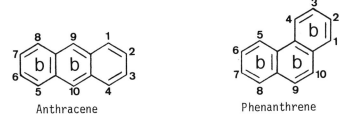

Fig. 5-2. Structural configuration of anthracene and phenanthrene.

leads to a high polarizability and that, as discussed in Chapter 1, this results in stronger adsorption of phenanthrene than of anthracene.

The addition of alkyl groups to the ring affects the chromatographic behavior only by steric hindrance. When alkyl groups are added, the size of the molecule is enlarged, and the retention is slightly increased. 2,3,6-Trimethylnaphthalene was found to have an R_F value in TLC approximately the same as for three-ring compounds (45). Similarly, the chromatographic behavior of partially hydrogenated PAH is influenced by the nonplanarity of the molecules, as the aromatic rings are distorted out of the normal planar arrangement. Anthracene has a higher R_F value than 9,10-dihydroanthracene in TLC (45).

Snyder (46) reported the relative elution order and potential compound class separability of PAH by alumina and silica gel. Alcoa F-20 alumina and Davison chromatography-grade silica were the adsorbents studied. When the adsorbents were partially deactivated by water, isotherm linearity increased while total capacity decreased. It was found also that the silica had a much greater adsorbed volume. The silica activity tended to decline at higher carbon numbers, causing shorter retention times. Alumina therefore gave greater separation of the parent aromatics than the silica. Alkyl substitutents alter the adsorption energies more for the silica, causing greater compound class overlap. This is because the solute cross section and, hence, the accessibility to the small adsorbent pores of silica are governed by the molecular geometry. Alumina was shown to be superior in the separation of aromatic hydrocarbon classes.

Extensive studies of PAH retention in adsorption chromatography have been made by Klemm (18, 47), Snyder (46, 48–50), and Popl (51–54). A more detailed discussion of this subject is given in Chapter 6.

The application of adsorption chromatography to aromatic samples was one of the earliest methods of separating PAH. Alumina was shown to be a better adsorbent than silica. The majority of work has been done using these two different adsorbents. Generally, for the most complete separations they were used in conjunction with preliminary separations on silica followed by a better resolving column of alumina.

Column chromatography provides a system of analysis which is nondestructive to the sample. However, there are many drawbacks to this method. Since flow rates are controlled by gravity or moderately low pressures and must be slow to achieve any resolution, analysis times are long. Some analyses take days by this method. The theory of adsorption chromatography is not complete enough to explain everything that happens. The process itself is not completely reproducible. It is seriously impaired by slight modifications in the adsorbents with traces of water or other contaminants in the sample to be analyzed. The method of separation is not adequate for trace analysis. The constituents can be lost or resolution may be inadequate for identification of all components.

Column chromatography using adsorbents led to the development of paper chromatography, thin-layer chromatography, and HPLC. The basic theories of adsorption are still used, yet better resolution and reproducibility can be achieved. Column chromatography is a good method for clean-up of a sample to prepare an aromatic fraction (see Chapter 4), and it is employed before many methods of chromatography that offer greater resolution.

III. PAPER AND THIN-LAYER CHROMATOGRAPHY

Early separations of PAC were done using classical column chromatographic procedures. Usually, column chromatography entailed at least two days of analysis time and fairly large samples. In an attempt to find less cumbersome methods, paper chromatographic and eventually thin-layer chromatographic (TLC) techniques were developed.

Paper chromatography usually requires less than a day of analysis time and only microgram sample quantities. As a natural extension of the development of paper chromatography, the possibility of finding a more easily accessible and much less expensive chromatographic adsorbent prompted the development of TLC. In this technique, a thin film of an appropriate adsorbent is bound to an inert backing and developed in the same manner as in paper chromatography. Though reproducibility of R_F values obtained by TLC is no better than in paper chromatography, other advantages, such as shorter run time (a few hours or less), sharper spots, and a broader spectrum of spray reagents, make this method more widely applicable to analytical problems.

A. Paper Chromatography

Paper chromatography of PAH was shown by Dubois (55) in 1959 to give better resolution than column chromatography on either alumina or silica gel, although the capacity of the paper techniques is usually smaller. Separations that often are difficult on a column often succeed on paper, e.g., benzo[a]pyrene (BaP) and benzo[e]pyrene (BeP); fluoranthene and benz[a]-anthracene (BaA); and BaP and benzo[ghi]perylene (BghiP). Furthermore, paper techniques are often able to separate compounds with practically identical UV absorption spectra, e.g., pyrene, 2-methylpyrene, and 4-methylpyrene (56).

Paper chromatography of PAH requires the use of hydrophobic paper.

The decisive breakthrough into this field was achieved in 1952 by Micheel and Schweepe, who were the first to use acetylated paper (57). Further refinements of acetylated papers and their applications have been described by Spotswood (39, 58). Although good separations are obtainable, paper chromatography suffers from the nonreliability of R_F values, unless the conditions of running the chromatograms are strictly controlled. The chief difficulty is due to the irreproducibility of the acetylated paper (59, 60), different batches having somewhat different characteristics. Van Duuren (61) exploited the principle of differing R_F values of PAH on differently acetylated paper to give selective, sequential-type separations.

Dubois et al. (55) have found that ascending chromatography is often more reproducible than descending chromatography. Using five solvent systems different from those of other authors, they established the R_F values found in Table 5-3 with a claimed reproducibility of $\pm 3\%$ for some PAH on Schleicher and Schull No. 2043b acetylated paper; a further claim of reasonable resolution at R_F differences of 0.05 was made.

Table 5-4 lists representative paper chromatographic systems as compiled by Schaad (2). Various applications of these systems are contained in the literature (80–88).

Disadvantages of paper chromatography include long separation times, irreproducible papers, often inadequate resolution, and poor quantification.

TABLE 5-3

Absolute R_F Values of Some PAH in
Toluene–Methanol–Water (1:10:1)
on Acetylated Paper

Compound	R_F Value
Acenaphthene	0.88
Fluorene	0.69
Anthracene	0.64
Phenanthrene	0.62
9-Methylanthracene	0.57
Pyrene	0.52
Fluoranthene	0.51
Benz[a]anthracene	0.43
Benzo[a]pyrene	0.41
Perylene	0.28
Chrysene	0.22
Benzo[ghi]perylene	0.18
Dibenz[a,h]anthracene	0.16
Benzo[e]pyrene	0.10
Anthanthrene	0.07

TABLE 5-4

Paper Chromatographic Systems

Paper[a]	Mobile phase	Method	Time	Ref.
Normal/DMF	Decalin	Decending		(24)
Acel/polystyrene	Methanol–petroleum ether	Ascending		(57)
Acel	Methanol–ether–water (4:4:1)	Ascending	12 hr	(58)
Acel	Ethanol–toluene–water (17:4:1)	Ascending	12 hr	(58)
Acel	Ethanol–benzene–water (12:6:1)	Ascending	12 hr	(58)
Normal	Acetic acid–water (2:3)	Rutter technique		(62, 63)
Normal	Methanol–water (9:1)	Ascending		(64)
Normal	Benzene–methanol–water (6:2:1)	Ascending		(65)
Acel	Methanol–ether–water (4:4:1)	Ascending	20 hr	(66)
Acel	Methanol–water–benzene (6:1:2)	Ascending	20 hr	(66)
Acel	Methanol–water–petroleum ether (20:1:5)	Ascending	20 hr	(66)
Acel	Methanol–toluene–water (6:1:1)	Ascending		(67)
Acel	Methanol–light petroleum–water (8:1:2)	Ascending		(67)
Acel	Methanol–chloroform (3:1)	Ascending	17 hr	(68)
Acel	Ethanol–toluene–water (30:5:1)	Ascending	16 hr	(69)
Normal/paraffin oil	Methanol–paraffin oil	Ascending		(70–76)
Normal/ABN	ABN (α-bromonaphthalene), acetic acid	Descending	2 hr	(77)
Normal	Ethanol–ammonia–water (20:1:4)	Descending	2 hr	(77)
Normal/DMF	Dimethylformamide–hexane	Descending	2 hr	(77, 79)
Acel	Methanol–toluene–water (20:3:2)	Ascending		(78)

[a] Acel refers to acetylated paper.

For these reasons, paper chromatography is rarely used in modern PAH separations.

B. Thin-Layer Chromatography

In 1959, Pavelka and D'Ambrosio (87) used an alumina plate and separated a mixture of seven PAH with cyclohexane as the developing solvent, but it was not until early 1964 that TLC was applied to any great extent to the separation of PAH mixtures. Some of the advantages of TLC are as follows:

1. Elution times are short, thus minimizing losses from volatility and instability.
2. The thin layer of inorganic (usually) material permits the use of aggressive universal indicators such as heat, concentrated mineral acids, alkaline permanganate solutions, and dichromate/sulfuric acid mixtures.

3. Usually spot density is great enough to be detected by direct fluorescence, fluorescence quenching, or color development.

4. Spots may be excised and extracted for identification by various spectroscopic methods.

5. Valuable information regarding possibilities for separations desired, choices of solvent systems, and choices of adsorbents can be obtained quickly and simply for subsequent applications to preparative column chromatography.

The principal disadvantage lies in the ever-present possibility of accelerating unwanted oxidation (air, ultraviolet) of sought compounds, or inducing rearrangements or isomerizations, when sensitive PAH are adsorbed onto an active surface such as alumina. Another disadvantage is that elution problems may occur if the compound is strongly bound to the adsorbent.

The use of TLC for the separation of PAH has been reviewed by several workers (2, 56, 88, 127). In general, it can be stated that there is no universal method or one method which is superior to another. The choice of a par-

TABLE 5-5

R_B Values of PAH (R_B of BaP = 1.00)

	Systems[a]		
Compound	1	2	3
Phenanthrene	1.13	3.74	1.99
Anthracene	1.14	3.33	1.99
Fluoranthene	1.09	2.92	1.89
Chrysene	1.10	—	1.75
Pyrene	1.25	3.16	1.72
Triphenylene	1.07	—	1.49
Benz[a]anthracene	1.03	2.70	1.47
11H-Benzo[b]fluorene	1.08	3.54	1.33
Benzo[e]pyrene	1.04	2.94	1.16
Perylene	0.19	2.86	1.14
Benzo[k]fluoranthene	0.98	2.40	1.03
Benzo[a]pyrene	1.00	1.00	1.00
Anthanthrene	0.71	2.17	0.70
Benzo[ghi]perylene	0.89	3.04	0.69
Dibenz[a,h]anthracene	0.74	2.92	0.66
Naphtho[1,2,3,4-def]chrysene	0.78	1.85	0.48
Benzo[rst]pentaphene	0.68	2.41	0.45
Coronene	0.46	2.87	0.37
Benzo[a]coronene	0.10	2.48	0.15
Dibenzo[h,rst]pentaphene	0.12	2.35	0.14

[a] See text.

ticular system or systems will depend upon the hydrocarbons to be separated and the chemical characteristics of the background material from which they are to be isolated. It is often found that where one system fails for a specific group of hydrocarbons, another may prove satisfactory. With a complex mixture, therefore, successive fractionations on different systems may be necessary. The three systems most commonly used are: (1) alumina with pentane–diethyl ether (19:1), (2) cellulose acetate with ethanol–toluene–water (17:4:4), and (3) cellulose with dimethylformamide–water (1:1). These three systems have been compared by Sawicki et al. (88), who obtained the data in Table 5-5. The retention values (R_B values) are the ratios of the distances travelled by the compounds to the distance travelled by BaP. Although alumina gave the best separation of PAH from other compound types, the PAH were poorly separated from each other. The cellulose system gave the widest range of retention values, and the separation of difficult pairs could often be improved with a slight change of water content in the mobile phase. The cellulose acetate system gave the best separation of the "benzopyrene" fraction obtained from alumina column chromatography— i.e., BaP is completely separated from BkF, BeP, and perylene.

Recoveries from cellulose and cellulose acetate are much higher (93–98%) than from alumina (50–80%). Poor recoveries are mainly attributed to photochemical changes taking place during chromatography. These reactions are accelerated by ultraviolet light and by the presence of chlorinated solvents as mobile phases. Inscoe (89) identified pyrene-1,6- and -1,8-dione among the numerous photooxidation products formed during the chromatography of pyrene on silica (see Chapter 1).

The cellulose thin-layer chromatographic systems are summarized in Table 5-6.

TABLE 5-6

Cellulose TLC Systems

Thin layer	Mobile phase	Reference
Acetylated cellulose	Methanol–ethanol–water (4:4:1)	(60, 92)
Acetylated cellulose	Ethanol–toluene–water (17:4:4)	(90)
Cellulose	Isooctane	(90)
Acetylated cellulose	Ethanol–dichloromethane–water (20:10:1)	(91, 96, 97)
Cellulose	N,N-Dimethylformamide–water (1:1)	(93)
Acetylated cellulose	Methanol–water	(94)
Acetylated cellulose	Various systems	(95)
Cellulose acetate	Ethanol–toluene–water (17:4:4)	(98)

Table 5-7 summarizes representative adsorbent-layer chromatography systems that involve the use of either alumina or silica gel.

Organic porous hydrocarbons like Porapak Q and related types have been found to be very suitable as lipophilic solid phases for thin-layer chromatography (117). The material is very convenient for the production of standard thin-layer plates, e.g., by cold-pressing the powdered polymer on plates of suitable material. Its hard, granular physical properties also make it easily wettable.

Many solvents in the polarity range from carbon tetrachloride up to methanol travel well along the layer in a reasonable time. Separation of

TABLE 5-7

TLC Systems

Thin Layer[a]	Mobile phase[b]	Ref.
Alumina	Pentane–ether (19:1)	(88, 98)
Silica	Methanol–ether–water (4:4:1)	(99)
Silica	Hexane–benzene (10:1)	(101)
Silica	Cyclohexane–pyridine (95:5)	(102)
Alumina	Pentane	(103)
Silica	Hexane–pyridine (30:1)	(104, 105)
Sil-CuSO$_4$	Benzene–methanol (95:5)	(106)
Sil-CuSO$_4$	Ethyl acetate–methanol–formic acid (80:10:10)	(106)
Alumina	Isooctane	(107)
Alum-acel (2:1)	Methanol–ether–water (4:4:1) (1st direction)	(108, 109)
Alum-acel (2:1)	Hexane–pentane–toluene (18:1:1) (2nd direction)	(108, 109)
Alum,sil	Hexane, CCl$_4$, CH$_2$Cl$_2$, or trichloroethane	(100)
Alum-acel	Pentane (1st direction)	(110, 111)
Alum-acel	Ethanol–toluene–water (17:4:4) (2nd direction)	(110, 111)
Alum-acel (1:1)	2-Dimensional hexane (1st direction)	(112)
Alum-acel (1:1)	Methanol–ether–water (4:4:1) (2nd direction)	(112)
Alum-acel (2:1)	Cyclohexane	(113)
Sil-alum-acel (1:1:1)	Isooctane (1st direction)	(114)
Sil-alum-acel (1:1:1)	Hexane–benzene (19:1) (2nd direction)	(114)
Sil-alum-acel (1:1:1)	Methanol–ether–water (4:4:1) (2nd direction)	(114)
Sil-gyp	Hexane–o-dichlorobenzene–pyridine (20:2:1)	(115)
Sil-cel	Hexane–o-dichlorobenzene–pyridine (20:2:1)	(115)
Alum-cel	Pentane–ether (19:1)	(115)
Alum-acel	Pentane–ether (19:1)	(115)
Alum,sil	Pentane–ether (19:1)	(115)
Alum,sil	n-Hexane	(115)
Alum,sil	Cyclohexane–benzene (4:1)	(115)
Silica	Hexane–o-dichlorobenzene–pyridine (20:2:1)	(115)
Silica	Benzene–CCl$_4$–acetic acid (50:75:0.6)	(116)

[a] Alum = alumina, sil = silica, cel = cellulose, acel = acetylated collulose, gyp = gypsum.
[b] 1st and 2nd direction refer to two-dimensional TLC.

aromatic hydrocarbons is due to the dispersion forces (solute–sorbent) which act not only according to the number of nuclei but also according to the shapes of the respective molecules (e.g., phenanthrene–anthracene). Very sharp separations can be obtained if the electron cloud of the hydrocarbon part of any molecule is disturbed and/or sterically shielded by a heteroatom.

Table 5-8 (*118*) summarizes the R_F values of some PAH on Porapak Q measured in five solvent systems.

Porapak T, a polar-substituted ethylvinylbenzene polymer, has also been used as an adsorbent (*119*). The polar groups have a positive electrostatic field and affect the adsorption selectivity, particularly toward molecules with localized double bonds. The selectivity thus generated enables substances that differ in structure by only one double bond to be separated.

Recently, a thin-layer plate coated with polyaminoundecanoic acid (polyamide) (*120*) was developed with a mixture of toluene–methanol (4:1) at room temperature. BaP, BghiP, benzo[*k*]fluoranthene, coronene, perylene, dibenz[*a*,*h*]anthracene, and 7,12-dimethylbenz[*a*]anthracene were all separated.

Magnesium oxide has been used in the past as a column-chromatography adsorbent. Work has shown that satisfactory thin layers can also be pre-

TABLE 5-8

R_F Values of Porapak Q[a]

Compound	S1	S2	S3	S4	S5
Naphthalene	0.89	0.81	0.74	0.43	0.51
2,6-Dimethylnaphthalene	0.87	0.70	0.74	0.27	0.42
1,5-Dimethylnaphthalene	0.85	0.77	0.73	0.27	0.42
Anthracene	0.85	0.73	0.48	0.16	0.22
Phenanthrene	0.82	0.74	0.70	0.26	0.33
Fluorene	0.80	0.70	0.66	0.27	0.36
Acenaphthene	0.82	0.77	0.68	0.32	0.35
1-Methylanthracene	0.85	0.73	0.47	0.15	0.23
2-Methylanthracene	0.85	0.73	0.51	0.15	0.22
Carbazole	0.90	0.83	0.85	0.62	0.65
Pyrene	0.75	0.70	0.61	0.16	0.28
Fluoranthene	0.75	0.64	0.47	0.05	0.10
Benzo[*b*]fluorene	0.75	0.61	0.48	0.05	0.12
Benzo[*a*]pyrene	0.65	0.47	0.34	0.05	0.06
Benzo[*e*]pyrene	0.69	0.59	0.42	0.06	0.10
Benzo[*ghi*]perylene	0.60	0.46	0.35	0.06	0.08

[a] S1 = ethylacetate, S2 = acetone, S3 = acetone–propanol, S4 = ethanol, S5 = propanol.

pared from magnesium hydroxide (*121*). This adsorbent bears a close resemblance in several respects to alumina, but the two systems are not equivalent; magnesium hydroxide/benzene has at least four important advantages over alumina/pentane–ether. First, the R_F difference between two given compounds is usually larger. For example, with alumina, the R_F difference between anthracene and benz[*a*]anthracene is 0.24, while with magnesium hydroxide the difference is 0.34. Second, magnesium hydroxide has a high capacity. A 50-mg sample was easily resolved by a magnesium hydroxide plate, whereas an alumina plate was only able to provide a preliminary fractionation. Third, results suggest that magnesium hydroxide is less sensitive to variations in experimental detail, such as activation times and temperature. Consequently, reproducibility of R_F values appears better. Fourth, the plates themselves have excellent mechanical strength. Disadvantages include longer analysis times (90 min as compared to 45 min on alumina), incompatibility with various spray reagents, and insolubility in some extracting solvents, making quantitative recovery difficult.

TLC adsorption procedures have been used in a number of applications (*122–126*). Sawicki (*127*) has reviewed their application to air-pollution research.

The ability of polycyclic hydrocarbons to form charge-transfer complexes can, in principle, form the basis of a method for their separation. Aromatic hydrocarbons, as donors of π-electrons, are capable of forming donor–acceptor complexes with substances having electron-acceptor properties. Franck-Neumann (*128*) was the first to point out the possibility of using these complexes in TLC. Berg and Lam (*59*) found that the separation of PAH on alumina or silica gel was improved by impregnating these adsorbents with an electron acceptor like 2,4,7-trinitrofluoren-9-one (*129*) or caffeine. The most useful mobile phases were aliphatic hydrocarbons containing about 1% of a polar solvent such as ether, acetic acid, or pyridine.

Short and Young (*130*) achieved improved separation by incorporation of 30% pyromellitic dianhydride into their silica gel stationary phase. In addition to improved separations, the formation of charge-transfer complexes usually results in a brightly fluorescent color which aids in visually identifying separated components. Other workers (*131–133*) have used other electron acceptors such as picric acid, chloranil, and trinitrobenzene.

Two-dimensional development in TLC has been used and is particularly valuable for mixtures of many components such as are normally found in isolated PAC fractions. Using this technique, the sample is applied to the plate about 3–4 cm from one corner and is developed as usual. The plate is then turned through 90° and development carried out in the second direction. The solvent is varied so that different separation effects are obtained in the second dimension.

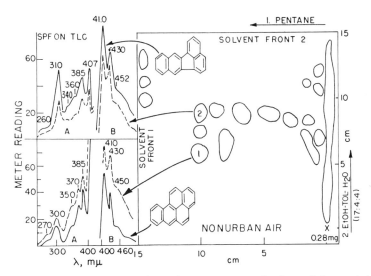

Fig. 5-3. Two-dimensional thin-layer chromatogram on alumina–cellulose acetate (2:1) of the benzene-soluble fraction of airborne particulates. [Reproduced with permission from E. Sawicki, T. W. Stanley, S. A. McPherson, and M. Morgan, *Talanta* **13**, 619 (1966). Copyright, Pergamon Press, Ltd.]

Figure 5-3 shows a two-dimensional, thin-layer chromatogram on alumina–cellulose acetate (2:1) of the benzene-soluble fraction of airborne particulates (*111*). *n*-Pentane was used as the first developing solvent, followed by ethanol/toluene/H_2O (17:4:4) in the second dimension. Benzo[*a*]pyrene and benzo[*k*]fluoranthene were identified by comparison with one-dimensional R_F values of standards.

The term *channel TLC* (*134–136*) is used to describe a procedure in which components that have to be fractionated on a thin plate are made to flow into narrow development channels to prevent the spreading of the chromatographic spot. In channel TLC, the chromatographic spot has a rectangular shape, and both its area and length are related to the concentration of the eluted species.

This method of analysis may be applied to PAH analysis, since these compounds have similar R_F values in controlled chromatographic operating conditions and are fluorescent under UV light. The procedure allows the determination of total PAH content in a mixture of hydrocarbons. Channel TLC is carried out by scoring a channel with a metal stylus and a template in a ready-to-use silica gel plate, and drawing two lines at an angle at the end of the channel, where a sample of the mixture to be analyzed is placed (*137*), as seen in Fig. 5-4.

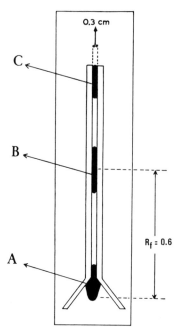

Fig. 5-4. Channel thin-layer chromatogram of a mixture containing (A) hydrophilic compounds, (B) PAH, and (C) alkanes. Elution solvent: hexane–benzene (1:1). [Reproduced with permission from: L. Zoccolillo and A. Liberti. *J. Chromatogr.* **120**, 485 (1976). Copyright, Elsevier Scientific Publishing Co.]

It has been shown that chromatogram development is sensitive to temperature changes, and in order to obtain reproducible results, it is necessary to operate at constant temperature. Plots of chromatographic spot lengths versus concentration for various PAH, such as phenanthrene, indeno-[1,2,3-*cd*]pyrene, BaP, and fluoranthene, are linear, but they do not coincide, as the molecular weight apparently influences spot size. By running an analysis of artificial mixtures of various compositions, it has been found that the standard deviation of the quantitation is ±0.02 when silica gel plates are used and ±0.17 when chromatograms are run on plates containing a fluorescent indicator.

Poor resolution and sensitivity are not inherent to TLC. They merely reflect the fact that TLC, as normally practiced, is not a high-performance (efficient) technique. When efforts are made to improve the efficiency of TLC, the resulting improvements in resolution and sensitivity make high-performance TLC fully comparable to HPLC (*138*). Two approaches to improved TLC performance are the use of small-particle TLC plates and the technique of programmed multiple development. A recent study (*139*) of "micro-HPTLC" suggests a bed efficiency of some 4200 theoretical plates in only 16 min. The chief limitation of these beds seems to be the small (nanoliter) sample volumes required for optimum results.

The second approach exploits the spot reconcentration which occurs whenever the solvent front crosses a spot on a TLC plate (*140, 141*). Loosely speaking, the lower edge of the spot has a chance to catch up with the upper edge. Programmed multiple development (PMD) is the repeated, same-direction, same-solvent development to gradually increasing distance. This technique has the ability to carry out a sufficient number of developments to assure compact spots while maintaining separation between spots. Sensitivity is a function of spot size, and any technique which produces more compact spots is expected to improve sensitivity. In this case, PMD provides an improvement factor of 5.

In conclusion, TLC and paper chromatography have been widely used for the separation of PAH. Even with more efficient and automated systems available, such as HPLC or GC, TLC separations are still extensively used. This can be attributed to, among other things, simplicity of operation and sample recovery possibilities. Many techniques using a variety of adsorbents and developing systems have now been established. Although no single system exists that will give superior separation of all possible components, some selective systems give very good separations for a particular matrix.

IV. GEL PERMEATION CHROMATOGRAPHY

The basis of gel permeation chromatography (GPC) is the selective retardation of solute molecules, due to the relative degree of penetration into solvent filled pores in the gel matrix. Molecules of smaller size permeate the gels much more deeply than larger molecules, and are thus retained longer on the column, effecting a separation in reverse order of most chromatographic systems, i.e., elution in order of decreasing size. The molecular parameter, then, which governs GPC separations is the hydrodynamic volume or "effective size" (*142*) of the molecule, placing GPC in an entirely different chromatographic class from those methods which depend on solubility, volatility, polarity, ionic characteristics, or adsorption. Gel permeation is more definitively known as *steric exclusion chromatography* (*143*), but *molecular sieving, gel filtration*, and *size exclusion chromatography* are also accepted terms.

Two gels have been used predominantly for the analysis of PAH by GPC: Bio-Beads SX series, and Sephadex LH-20. The Bio-Beads are a neutral, porous, styrene–divinylbenzene copolymer, and exhibit relatively little swelling in organic solvents because of the rigid polystyrene structure.

Sephadex LH-20 is a propylhydroxylated dextran (*144*) (dextran: a poly-saccharide containing more than 90% α-1,6-glycosidic linkages; see Fig. 5-5), formed by the reaction of propylepoxide with the free hydroxyls in Sephadex G-25. The ether and hydroxyl groups in the LH-20 make this a polar gel (*145*). The G values used in describing the Sephadex gels are a measure of the amount of cross-linking in the polymer.

Two mechanisms describing the elution of the PAH on these gels have been suggested: a steric exclusion effect, as previously described, and an adsorption mechanism.

In adsorption gel chromatography, the PAH elute in order of increasing ring number, just the reverse of steric exclusion elution order. Streuli (*146*) suggests that because the PAH are electron-rich bases, the adsorption effect can be understood as the formation of Lewis acid–base complexes through the relatively dense and localized π-clouds of the PAH and suitable sites in the gel matrices: the benzene rings in the Bio-Beads and the ether and hydroxyl groups in LH-20.

The separation mechanism is determined largely by the choice of solvent. PAH eluted with tetrahydrofuran (THF), dimethylformamide (DMF), and dimethylaniline (DMA) on Bio-Beads elute in order of decreasing ring

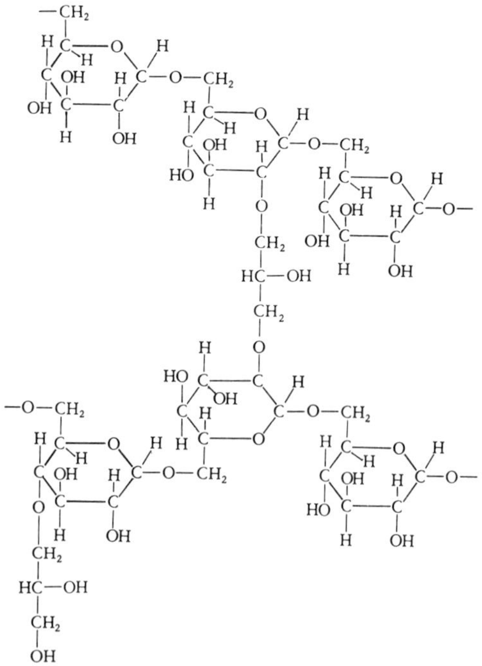

Fig. 5-5.   Chemical structure of Sephadex LH-20.

number (steric exclusion), while methylene chloride, benzene, acetone, and, very slightly, methyl ethyl ketone show reversed elution order (adsorption effect) on the same gel. The PAH are separated according to the steric exclusion principle on Sephadex LH-20 when eluted with DMF, THF, and DMA, but elute according to adsorptive effects on LH-20 in methanol, isopropanol, acetone, and acetonitrile. A look at the relative polarities and solvating strengths for PAH of these solvents will aid the understanding of the different effects they have in the two gels (*147*).

In Table 5-9, the first column, (a), is a measure of the polarity of each solvent, with values increasing with polarity. The next column, (b), is a measure of the ability of the solvent to interact with solute molecules via dispersion or London forces. Larger values indicate preferential reaction with highly polarizable molecules, i.e., aromatics.

Column (c) is a measure of the solubility of toluene in each solvent; the larger the value, the greater the solubility. The fourth column describes which mechanism is favored with which gel in that solvent. The solvents can be divided into three groups which indicate roughly their solvating strength for PAH: (1) the poor solvents are polar (methanol, isopropanol, and acetonitrile); (2) the moderate solvents are slightly less polar (DMF, acetone, and methyl ethyl ketone); (3) the good solvents are even less polar (benzene, THF, and methylene chloride). The decrease in polarity from poor to good solvents is paralleled by increasing solubility of toluene as in (c), and more approximately by an increase in the preferential interaction of the solvents with the highly polarizable molecules as in (b). It is apparent from Table 5-9 that only empirical rules can be formulated for the choice of solvents with Bio-Beads; the trends are somewhat clearer for LH-20, and suggest that the

TABLE 5-9

Commonly Used Solvents in GPC of PAH[a]

Solvent	(a)	(b)	(c)	Mechanism
Methanol	12.9	6.2	1,487	Adsorp., LH-20
Isopropanol			1,566	Adsorp., LH-20
Acetonitrile	11.8	6.5	2,666	Adsorp., LH-20
Dimethylformamide	11.5	7.9	3,840	GPC, LH-20, Bio-Beads
Acetone	9.4	6.8	4,121	Adsorp., LH-20, Bio-Beads
Methyl ethyl ketone			5,122	Adsorp., Bio-Beads
Benzene	9.2	9.2	7,930	Adsorp., Bio-Beads
Tetrahydrofuran	9.1	7.6	10,800	GPC, LH-20, Bio-Beads
Methylene chloride	9.6	6.4	11,100	Adsorp., Bio-Beads

[a] See text for descriptions of columns (a), (b), and (c).

polar solvents with less affinity for the PAH (as indicated by a value in (b) <7.0) will favor the adsorption mechanism, while the less polar, better solvents for PAH will suppress (148, 149) the adsorption effects in favor of size-exclusion chromatography. This effect can be understood in terms of the degree of solvation of each PAH molecule: the more highly solvated, the less chance for π-interactions with the LH-20 gel matrix. Detailed descriptions of the applications of Bio-Beads and Sephadex LH-20 to PAH separations are given in the following sections.

A. Bio-Beads

Bio-Beads in THF will supposedly elute the PAH in order of decreasing ring number—that is, by molecular sieving where the larger molecules do not penetrate the gel pores deeply, if at all, and are rapidly washed through the column while the smaller molecules are retained in the gel matrix. This explanation of the behavior of PAH on Bio-Beads in THF cannot wholly account for the results many authors have reported. Edstrom and Petro (150) show that while pentacene eluted first, benzo[a]coronene, benzo[ghi]-perylene, anthanthrene, and coronene all eluted about the same time as benzene. They explain that the flat, highly condensed molecules have a smaller effective size, thus enabling them to permeate the gel more thoroughly. This explanation holds well when comparing these molecules with the 4- and 5-ring cata-condensed PAH; however, the explanation fails when comparing the effective size of coronene with benzene: the mean molecular size ($\bar{L}$) used by Popl et al. (142) is 10.18 Å for coronene and 6.54 Å for benzene. Obviously, coronene and benzene will not permeate the styrene–divinylbenzene pores to the same degree, and some other explanation must be sought for the apparently anomalous elution behavior of the peri-condensed aromatics.

Hendrickson and Moore (151) described several forces which could modify elution volumes on styrene–divinylbenzene polymer: (1) changes in widths of the pores because different solvents swell the gel to different extents; (2) solvent–solute associations, i.e., the solvated molecule demonstrates changes in effective volume or size with different solvents; (3) solute dimerization; and (4) adsorption on or into the gel structure. They suggested that solvent–solute interactions (in THF, see Table 5-9) should be strong enough to suppress the adsorptive effects, which should be very slight because the Bio-Beads are a neutral polymer. There are, however, benzene rings present in the polymer, and the authors did not specifically consider the possible effect of the π-electrons of these benzene rings on the dense, localized π-electron clouds of the PAH.

Minarik et al. (152) did attempt to measure the adsorption coefficients of benzene, 2-methylnaphthalene, naphthalene, and phenanthrene on Bio-Beads eluted with THF. They used a gel with excessively large pore diameters in order to eliminate sieving action, and reported that sorption was so low as to be immeasurable and hence negligible. One should note, however, that their observations were generalized from three low-molecular-weight cata-condensed aromatics to the whole class of PAH, and they did not report on the elution behavior of the more highly condensed, electron-rich peri-aromatics.

Klimisch and Reese (*153*) did a thorough study of the elution behavior of PAH in the THF/styrene–divinylbenzene system. They noted that molecular sieving effects predominated for alkanes and cata-aromatics, while the peri-aromatics eluted strictly according to adsorption principles, and the elution of cata-peri-aromatics (dibenzopyrenes) showed the result of both exclusion and adsorption effects. The elution order of the cata-aromatics fits the theory of steric exclusion chromatography, but the reverse order of elution of the peri-aromatics suggests strongly that other mechanisms can be considered with THF/Bio-Beads, and that both sieving and adsorption takes place concurrently. Klimish and Ambrosius (*154*) reported that the peri-condensed aromatics complexed with the gel matrix under the same conditions that the cata-aromatics eluted according to size exclusion principles. This anomalous elution behavior of the peri-systems can be explained by the increased adsorption caused by the more localized denser π-electron clouds of the peri-aromatics available for $\pi–\pi$ complexing.

The same authors eliminated the adsorption of the peri-aromatics by the use of DMA as the eluent. The charge-transfer complexes formed between the PAH and the solvent are favored over PAH–gel complexation, and all the PAH, cata- and peri-condensed, eluted in order of decreasing ring number: coronene, benzo[*ghi*]perylene, naphthacene, pyrene, anthracene, naphthalene, and benzene.

Klimisch and Reese (*153*) showed that the elution volume of the alkylated cata-aromatics decreased regularly as the alkyl chain length was increased by one carbon atom at a time, explaining that the effective size of the molecule grew such that elution occurred earlier according to steric exclusion. Edstrom and Petro (*150*) noted similar changes in elution volume as the parent compounds were methylated in different positions.

The polar, protic solvents, methanol and isopropanol, gave very poor separations of the PAH on Bio-Beads (*155*), but the elution order, except for perylene and pyrene (which eluted after benzene), was generally in order of decreasing ring number. These solvents are not suggested for analysis of PAH by gel filtration techniques.

The approach to Bio-Beads in adsorption-promoting solvents must be made with caution gained from the realization of multiple effects occurring in exclusion-promoting solvents: The explanation of the elution order of

PAH or their alkylated derivatives may include not only adsorption effects but also steric exclusion effects.

Although much work has been done on the chromatography of PAH on the system of benzene/Bio-Beads (156–159), the most thorough study was done by Popl et al. (142). They used the mean molecular size $\bar{L}$ as suggested by Giddings et al. (160) to characterize each solute molecule in order to describe the adsorption of PAH on the gel. The elution order followed roughly an increase in $\bar{L}$ from naphthalene ($\bar{L} = 7.75$ Å) to ovalene ($\bar{L} = 12.39$ Å) showing the effect of adsorption on or into the gel. The larger the molecule, the greater the amount of adsorption due to the π–π complexes. Inspection of their data reveals that (1) isomers elute in order of decreasing $\bar{L}$ parameters; (2) pyrene elutes after the four-ring isomers; and (3) the peri-condensed aromatics are all retained relatively longer in the gel than the other compounds of equivalent $\bar{L}$ value (e.g., ovalene, $\bar{L} = 12.39$ Å, eluted 4.4 ml after decacyclene, $\bar{L} = 12.54$ Å). The explanation of all three observations is a function of $\bar{L}$ and π-electron density: the more compact the molecule (in order of decreasing $\bar{L}$ parameter), the denser the π-electron cloud and the longer the retention due to adsorption; and the more compact the molecule, the greater the permeation in the gel. On the basis of their data, the authors formulated a linear equation showing the dependence of the elution volume on the parameter $\bar{L}$ and on the number of aromatic carbon atoms, and, comparing it with an equation suggested by Snyder (148) for adsorption chromatography, noted that the difference between the two equations could be attributed to the contribution of exclusion (which has a reverse effect on retention volume) to the total adsorption energy term. In other words, the retention volume of PAH on benzene/Bio-Beads is a function of both exclusion and adsorption.

The effect of the two mechanisms operating simultaneously is also illustrated by the elution behavior of several alkylated PAH (142). Adding a methyl group to naphthalene decreased the elution volume 0.7 ml from the elution volume of the parent compound. 1-n-Propylnaphthalene eluted 3 ml earlier than naphthalene, and 1-n-amylnaphthalene almost 4.5 ml earlier. The same molecular sieving effect could be seen with the methyl-, ethyl-, n-propyl-, and n-amylphenanthrenes. The alkylated PAH followed the steric exclusion chromatography of the n-alkanes (142, 152) very strictly. For example, the retention volumes of two n-C_5-alkyl PAH are within 0.9 ml of the retention volume of n-pentane. The adsorptive effects of the PAH part of the molecule played an insignificant role in determining elution behavior, except for the alkylated peri-condensed perylene and highly condensed triphenylene. While 9-isoamylphenanthrene eluted only 0.9 ml before n-pentane, 2-n-octyltriphenylene eluted 1.1 ml after a C_8 alkane, and 3-n-hexylperylene eluted amost 6 ml after n-hexane. The elution of PAH and their alkylated derivatives on benzene–divinylbenzene is again a function of both adsorption and steric exclusion.

Asche and Oelert (*161*) chromatographed the PAH in the system of methylene chloride/Bio-Beads, using gels of different pore size. The cata-aromatics exhibited gel permeation behavior, while the peri-aromatics eluted in order of increasing ring number on the smaller pore gels, S-X8 and S-X12. All of the aromatics eluted by adsorption principles on the largest pore gels, S-X2 and S-X3. The larger pores eliminated all steric exclusion effects, and, due to the deeper penetration of the PAH into the gel structure, provided more active sites, thus enhancing the adsorption of the cata-aromatics (*162*).

Stedman *et al.* (*163*) eluted five PAH in order of increasing ring number with acetone on Bio-Beads without any anomalous elution behavior.

B. Sephadex LH-20

Even though Sephadex LH-20 is more polar than the Bio-Beads, the same solvents which favor gel filtration in the latter gel also favor it in LH-20. THF and DMF have been used most widely, and DMA has found some application. PAH have a high affinity to the ether and hydroxyl active sites in the LH-20 gel matrix, thus requiring the use of solvents which strongly solvate the PAH solute molecules in order to eliminate $\pi-\pi$ bonding. THF and DMF both have reasonably high values in the (b) column of Table 5-9, indicating relatively good solvating strength for PAH.

Streuli (*164*) studied the elution behavior of several cata-condensed aromatics on the system of THF/LH-20, and reported that the solvent eliminated π-bonding between the solute and the gel, favoring steric exclusion separations. It must be noted that he did not include any of the peri-condensed aromatics in his work.

Streuli (*164*) discussed the use of DMF with LH-20. The DMF totally masks the solute, thus preventing solute–gel interactions and favoring molecular sieving. Klimisch and Ambrosius (*154*) used DMA to achieve steric exclusion chromatography on LH-20 of both cata- and peri-aromatics. They attributed the strong solute–solvent interactions to charge-transfer complexes between the PAH and DMA.

The study of adsorption on Sephadex LH-20 reveals a more straightforward situation. No molecular sieving is present, and the peri-aromatics are not retained in the gel significantly longer than their cata-condensed isomers. The solvents most commonly used are methanol, isopropanol, and acetonitrile, but the first two are used predominantly. These three solvents do not solvate the PAH well (*165*) (see Table 5-9); they are very polar and do not interact with highly polarizable molecules. Therefore, theoretically, the solute molecules are free to bind with the active sites in the gel matrix. The only problem is that the polar solvents should saturate the polar gel sites, masking the solute–gel interactions; but this effect does not seem to occur. Possibly, the π-clouds of the gel and the PAH act "through" the layers of orientated solvent molecules.

Wilk *et al.* (*165*) published one of the first papers on adsorption gel chromatography of PAH in isopropanol. They obtained good separations between benzo[*e*]- and benzo[*a*]pyrene and the other cata- and peri-aromatics. No anomalous elution behavior was exhibited by either class of aromatics. Streuli (*146, 164*) showed a linear correlation between the adsorption values for the PAH and their respective resonance energies. The resonance energies are a measure of the π-electron cloud density. Because he worked only with the cata-condensed aromatics, one would have to be careful generalizing his statements to include the peri-condensed aromatics, although several authors have reported that pyrene elutes in the area of fluoranthene (*166, 167*), and possibly the other peri-aromatics could be assumed to act similarly. Klimisch and Reese (*155*) compared the elution of PAH in methanol and isopropanol on LH-20, and observed that the elution volumes are greater in methanol, indicating stronger adsorption of the PAH to the gel. This behavior leads to the conclusion that methanol separates the smaller PAH better, while isopropanol is preferred for the separation of the higher molecular weight PAH. They suggest the use of small beads with small pores eluted with an alcohol to optimize the separations of PAH.

Klimisch (*145*) and Jones *et al.* (*168*) applied adsorption gel chromatography on LH-20 to the clean-up and fractionation steps necessary to obtain a PAH-rich sample from cigarette-smoke condensate, and a shale oil and coal oil, respectively. Following fractionation of the lipophilic subfraction into nonpolar and polar subfractions with THF, the nonpolar subfraction was chromatographed with isopropanol. The aliphatics were eluted first, and the aromatics, by adsorption, followed after. After the two-ring PAH had eluted, the solvent was changed to THF to wash the remainder of the PAH rapidly off the column. Klimisch effected a total separation of the aliphatics from the aromatics, a difficult separation to achieve by any other adsorption procedure, by using the same chromatographic system: LH-20 eluted with isopropanol. The aliphatics elute first from the column in order of decreasing molecular weight (gel filtration), while the aromatics are retained by adsorptive forces. The aromatics are then eluted in order of increasing ring number.

According to the data given by Oelert (*167*), the methylated PAH elute only shortly before their respective parent compounds, by approximately

3 ml per carbon atom in the sidechain. Since exclusion effects do not appear to be present with LH-20 under adsorption conditions, this elution behavior of methylated PAH could be attributed to sterically hindered adsorption effects (*156*).

Bergmann *et al.* (*169*) showed the relationship of relative solubilities of PAC in various solvents to the separation mechanism. Separations of cata- and peri-condensed PAH, and N- and S-heterocycles, were effected by both steric exclusion and adsorption when THF was used as the eluant. On the other hand, elution strictly according to size exclusion was obtained using 1,2,4-trichlorobenzene as the eluant. The authors suggested that because PAH are more soluble in TCB than in THF, the adsorption process is practically eliminated. The N- and S-heterocycles also eluted in order of decreasing molar volume.

Snook (*158*) discussed the elution order of O-, S-, and N-heterocycles in the system benzene/Bio-Beads in terms of the adsorption mechanism. The elution of the heterocycles relative to their PAH analogs is a function of the ring current or π-electron density of each compound.

Chamberlain *et al.* (*170*) obtained both a purified aromatic ketone fraction and an aromatic phenolic fraction from the ether-soluble neutral fraction of cigarette-smoke condensate. Chromatography with both benzene/Bio-Beads and chloroform/LH-20 were necessary for the purification of the two O-heterocycle fractions.

Gel permeation chromatography of PAC, by steric exclusion and/or adsorption, is an effective tool for the fractionation of high- and low-molecular-weight aromatic mixtures and for the clean-up and analysis of trace amounts of PAC in complex matrices. The separation mechanism on Bio-Beads and Sephadex LH-20 is a function of the solvent, the solute, and the gel pore size.

The principles of steric exclusion chromatography are practically the same on both Bio-Beads and LH-20: the same solvents are used for both gels, but the presence of adsorption effects on the polystyrene gel does differentiate the two systems. Adsorption gel chromatography, on the other hand, is considerably different for each gel: solvents of totally different polarity and solvating strength for PAH are required; steric exclusion effects play a significant role in separations on Bio-Beads; the increased π-cloud densities of the peri-aromatics cause anomalous elution of those compounds on Bio-Beads with respect to their cata-condensed isomers, while all the PAC elute practically according to molecular weight and/or size on LH-20; and the optimum Bio-Beads adsorption system requires large pores but small beads, while small pores optimize the LH-20 adsorption system.

REFERENCES

1. E. Sawicki, *Chemist-Analyst* **53,** 24, 28, 56, 88 (1964).

2. R. E. Schaad, *Chromatogr. Rev.* **13,** 61 (1970).

3. R. E. Leitch and J. J. De Stefano, *J. Chromatogr. Sci.* **11,** 105 (1973).

4. C. H. Giles and I. A. Easton, *Adv. Chromatogr.* **3,** 67 (1966).

5. H. Brockmann and H. Schodder, *Ber.* **74B,** 73 (1941).

6. H. L. Falk and P. E. Steiner, *Cancer Res.* **12,** 30 (1952).

7. P. Wedgwood and R. L. Cooper, *Analyst* (*London*) **78,** 170 (1953).

8. P. Kotin, H. L. Falk, and M. Thomas, *AMA Arch. Ind. Hyg. Occup. Med.* **9,** 164 (1954).

9. P. Kotin, H. L. Falk, P. Mader, and M. Thomas, *AMA Arch. Ind. Hyg. Occup. Med.* **9,** 153 (1954).

10. R. L. Cooper, *Analyst* (*London*) **79,** 573 (1954).

11. R. L. Cooper and A. J. Lindsey, *Br. J. Cancer* **9,** 304 (1955).

12. M. Kuratsune, *J. Natl. Cancer Inst.* **16,** 1485 (1956).

13. A. J. Lindsey, E. Pash, and J. R. Stanbury, *Anal. Chim. Acta* **15,** 291 (1956).

14. H. J. Cahnmann and M. Kuratsune, *Anal. Chem.* **29,** 1312 (1957).

15. M. J. Lyons and H. Johnston, *Br. J. Cancer* **11,** 554 (1957).

16. H. R. Bentley and J. G. Burgan, *Analyst* (*London*) **83,** 442 (1958).

17. B. T. Commins, *Analyst* (*London*) **83,** 386 (1958).

18. L. H. Klemm, D. Reed, L. A. Miller, and B. T. Ho, *J. Org. Chem.* **24,** 1468 (1959).

19. G. E. Moore and M. Katz, *Int. J. Air Pollut.* **2,** 221 (1960).

20. E. Sawicki, W. Elbert, T. W. Stanley, T. R. Hauser, and F. T. Fox, *Anal. Chem.* **32,** 810 (1960).

21. P. Stocks, B. T. Commins, and K. V. Aubrey, *Int. J. Air Water Pollut.* **4,** 141 (1961).

22. L. R. Snyder, *Anal. Chem.* **33,** 1527 (1961).

23. G. J. Cleary, *J. Chromatogr.* **9,** 204 (1962).

24. G. Grimmer and A. Hildebrandt, *J. Chromatogr.* **20,** 89 (1965).

25. A. Zdrojewski, L. Dubois, G. E. Moore, R. S. Thomas, and J. L. Monkman, *J. Chromatogr.* **28,** 317 (1967).

26. B. B. Chakraborty and R. Long, *Environ. Sci. Technol.* **1,** 828 (1967).

27. E. Sawicki, *Health Lab. Sci.* **7,** 45 (1970).

28. E. Sawicki, *Health Lab. Sci.* **7,** 31 (1970).

29. A. Candeli, V. Mastrandrea, G. Morossi, and S. Toccaceli, *Atmos. Environ.* **8,** 693 (1974).

30. A. Liberti, G. Morozzi, and L. Zoccolillo, *Ann. Chim.* **65,** 573 (1975).

31. R. P. W. Scott and P. Kucera, *J. Chromatogr. Sci.* **12,** 473 (1974).

32. H. J. Cahnmann, *Anal. Chem.* **29,** 1307 (1957).

33. D. Hoffmann and E. L. Wynder, *Anal. Chem.* **32,** 295 (1960).

34. I. Schmeltz, R. L. Stedman, and W. J. Chamberlain, *Anal. Chem.* **36,** 2499 (1964).

35. D. E. Hirsch, R. L. Hopkins, H. J. Coleman, F. O. Cotton, and C. J. Thompson, *Anal. Chem.* **44,** 915 (1972).

36. D. Hoffmann and E. L. Wynder, *Cancer* **27,** 848 (1971).

37. W. H. Griest, H. Kubota, and M. R. Guerin, *Anal. Lett.* **8,** 949 (1975).

38. D. Hoffmann and E. L. Wynder, *Identif. Meas. Environ. Pollut. Symp.,* Ottawa, *1971,* p. 9.

39. T. M. Spotswood, *J. Chromatogr.* **33,** 101 (1960).

40. M. Godlewicz, *Nature* (*London*) **174,** 134 (1954).

41. T. Balint, *Acta Chim. Acad. Sci. Hung.* **31,** 17 (1962).

42. N. P. Buu-Hoi and P. Jacquignon, *Experientia* **13,** 375 (1957).

43. R. Tye and Z. Bell, *Anal. Chem.* **36,** 1612 (1964).

References

153

44. P. H. Berthold, *Erdoel. Kohle* **19**, 21 (1966).
45. H. J. Petrowitz, *in* "Progress in Thin-Layer Chromatography and Related Methods" (A. Niederwieser and G. Pataki, eds.), Vol. III, p. 1. Ann Arbor, Michigan, 1972.
46. L. R. Snyder, *Anal. Chem.* **33**, 1535 (1961).
47. L. H. Klemm, D. Reed, and C. D. Lind, *J. Org. Chem.* **22**, 739 (1957).
48. L. R. Snyder, *J. Chromatogr.* **5**, 430 (1961).
49. L. R. Snyder, *J. Chromatogr.* **6**, 22 (1961).
50. L. R. Snyder, "Principles of Adsorption Chromatography." Marcel Dekker, New York, 1968.
51. M. Popl, *J. Chromatogr.* **53**, 233 (1970).
52. M. Popl, J. Mostecky, and V. Dolansky, *J. Chromatogr.* **59**, 329 (1971).
53. M. Popl, J. Mostecky, and M. Kuras, *Anal. Chem.* **43**, 518 (1971).
54. M. Popl, J. Mostecky, and V. Dolansky, *J. Chromatogr.* **91**, 649 (1974).
55. L. Dubois, A. Corkery, and J. L. Monkman, *Int. J. Air Pollut.* **2**, 236 (1960).
56. F. A. Gunther and F. Buzzetti, *Residue Rev.* **9**, 90 (1964).
57. F. Micheel and J. Schweepe, *Naturwissenschaften* **39**, 380 (1952).
58. T. M. Spotswood, *J. Chromatogr.* **2**, 90 (1959).
59. A. Berg and J. Lam, *J. Chromatogr.* **16**, 157 (1964).
60. G. M. Badger, J. K. Donnelly, and T. M. Spotswood, *J. Chromatogr.* **10**, 397 (1963).
61. B. L. Van Duuren, *J. Natl. Cancer Inst.* **21**, 1, 623 (1958).
62. B. D. Tebbens, J. F. Thomas, and M. Mukai, *Arch. Ind. Health* **14**, 413 (1956).
63. J. F. Thomas, B. D. Tebbens, M. Mukai, and E. D. Sanborn, *Anal. Chem.* **29**, 1835 (1967).
64. A. Pietzsch, *Pharmazie* **12**, 24 (1957).
65. W. Kutscher, R. Tomingas, and H. P. Weisfeld, *Arch. Hyg. Bakteriol.* **151**, 656 (1967).
66. T. Wieland and W. Kracht, *Angew. Chem.* **69**, 172 (1957).
67. E. D. Bergmann and T. Grunwald, *J. Appl. Chem.* **7**, 15 (1957).
68. W. Kracht, Ph.D. Dissertation, University of Frankfurt/Main, 1958.
69. E. Deschner, Ph.D. Dissertation, University of Karlsruhe, 1960.
70. E. Maly, *Prac. Lek.* **12**, 347 (1960).
71. E. Maly, *Nature (London)* **181**, 698 (1958).
72. E. Maly, *J. Chromatogr.* **7**, 422 (1962).
73. E. Maly, *Prac. Lek.* **13**, 67 (1961).
74. E. Maly, *J. Chromatogr.* **40**, 190 (1969).
75. E. Maly, *Mikrochim. Acta*, 800 (1971).
76. E. Hluchan, M. Jenik, and E. Maly, *J. Chromatogr.* **91**, 531 (1974).
77. J. Gasparic, *Mikrochim. Acta*, 681 (1958).
78. G. Lindtedt, *Atm. Environ.* **2**, 1 (1968).
79. D. S. Tarbell, E. G. Brooker, A. Vanterpool, W. Conway, C. J. Claus, and T. J. Hall, *J. Am. Chem. Soc.* **77**, 767 (1955).
80. D. Hoffmann and E. L. Wynder, *Cancer* **13**, 1062 (1960).
81. C. I. Ayres and R. E. Thornton, *Beitr. Tabakforsch.* **3**, 285 (1965).
82. W. Graf and H. Diehl, *Arch. Hyg. Bakteriol.* **150**, 49 (1966).
83. J. W. Howard, R. T. Teague, R. H. White, and B. E. Fry, *J. Assoc. Off. Agr. Chem.* **49**, 595 (1966).
84. W. Lijinsky and P. Shubik, *Food Cosmet. Toxicol.* **3**, 145 (1965).
85. H. J. Davis, L. A. Lee, and T. R. Davidson, *Anal. Chen.* **38**, 1752 (1966).
86. F. Micheel and W. Schminke, *Angew. Chem.* **69**, 334 (1957).
87. F. Pavelka and A. D'Ambrosio, "Centro Provinciale per lo Studio Sugli Inquinamenti Atmosferici," p. 111. Amministrazione Provinciale Di Milano, Italy, 1959.

88. E. Sawicki, T. W. Stanley, W. C. Elbert, and J. D. Pfaff, *Anal. Chem.* **36**, 497 (1964).

89. M. N. Inscoe, *Anal. Chem.* **36**, 2505 (1964).

90. R. H. White and J. W. Howard, *J. Chromatogr.* **29**, 108 (1967).

91. R. E. Schaad, *Microchem. J.* **15**, 208 (1970).

92. T. Wieland, G. Luben, and H. Determan, *Experientia* **18**, 432 (1962).

93. C. G. Smith, C. A. Nau, and C. H. Lawrence, *Am. Ind. Hyg. Assoc. J.* **29**, 242 (1968).

94. L. Toth, *J. Chromatogr.* **50**, 72 (1970).

95. H. Woidich, W. Pfannhauser, G. Blaicher, and K. Tiefenbacher, *Chromatographia* **10**, 140 (1977).

96. J. Klimisch and E. Kirchheim, *Chromatographia* **9**, 119 (1976).

97. R. Schaad, R. Bachman, and A. Gilgen, *J. Chromatogr.* **41**, 121 (1969).

98. E. Sawicki, T. R. Stanley, J. D. Pfaff, and W. C. Elbert, *Chemist-Analyst* **53**, 7 (1964).

99. H. J. Petrowitz, *in* "Dunnschicht-Chromatographie" (K. Randerath, ed.), p. 219. Verlag Chemie, Weinheimbergstr., 1962.

100. N. Kucharczyk, J. Fohl, and J. Vymetal, *J. Chromatogr.* **11**, 55 (1963).

101. J. Pavlu, *Acta Univ. Carol. Med.* **12**, 225 (1966).

102. U. Pfetsch and K. Potzl, *Zentrabl. Arbeitsmed.*, 78 (1970).

103. T. W. Stanley, M. J. Morgan, and J. E. Meeker, *Anal. Chem.* **39**, 1327 (1967).

104. L. V. S Hood and J. D. Winefordner, *Anal. Chim Acta* **42**, 199 (1968).

105. H. Matsushita, Y. Suzuki, and H. Sakabe, *Bull. Chem. Soc. Jpn.* **36**, 1371 (1963).

106. H. Wagner and H. Lehmann, *Fresenius Z. Anal. Chem.* **291**, 366 (1978).

107. G. Biernoth, *J. Chromatogr.* **36**, 325 (1968).

108. M. Kohler, H. Golder, and R. Schiesser, *Fresenius Z. Anal. Chem.* **206**, 406 (1964).

109. L. E. Stromberg and G. Widmark, *J. Chromatogr.* **47**, 27 (1970).

110. G. Chatot, W. Jequier, M. Jay, and R. Fontanges, *J. Chromatogr.* **45**, 415 (1969).

111. E. Sawicki, T. W. Stanley, S. A. McPherson, and M. Morgan, *Talanta* **13**, 619 (1966).

112. M. Kohler, *Staub* **25**, 67 (1965).

113. D. F. Bender, *Environ. Sci. Technol.* **2**, 204 (1968).

114. H. Kunte, *Arch. Hyg. Bakteriol.* **151**, 193 (1967).

115. H. Matsushita, H. Hayashi, K. Nozaki, and Y. Suzuki, *Ind. Health* **3**, 126 (1965).

116. Th. A. Kouimtzis and I. N. Papadoyannis, *Anal. Chim. Acta* **96**, 203 (1978).

117. J. Janak, *Chem. Ind.*, 1137 (1967).

118. J. Janak and V. Kubecova, *J. Chromatogr.* **33**, 132 (1968).

119. V. Martinu and J. Janak, *J. Chromatogr.* **65**, 477 (1972).

120. G. Bories, *J. Chromatogr.* **130**, 387 (1977).

121. L. K. Keefer, *J. Chromatogr.* **31**, 390 (1967).

122. E. Sawicki, T. W. Stanley, W. C. Elbert, J. Meeker, and S. McPherson, *Atm. Environ.* **1**, 131 (1967).

123. G. Chatot, M. Castegnaro, J. L. Roche, and R. Fontanges, *Chromatographia* **3**, 507 (1970).

124. R. Tomingas, G. Voltmer, and R. Bednarik, *Sci. Total Environ.* **7**, 261 (1977).

125. E. Sawicki and J. D. Pfaff, *Anal. Chim. Acta* **32**, 521 (1965).

126. R. C. Pierce and M. Katz, *Anal. Chem.* **47**, 1743 (1975).

127. C. R. Sawicki and E. Sawicki, *in* "Progress in Thin-Layer Chromatography and Related Methods" (A. Niederwieser and G. Pataki, eds.), Vol. III, p. 233. Ann Arbor Sci. Publ., Ann Arbor, Michigan, 1972.

128. M. Franck-Neumann and P. Jossang, *J. Chromatogr.* **14**, 280 (1964).

129. G. H. Schenk, P. W. Vance, J. Pietrandrea, and C. Mojzis, *Anal. Chem.* **37**, 371 (1965).

130. G. D. Short and R. Young, *Analyst (London)* **94**, 259 (1969).

131. V. Libickova, M. Stuchlik, and L. Krasnec, *J. Chromatogr.* **45**, 278 (1969).

132. R. G. Harvey and M. Halonen, *J. Chromatogr.* **25**, 294 (1966).
133. H. Kessler and E. Muller, *J. Chromatogr.* **24**, 469 (1966).
134. I. Berthold, *Fresenius Z. Anal. Chem.* **240**, 320 (1968).
135. G. Goretti, A. Liberti, and B. M. Petronio, *Ann. Chim.* **64**, 653 (1973).
136. G. Goretti, A. Liberti, and B. M. Petronio, *Riv. Ital. Sostanze Grasse* **52**, 165 (1975).
137. L. Zoccolillo and A. Liberti, *J. Chromatogr.* **120**, 485 (1976).
138. T. H. Jupille, *J. Am. Oil Chem. Soc.* **51**, 179 (1977).
139. J. Riphahn and H. Halpaap, *J. Chromatogr.* **112**, 81 (1975).
140. T. H. Jupille and J. A. Perry, *J. Chromatogr. Sci.* **13**, 163 (1975).
141. T. H. Jupille and J. A. Perry, *J. Chromatogr.* **99**, 231 (1974).
142. M. Popl, J. Fähnrich, and M. Stejskal, *J. Chromatogr. Sci.* **14**, 537 (1976).
143. V. F. Gaylor and H. L. James, *Anal. Chem.* **50**, 29R (1978).
144. J. Sjövall, E. Nyström, and E. Haahti, *Adv. Chromatogr.* **6**, 119 (1968).
145. H.-J. Klimisch, *Fresenius Z. Anal. Chem.* **264**, 275 (1973).
146. C. A. Streuli, *J. Chromatogr.* **56**, 219 (1971).
147. L. R. Snyder and J. J. Kirkland, "Introduction to Modern Liquid Chromatography." Wiley, New York, 1974.
148. L. R. Snyder, "Principles of Adsorption Chromatography." Marcel Dekker, New York, 1968.
149. D. H. Freeman, *Anal. Chem.* **44**, 117 (1972).
150. T. Edstrom and B. A. Petro, *J. Polym. Sci. Part C* **21**, 171 (1968).
151. J. G. Hendrickson and J. C. Moore, *J. Polym. Sci. Part A*, **4**, 167 (1966).
152. M. Minarik, J. Coupek, and R. Komers, *Coll. Czech. Chem. Commun.* **43**, 2540 (1978).
153. H.-J. Klimisch and D. Reese, *J. Chromatogr.* **67**, 299 (1972).
154. H.-J. Klimisch and D. Ambrosius, *J. Chromatogr.* **94**, 311 (1974).
155. H.-J. Klimisch and D. Reese, *J. Chromatogr.* **80**, 266 (1973).
156. M. E. Snook, *Anal. Chim. Acta* **81**, 423 (1976).
157. R. F. Severson, M. E. Snook, R. F. Arrendale, and O. T. Chortyk, *Anal. Chem.* **48**, 1866 (1976).
158. M. E. Snook, *Anal. Chim. Acta* **99**, 299 (1978).
159. M. E. Snook, W. J. Chamberlain, R. F. Severson, and O. T. Chortyk, *Anal. Chem.* **47**, 1155 (1975).
160. J. C. Giddings, E. Kucera, C. P. Russell, and M. N. Myers, *J. Phys. Chem.* **72**, 4397 (1968).
161. W. Asche and H. H. Oelert, *J. Chromatogr.* **106**, 490 (1975).
162. D. Eaker and J. Porath, *Sep. Sci.* **2**, 507 (1967).
163. R. L. Stedman, R. L. Miller, L. Lakritz, and W. J. Chamberlain, *Chem. Ind.*, 394 (1968).
164. C. A. Streuli, *J. Chromatogr.* **56**, 225 (1971).
165. M. Wilk, J. Rechlitz, and H. Bende, *J. Chromatogr.* **24**, 414 (1966).
166. R. Gladen, *Chromatographia* **5**, 236 (1972).
167. H. H. Oelert, *Fresenius Z. Anal. Chem.* **244**, 91 (1969).
168. A. R. Jones, M. R. Guerin, and B. R. Clark, *Anal. Chem.* **49**, 1766 (1977).
169. J. G. Bergmann, L. J. Duffy, and R. B. Stevenson, *Anal. Chem.* **43**, 131 (1971).
170. W. J. Chamberlain, M. E. Snook, J. L. Baker, and O. T. Chortyk, *Anal. Chim. Acta* **111**, 235 (1979).

6

High-Performance
Liquid Chromatography

I. INTRODUCTION

Very few instrumental methods parallel the enormous expansion of high-performance liquid chromatography (HPLC) during the last 10 years. Following the example of gas chromatography in both the accessibility of theoretical treatment and instrumentation, modern LC has undergone very rapid development. While prediction of higher separation efficiencies with particles of smaller size and the consequent use of high inlet pressure were straightforward considerations (*1*), the experimental investigations of Hamilton (*2*), Horvath (*3*), Kirkland (*4*), and others laid the foundation for the rapid expansion of the method.

The earlier developments in modern LC in the late 1960s were closely associated with improvements in column technology. While "classical" packings were found to perform only marginally under the conditions of high pressure, a need for mechanically stable packings with a limited inner porosity was soon realized. The latter requirement is due to the slow diffusion in the liquid phase of solute molecules in and out of the porous structure of the usual adsorbents. Since the speed of analysis in chromatography could be theoretically improved by the design of superfically porous packings with an impermeable core, Horvath *et al.* (*3*, *5*) developed techniques for the preparation of spherical beads coated with small particles of the chromatographic media. Additional approaches have been made to the preparation of ion-exchange and adsorbent-coated materials (typically, 20–30 μm particles) which became widely known as "pellicular" or "controlled surface porosity" packings. Such packings possess favorable mass-transfer properties.

Superficially porous materials are still used in the practice of modern LC. While they provide moderately high column efficiencies at medium inlet pressures, these materials are easy to pack and use in analytical work. However, certain sample-size limitations occur, i.e., the pellicular columns are easy to overload.

Simple theoretical analysis of the particle size optimization in LC by Knox and Saleem (6) indicated that further efficiency improvements were feasible with ordinary high-pressure equipment. The particle size is implicated in two constants (A and C) in the general chromatographic equation [Eq. (1)]:

$$H = L/N = A + B/u + Cu \tag{1}$$

where H = height equivalent to a theoretical plate, L = column length, N = number of theoretical plates, A = flow nonuniformity, B = longitudinal diffusion, C = resistance to mass transfer, and u = linear velocity of the mobile phase. To incorporate particle-size considerations, Eq. (1) can be rearranged (7, 8) to read

$$h = B/v + Av^{0.33} + Cv \tag{2}$$

where the reduced plate height, $h = H/d_p$, and the reduced velocity, $v = u(d_p/D_M)$ (d_p = particle size; D_M = solute diffusion coefficient in the mobile phase). The plots of reduced values are now generally recognized to be the best measure of the kinetic performance of chromatographic columns.

While it follows from the above that a significant decrease in plate height (and a corresponding increase in column efficiency) can be obtained through a reduction of the particle size, certain practical aspects must be considered first; the specific column permeability, K_0, is a function of the particle size, and a higher pressure must follow reduction of H through decreased d_p:

$$u = \frac{K_0 d_p^2}{\eta} \frac{\Delta p}{L} \tag{3}$$

where Δp is the pressure gradient, and η is the viscosity of the mobile phase. Knox and Saleem (6) calculated that the theoretical optimum for d_p with the pressure limitations of 100–200 bar should be around 2 μm.

The second practical limitation lies in packing difficulties with very small particles that tend to agglomerate. Whereas conventional dry packing techniques work well with superficially porous materials and particles above 25 μm, they are of limited use with smaller particles. Most of these problems were largely overcome by the development of slurry-packing procedures (9, 10). While variations in packing techniques in different laboratories are bound to produce somewhat different results, prepacked and tested LC columns are now available from a number of manufacturers.

Third, limitations occur when working with excessive pressures, and columns packed with particles smaller than 5 μm may be inconvenient. In addition, subtle problems may arise from the heat of friction (11), causing temperature and viscosity gradients inside the column. Any practical consequences of this effect are, however, not known at present.

Advances in column packing technology and effective utilization of small porous particles since the early 1970s have resulted in significant improvements in column performance. Partially or totally porous spherical particles are now preferred in many applications. While the utilization of 5-μm or smaller particles is still under development, 10-μm columns are now commonly employed. Thus, columns of different selectivities that yield efficiencies above 5,000 theoretical plates are widely used. General aspects of modern liquid chromatography have been reviewed in a number of recently published books.

Modern LC instrumentation, while expensive, is suitable for most analytical work. Typical commercial instruments operate with inlet pressures up to 300 atm. Although detection and characterization of LC effluents remain less than ideal compared to the capabilities of GC methods, the available methodology is quite sufficient for many classes of compounds.

Methods for the separation and detection of PAC and related compounds by LC have been under development for some time. Fortunately, there are no severe deficiencies with respect to detection, since the most versatile LC detectors, UV and fluorescence monitors, lend themselves naturally to effective analyses of such compounds. While the merits of high-performance LC in the fractionation of complex PAC mixtures and sample clean-up are beyond dispute, the method also now successfully competes with GC as an analytical tool. Lack of resolution may frequently be less critical here due to the selectivity of LC spectroscopic detectors.

II. CHROMATOGRAPHIC COLUMNS

Numerous types of chromatographic media have been used for both classical columns and thin layers in the separation of PAC (Chapter 5). With the wider use of high-performance LC, many of these materials have lost their primary importance for analytical purposes. This is partly due to the fact that some very useful packings (e.g., lipophilic gels) are unsuitable for high-pressure work, and partly due to a less distinct need for a whole range of chromatographic materials of varying selectivity. In the latter case, column efficiency seems to be a better general substitute. Nevertheless, the stationary phases used previously in classical separations may continue to be of use as the means of sample purification and fractionation. Since the microparticulate (totally porous) packings that are now being increasingly applied in modern LC have good sample capacity, high-performance LC may even compete with conventional column chromatography as a more effective and faster fractionation method. However, its present strength lies primarily in analytical applications.

Analytical separations of PAC have been performed with both adsorbents and various chemically bonded stationary phases. Both cases are discussed below, with selected illustrations.

Column choice in chromatography of PAH is determined by the properties of mobile phases and solubility considerations. For the amounts chromatographed, common PAH are adequately soluble in both moderately polar and nonpolar solvents, thus allowing a wide choice of chromatographic conditions. These considerations apply for up to six- and seven-ring systems. The solubilities of high-molecular-weight components in conventional solvents are decreased. Proper combinations of mobile phases and column materials for the larger PAH molecules have yet to be devised.

A. Classical Adsorbents: Silica and Alumina

Alumina and silica particles have been among the most widely used column materials for the separation of PAH for many years. The recent availability of these adsorbents in both pellicular and microparticulate form is likely to maintain further interest in both types of column materials. Since the sensitivity of current LC detectors permits chromatography of PAH in the linear range of their respective adsorption isotherms, there may be a number of reasons for the utilization of the selective adsorption processes for analytical purposes. Hydrocarbon mobile phases are usually used.

Numerous attempts were made in the past to understand the processes of liquid–solid chromatography, in which aromatic hydrocarbons have frequently been used as model solutes. Earlier work on these correlations is summarized in the book by Snyder (12).

In general, the retention of aromatic hydrocarbons increases with molecular weight and the number of aromatic rings. Increases in retention are also noticable with greater lengths of aliphatic side chains in alkyl-substituted PAH, but these appear less predictable (13, 14). According to Snyder (12), the retention volume of an adsorbed solute, R^0, is related to the adsorption energy, S^0, and other molecular parameters:

$$\log R^0 = \log V_a + \alpha(S^0 - A_s \varepsilon^0) \tag{4}$$

where V_a = adsorbent surface volume, α = adsorbent activity function, A_s = relative size of the adsorbate (sample) molecule, and ε^0 = polarity factor. For pentane as a mobile phase, ε^0 is equal to zero [values for other solvents are tabulated (*12*)], and the equation becomes

$$\log R^0 = \log V_a + \alpha S^0 \tag{5}$$

Also, the adsorption energy, S^0, can be expressed as

$$S^0 = \sum_{i} Q_i^0 + \sum_{j} q_j^0 \tag{6}$$

where Q_i^0 are the additive contributions of individual groups within a molecule to the overall adsorption energy and q_j^0 account for possible interactions of such groups. There is a similarity between this concept and the retention rules derived by Martin for partition chromatography (*15*). Tabulation of the corresponding values (*12*) permits comparison of calculated and experimental retention as demonstrated in Fig. 6-1. Although this case relates to thin-layer chromatography of PAH on alumina, similar relationships are valid for chromatography in columns.

Systematic retention studies with alkylbenzenes (*13*) and alkylated di- and tricyclic systems (*14*) were carried out by Popl and co-workers. Whereas good agreement was found with straight-chain alkylated benzenes, retention

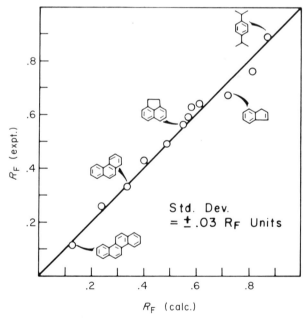

Fig. 6-1. Comparison of experimental and calculated R_F values for various aromatic hydrocarbons in the pentane–alumina TLC system. [Reproduced with permission from L. R. Snyder, *in* "Chromatography" (E. Heftmann, ed.), p. 75. Van Nostrand-Reinhold, New York, 1967.]

of solutes with branched chains was less predictable. Steric hindrance to adsorption of an aromatic nucleus was suggested as an explanation. Similar observations were made with alkyl-substituted naphthalenes and phenanthrenes (*14*), but the presence of a second alkyl group caused significant deviations in experimental data from the theoretical predictions.

Certain predictions can also be made with somewhat larger polycyclic molecules. Snyder (*12*) proposed that an empirical relationship exists between the adsorption energy, S^0, and the number of "aromatic carbon" atoms in a molecule nC_A:

$$S^0 = 0.31 \cdot nC_A \tag{7}$$

Obviously, the adsorption process discriminates aromatic from aliphatic carbon atoms, but the relationships can be quite complicated. The shape of the molecule is of primary importance in all adsorption processes. Also, the acid–base interactions may have a profound influence on the values of S^0 and the solute retention.

In their systematic study, Popl *et al.* (*16*) established that Snyder's empirical formula is strictly followed in the pentane–alumina system for one- to three-ring compounds. As demonstrated in Fig. 6.2, deviations exist for the four-ring planar molecules, and nonplanar PAH compounds demonstrate further significant deviation. Thus, lower values of the adsorption energy were found for terphenyls, and similar observations were made for other nonplanar molecules such as 9,10-dihydroanthracene, dibenzocycloheptane, and dibenzocyclooctane.

Acid–base interactions can also play an important role in adsorption chromatography. Adsorption energy values were found to be significantly increased for PAH possessing acidic hydrogen atoms (e.g., indene, fluorene, and benzofluorenes) and low ionization potentials (e.g., anthracene, naphthacene, and pentacene) on alumina adsorbents (*16*). Depending on the surface acidity of alumina, retention of certain solutes can be drastically affected, as exemplified in Fig. 6-3. While this phenomenon can be used with advantage in both fractionation and analytical work, caution should be exercised in standardization of adsorbents with respect to their acidity.

It is hardly surprising that alumina and silica adsorbents exhibit significantly different adsorption properties for PAH molecules (see Chapter 5). Their distinctly different surface chemistry and acidities are bound to be reflected in the strength of solute–sorbent interactions. Although the surface chemistry of alumina is at present less well understood than that of silica, the adsorption of sample molecules on alumina is believed to be of predominantly electrostatic character, whereas hydrogen bonding is undoubtedly involved in many interactions of molecules with siliceous surfaces.

Different retention mechanisms on the two adsorbents can be noticed even

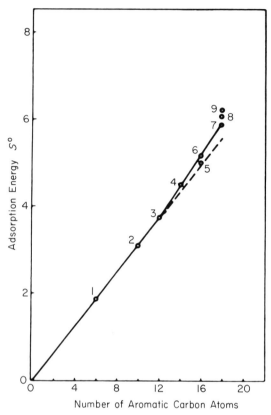

Fig. 6-2. Dependence of adsorption energy on the number of aromatic carbon atoms. Broken line, theoretical values; solid line, experimental values. 1, Benzene; 2, naphthalene; 3, biphenyl; 4, phenanthrene; 5, pyrene; 6, fluoranthene; 7, triphenylene; 8, benz[*a*]anthracene; 9, chrysene. [Reproduced with permission from M. Popl, V. Dolansky, and J. Mostecky, *J. Chromatogr.* **91**, 649 (1974). Copyright, Elsevier Scientific Publishing Co.]

with the small alkylated aromatic hydrocarbons. Martin *et al.* (*17*) observed that the presence of side chains in alkylbenzenes increased the retention values more for silica gel than for alumina. Similarly, alkyl groups were found to have different effects on alumina as compared to silica for a series of substituted naphthalenes and anthracenes (*14*).

Acid–base interactions are most apparent during the separation of azaarenes on silica. Here, accessibility of the nitrogen atom(s) is of primary importance. During their work on thin layers of silica gel, Engel and Sawicki (*18*) found that compounds with hindered nitrogen atoms migrated at a faster rate in the pentane–ether solvent system. Dong and Locke (*19*) have verified this phenomenon in adsorption HPLC with microparticulate silica.

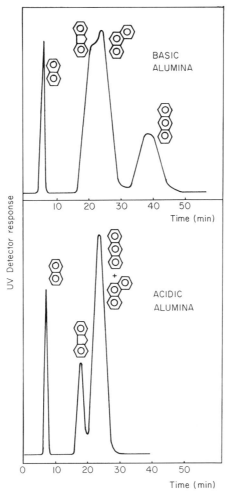

Fig. 6-3. Chromatograms of a standard PAH mixture on two different types of alumina. [Reproduced with permission from M. Popl, V. Dolansky, and J. Mostecky, *J. Chromatogr.* **91,** 649 (1974). Copyright, Elsevier Scientific Publishing Co.]

Again, the extent of separation was strongly dependent on how well the nitrogen electron lone-pair was shielded from interaction with the hydroxyl groups of the silica surface. For example, benzo[*h*]quinoline, benz[*c*]acridine, and dibenz[*a,h*]acridine were only weakly retained.

Specific interactions of nitrogen-containing PAC with the adsorbent surface can further be enhanced by impregnation techniques. Pyridine homologs were successfully separated on thin layers of silica impregnated with silver oxide (*20*). A superficially porous siliceous packing coated with silver ions was also found effective in separating azaarenes (*21*), and there is no particular reason why a similar technique could not be applied to small-particle column technology. According to Vivilecchia *et al.* (*21*), compounds are separated by the mechanism of donor–acceptor complexing. The silver ion acts as a Lewis acid toward the basic heterocyclic nitrogen atom. Thus, the elution order of azaaromatics in such a system can be explained in terms of compound basicity as well as steric accessibility of the nitrogen lone-pairs. Even though compound basicity is an important parameter in the separation process, steric factors can easily override basicity considerations. In such cases, the compound's retention can be dramatically reduced in spite of its appreciable basicity.

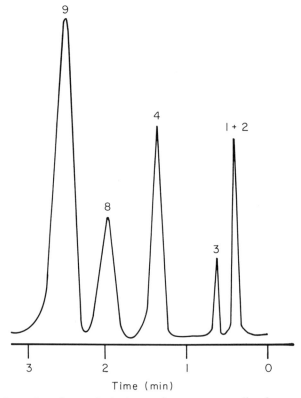

Time (min)

Fig. 6-4. Separation of a standard mixture of azaarenes on a silver-impregnated porous-layer bead column. 1, Benzo[*h*]quinoline; 2, dibenzo[*a,c*]phenazine; 3, phenazine; 4, acridine; 8, benzo[*c*]quinoline; 9, benzo[*f*]quinoline. [Reproduced with permission from R. Vivilecchia, M. Thiebaud, and R. W. Frei, *J. Chromatogr. Sci.* **10**, 411 (1972). Copyright, Preston Technical Abstracts Co.]

Due to the adsorption of azaarenes, chromatographic peaks of these compounds exhibit tailing; addition of a polar modifier into the hydrocarbon mobile phase usually changes this. Useful separations of azaaromatics were obtained with both unmodified and silver-impregnated silica (*19, 21*). An example is shown in Fig. 6-4.

Adsorbent selectivity for different alkyl groups within the aromatic molecule is one of the most useful attributes of these column materials. Although interpretation of the role of such groups in the overall adsorption

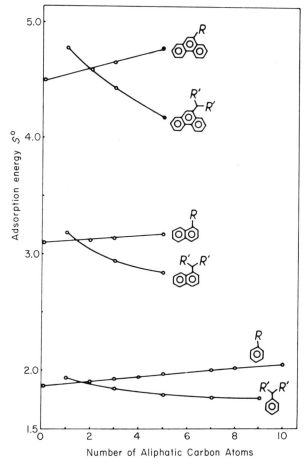

Fig. 6-5. Plots of adsorption energy versus the number of aliphatic carbon atoms for *n*-alkyl aromatics and alkyl aromatics symmetrically branched on the α-carbon atom. [Reproduced with permission from M. Popl, V. Dolansky, and J. Mostecky, *J. Chromatogr.* **91**, 649 (1974). Copyright, Elsevier Scientific Publishing Co.]

effect is far from being complete, the practical utility of these selective processes is not diminished. However, it is advantageous to an analyst to consider certain retention rules for the purposes of both fractionation and analysis. As already indicated, alumina and silica show quite different effects for alkylated aromatics. Systematic studies by Popl et al. (13, 14, 16) reveal certain retention regularities in the system pentane–alumina. Adsorption energies increase in a linear fashion with the length of the side chain, but the slope is dependent on the type of aromatic molecules. Multiple substituents do not always contribute to the overall adsorption energy in an additive fashion. In contrast to the results obtained with silica, branching tends to decrease retention of planar PAH molecules (see Fig. 6-5); steric hindrance is deemed responsible (16). Surprisingly, certain cycloaromatics were observed to have approximately the same retention as polymethylaromatics with similar carbon numbers. Elucidation of the phenomena which govern retention of alkylated aromatics in adsorption chromatography remains an interesting theoretical problem.

Quite importantly, different positions of alkyl groups frequently give rise to changes in chromatographic mobility. Again, steric considerations can be offered as an explanation of this behavior. This type of selectivity is much less frequent in partition chromatographic systems, and considerably higher column efficiencies are needed to resolve alkylated isomeric compounds (e.g., in capillary GC). Table 6-1 shows some selected examples. Index values in Table 6-1 are related to the compound retention (16) by the formula

$$\log I_x = \log I_n + \frac{\log R_x - \log R_n}{\log R_{n+1} - \log R_n} \tag{8}$$

where x, n, and $(n + 1)$ denote the unknown compound, and the bracketing standards, respectively. This is a treatment similar to the well-known Kovats retention index system (22). In the system above, used by Popl and co-workers, a retention index of 10 is assigned to benzene, 100 to naphthalene, 1000 to phenanthrene, 10,000 to benz[a]anthracene, etc.

B. Less Common Sorbents

Different types of column materials (other than the "classical adsorbents") have often been sought for selective separations of certain PAC. Klimisch (23) employed cellulose acetate and polyamide columns to accomplish quantitative determination of isomeric benzopyrenes and coronene. Both materials could withstand moderately high pressures. A polyamide packing (24) and spherical cellulose-based materials (25) may improve results considerably.

TABLE 6-1

Retention of Isomeric Alkylated Aromatics on Alumina[a]

Solute	Number of carbon atoms	Retention index
1,2,4-Trimethylbenzene	9	16.8
1,2,3-Trimethylbenzene	9	21.8
1,2,4,5-Tetramethylbenzene	10	21.8
1,2,3,4-Tetramethylbenzene	10	27.3
1-Methylindane	10	15.9
5-Methylindane	10	16.8
1-Methylnaphthalene	11	114
2-Methylnaphthalene	11	130
1,4-Dimethylnaphthalene	12	95
1-Ethylnaphthalene	12	102
1,5-Dimethylnaphthalene	12	110
2-Ethylnaphthalene	12	124
1,7-Dimethylnaphthalene	12	150
1,3-Dimethylnaphthalene	12	153
1,2-Dimethylnaphthalene	12	180
2,7-Dimethylnaphthalene	12	200
2,3-Dimethylnaphthalene	12	210
1-Phenylnaphthalene	16	268
2-Phenylnaphthalene	16	1394
9-Methylphenanthrene	15	1500
1-Methylphenanthrene	15	1520
3-Methylphenanthrene	15	1630
2-Methylphenanthrene	15	1680
9-Ethylphenanthrene	16	1120
3,6-Dimethylphenanthrene	16	1700

[a] According to M. Popl *et al.* (*16*).

A macroporous polystyrene gel was used for the separation of PAH by Popl *et al.* (*26*). This spherical cross-linked material can be used at least at inlet pressures up to 200 atm. Good column efficiencies were obtained at low flow rates, but the plate-height values rose quite steeply at higher mobile-phase velocities. Although the column material with a water–methanol–ether mobile phase behaved in several respects similarly to a reversed-phase packing (see below), some selective retention was observed for certain PAH solutes. Mixed adsorption and partition processes are likely to occur with this and similar types of column materials.

Regrettably, many potentially selective column materials are not fully compatible with the high-pressure conditions of high-performance LC. A possible solution to this problem lies in the deposition of such materials on the surface of a mechanically stable core (e.g., spherical silica or glass beads).

Although this technology has not been adequately developed at this time, both Benton 34 (a modified clay material) (27) and carbon black (28) were deposited on the surface of siliceous packings and used for chromatography of aromatic hydrocarbons. A resemblance of the latter to reversed phases was noticed.

C. Chemically Bonded Stationary Phases

Although siliceous adsorbents belong among the most utilized sorption media, the necessity for their modification has long been obvious. Chemical adsorbents frequently possess unique selectivity toward the chromato-graphed solutes. Unfortunately, there is only limited range in which the column selectivity can be adjusted (e.g., through controlled addition of water). In spite of the overall utility of adsorbents (both silica and alumina), the necessary adjustments of the system selectivity are generally somewhat troublesome and irreproducible.

One serious problem is the limited reproducibility of compound retention. Due to the trace amounts of water in the mobile phases, an unwanted "modification" of adsorbents frequently occurs. Subsequently, retention of various solutes can vary widely during repeated runs, and the column equilibration becomes a time-consuming process. This is particularly true when using gradient elution (going from a less polar to a more polar solvent), since traces of water and other polar impurities are not easy to remove from the solvents used.

If the adsorbents are too active, irreversible adsorption and tailing may occur. The common remedies to cure this problem (addition of a "modifier" or the geometrical modification of the adsorbent surface) frequently result in the loss of specific surface area and adsorbent capacity.

Liquid–liquid partition chromatography has enjoyed considerably less popularity in spite of its potential for selectivity. This is obviously due to the fact that solvent programming techniques cannot be used for the con-venient adjustment of retention. Coated stationary phases are easily stripped off the solid supports with continuous flow of solvents. To a minor degree, this applies even for relatively immiscible combinations of the mobile and stationary phases. Thus, it is hardly surprising that only a limited number of papers deal with liquid–liquid chromatography of PAH, and most of these investigations were carried out prior to the wider availability of chemically bonded phases. Separations of PAH in isooctane–oxydipropionitrile (29) and octane–polyethylene glycol (30) systems can be cited as representative examples of these earlier attempts. The development of chemically bonded stationary phases was the major important step in expanding the capabilities of liquid chromatography, in general. While the many aspects of bonded phases are treated in a specialized book (*31*), the following account will deal specifically with bonded-phase chromatography of PAH.

D. Reversed-Phase Columns

As early as 1950, Howard and Martin (*32*) reported the modification of a siliceous material with a hydrophobic partitioning layer. This modification was made to improve chromatographic separation of fatty acids; a polar mobile phase was employed and the now common term of "reversed-phase chromatography" was then used for the first time. The systematic studies of Kiselev and co-workers [for a review, see reference (*33*)] have further extended our knowledge of various surface reactions that may lead to chromatographic packings of different selectivities. Although various surface derivatization techniques have great utility far beyond chromatographic applications, the rapid developments of high-performance LC in recent years have primarily helped in promoting this important technology. As a result, both polar and nonpolar bonded phases are now available as superficially porous as well as microparticulate column materials.

Because of its great universal utility in separations of both PAC and their metabolic products, reversed-phase chromatography will be described first. In this mode of chromatography, the solutes are retained by the hydrophobic (or, generally, less polar) stationary phase in the order of decreasing polarity. However, with commonly used columns, such as those with bonded long-chain silicone polymers, a mixed retention mechanism is clearly suggested (*34*). Besides hydrophobic-phase "solubility effects," some residual adsorption and modifying effects of the mobile phase on the properties of the stationary phase (i.e., production of a "binary phase" through the entrapment of some solvent molecules into the surface structures) are all likely to occur. A better understanding of the retention properties of bonded phases is being currently sought in a number of laboratories. While the separation mechanism is complex, the general utility of this type of chromatography is widely recognized.

Probably the first successful application of a reversed-phase high-performance column to PAH analysis is that of Schmit *et al.* (*34*), who showed separations of both standard compounds and the components of an autoexhaust condensate. The chromatogram obtained in this work is reproduced in Fig. 6-6, showing that retention increases with the number of rings. Some resolution of PAH compounds with different ring fusion is also evident.

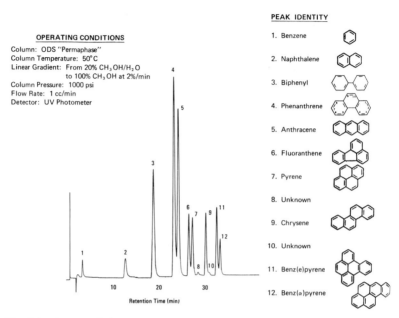

Fig. 6-6. A reversed-phase separation of standard PAH compounds using linear gradient elution. [Reproduced with permission from J. A. Schmit, R. A. Henry, R. C. Williams, and J. F. Kieckman, *J. Chromatogr. Sci.* **9**, 645 (1971). Copyright, Preston Technical Abstracts Co.]

Considering that this chromatogram was obtained with only a 1-meter column packed with 30-μm particles, the separation is quite impressive. Obviously, the use of gradient elution was helpful.

It will be evident from the following discussion that there is no particular difference in the basic retention order of parent PAH on reversed phases, common adsorbents, or conventional (polar) stationary phases. However, the hydrophobic interaction is highly important in the retention of alkylated PAH. Thus, compounds with sidechains are more soluble in hydrophobic stationary phases and are retained longer in reversed-phase systems.

Sleight (35) studied the retention of a number of PAH on an octade-cylsilane-bonded pellicular column using methanol–water as the mobile phase. A retention plot for the parent PAH gave a linear relationship, extending from benzene to the benzopyrene isomers. Such a relationship fits a simple equation

$$\log k = a + bC_N \tag{9}$$

where k = capacity ratio, C_N = carbon number, and a and b are constants.

Alkyl substitution in reversed-phase chromatography causes an appreciable increase of retention due to the decreased solubility in the polar mobile phase. Consequently, this type of chromatography should be preferable in cases where compounds with different side-chain lengths and numbers of alkyl groups are to be separated. Retention data obtained by Sleight (35) demonstrate the degree of solute retention with increasing number of aliphatic carbon atoms (Fig. 6-7). Similarly, a simple equation

$$\log k = A + BC'_N \tag{10}$$

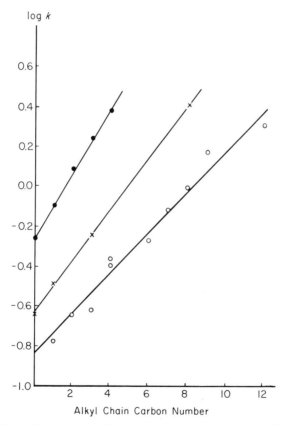

Fig. 6-7. Plot of $\log k$ against alkyl side-chain carbon number for alkylbenzenes ($\bigcirc$), naphthalenes ($\times$), and phenanthrenes ($\bullet$). Chromatographic system: methanol–water (7:3)/ octadecylsilane–Zipax; temperature, 60°C. [Reproduced with permission from R. B. Sleight, *J. Chromatogr.* **83**, 31 (1973). Copyright, Elsevier Scientific Publishing Co.]

is applicable to this case (C_N' is the total number of alkyl carbons). The constants A and B will change with the type of aromatic structure.

The mobile-phase composition has an effect on retention that is appreciably different for the parent and substituted PAH. Selected examples (*35*) are presented in Table 6-2. In some instances, even the elution order can be reversed.

As observed by Schmit *et al.* (*34*), increasing the temperature in a reversed-phase system can result in an increase in both column efficiency and solute retention. Thus, both the solvent composition and column temperature can be used as important variables.

Microparticulate columns for reversed-phase chromatography have become popular for PAH separations. Applications of such systems to the analysis of air-pollution samples (*36–38*), used engine oils (*39*), and coal-tar pitch (*40*) can be cited as representative examples. The utility of such columns was also recognized in work with azaarenes (*29*) and the oxygenated metabolites of PAH (*41*).

The majority of analyses of PAC by HPLC are performed using octadecylsilane (C_{18}) reversed-phase packing materials. Numerous C_{18} materials are commercially available; however, they differ significantly in their selectivity for PAH. A number of these packing materials were evaluated and compared by Wise *et al.* (*42*). Retention data for over 80 PAH on two C_{18} columns with different selectivities were reported.

A recent study involving the analysis of high-molecular-weight PAC in carbon black on a reversed-phase column has resulted in the separation

TABLE 6-2

Dependence of Retention on Solvent Composition for Alkylated and Nonalkylated Aromatic Hydrocarbons in Two Reversed-Phase Systems.[a,b]

	k value	
Compound	Methanol–water (7:3)	Methanol–water (6:4)
1,3-Diisopropylbenzene	0.54	3.63
Anthracene	0.67	2.82
2-Isopropylnaphthalene	0.57	3.03
Phenanthrene	0.55	2.42
2-Methylnaphthalene	0.32	1.45
Biphenyl	0.34	1.19

[a] According to Sleight (*35*).

[b] Temperature 60 °C.

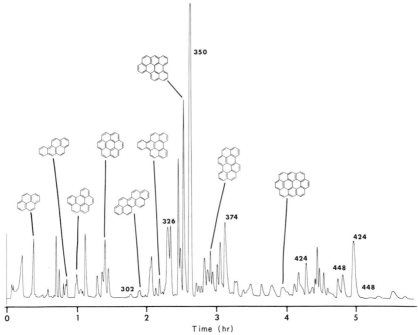

Fig. 6-8. HPLC chromatogram of high-molecular-weight fraction extracted from carbon black on a 25 cm × 4.6 mm i.d. Vydac 201TP C_{18} reversed-phase column. Solvent programming was 50:50 water–acetonitrile for 15 min. then to 100% acetonitrile during 70 min. to 100% ethyl acetate during 130 min, and to 100% methylene chloride during the last 60 min. [Reproduced with permission from P. A. Peaden, M. L. Lee, Y. Hirata, and M. Novotny, *Anal. Chem.* **52,** 2268 (1980). Copyright, American Chemical Society.]

shown in Fig. 6-8. Compounds up to molecular weight 448 were successfully chromatographed (*43*).

Due to the excellent reproducibility and stability of bonded reversed-phase columns, they now play a major role in LC analysis of PAC. The unique selectivity for alkylated aromatics is an additional feature of practical importance.

E. Polar Columns

Systems that employ a bonded polar stationary phase and a nonpolar mobile phase are also applicable to the separation of PAC. Naturally, the mechanism of retention is likely to be quite complex in such cases. Interactions between the π-electron system of PAC solutes and various polar surface structures must be taken into account.

Basing their work on the previously reported use of oxydipropionitrile as

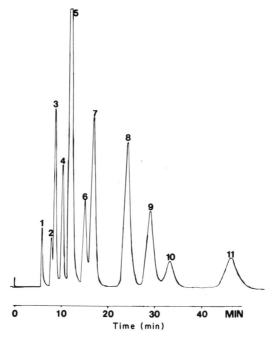

Fig. 6-9. Chromatogram of standard PAH on an aminosilane-bonded silica. Mobile phase: *n*-hexane at room temperature. Peaks 1–11 are benzene, naphthalene, biphenyl, fluorene, anthracene, pyrene, benzofluorene, triphenylene, benzo[*a*]pyrene, perylene, and dibenzanthracene. [Reproduced with permission from M. Novotny, *in* "Bonded Stationary Phases in Chromatography" (E. Grushka, ed.), Ann Arbor Science Publishers, 1974, p. 221.]

a coated stationary phase for PAH separations (*29*), Novotny *et al.* (*44*) used a bonded version of this phase (Fig. 6-9). Since fractionation of a mixture prior to capillary GC and NMR analysis was the major objective, the use of a nonpolar hydrocarbon mobile phase had the advantage of easy solvent removal. Similar objectives were also fulfilled by a bonded material synthesized in the same laboratory from silica microparticles and a triethoxyaminosilane (*45*). Again, the individual PAH were eluted in the order of increasing molecular weight. Recently, a similar type of material was also evaluated by Wise *et al.* (*46*), who advocated the complementary use of a polar amino phase followed by a reversed-phase column for effective resolution of complex mixtures.

Minor selective effects were observed with the amino phase (*46*). In another report, a phthalimidopropylsilane-bonded packing (*47*) was also found useful in certain applications. For instance, the resolution of benzo[*k*]-fluoranthene and perylene was possible, a separation of some importance in the analysis of certain foodstuffs. The reversed elution order of benzo[*a*]-

pyrene and benzo[e]pyrene was an additional unusual effect of this phase.

A different polar phase was also prepared by Ray and Frei (48) by reacting p-nitrophenyl isocyanate with surface silanol groups of a pellicular material (Corasil I). This phase was specifically intended for use in the separation of electron-donor compounds, such as azaarenes. Blumer and Zander (49) described the use of a similar bonded nitrophenyl phase for the selective retention and separation of nitrogen-containing constituents in coal tar pitch. The mechanism of retention was suggested to be the formation of donor–acceptor complexes.

The separation of high-molecular-weight PAH up to molecular weight 600 has also been reported using the bonded nitrophenyl phase (49).

III. GRADIENT ELUTION TECHNIQUES

Chromatographic resolution of PAC mixtures can be related to the number of theoretical plates N, column selectivity α, and the solute capacity ratio k. Resolution R of a solute pair in a chromatogram is related to these variables by

$$R = \frac{N^{1/2}}{4}\left(\frac{\alpha - 1}{\alpha}\right)\left(\frac{k}{1 + k}\right) \tag{11}$$

Since N and α are more or less fixed by our capabilities to prepare efficient and selective column packings, the solute retention (and resolution) is commonly controlled by the choice of phase systems. Since the capacity ratio, k, is directly proportional to the thermodynamic distribution coefficient, the mobile-phase composition and temperature are the most important variables in the optimization of separation processes.

It is common experience that under isocratic conditions (constant mobile-phase composition), solutes with large k values can be eluted only after an appreciable time period. This may often lead to poor separation of certain components in a complex mixture, excessive analysis times, and difficult detection of the late mixture constituents. These problems can be considerably reduced by the employment of certain programming (gradient) techniques. The situation is similar to that encountered in gas chromatography, where temperature programming is the most frequently used means to control solute retention.

In liquid–liquid (partition) chromatography, solvent programming is prohibitive. With the exception of a limited number of selected solutes, there is little prospect that conventional partition liquid chromatography will be of general utility because of the limited miscibility of the two phases. Immiscibility of the two phases is almost synonymous with a great difference in polarity and, consequently, "the sample components tend to have partition coefficients in such a system tending either to zero or infinity (50)." These problems are virtually nonexistent in bonded-phase chromatography, where solvent gradient techniques can easily be used.

To optimize the capacity ratios of PAC in a mixture, the solvent composi-

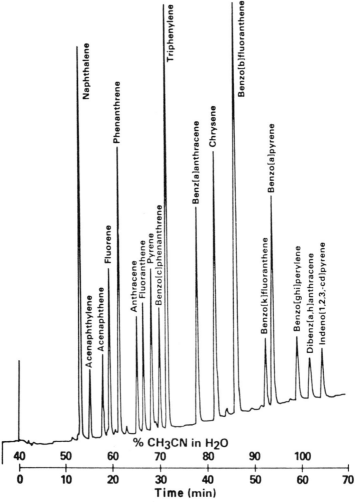

Fig. 6-10. Reversed-phase liquid chromatographic separation of priority-pollutant PAH. Column: Vydac 201TP reversed phase. Detection: UV absorbance at 256 nm. Condition: linear gradient from 40–100% acetonitrile in water at 1% min^{-1} and 1 ml min^{-1}. [Reproduced with permission from S. A. Wise, W. J. Bonnett, and W. E. May, *in* "Polynuclear Aromatic Hydrocarbons: Chemistry and Biological Effects" (A. Bjørseth and A. J. Dennis, eds.) p. 791. Battelle Press, Columbus, Ohio, 1980].

tion may be gradually changed in both adsorption and bonded-phase chromatography. In solvent programming, the solvent polarity is usually increased in accord with the eluotropic series (12). Alternatively, incremental gradient elution techniques, as proposed by Scott and Kucera (51), can be used for mixtures of compounds with widely differing polarities.

Typical reversed-phase separations are carried out with the systems of methanol–water or acetonitrile–water. In these "solubility separations," the PAH of increasing molecular weight or hydrophobicity are gradually eluted with an increasing amount of the organic component in the mobile phase. Examples of gradient elution of PAH mixtures are now widely documented throughout the literature. It should be noted that the mobile-phase composition can change the order of elution.

A particularly impressive separation of the priority pollutant PAH was obtained by Wise et al. (42) using a linear gradient from 40–100% acetonitrile in water on a C_{18} reversed-phase packing. This chromatogram is shown in Fig. 6-10.

Although additional gradient techniques have neither been applied to PAC separations nor to other complex mixtures, their potential should briefly be mentioned. In particular, they include temperature or flow programming and the so-called "coupled columns" (52). The resolving power of various gradient methods was evaluated by Snyder (52).

Since both solubilities and adsorption energies are temperature-dependent, temperature programming can be used to change retention of PAC solutes. However, the general utility of this approach is considerably less than in gas chromatography.

In flow programming, since the solvent flow is programmed by a gradual increase in the inlet pressure, certain equipment limitations exist. In addition, since there is a general increase of the plate height with mobile-phase flow, a decrease in k values is only possible with some sacrifice of column efficiency.

In conclusion, solvent gradient methods are currently the most useful means of effecting separation, and their wide use is fully justified. There are a number of ways to mix solvents for gradient elution (53). However, modern analytical instruments typically utilize systems containing two pulseless pumps and sophisticated, electronically controlled solvent programmers.

IV. DETECTION SYSTEMS

The availability of both sensitive general-purpose and selective detectors remains among the most serious problems of modern liquid chromatography, but there is much less difficulty with the analysis of PAC. Aromatic molecules are strong absorbers of UV light. In addition, some PAC yield intense fluorescence. UV monitors and fluorescence detectors are the most widely used detectors for high-performance liquid chromatography.

A. UV Detection

Flow-cell photometers and spectrophotometers have been widely used since the inception of high-performance liquid chromatography. Over the years of application, their technology has improved considerably. UV monitors with cell volumes less than 10 μl are now available from a number of commercial sources. The small cell volumes are particularly desirable in work with high-efficiency microparticulate columns. Because of the wide utilization of liquid chromatography in the analysis of biochemicals and various pharmaceutical compounds, the less expensive commercial UV detectors are confined to fixed-wavelength operation within the range 250–280 nm.

Since PAC strongly absorb in the above spectral region, the UV monitor can be considered a "universal detector" for this class of compounds. Of course, aromatic solvents cannot be considered as potential mobile phases. This is no particular problem, since a variety of suitable solvents are available for both adsorption and bonded-phase chromatography which have minimal absorptivities at the wavelengths of detection. Only small baseline drifts due to minor changes of absorbance or refractive index may be occasionally observed during gradient elution.

Since the Beer–Lambert law is generally obeyed by PAC in these LC detectors, the responses are predictable and the calibration plots are linear over an extensive concentration range. Although the UV absorption of various PAH and related compounds differ somewhat at fixed wavelength, typical sensitivities in the lower nanogram range are feasible.

The application of spectrophotometric detectors in modern liquid chromatography has resulted in improvements in both selectivity and sensitivity of detection. Both aspects are of utmost importance in the analysis of PAC. Selective monitoring devices, commonly referred to as "variable wavelength detectors," are now being increasingly used.

Due to the great complexities of various PAC mixtures, a complete resolution of individual compounds is clearly not feasible. Capillary GC comes closest to accomplishing this goal, whereas the present resolving power of high-performance LC is limited. Consequently, it is advantageous to use selective detectors. Since various PAC have different absorptivities at a given wavelength, the detector can be tuned for maximum selectivity for specific compounds. Published spectra (e.g., references in Chapter 9) can be of use in selecting the wavelength at which the greatest absorption occurs, while discriminating against known interfering compounds.

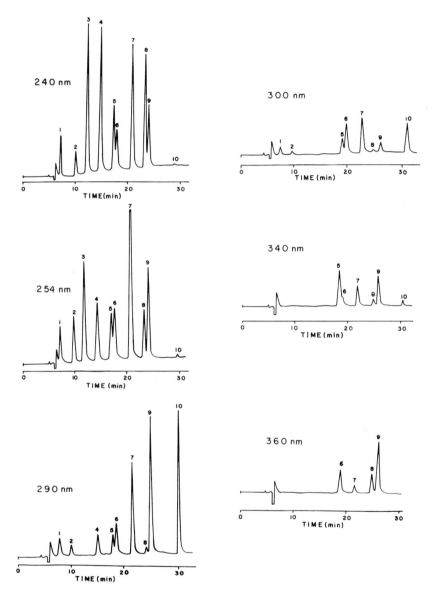

Fig. 6-11. Detection of PAH at different wavelengths. Order of elution: (1) benzene (37.9 μg), (2) naphthalene (1.55 μg), (3) biphenyl (1.50 μg). (4) phenanthrene (0.65 μg), (5) fluoranthene (1.00 μg), (6) pyrene (0.65 μg), (7) chrysene (0.65 μg), (8) perylene (1.00 μg), (9) benzo[*a*]pyrene (0.35 μg), and (10) coronene (0.91 μg). [Reproduced with permission from A. M. Krstulovic, D. M. Rosie, and P. R. Brown, *Anal. Chem.* **48**, 1383 (1976). Copyright, American Chemical Society.]

Detection of PAH at different wavelengths was studied by Krstulovic *et al.* (*54*). Chromatograms obtained with a standard mixture are shown in Fig. 6-11. Through the proper wavelength selection (290 nm), benzo[a]-pyrene can be monitored without interference from perylene. Similarly, pyrene can be quantified in the presence of fluoranthene at 360 nm. In an application of reversed-phase chromatography and variable-wavelength detection, Boden (*40*) observed that benzo[a]pyrene can be measured in coal tar with ease, in spite of the obvious sample complexity.

While optimizing detector response by selecting a wavelength corresponding to a higher absorbance, the overall sensitivities can be significantly increased. Table 6-3 shows detection limits determined for several common PAH by Krstulovic *et al.* (*38*). Naturally, sensitivity values will differ among different analytical systems, commercial detectors, etc. Detection limits are particularly dependent on the noise levels produced by various types of high-pressure pumps; in this respect, pulseless syringe-type pumps appear most suitable for sensitive analytical determinations.

Although a UV absorption spectrum of an isolated fraction may be identical with the spectrum obtained from an authentic PAC, this cannot be taken as complete structural proof. Useful information, however, can be derived from it. Thus, the possibility of acquiring spectral data directly under LC detection conditions deserves some attention. While the identification power of UV spectroscopy will be discussed in Chapter 9, we will briefly discuss here the possibilities of "on-line" investigations.

Spectrophotometric (or spectrofluorimetric) flow-cell devices lend them-

TABLE 6-3

Detection Limits of Some PAC[a]

Compound (wavelength in nm)	Amount (ng)
Benzene (254)	0.04
Naphthalene (254)	0.11
Biphenyl (254)	0.36
Phenanthrene (254)	0.16
Fluoranthene (254)	0.38
Pyrene (300)	0.71
Chrysene (254)	0.01
Perylene (254)	0.52
Benzo[a]pyrene (254)	0.05
Dibenz[a,c]anthracene (300)	0.22
Coronene (290)	0.29

[a] From Krstulovic *et al.* (*38*).

selves quite naturally to such qualitative studies. As has already been pointed out (54), the ratios of peak areas detected at different wavelengths can yield certain quantitative information. Simultaneous detection at two different wavelengths (with two detectors in series) provides such an opportunity. However, coeluting "impurities" with similar absorption properties remain a problem.

Certain approaches to the acquisition of whole absorption spectra during the elution of chromatographic peaks have been under development. While these devices have not been applied yet to PAC analysis, their potential is clearly suggested. Spectral recording can be achieved through rapid scanning with an oscillating mirror (55), the use of a silicon target vidicon tube (56), or a computer-controlled solid-state photodiode array system (57, 58). It is important for such developments that the spectral information is obtained without appreciable loss of sensitivity.

B. Fluorescence Detection

Although fluorimetric detectors have earlier been considered as generally less useful devices than UV monitors, the recent advances in labeling (pre- or postcolumn derivatization) of biologically important molecules (59, 60) have significantly influenced developments in both instrumentation and application of such measurements. Consequently, several micro-cell filter-type monitors and spectrofluorimeters (variable wavelength devices) are now commercially available.

Although much of what has been said in the previous section about UV detection applies here as well, the basic difference lies in the detector selectivity. All PAC absorb in the UV spectral region, but only some are fluorescent. Structural features of PAC related to their fluorescence spectra are discussed in Chapter 9.

Although fluorimetric detection overlooks a number of PAC in complex mixtures, this type of selectivity can be a bonus in certain analyses. This is, for example, demonstrated in a publication by Ogan et al. (61), where 16 selected components of environmental significance were monitored in a fairly complex sample matrix; if a UV detector were used, inadequate chromatographic resolution would not allow reliable quantitation. Sensitivity is also a unique and most welcome attribute of this type of detection. The detection limits are at least an order of magnitude lower than in UV detection; sensitivities between 10–100 pg have typically been reported. In fact, such sensitivity is unfortunately well ahead of capabilities of the most powerful identification tools.

There appears to be much potential for further sensitivity improvements. In a recent study involving laser-induced fluorescence of PAH molecules,

Richardson and Ando (*62*) reported the limits of detection in the low parts-per-trillion range. Further technological advances may result in even lower levels. A report by Richardson *et al.* (*63*) describes the use of a pulsed laser as an excitation source for an LC detector; if the delay between excitation and detection is sufficiently long, only fluorophores with long fluorescence lifetimes are monitored. This approach provides another dimension in selectivity.

Fine spectral tuning in fluorimetric detection offers some possibilities for identification work at extremely low concentration levels of PAC. The effect of wavelength selectivity is dramatically demonstrated in Fig. 6-12 (*39*) by the appearance or almost total disappearance of several chromatographic peaks, even though compound identification was not an objective of that study.

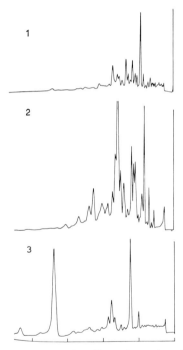

Fig. 6-12. Chromatograms of an engine oil sample monitored by (1) a UV detector at 254 nm; (2) fluorimetric detector at $\lambda_{ex} = 275$ nm and $\lambda_{em} = 320$ nm; and (3) fluorimetric detector at $\lambda_{ex} = 360$ nm and $\lambda_{em} = 460$ nm. Column: 25-cm bonded-phase microparticulate packing; solvent, methanol–water (9:1). [Reproduced with permission from B. B. Wheals, C. G. Vaughan, and M. J. Whitehouse, *J. Chromatogr.* **106,** 109 (1975). Copyright, Elsevier Scientific Publishing Co.]

V. CAPILLARY AND MICROBORE COLUMNS

Just as with many other complex mixtures, the number of PAC isomers increases greatly with molecular weight. Since volatility is substantially reduced for larger PAC molecules, high-resolution GC is of limited value for compounds containing more than six rings.

While LC becomes a natural method of choice (volatility is irrelevant), limitations exist concerning its practical resolving power. Attempts to increase the column length have met with little success. This is mainly due to technical problems encountered in obtaining low-volume connections of columns connected in series, as well as difficulties generated by heats of friction that apparently distort the chromatographic zones (11). Recently developed microbore columns (64, 65) and LC capillary columns (66–68) may successfully avoid the above problems.

Microbore columns, originated by Scott and Kucera (64, 65), are a logical extension of the commonly used LC columns. Specialized technology is necessary to pack columns with 1 mm i.d. Since these columns are operated at low flow rates (to obtain low plate-height values), modified sampling and detector units must be employed—i.e., their volumes must be substantially reduced. With typical flow rates of 40–50 μl/min, sample volumes of approximately 0.2 μl and a detector with 1 μl cell volume were employed (65). While ordinary HPLC columns seldom exceed 10^3 theoretical plates, efficiencies between 10^5 and 10^6 plates have been demonstrated with microbore columns.

The development of capillary liquid chromatography has grown from theoretical considerations (69) of the potential of gas and liquid chromatographic methods. In a diffusion-controlled chromatographic process, a reduction of the column radius (for LC) is essential to counterbalance the four to five orders of magnitude slower diffusion rate of solute molecules in the liquid mobile phase. The plate height, H, is given by

$$H = (6R^2 - 16R + 11)\frac{r^2 u}{24D_M} \tag{12}$$

where $R = 1/(1 + k)$, r = column radius, u = mobile-phase linear velocity, and D_M = solute diffusion coefficient in the mobile phase (1).

Experimental investigations by Tsuda and Novotny (70) as well as the recent detailed theoretical analysis of capillary LC by Knox and Gilbert (71) indicate that very efficient separations can be achieved by substantially reducing the column radius down to about 20 μm. Obviously, such columns require a decrease in both sampling and detector volumes, the latter being estimated (71) to be 10 nl. While the necessary trend toward miniaturization

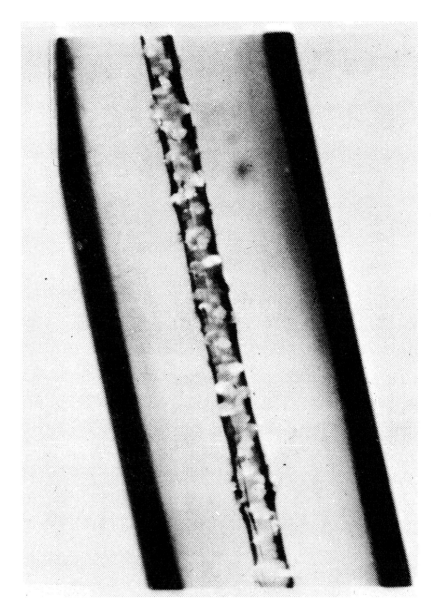

Fig. 6-13. Micrograph of a section of alumina-packed microcapillary. Column inner diameter, 75 μm; average particle size, 30 μm. [Reproduced with permission from T. Tsuda and M. Novotny, *Anal. Chem.* **50,** 271 (1978). Copyright, American Chemical Society.]

of LC has been initiated by Ishii *et al.* (*72*), capillary LC in open tubes has not been sufficiently developed for complex mixture analysis.

Another type of column, developed by Tsuda and Novotny (*66*), is the glass packed microcapillary column. As shown in Fig. 6-13, the adsorbent particles are embedded into the thick glass wall; this is achieved by drawing larger glass tubes, tightly prepacked with particles of the desired size. Such columns are subsequently modified to provide chemically bonded stationary phases (*73*) of various selectivities.

Specially designed equipment is needed for capillary HPLC, including submicroliter sampling systems and detectors. Ultraviolet (*67, 68*), fluorescence (*68*), and electrochemical (*74*) detectors have been modified to accommodate cells of approximately 0.1 μl volume. These are needed to avoid band-broadening due to extracolumn factors. Flow rates through packed microcapillaries are of the order of 1 μl/min.

Some preliminary results for PAC have been obtained (*68, 75*) using microcapillary LC. In order to resolve complex mixtures, stepwise gradient techniques have been investigated (*68*). Figure 6-14 demonstrates the separation

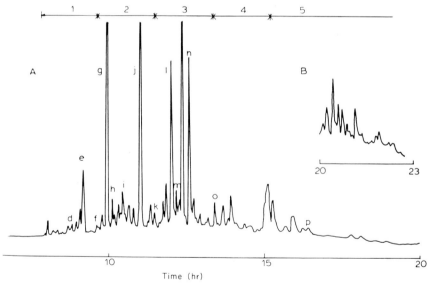

Fig. 6-14. Chromatogram of the aromatic fraction of coal tar on a reversed-phase packed microcapillary column (55 meter × 70 μm i.d., packed with 30 μm basic alumina–octadecylsilane.) Detection, 0.1 μl fluorimetric detector. Components: d, fluorene; e, phenanthrene; f, anthracene; g, fluoranthene; h, pyrene, i, benz[*a*]anthracene; k, chrysene; l, benzo[*e*]pyrene; m, perylene; n, benzo[*a*]pyrene; o, dibenzo[*ghi*]perylene; and p, coronene. [Reproduced with permission from Y. Hirata and M. Novotny, *J. Chromatogr.* **186**, 521 (1980). Copyright, Elsevier Scientific Publishing Co.]

of the aromatic fraction of coal tar (68), using a 55 m × 70 μm (i.d.) capillary packed with 30-μm alumina particles, modified (73) as a reversed-phase column. A fluorimetric detector was used. Long analysis times are typical for this chromatographic technique.

Both packed microcapillaries and microbore columns offer some potential for the separation and identification of nonvolatile PAC.

REFERENCES

1. J. C. Giddings, "Dynamics of Chromatography." Marcel Dekker, New York, 1965.
2. P. B. Hamilton, *Adv. Chromatogr.* **2**, 3 (1966).
3. C. Horváth, B. Preiss, and S. R. Lipsky, *Anal. Chem.* **39**, 1422 (1967).
4. J. J. Kirkland, *J. Chromatogr. Sci.* **7**, 7 (1969).
5. I. Halász and C. Horváth, *Anal. Chem.* **35**, 499 (1963).
6. J. H. Knox and M. Saleem, *J. Chromatogr. Sci.* **7**, 614 (1969).
7. G. J. Kennedy and J. H. Knox, *J. Chromatogr. Sci.* **10**, 549 (1972).
8. M. Martin, C. Eon, and G. Guiochon, *J. Chromatogr.* **99**, 357 (1974).
9. J. J. Kirkland, *J. Chromatogr. Sci.* **10**, 593 (1972).
10. R. E. Majors, *Anal. Chem.* **44**, 1722 (1972).
11. I. Halász, R. Endele, and J. Asshauer, *J. Chromatogr.* **112**, 37 (1975).
12. L. R. Snyder, "Principles of Adsorption Chromatography." Marcel Dekker, New York, 1968.
13. M. Popl, J. Mostecky, V. Dolansky, and M. Kuras, *Anal. Chem.* **43**, 518 (1971).
14. M. Popl, V. Dolansky, and J. Mostecky, *Collect. Czech. Chem. Commun.* **39**, 1836 (1974).
15. A. J. P. Martin, *Biochem. Soc. Symp.* **3**, 4 (1949).
16. M. Popl, V. Dolansky, and J. Mostecky, *J. Chromatogr.* **91**, 649 (1974).
17. M. Martin, L. Loheac, and G. Guiochon, *Chromatographia* **5**, 33 (1972).
18. C. R. Engel and E. Sawicki, *J. Chromatogr.* **31**, 109 (1967).
19. M. Dong and D. C. Locke, *J. Chromatogr. Sci.* **15**, 32 (1977).
20. S. Taback and M. R. M. Verzola, *J. Chromatogr.* **51**, 334 (1970).
21. R. Vivilecchia, M. Thiebaud, and R. W. Frei, *J. Chromatogr. Sci.* **10**, 411 (1972).
22. E. Kováts, *Helv. Chim. Acta* **41**, 1915 (1958).
23. H.-J. Klimisch, *J. Chromatogr.* **83**, 11 (1973).
24. F. M. Rabel, *Anal. Chem.* **45**, 957 (1973).
25. Commercial literature, Pharmacia, Uppsala, Sweden.
26. M. Popl, V. Dolansky, and J. Coupek, *J. Chromatogr.* **130**, 195 (1977).
27. D. W. Grant, R. B. Meiris, and M. G. Hollis, *J. Chromatogr.* **99**, 721 (1974).
28. H. Colin and G. Guiochon, *J. Chromatogr.* **126**, 43 (1976).
29. K.-P. Hupe and H. Schrenker, *Chromatographia* **5**, 44 (1972).
30. R. E. Jentoft and T. H. Gouw, *Anal. Chem.* **40**, 923 (1968).
31. E. Grushka (ed.), "Bonded Stationary Phases in Chromatography." Ann Arbor Sci. Publ., Ann Arbor, Michigan, 1974.
32. G. A. Howard and A. J. P. Martin, *Biochem. J.* **56**, 532 (1950).
33. A. V. Kiselev and Y. I. Yashin, "Gas-Adsorption Chromatography." Plenum, New York, 1969.
34. J. A. Schmit, R. A. Henry, R. C. Williams, and J. F. Dieckman, *J. Chromatogr. Sci.* **9**, 645 (1971).
35. R. B. Sleight, *J. Chromatogr.* **83**, 31 (1973).

36. M. Dong, D. C. Locke, and E. Ferrand, *Anal. Chem.* **48**, 368 (1976).
37. M. A. Fox and S. W. Staley, *Anal. Chem.* **48**, 992 (1976).
38. A. M. Krstulovic, D. M. Rosie, and P. R. Brown, *Am. Lab.* **9**, 11 (1977).
39. B. B. Wheals, C. G. Vaughan, and M. J. Whitehouse, *J. Chromatogr.* **106**, 109 (1975).
40. H. Boden, *J. Chromatogr. Sci.* **14**, 391 (1976).
41. J. K. Selkirk, R. G. Croy, and H. V. Gelboin, *Science* **184**, 169 (1974).
42. S. A. Wise, W. J. Bonnett, and W. E. May, *in* "Polynuclear Aromatic Hydrocarbons: Chemistry and Biological Effects" (A. Bjørseth and A. J. Dennis, eds.), p. 791. Battelle Press, Columbus, Ohio, 1980.
43. P. A. Peaden, M. L. Lee, Y. Hirata, and M. Novotny, *Anal. Chem.* **52**, 2268 (1980).
44. M. Novotny, M. L. Lee, and K. D. Bartle, *J. Chromatogr. Sci.* **12**, 606 (1974).
45. M. Novotny, *in* "Bonded Stationary Phases in Chromatography" (E. Grushka, ed.), p. 199. Ann Arbor Sci. Publ., Ann Arbor, Michigan, 1974.
46. S. A. Wise, S. N. Chesler, H. S. Hertz, L. R. Hilpert, and W. E. May, *Anal. Chem.* **49**, 2306 (1977).
47. D. C. Hunt, P. J. Wild, and N. T. Crosby, *J. Chromatogr.* **130**, 320 (1977).
48. S. Ray and R. W. Frei, *J. Chromatogr.* **71**, 451 (1972).
49. G. P. Blumer and M. Zander, *Erdol Kohle Erdgas Petrochem. Suppl. Compend 78/79*, p. 1472, 1978.
50. H. N. M. Stewart and S. G. Perry, *J. Chromatogr.* **37**, 97 (1968).
51. R. P. W. Scott and P. Kucera, *Anal. Chem.* **45**, 749 (1973).
52. L. R. Snyder, *J. Chromatogr. Sci.* **8**, 692 (1970).
53. L. R. Snyder, *in* "Chromatography" (E. Heftmann, ed.), Chapter 5. Van Nostrand-Reinhold New York, 1967.
54. A. M. Krstulovic, D. M. Rosie, and P. R. Brown, *Anal. Chem.* **48**, 1383 (1976).
55. M. S. Denton, T. P. DeAngelis, A. M. Yacynych, W. R. Heineman, and T. W. Gilbert, *Anal. Chem.* **48**, 20 (1976).
56. A. E. McDowell and H. L. Pardue, *Anal. Chem.* **49**, 1171 (1977).
57. R. E. Dessy, W. D. Reynolds, W. G. Nunn, C. A. Titus, and G. F. Moler, *J. Chromatogr.* **126**, 347 (1976).
58. M. J. Milano, S. Lam, and E. Grushka, *J. Chromatogr.* **125**, 315 (1976).
59. S. Udenfriend, S. Stein, S. Bohlen, P. Dairman, W. Leimgruber, and M. Weigele, *Science* **178**, 871 (1972).
60. J. F. Lawrence and R. W. Frei, "Chemical Derivatization in Liquid Chromatography." Elsevier, Amsterdam, 1976.
61. K. Ogan, E. Katz, and W. Slavin, *Anal. Chem.* **51**, 1315 (1979).
62. J. H. Richardson and M. E. Ando, *Anal. Chem.* **49**, 955 (1977).
63. J. H. Richardson, K. M. Larson, G. R. Hangen, D. C. Johnson, and J. E. Clarkson, *Anal. Chim. Acta* **116**, 407 (1980).
64. R. P. W. Scott, *Analyst (London)* **103**, 37 (1978).
65. R. P. W. Scott and P. Kucera, *J. Chromatogr.* **169**, 51 (1979).
66. T. Tsuda and M. Novotny, *Anal. Chem.* **50**, 271 (1978).
67. T. Tsuda, K. Hibi, T. Nakanishi, T. Takeuchi, and D. Ishii, *J. Chromatogr.* **158**, 227 (1978).
68. Y. Hirata and M. Novotny, *J. Chromatogr.* **186**, 521 (1980).
69. J. C. Giddings, *Anal. Chem.* **36**, 1890 (1964).
70. T. Tsuda and M. Novotny, *Anal. Chem.* **50**, 632 (1978).
71. J. H. Knox and M. T. Gilbert, *J. Chromatogr.* **186**, 405 (1980).
72. D. Ishii, K. Asai, K. Hibi, T. Jonokuchi, and M. Nagaya, *J. Chromatogr.* **144**, 157 (1977).
73. Y. Hirata, M. Novotny, T. Tsuda, and D. Ishii, *Anal. Chem.* **51**, 1807 (1979).
74. Y. Hirata, P. Lin, M. Novotny, and R. M. Wightman, *J. Chromatogr. Biomed. Appl.* **181**, 287 (1980).

7

Gas Chromatography

I. INTRODUCTION

The separation of smaller aromatic hydrocarbon molecules was commonly done using gas chromatography in the very early stages of the development of this method. Thus, numerous papers on the subject can be found in the pertinent literature of the late 1950s and throughout the 1960s, including group separations (alkane–aromatic, aromatic–naphthene, and aromatic–alkane hydrocarbon groups), resolution of alkylated benzenes, the "classical" separation of xylene isomers, etc. In many of these determinations, stationary-phase selectivity was generally sought since only low resolution was available at that time.

It was not until gas chromatography was fully developed as an analytical method that its potential as an effective means of PAC analysis was unquestionably established. Several factors appear responsible for this situation. PAC of interest to the analyst cover a considerable volatility range. Whereas the boiling point of naphthalene is 218°C, the value for coronene is 525°C. Thus, high column temperatures are needed to bring about the elution of heavier PAC. Correspondingly, only the stationary phases of greatest thermostability have been of interest here.

Although the lack of functional groups on PAC and their hydrocarbon character could imply chromatographic stability, common experience shows otherwise. Irreversible adsorption phenomena and peak tailing are frequently met with insufficiently inert chromatographic columns. These effects are particularly obvious when working with small samples. The affinity of PAC for adsorptive siliceous surfaces are most likely to be the primary cause of such phenomena.

The advances in technology of chromatographic columns have played a substantial role in the feasibility of GC for PAC analysis. The reduction of

188

liquid-phase loading (1), the consequent improvements in quality of the solid supports, and the availability of new thermostable silicone phases in the early 1960s all contributed to new directions in PAC analysis.

As the molecular weight and ring number increase within the series of PAC compounds, the number of possible isomers also becomes significantly greater. Thus, it soon became obvious that the complexity of most PAC mixtures created resolution requirements beyond the capability of conventional packed columns. Although capillary columns had been successfully applied to many petrochemical problems (2), several earlier reports of their use in PAH separations (2–5) were exceptional rather than typical of such efforts. It was not until much later (6–8) that the capabilities of glass capillary columns in GC analysis of products of combustion were clearly indicated. Today, the glass capillary column assumes an important role in PAC analytical methodology. In addition to the great resolving power and inertness of glass capillary columns, thin-film coatings have also significantly reduced the temperatures needed to elute higher molecular weight PAC.

However, the volatility range in which various PAC can be chromatographed seldom exceeds that of six-ring structures. There is a point where further increases in analysis temperature are hardly profitable. There are many indications that chromatography in the liquid phase is a better choice for analysis of the very heavy mixture components. In addition, so-called "supercritical-fluid chromatography" (9) has been investigated mostly with large PAH as model substances; the method has not yet been developed to its full analytical potential so far. Since modern LC does not have the resolution capabilities of capillary GC, how are we to cope with the obvious sample complexity in this volatility region? This question presents a great challenge for future investigations in this research area.

Even the best separations are worthless without the identification of individual mixture components. Although some identification power is inherent in chromatographic retention data, additional information can be obtained by the use of selective detectors and ancillary techniques. Fortunately, gas chromatography lends itself quite naturally to coupling with ancillary tools, of which mass spectrometry is the most important.

II. STATIONARY PHASES

Since typical PAC mixtures contain a great number of isomers, adequate chromatographic resolution is imperative for both identification and quantification purposes. Such resolution is theoretically feasible by increasing the column efficiency or by choosing a selective stationary phase. Generally, combination of both factors is sought in chromatographic practice. The number of theoretical plates, N_{req}, required for adequate resolution of two adjacent peaks (98% separation of the peak areas) is related to column selectivity (relative retention, α) and to the capacity ratio, k, according to the well-known equation (10)

$$N_{req} = 16 \left(\frac{\alpha}{\alpha - 1}\right)^2 \left(\frac{k + 1}{k}\right)^2 \tag{1}$$

It is apparent that for values of k typically used in GC analysis and for values of α lower than 1.05, the use of capillary columns is mandatory. Values of α can be related to the difference in Gibbs free energy between two compounds according to (11)

$$\Delta(\Delta G) = \Delta G_2 - \Delta G_1 = RT \ln \alpha \tag{2}$$

Such differences can be small for very similar isomers. However, exploitations of very selective interactions of solute molecules with the stationary phase have been widely documented. While the column selectivity alone can frequently be instrumental in resolving pairs of compounds with very small differences in their chemical structure, it is of only limited value in chromatography of complex mixtures. This argument was appropriately expressed by Giddings (12) many years ago, in that "changes in selectivity may do little more than scramble the already crowded chromatogram, with new overlaps replacing the old; the only certain means of improving resolution is through an increase in the number of plates."

The importance of isomer resolution in the analysis of PAC cannot be overemphasized. Besides the well-documented difference in the toxicity of the benzopyrenes, as discussed in Chapter 3, the position of methyl-group substitution is now known to play a crucial role in the carcinogenic properties of numerous PAC molecules. Thus, the ideal column for the analysis of a complex mixture should have *both* a high number of theoretical plates (e.g., a wall-coated capillary column) and an adequate selectivity. Only when the sample is fractionated into groups of isomeric compounds prior to GC analysis will selective columns with only moderate efficiencies be sufficient.

However, it should be pointed out that column selectivity is frequently incompatible with the practical temperatures required for analysis. Most selective (polar) column partitioning systems have only limited temperature stability. The liquid crystalline phases discussed below are somewhat exceptional in this respect. Also, gas–solid (adsorption) chromatography is not always a viable alternative because solute-adsorption coefficients do not permit PAC of higher molecular weight to elute at reasonable column temperatures.

Many of the examples below represent compromises made between compound resolution, column thermal stability, stationary-phase bleeding, detection conditions, etc. The relative merits of GC with solid adsorbents, inorganic salt phases, nematic liquid crystals, and the conventional liquid phases will now be discussed. This discussion will be limited mostly to studies with conventional packed columns, whereas the merits of capillary GC (which can frequently solve many of the above problems simultaneously) will be treated subsequent to this discussion.

Due to the limited availability of thermostable phases during the first decade of gas chromatography and the insufficiently developed technology of thin-film coatings, PAC separations seldom went beyond the elution of phenanthrene and anthracene. The effective chromatography of higher homologs was seldom possible. Also, prior to the utilization of ionization detectors, aspects of column inertness were generally overlooked.

Some of the stationary phases employed earlier in the GC of PAC have preserved their utility to the present date. However, this statement is only generally true, since many commercial phases currently available are very refined derivatives of their predecessors. Silicone polymers and Apiezon phases are most commonly mentioned in the early publications on chromatography of PAH. Among the first investigations, the analytical studies of Dupire and Botquin (13, 14) Gudzinowicz and Smith (15), Wood (16), and Lijinsky et al. (17) should be mentioned. Additional references (18) were reviewed in 1970.

It is of historical interest to show (17) here a chromatogram (Fig. 7-1) representing what might have been considered in 1963 a very successful analysis, resulting from a combination of two major advances: (1) the availability of thermostable phases and (2) the use of an ionization detector.

The liquid stationary phases most commonly employed in PAC gas chromatography are listed in Table 7-1. Most of these phases can be readily categorized into one of the following types: silicone fluids or gums with variable polarity, polyphenyl ethers, and carborane polymers. Apiezons (hydrocarbon vacuum grease materials) are somewhat exceptional; although otherwise very useful stationary phases, Apiezons are at present less popular due to their batch-to-batch variations.

The exceptionally good thermal stability of silicones gives this class a preferential position among the numerous stationary phases. The selectivity characteristics of the individual silicones have not been extensively studied for PAH separations. It should be mentioned that even minor changes in selectivity may be helpful in capillary GC, although they make no visible difference in work with packed columns. For example, among the silicone phases, SE-52 (a phenylmethylsilicone gum) brings about somewhat better

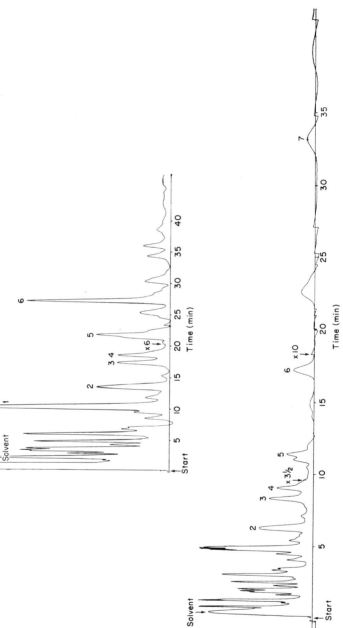

Fig. 7-1. Chromatograms of coal-tar sample (250 mg/5 ml benzene, 5 μl aliquot) on an SE-30 packed column at 160°C (upper) and 180°C (lower). Key: (1) phenanthrene/anthracene; (2) carbazole; (3) fluoranthene; (4) pyrene; (5) 1-methyl or 3-methylpyrene; (6) benz[a]-anthracene/chrysene; and (7) benzo[a]pyrene. (Reproduced with permission from W. Lijinsky, I. Domsky, G. Mason, H. Y. Ramaki, and T. Safavi, *Anal. Chem.* **35**, 952. (1963). Copyright, American Chemical Society.)

TABLE 7-1

Liquid Stationary Phases for PAH Analyses

Stationary phase		Estimated[a] upper temperature limit (°C)
Chemical type	Commercial name	
Methylsilicone fluid	OV-101 or SP-2100	320
Methylsilicone gum	SE-30 or OV-1	350
Methylphenylsilicone gum	SE-52 (5% phenyl substitution)	320
	SE-54 (1% vinyl and 5% phenyl substitution)	320
Methylphenylsilicone fluid	OV-3 (10% phenyl substitution)	290
	OV-17 (50% phenyl substitution)	300
	SP-2250 (50% phenyl substitution)	300
Carborane/silicone polymer	Dexsil 300 (methyl substitution)	400
	Dexsil 400 (methyl, phenyl substitution)	400
Polyphenyl ether	Polysev	250
Polymetaphenoxylene	Poly-MPE	350
Polyphenylether sulfone	Poly-S-179	400

[a] Estimations are approximate; the actual thermal stabilities are dependent on a particular column technology, solid support modification, carrier gas purity, etc. Also, there is frequently a difference between packed and capillary columns.

resolution of the phenanthrene–anthracene, benz[a]anthracene–chrysene, and benzo[a]pyrene–benzo[e]pyrene pairs than SE-30 (methylsilicone); however, the latter phase possesses a better temperature stability. Only very minor selectivity effects are generally gained for PAH separations with more polar silicone phases. This statement is generally supported by a recent study of Lysyuk and Korol (*19*), based on the thermodynamic considerations of retention data obtained with 16 model PAC on different stationary phases.

The availability of carborane polymers as chromatographic liquid phases in the early 1970s has generated interest in high-temperature separations of hydrocarbon materials. The great temperature stability of these polymers is a result of the structural combination of the ordinary silicones with carborane "cages" incorporated into the polymer chains. Normal alkanes up to C_{55} were analyzed at temperatures as high as 350°C (*20*) on carborane columns. Subsequently, many groups applied similar phases to the analysis of PAH. Commercial products under the names "Dexsil" and "Pentasil" are based on polycarborane–siloxane (semiorganic) polymers with differently incorporated carborane "cages" in the silicone structures. In addition, functional groups within the siloxane part of the macromolecules may be varied.

Since irreversible adsorption may frequently be severe in work with PAC, deactivation of the solid support must be highly effective. The quality control of the commercially available solid supports and ready-to-pack column materials is now generally adequate, and the diatomaceous materials are usually silanized prior to coating. It should be borne in mind that the feasibility of high-temperature operation (above 300°C) is only limited by the phase stability. The surface modification groups (e.g., silanized structures) are not lost at temperatures up to 400°C (21).

Although 1% or higher phase loadings are typically used with the diatomaceous supports, the use of lightly loaded columns seems worth considering. For example, columns containing glass beads coated with thin films of silicone phases were found by Bhatia (22) to be superior to conventional packed columns. An efficient separation of a PAH standard mixture (with coronene being eluted at 290°C) on a solid support modified with a thin layer of linear polyethylene polymer was reported by Aue et al. (23). The resulting chromatogram is shown in Fig. 7-2. A similar column material was also used by Karasek et al. (24) to analyze the PAH content of airborne particulates.

A. Inorganic Salt Stationary Phases

Attempts to employ inorganic salts or eutectic mixtures as stationary phases for PAH separations originated from earlier efforts to eliminate column bleed problems. In addition, numerous studies had been previously reported on the adsorption chromatography of other compounds with salt-modified adsorbents. Solomon (25) reported the GC of quaterphenyls and hexaphenyls on supports impregnated with LiCl, CsCl, and $CaCl_2$ within the temperature range of 200–500°C. Separations of PAH under temperature programming conditions were also described by Chortyk et al. (26). No bleeding was observed with LiCl columns (melting point, 614°C), while fair separations up to coronene were achieved.

Several years later Gump (27) presented a more systematic study of PAH chromatography on inorganic salts. Chlorides of Rb, Li, Ca, Sr, Ba, Mg, and Cs were evaluated, of which LiCl and RbCl showed most promise. The effects of salt amount, type of solid support, and temperature of firing (above the respective melting points) were investigated with model PAH compounds in order to establish optimum conditions. An example of a chromatographic run with such a packing is shown in Fig. 7-3. Note the partial resolution of benzo[e]pyrene and benzo[a]pyrene. Although a sample quantity of the

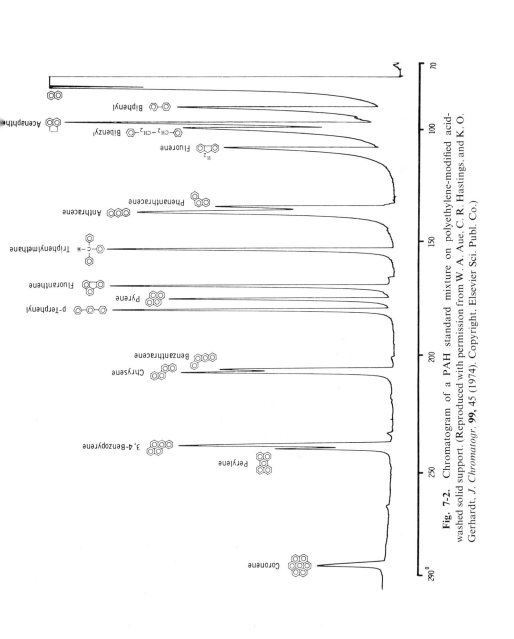

Fig. 7-2. Chromatogram of a PAH standard mixture on polyethylene-modified acid-washed solid support. (Reproduced with permission from W. A. Aue, C. R. Hastings, and K. O. Gerhardt, *J. Chromatogr.* **99**, 45 (1974). Copyright, Elsevier Sci. Publ. Co.)

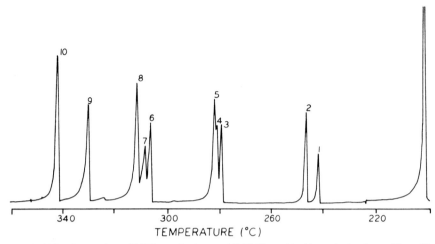

Fig. 7-3. Separation of PAH standards on a 20% lithium chloride packed column. Key: (1) pyrene; (2) fluoranthene; (3) triphenylene; (4) chrysene; (5) benz[a]anthracene; (6) benzo[e]-pyrene; (7) benzo[a]pyrene; (8) benzo[k]fluoranthene; (9) benzo[ghi]perylene; and (10) dibenz-[a,h]anthracene. (Reproduced with permission from B. H. Gump, *J. Chromatogr. Sci.* **7**, 755 (1969). Copyright, Preston Tech. Abstr. Co.)

order of micrograms was used here, peak tailing still occurs. This is, most likely, due to adsorption.

The mechanism of retention on inorganic salts was briefly investigated in Gump's work (27). For a number of PAH, weak bonding forces and certain reversals in elution were observed. The less symmetrical compounds such as fluoranthene, benz[a]anthracene, or benzo[k]fluoranthene are retained more strongly than their more symmetrical isomers (pyrene, chrysene, or perylene). Furthermore, there are some differences in retention on the various inorganic salts.

B. Graphitized Carbon Black

Although gas–solid chromatography is clearly not feasible in the case of PAH analysis with most column adsorption materials, graphitized carbon black is a unique adsorbent. Its properties have been described by Kiselev (28). The homogeneous nature of the surface, the large surface area, and the reversible solute–sorbent interactions with a great number of compounds are the fundamental reasons why the various forms of this adsorbent have been extensively studied by a number of chromatographers. The most distinct advantage of gas adsorption chromatography in PAH analysis is the remarkable column selectivity for various geometrical isomers.

Earlier attempts to use graphitized carbon black for the separation of biphenyls and terphenyls (29) and for fused-ring PAH compounds (30) met with limited success. While the resolution of isomers is frequently greater than for the conventional (nonpolar) liquid stationary phases, column efficiencies are generally too low to be useful in the analysis of complex mixtures. Similar conclusions can also be drawn from later publications by Frycka (31, 32). In addition, the required column temperatures are obviously too high (typically, much higher than 300°C).

Perhaps the best separations obtained on carbon black were demonstrated by Vidal-Madjar et al. (33). Figure 7-4 shows an impressive resolution of the phenanthrene and anthracene peaks. The column used was a 20-meter capillary coated with an adsorptive carbon-black layer by a static coating procedure. An attempt was also made to separate larger PAH. However, a temperature of 585°C was needed to elute dibenz[a,h]anthracene from an 8.5-meter column.

While it is obvious that the specific interactions with the carbon black surface are beneficial for the resolution of isomers, excessive analysis temperatures are a serious limitation. This problem might be reduced through

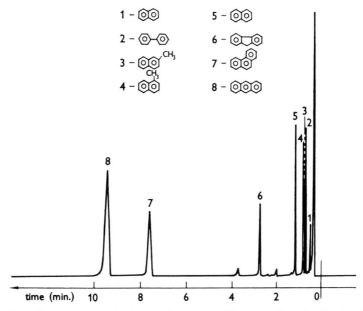

Fig. 7-4. Chromatogram of a PAH mixture on a 20-meter adsorption (carbon black) capillary column; temperature, 320°C; hydrogen inlet pressure, 1.7 atm. (Reproduced with permission from C. Vidal-Madjar, J. Ganansia, and G. Guiochon, in "Gas Chromatography— 1970," pp. 20–34. Institute of Petroleum, London, 1971.)

the use of a liquid-modified adsorbent (a method sometimes referred to as "gas–liquid–solid chromatography"). As demonstrated by Bruner *et al.* (*34*), a carbon black modified with a 1.5% coating of a polyphenyl ether phase shows decreased retention, while some unique selectivity is preserved. Micropacked columns with efficiencies up to 30,000 theoretical plates can be produced. Even more interesting possibilities of maximizing both efficiency and selectivity may be derived from the reported technology of the carbon black version of support-coated open tubular columns (*35*).

C. Liquid Crystalline Phases

Utilization of liquid crystals as stationary phases of remarkable selectivity has been of interest for some time (*36, 37*). Nematic liquid crystalline phases that exhibit an ordered molecular arrangement within a certain temperature range are particularly useful in the separation of isomeric compounds.

Certain aspects of the separation of PAH on liquid crystalline phases were studied by several groups. The primary cause of resolution appears to be the length-to-breadth ratio of different PAH molecules: the rod-like molecules are strongly retained due to the greater probability of charge-transfer interactions. These interactions are sufficiently strong to override volatility and often reverse the expected order of elution. For example, 1-methyl-naphthalene is eluted on such column substrates before 2-methylnaphthalene, a sequence opposite to that of their boiling points (*38, 39*).

The length-to-breadth differences lead also to selective separations of pairs of geometrical isomers such as phenanthrene and anthracene, benzo[*e*]-pyrene and *benzo*[*a*]pyrene, and others. The resolution of such isomers is remarkably good. An example is shown in Fig. 7-5, where a synthetic mixture of 16 PAH was chromatographed on nematic *N,N'*-bis(*p*-methoxybenzyli-dene)-α,α'-di-*p*-toluidine (*38*).

The preparation and retention properties of even more stable liquid crystalline phases were described in subsequent publications by Janini *et al.* (*40, 41*). For example, separation of the heavy C_{24} PAH at 290°C is illustrated in Fig. 7-6. Again, a remarkable phase selectivity is still evident.

The key to successful separations of PAH on liquid crystalline phases has been the development of thermostable phases by Janini and co-workers (*38, 40, 41*). The achievement of thermal stability has been most necessary for the reduction of column bleed, a problem often encountered in GC/MS, and establishment of long column lifetimes. Solute retention on the liquid crystalline phases is strongly dependent on temperature. Figure 7-7 demonstrates that the retention of both phenanthrene and anthracene decreases with increases in column temperature until the smectic-nematic phase transition is reached, whereupon retention times suddenly become longer.

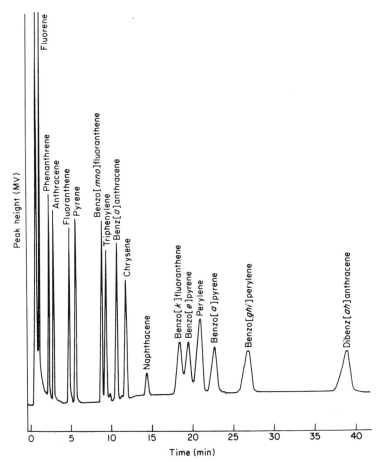

Fig. 7-5. Sixteen PAH of wide molecular weight range, separated on a liquid crystal stationary phase; temperature program 4°C/min, from 185 to 265°C. (Reproduced with permission from G. M. Janini, K. Johnston, and W. L. Zielinski, Jr., *Anal. Chem.* **47,** 670 (1975). Copyright, American Chemical Society.)

Later, retention decreases again due to the fact that compound volatility becomes more important. Resolution of geometrical isomers is also a function of the column temperature.

Retention of azaarenes on a nematic crystal column was also investigated (*42*). Here, in addition to molecular shape considerations, contributions of polarity and basicity are involved.

For the toxicologically important separations of various alkyl-substituted PAH, liquid crystalline phases may be effective. Thus, Wasik and Chesler

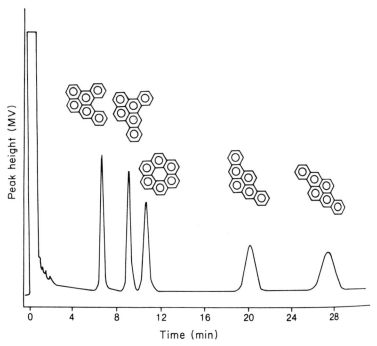

Fig. 7-6. High-temperature separation of five C_{24} PAH on a liquid crystal phase column. [Reproduced with permission from G. M. Janini, G. M. Muschik, J. A. Schroer and W. L. Zielinski, Jr., *Anal. Chem.* **48,** 1879 (1976). Copyright, American Chemical Society.]

(*39*) demonstrated that methyl- and ethyl-substituted naphthalenes, as well as all possible dimethylnaphthalene isomers, can be completely separated with the *N,N'*-bis(*p*-methoxybenzylidene)-α,α'-bi-*p*-toluidine phase. Although the order of elution is reversed when compared to the boiling-point sequence, the resolution obtained with the liquid crystal is considerably better than with other stationary phases. Interestingly, selective effects were also observed at temperature below the nematic region, but the column efficiency was lower.

Due to the hitherto unmatched selectivity, the liquid crystalline phases certainly offer some unique analytical possibilities in PAH analysis. However, it is illusory to expect that the selectivity alone can be utilized in an effective resolution of the numerous compounds encountered in complex PAH mixtures. On the other hand, the development of capillary column technology with the liquid crystal stationary phases would be eminently worthwhile. In general, the insufficient long-term stability of these phases is a serious drawback in this approach.

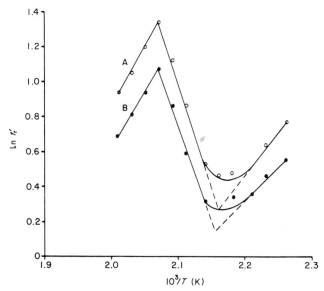

Fig. 7-7. Retention behavior of anthracene (A) and phenanthrene (B) on the N,N'-bis-(p-butoxybenzylidene)-α,α'-bi-p-toluidine stationary phase in the temperature range 170°–225°C, expressed as logarithm of the corrected retention time (related to benzene) vs. reciprocal absolute temperature. [Reproduced with permission from G. M. Janini, G. M. Muschik, and W. L. Zielinski, Jr., *Anal. Chem.* **48**, 809 (1976). Copyright, American Chemical Society.]

III. CAPILLARY GAS CHROMATOGRAPHY

Due to the extraordinarily great complexity of environmental PAC mixtures, capillary GC becomes a mandatory approach whenever detailed analyses are needed. Recent work demonstrates that capillary GC can be used today as a routine analytical tool.

Several advantages can be derived from using capillary GC methods. Their importance in the resolution of isomers has already been pointed out. As it will be shown, even such minor structural differences as the position of alkyl groups can be readily distinguished by efficient capillary columns. Another important advantage lies in the combination of capillary GC with mass-spectroscopic techniques; unless the individual mixture components are well resolved, difficulties may be experienced in using a mass spectrometer for structural elucidation. Although many isomers provide nearly identical mass spectra, the presence of certain compounds can be positively established from additional information on their chromatographic mobility. Thus, the most reliable structural assignments will be dependent on the column's resolving power and the precision of retention data.

Although capillary columns were previously regarded as unsuitable for trace analysis, this consideration is no longer valid today. Above all, sample splitting is no longer a necessary condition for proper capillary column use. Direct sampling procedures are now readily available, utilizing effectively large sample aliquots. It has been shown in many instances that capillary columns are associated with an increased signal-to-noise ratio as far as detection is concerned. Trace amounts are now generally detected with comparable precision to that of packed-column GC. Improvements in the overall analytical performance are also due to the greater column inertness available with today's capillary columns.

Due to different flow conditions in capillary GC, it should not be forgotten that various detection and ancillary techniques must be adapted to meet such changes. Only then can the maximum performance be expected. On the other hand, the use of capillary columns can frequently result in better operation of the ancillary equipment. For example, a reduced contamination problem is encountered in capillary GC/MS due to lower column flow rates and, subsequently, less bleeding of the stationary phase into the mass spectrometer.

Finally, lower temperatures are necessary in capillary GC for the elution of chromatographed samples. This temperature reduction can be quite substantial, leading to fewer decomposition problems and shorter analysis times. Most importantly, the range of analyzed compounds can be significantly extended by using thin-film, inert capillary columns. Whereas elution of coronene with packed columns may need temperatures over 300°C, a short glass capillary column can perform the same function at 240°C (8).

Although capillary GC has been, in the past, a privilege of only a few laboratories, reliable instrumentation is now readily available. Many examples of PAC separations with capillary columns have now appeared throughout the literature. In fact, there are reasons to believe that GC of PAC mixtures with packed columns may soon be of historical interest only.

A. Glass Capillary Columns

Although the first glass capillary columns were prepared for the analysis of light petroleum samples as early as 1960 (43, 44), many years of research were necessary to arrive at the glass columns of high quality that are used in today's analytical work. Various aspects of modern capillary GC have recently been reviewed (45–48).

As a result of these developments, glass capillary columns of high efficiency and surface inertness are now readily available for the separation of PAC compounds. There is every indication that the glass capillary column is rapidly becoming an indispensable tool in this area.

Although a few applications to PAH of stainless steel capillary columns have been reported (*2, 3, 49–51*), glass has the distinct advantage of inertness. Since the technology of glass capillary columns has been considerably advanced by better understanding of the surface chemistry of glass, such columns are now available with considerably higher plate numbers than those prepared from other tubing materials. Whereas thin-film columns are being frequently sought for the separation of heavier PAC compounds, the film thickness can be varied during column preparation as needed for different applications.

Glass capillary columns for most PAC analytical work need not be longer than 10–25 m, with the internal diameters around 0.2–0.3 mm. Figure 7-8 shows a plot (*52*) of column resolution (as measured by the separation number) versus column length for a series of glass capillaries with 0.30-mm *i.d.* and 0.25-μm film thickness of SE-52. When the separation number data are plotted against the square root of length, a nearly linear plot is obtained. This is in accordance with chromatographic theory; column resolution is proportional to the square root of length. It is informative to note, however, that the loss in resolution in going from a 30-m- to a 15-m-long column is minimal. Relatively short lengths, on the order of 15 m, are a good compromise between overall efficiency and the elution of heavier PAC. A high number of theoretical plates per unit column length is the fundamental key to a successful analysis. Typical columns should have around 3,000 theoretical plates per meter.

An example of a capillary separation on a 12-m capillary column is shown in Fig. 7-9. As pointed out earlier, a good qualitative evaluation of the resolving power of a chromatographic column is the observation of the

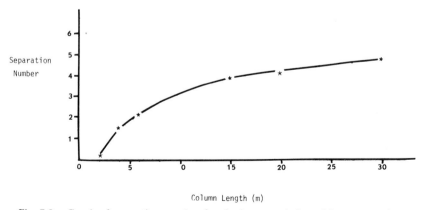

Column Length (m)

Fig. 7-8. Graph of separation number for the isomer pair benzo[*e*]pyrene–perylene vs. column length.

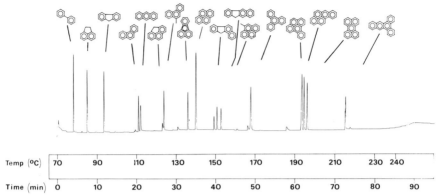

Temp (°C)	70	90	110	130	150	170	190	210	230 240

| Time (min) | 0 | 10 | 20 | 30 | 40 | 50 | 60 | 70 | 80 | 90 |

Fig. 7-9. Gas chromatogram of standard PAH on a 12 m × 0.28 mm i.e. glass capillary coated with SE-52 elastomer (a dry sampling technique was used). [Reproduced with permission from M. L. Lee, D. L. Vassilaros, L. V. Phillips, D. M. Hercules, H. Azumaya, J. W. Jorgenson, M. P. Maskarinec, and M. Novotny, *Anal. Lett.* **12**, 191 (1979). Copyright, Marcel Dekker, Inc.]

resolution of the pairs phenanthrene–anthracene, benz[a]anthracene–chrysene, and benzo[a]pyrene–benzo[e]pyrene. A recent study (52) illustrates the separations obtained of coal-tar PAH on 6-m, 4-m, and 2-m capillary columns. Inspection of the chromatograms shows remarkable retainment of resolution and significantly shorter analysis times.

Elution of heavier aromatics can also best be accomplished with shorter columns. A chromatogram (Fig. 7-10) of the extract of airborne particulates displays PAH from tricyclic compounds up to at least coronene (seven aromatic rings) at relatively low temperatures. This separation was obtained with a column only 11 m long (8), but over 120 extract components were identified by means of combined GC/MS. However, it should be pointed out that shorter capillary columns have somewhat reduced sample capacities.

The separations illustrated in Figs. 7-9 and 7-10 were performed with glass capillary columns coated with a silicone elastomer SE-52. Although other thermostable phases could be used, incorporation of some aromatic moieties into the polymeric phase (as present in SE-52, SE-54, and OV-3 phases) appears beneficial. It has been our experience that such columns can provide reliable service for many months, provided that adequate sample clean-up is done and excessively high column temperatures are used infrequently.

Recent investigations (53) indicate that thermal stability of a column is also very dependent on the surface characteristics of glass. Depolymerization of common stationary phases due to the presence of various glass ingredients is strongly suspected to be the primary cause of column bleed at high tem-

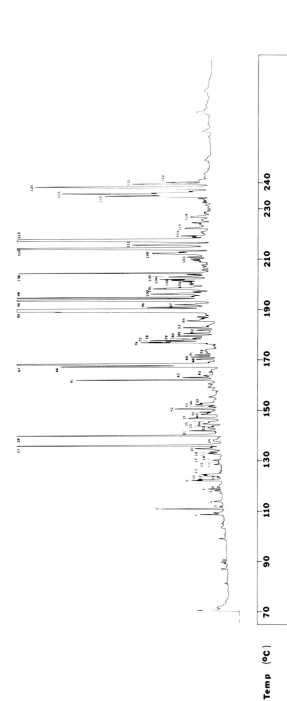

Fig. 7-10. Capillary gas chromatography of the total PAH fraction of air-particulate matter. Conditions: 11 m × 0.26 mm i.d. glass capillary column coated with SE-52. [Reproduced with permission from M. L. Lee, M. Novotny, and K. D. Bartle, *Anal. Chem.* **48**, 1566 (1976). Copyright, American Chemical Society.]

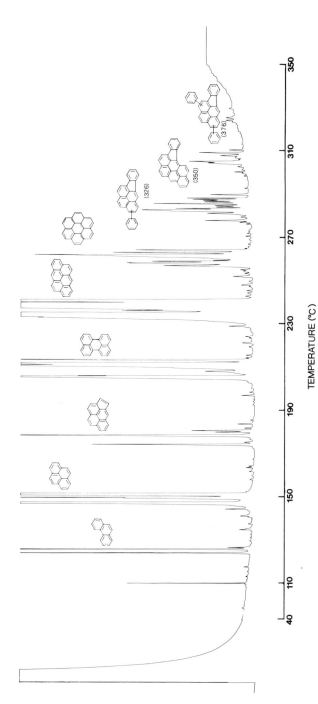

TEMPERATURE (°C)

Fig. 7-11. High-temperature capillary column gas chromatogram of the PAH extract from carbon black. Chromatographic conditions: 15 m × 0.27 mm i.d. glass column coated with SE-52, temperature programmed from 40° to 110°C at 10 C/min and then from 110° to 350°C at 2°C/min.

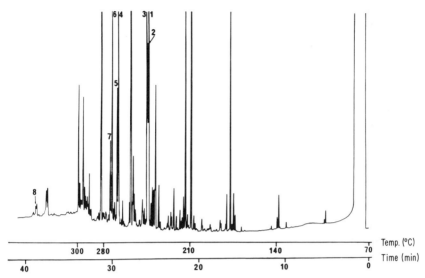

Fig. 7-12. Chromatogram of a PAH sample derived from cyclone dust collected at an aluminum plant. Conditions: 20 m × 0.22 mm i.d. Pyrex glass capillary column coated with a bonded methylpolysiloxane phase; carrier gas, hydrogen. Keys: (1) benz[*a*]anthracene; (2) chrysene; (3) triphenylene; (4) benzo[*b*]fluoranthene; (5) benzo[*k*]fluoranthene; (6) benzo[*e*]-pyrene; (7) benzo[*a*]pyrene; and (8) coronene. [Reproduced with permission from L. Blomberg and T. Wannman, *J. Chromatogr.* **186**, 159 (1980). Copyright, Elsevier Sci. Publ. Co.

peratures. As shown in Fig. 7-11 (*54*), an adequately pretreated glass surface extends the temperature range of a stationary phase that could previously be used to only 260°C. The chromatogram was obtained with an extract of carbon black which is known to contain numerous higher molecular weight PAC.

Although chemically bonded stationary phases have been primarily a domain of liquid chromatography, recent work (*55, 56*) shows their potential in capillary GC work with PAC. Nonpolar silicone polymers are coated and attached to a suitably pretreated glass capillary wall by heating which induces cross-linking. A polymeric film of this kind is nonextractable with conventional organic solvents. It can also withstand very high temperatures, as demonstrated in Fig. 7-12. A distinct advantage of this column technology (compared to coated films) is that large solvent injections cause little harm to these columns.

B. Injection Techniques

Sampling aspects acquire special meaning in capillary GC. Firstly, the amount of sample required for effective capillary GC work is considerably smaller than in conventional GC. The introduction of such a sample at the capillary column inlet has been a major technical problem. Secondly, difficulties may rise from sample transfer from a preconcentration medium into the column itself. And, thirdly, sample clean-up is more critical in capillary GC, since experience shows that deposition of nonvolatile material of crude samples may reduce the effective column lifetime.

The lack of sampling techniques had for years excluded capillary GC from the area of trace analysis. In order to inject small samples onto a capillary column rapidly enough, splitting has conventionally been used. In such a sampling procedure, relatively concentrated sample solutions are injected into a hot injector zone, the sample is completely vaporized, "homogenized," and divided unequally between the capillary inlet and a vent. Thus, only a small fraction (typically, less than 1%) of the sample is utilized for analysis. Obviously, such an approach is unacceptable for trace analysis as most of the sample is wasted. Unless concentrated PAC solutions are available for analysis, this procedure is not recommended. Splitting techniques may also result in less quantitative results due to the mass discrimination inevitable in most splitter designs.

The development of splitless sample injection techniques has significantly extended the scope of capillary GC applications. As shown by Grob and Grob (57, 58), chromatographic resolution is hardly affected if a dilute sample is injected into the column kept at low temperature. While the more volatile excess solvent passes through it, trace (less volatile) solutes are desorbed from the hot injector into the first section of the column and effectively concentrated in a narrow band. The subsequent warm-up of the column will initiate migration of the sample molecules. Depending on the temperature of operation, the solvent band can have a desirable concentration effect. Since it also partially condenses inside the capillary column, it will efficiently concentrate trace organics on the solvent tail part (58). Thus, the sample band is effectively "sandwiched" between the hot injector and the solvent peak zones. Up to several microliters of dilute samples can be injected without problems using this procedure even if they are in the vicinity of a large solvent peak (59).

Another very attractive sampling alternative, described more recently (60), is the direct introduction of small liquid samples directly (without prior vaporization) onto a capillary column.

Sample preconcentration is distinctly advantageous in capillary GC. The general approaches to isolation of PAC fractions from complex matrices have already been described in Chapter 4. It should be added that in the trace analysis of such compounds by capillary GC, it is desirable to remove as many as possible of the heavy mixture components which would otherwise gradually deposit in the first section of the capillary column and cause undesirable interactions and peak spreading with new samples. Such problems are frequently observed, and the column efficiency can often be regained after removing the contaminated section of the column. A better alternative is to prevent the occurrence of this problem by more effective sample clean-up or retaining the nonvolatile sample portion by a sampling precolumn method (61).

Many researchers have attempted the direct preconcentration of PAH and related compounds from both air and water pollution samples. It would be quite desirable to have a selective sampling tube for PAH with subsequent quantitative thermal or solvent sample stripping for direct capillary GC or LC analysis. As already discussed in Chapter 4, different approaches based on bonded stationary phases, polyurethane foams, synthetic polymers such as Tenax GC, or XAD resins have been tried with varying degrees of success.

Whether the final PAC solution is obtained through a complex partitioning process, chromatographic clean-up procedure, or solvent stripping from an adsorbent concentration system, a common problem exists: while the typical volume of such a sample is 50 μl or more, no more than a few microliters can be injected onto a capillary column. Thus, only a small sample aliquot is utilized even in the cases where detection sensitivity is a problem. As shown by Novotny et al. (61, 62), a precolumn procedure can overcome this limitation. A small amount of a deactivated solid support (no more than a few milligrams) is packed into the injector port glass liner. Large injections of a dilute sample can be introduced on the precolumn material, and the volatile solvent removed. Following the introduction of the sample, the precolumn is inserted back into the hot injector and the sample is trapped at the column inlet which is held at a sufficiently low temperature. After the sample is quantitatively trapped, the column temperature is elevated to perform the usual analysis. When using volatile solvents, this technique is quantitative for PAH with three rings and above.

A major disadvantage of capillary GC is the sample capacity. Maximum permissible amounts per components are of the order of 10^{-7} g with thin-film glass capillary columns. When the total sample size (of all mixture components) becomes too high, additional sample fractionation is worth considering, such as a more extensive utilization of LC methods (6), or switching an effluent from one GC column to another (63).

C. Instrumental Aspects of Capillary GC

Although capillary GC is a method of extraordinarily high resolving power, its advantages can only be utilized with high-quality equipment and sufficient technical expertise. In particular, the design of the gas chromato-graph is more critical for capillary GC than it is for conventional packed-column work. Most importantly, special injection techniques are needed. The main additional requirement concerns minimum dead volumes in the injector and detector parts. It should be pointed out that the present-day gas chromatographs are relatively well-equipped for capillary GC work. Also, many older instruments can be modified to function adequately.

Although the fragility of glass is a disadvantage, the necessary skills to handle, install, and use such columns can be learned in a short time. In order to take advantage of the inertness of glass capillary columns, the inlet system and transfer lines to detectors, connection points, etc. must also be designed to minimize sample decomposition. The flexibility of the recently developed fused silica columns (*64*) is likely to improve such conditions.

The effect of carrier-gas velocity on the chromatography of PAH has recently been studied (*54*). It was found that the optimum velocity of helium is close to 50 cm/sec. The retention times of PAH are affected considerably by the flow rate; compounds are eluted at lower temperatures under tem-perature programming conditions when using greater velocities.

Schomburg, *et al.* (*53*) recently described the influence of nitrogen and hydrogen as carrier gases on the elution of coal tar PAH. The retention times, and hence elution temperature, were considerably lower using hydrogen.

Although flow programming has been described (*65*) in capillary GC, temperature programming is most frequently used. With the exception of wide-bore (0.7 mm i.d.) capillary columns, the inlet pressure rather than the column flow is regulated. With the typical capillary columns used in analyti-cal work (0.2–0.3 mm i.d.), the helium flow is around 1 cm^3/min. However, flow conditions must often be adjusted in relation to the optimum perform-ance of certain detectors and ancillary tools.

D. PAC Structure versus Retention

The significance of retention values was recognized early in the years following the inception of gas chromatography. Among the early attempts of many workers to standardize retention measurements and qualitatively utilize log–log plots of the solute retention on two stationary phases of different polarity, the general retention index system, as introduced by Kováts (*66*), has met with widest acceptance. In the Kováts retention index system, normal alkanes are used as reference substances. The retention of a compound can be studied under either isothermal or programmed-tem-perature conditions.

Since retention data were shown to be crucially important in the assign-ments of structure, numerous studies have been devoted to the subject. A review of these earlier attempts is given by Schomburg (*67*). The availability of combined gas chromatography/mass spectrometry since the middle 1960s has somewhat decreased the popularity of this approach. However, as researchers increasingly learn that there are many cases where a mass spectrometer falls short in its identification power, retention data will become more significant.

It is the current view that *combined data* obtained from both GC retention and mass spectral information are of the utmost importance. With small sample amounts, these may frequently be the only means of identification. This consideration is particularly valid for isomeric compounds where mass spectra may be nearly identical, yet differences in the chromatographic mobility are distinct. Isomers encountered in PAC analyses fit this category extremely well.

The identification power of retention data is strongly dependent on two factors: (a) definite differences in peak position for the isomers in question (preferably, on more than one stationary phase), and (b) reproducibility of retention measurements. The first factor is directly related to the ability of a chromatographic column to distinguish such isomers, i.e., its resolving power. Thus, retention measurements acquire a special meaning in capillary GC. As will be demonstrated below, glass capillary columns are capable of resolving a number of close isomers that yield similar mass spectra.

Interlaboratory comparison of retention data has been of some concern for a number of years. The main factors affecting reproducibility of retention are temperature and flow control, chemical nature and stability of the liquid stationary phase, and the surface chemistry of a given solid support or capillary wall.

It was unfortunate that earlier attempts to utilize retention data for qualitative purposes were made when both reliable instrumentation and column technology had not been sufficiently developed. However, more recent advances in both directions now make high-precision measurements feasible. In particular, attempts to arrive at standard stationary phases and well-defined surfaces (*68*) were crucial to this development. It should be pointed out that different surfaces coated with an identical liquid phase can have vastly different retention properties (*69*).

Due to the currently available resolving power of glass capillary columns, high-precision chromatographic data at the level of reproducibility of one retention index unit or less should be routinely sought. As discussed below, there are indications that this is now feasible.

Although a combination of mass spectra and peak retention values in capillary GC can lead in most cases to the positive identification of a PAC compound, this approach has been thus far utilized only to a limited degree. This is undoubtedly due to the unavailability of various standard PAC isomers. Whereas a match with a reference compound is ultimately needed, there is some hope that systematic retention studies may eventually lead to establishing some general rules for the prediction of retention data.

When comparing retention of various PAH compounds, two major isomer groups are of interest: parent PAH with different ring fusion, and alkylated PAH with different chain length and position of alkyl groups. With the exception of the liquid crystalline phases discussed above, selective retentions within the PAH series are minimal and the elution generally follows boiling-point considerations. Although there might be minor changes in the column selectivity (often not negligible from the analysis point of view), the order of elution is generally preserved with stationary phases of different polarity.

As shown in Fig. 7-13, the retention of a PAH increases nearly linearly

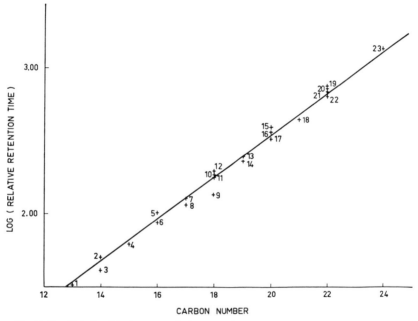

Fig. 7-13. Relationship between relative retention time and carbon number for selected PAH on SE-30 methylsilicone stationary phase. Key: (1) fluorene; (2) phenanthrene; (3) 2-methylfluorene; (4) 3-methylphenanthrene; (5) pyrene; (6) fluoranthene; (7) 16,17-dihydro-15H-cyclopenta[a]phenanthrene; (8) benzo[a]fluorene; (9) 17-methyl-16,17-dihydro-15H-cyclopenta[a]phenanthrene; (10) chrysene; (11) benzo[ghi]fluoranthene; (12) naphthacene; (13) 2-methylchrysene; (14) 1-methylchrysene; (15) perylene; (16) benzo[e]pyrene; (17) benzo[b]fluoranthene; (18) 3-methylcholanthrene; (19) indeno[1,2,3-cd]pyrene; (20) picene; (21) benzo[ghi]perylene; (22) anthanthrene; and (23) coronene. [Reproduced with permission from J. R. Wilmshurst, *J. Chromatogr.* **17**, 50 (1965). Copyright, Elsevier Sci. Publ. Co.]

with molecular weight. Most plotted values (49) show only minor deviation from the expected straight line. However, it should be pointed out that aliphatic carbons in the PAH molecules contribute less to the overall retention than the aromatic carbons. Thus, the alkylated PAH show some departure from the linear plot.

Differentiation of isomers that have different toxicological properties is of the utmost importance and capillary GC appears to be an ideal method. Among the parent PAH molecules, most compounds with different molecular shapes can be distinguished. Although differences are frequently minor, somewhat longer retention tends to be associated with more "extended" molecules of the same ring number and molecular weight. Thus, phenanthrene is eluted slightly before anthracene, benzo[j]fluoranthene before benzo[k]fluoranthene, and three isomers with molecular weight 252 are eluted in the following order: benzo[e]pyrene < benzo[a]pyrene < perylene. Hydro derivatives of PAH are generally eluted before the parent compounds.

Until recently, there has been a general lack of chromatographic data on isomeric PAH compounds due to the unavailability of reference compounds. In view of the importance of distinguishing isomers and the potential of capillary GC in this direction, the following discussion involves glass capillary columns only. Since the resolution of packed columns is inadequate, the earlier studies on the subject are of little use (e.g., relative retention data of 60 alkyl phenanthrenes (70) gathered on different stationary phases).

Since alkylated PAH isomers include derivatives with different alkyl carbon chain lengths, positions of alkyl groups within a molecule, multiple substitution, etc., chromatographic resolution is very important. Currently available spectroscopic techniques do not readily identify such molecular differences. The only exception is NMR (see Chapter 10), but there are sample-size limitations.

Surprisingly, a minor change in a PAH molecule such as a difference in position of an alkyl group frequently gives rise to distinguishable chromatographic behavior in capillary GC. This was demonstrated by Wilmhurst (49) and Lee et al. (6, 7) for methylated fluorenes, phenanthrenes, pyrenes, and chrysenes. Retention data within the series of alkylated benzenes are also available (71, 72). A significant extension of data on isomeric PAH has been the subject of a recent study by Lee et al. (73). Retention data have been gathered and standardized for over 200 PAC on glass capillary columns (see Table 7-2).

In the retention studies of 209 PAH and related compounds (73), changes of flow rate, different temperature programming rates, the stationary-phase film thickness, column length, etc., were all found to have adverse effects on retention reproducibility when n-alkanes were used as reference compounds. However, on choosing appropriate PAH as reference compounds such

TABLE 7-2

PAH Retention Indices

Compound no.	Compound name	Average index	Standard deviation
1.	1,2-Dihydronaphthalene	197.01	0.07[c]
2.	1,4-Dihydronaphthalene	197.01	0.07[c]
3.	Tetralin	197.04	0.05[d]
4.	Naphthalene	200.00	
5.	Benzo[b]thiophene	201.47	0.05[f]
6.	Indoline	204.74	0.04[c]
7.	Indole	205.26	0.08[d]
8.	Quinoline	209.70	—[b]
9.	Isoquinoline	215.61	1.01[d]
10.	2-Methylnaphthalene	218.14	0.28[h]
11.	2-Methylbenzo[b]thiophene	218.74	0.04[c]
12.	Azulene	219.95	0.24[c]
13.	Quinoxaline	220.37	0.05[d]
14.	3-Methylbenzo[b]thiophene	221.02	0.03[c]
15.	1-Methylnaphthalene	221.04	0.25[e]
16.	8-Methylquinoline	223.02	0.06[e]
17.	1,2,3,4-Tetrahydroquinoline	225.97	0.10[c]
18.	6-Methylquinoline	229.82	0.03[c]
19.	1,2,2a,3,4,5-Hexahydroacenaphthylene	232.70	—[b]
20.	Biphenyl	233.96	0.24[d]
21.	2-Ethylnaphthalene	236.08	0.16[c]
22.	1-Ethylnaphthalene	236.56	0.14[e]
23.	3-Methylindole	236.66	—[b]
24.	2-Methylindole	237.42	0.21[f]
25.	2,6-Dimethylnaphthalene	237.58	0.17[f]
26.	2,7-Dimethylnaphthalene	237.71	0.07[d]
27.	5-Ethylbenzo[b]thiophene	238.46	0.37[d]
28.	2-Methylbiphenyl	238.77	0.04[c]
29.	1,3-Dimethylnaphthalene	240.25	0.16[e]
30.	1,4-Naphthoquinone	240.82	0.04[c]
31.	1,7-Dimethylnaphthalene	240.66	0.25[c]
32.	1,6-Dimethylnaphthalene	240.72	0.09[d]
33.	2,2′-Dimethylbiphenyl	241.94	—[b]
34.	2,6-Dimethylquinoline	242.43	0.29[e]
35.	2,3-Dimethylnaphthalene	243.55	0.19[c]
36.	1,4-Dimethylnaphthalene	243.57	0.16[f]
37.	1,5-Dimethylnaphthalene	244.98	0.16[c]
38.	Diphenylmethane	243.35	0.11[f]
39.	Acenaphthylene	244.63	0.19[g]
40.	2,2′-Bipyridyl	245.48	0.27[c]
41.	1,2-Dimethylnaphthalene	246.49	0.30[e]
42.	1,8-Dimethylnaphthalene	249.52	—[b]
43.	2-Ethylbiphenyl	250.85	—[b]

(Continued)

TABLE 7-2 (*Continued*)

Compound no.	Compound name	Average index	Standard deviation
44.	Acenaphthene	251.29	0.14[g]
45.	4-Methylbiphenyl	254.71	0.17[d]
46.	3-Methylbiphenyl	254.81	0.15[f]
47.	2,3-Dimethylindole	255.48	—[b]
48.	Dibenzofuran	257.17	0.05[c]
49.	2-Methyl-1,4-naphthoquinone	259.23	0.11[c]
50.	2,3,6-Trimethylnaphthalene	263.31	0.12[c]
51.	1-Methylacenaphthylene	265.24	0.02[d]
52.	2,3,5-Trimethylnaphthalene	265.90	0.14[e]
53.	Dibenzo-*p*-dioxin	267.27	0.20[h]
54.	Fluorene	268.17	0.15[g]
55.	*trans*-1,2,3,4,4a,9a-Hexahydrodibenzothiophene	269.67	0.37[f]
56.	*cis*-1,2,3,4,4a,9a-Hexahydrodibenzothiophene	271.39	0.27[f]
57.	3,3′-Dimethylbiphenyl	271.87	—[b]
58.	9-Methylfluorene	272.38	0.17[c]
59.	2,3,5-Trimethylindole	272.57	—[b]
60.	4,4′-Dimethylbiphenyl	274.59	—[b]
61.	5*H*-Indeno[1,2-*b*]pyridine	279.31	0.19[c]
62.	Xanthene	280.48	0.22[c]
63.	9,10-Dihydroanthracene	284.89	0.19[e]
64.	9-Ethylfluorene	284.99	0.20[c]
65.	9,10-Dihydrophenanthrene	287.09	0.16[c]
66.	1,2,3,4,5,6,7,8-Octahydroanthracene	287.69	0.20[c]
67.	2-Methylfluorene	288.21	0.15[d]
68.	1-Methylfluorene	289.03	0.04[d]
69.	1,2,3,4,5,6,7,8-Octahydrophenanthrene	292.03	0.06[c]
70.	1,2,3,4-Tetrahydrodibenzothiophene	294.30	0.12[c]
71.	9-Fluorenone	294.79	0.26[d]
72.	Dibenzothiophene	295.81	0.03[g]
73.	1,2,3,4-Tetrahydrophenanthrene	297.21	—[b]
74.	Phenanthrene	300.00	
75.	Anthracene	301.69	0.08[g]
76.	Benzo[*h*]quinoline	302.22	0.11[d]
77.	9,10-Dihydroacridine	304.33	—[b]
78.	Acridine	304.50	0.05[c]
79.	1,2,3,4-Tetrahydrocarbazole	306.76	0.19[d]
80.	Phenanthridine	308.79	0.28[d]
81.	Benzo[*f*]quinoline	309.25	0.19[d]
82.	Carbazole	312.13	—[b]
83.	9-Ethylcarbazole	313.97	0.13[d]
84.	1-Phenylnaphthalene	315.19	0.05[c]
85.	1,2,3,10b-Tetrahydrofluoranthene	316.37	0.14[d]
86.	9-*n*-Propylfluorene	318.01	0.20[c]

(*Continued*)

TABLE 7-2 (*Continued*)

Compound no.	Compound name	Average index	Standard deviation
87.	3-Methylphenanthrene	319.46	0.12^e
88.	2-Methylphenanthrene	320.17	0.12^d
89.	3-Methylbenzo[f]quinoline	320.77	0.24^d
90.	2-Methylanthracene	321.57	0.12^c
91.	o-Terphenyl	321.99	0.16^d
92.	4H-Cyclopenta[def]phenanthrene	322.08	0.15^d
93.	9-Methylphenanthrene	323.06	0.24^d
94.	4-Methylphenanthrene	323.17	—^{b}
95.	1-Methylanthracene	323.33	—^{b}
96.	1-Methylphenanthrene	323.90	0.08^c
97.	2-Methylacridine	324.46	0.05^c
98.	9-n-Butylfluorene	328.99	0.37^d
99.	9-Methylanthracene	329.13	0.19^d
100.	4,5,9,10-Tetrahydropyrene	329.69	0.24^c
101.	4,5-Dihydropyrene	330.01	0.02^d
102.	Thianthrene	330.13	0.26^e
103.	Anthrone	330.53	0.67^c
104.	2-Phenylnaphthalene	332.59	0.14^e
105.	9-Ethylphenanthrene	337.05	0.08^d
106.	2-Ethylphenanthrene	337.50	—^{b}
107.	3,6-Dimethylphenanthrene	337.83	0.14^d
108.	2,7-Dimethylphenanthrene	339.23	0.16^d
109.	1,2,3,6,7,8-Hexahydropyrene	339.38	0.17^c
110.	6-Phenylquinoline	342.45	0.40^c
111.	Fluoranthene	344.01	0.16^g
112.	9-Isopropylphenanthrene	345.78	0.28^c
113.	1,8-Dimethylphenanthrene	346.26	0.27^c
114.	2-Phenylindole	347.47	0.07^c
115.	Indeno[1,2,3-ij]isoquinoline	347.57	0.19^c
116.	9-n-Hexylfluorene	348.54	0.10^c
117.	9-n-Propylphenanthrene	350.30	0.17^c
118.	Pyrene	351.22	0.08^d
119.	9,10-Dimethylanthracene	355.49	0.03^c
120.	Benzo[lmn]phenanthridine	358.53	0.28^c
121.	9-Methyl-10-ethylphenanthrene	359.91	0.19^c
122.	m-Terphenyl	360.73	0.03^c
123.	Benzo[kl]xanthene	361.38	0.15^c
124.	4H-Benzo[def]carbazole	364.22	0.12^d
125.	p-Terphenyl	366.10	0.16^g
126.	Benzo[a]fluorene	366.74	0.13^g
127.	11-Methylbenzo[a]fluorene	367.04	0.14^d
128.	9,10-Diethylphenanthrene	367.97	0.15^d
129.	1-Methyl-7-isopropylphenanthrene	368.67	0.19^c
130.	Benzo[b]fluorene	369.39	0.15^c

(*Continued*)

TABLE 7-2 (*Continued*)

Compound no.	Compound name	Average index	Standard deviation
131.	4-Methylpyrene	369.54	0.13[c]
132.	2-Methylpyrene	370.15	0.44[d]
133.	4,5,6-Trihydrobenz[de]anthracene	370.86	—[b]
134.	1-Methylpyrene	373.55	0.11[c]
135.	3,5-Diphenylpyridine	373.79	0.18[e]
136.	5,12-Dihydronaphthacene	381.56	—[b]
137.	9,10-Dimethyl-3-ethylphenanthrene	381.85	0.18[e]
138.	9-Phenylcarbazole	382.09	—[b]
139.	1-Ethylpyrene	385.35	—[b]
140.	2,7-Dimethylpyrene	386.34	0.06[d]
141.	1,2,3,4,5,6,7,8,9,10,11,12-Dodecahydrotriphenylene	386.36	0.29[e]
142.	11-Benzo[a]fluorenone	386.41	0.06[d]
143.	1,1'-Binaphthyl	388.38	0.23[d]
144.	Benzo[b]naphtho[2,1-d]thiophene	389.26	0.16[e]
145.	Benzo[ghi]fluoranthene	389.60	0.06[d]
146.	Benzo[c]phenanthrene	391.39	0.40[c]
147.	Benz[c]acridine	392.50	0.13[e]
148.	9-Phenylanthracene	396.38	—[b]
149.	Cyclopenta[cd]pyrene	396.54	0.08[d]
150.	Benz[a]anthracene	398.50	0.08[c]
151.	Benz[a]acridine	398.74	—[b]
152.	Chrysene	400.00	
153.	Triphenylene	400.00	0.01[c]
154.	Benzo[a]carbazole	401.81	—[b]
155.	1,2'-Binaphthyl	405.35	0.50[c]
156.	7-Benz[de]anthrone	406.54	0.34[d]
157.	9-Phenylphenanthrene	406.90	0.14[c]
158.	Naphthacene	408.30	0.22[d]
159.	Benzo[b]carbazole	410.12	—[b]
160.	11-Methylbenz[a]anthracene	412.72	—[b]
161.	2-Methylbenz[a]anthracene	413.78	0.44[c]
162.	1-Methylbenz[a]anthracene	414.37	0.17[c]
163.	1-n-Butylpyrene	414.87	0.12[d]
164.	1-Methyltriphenylene	416.32	—[b]
165.	9-Methylbenz[a]anthracene	416.50	0.20[d]
166.	3-Methylbenz[a]anthracene	416.63	—[b]
167.	9-Methyl-10-phenylphenanthrene	417.16	—[b]
168.	8-Methylbenz[a]anthracene	417.56	0.04[c]
169.	6-Methylbenz[a]anthracene	417.57	0.30[c]
170.	3-Methylchrysene	418.10	0.17[d]
171.	5-Methylbenz[a]anthracene	418.72	—[b]
172.	2-Methylchrysene	418.80	—[b]
173.	12-Methylbenz[a]anthracene	419.39	0.45[d]
174.	4-Methylbenz[a]anthracene	419.67	—[b]

(*Continued*)

TABLE 7-2 (*Continued*)

Compound no.	Compound name	Average index	Standard deviation
175.	5-Methylchrysene	419.68	—[b]
176.	6-Methylchrysene	420.61	0.04[c]
177.	4-Methylchrysene	420.83	0.16[c]
178.	2,2′-Biquinoline	421.12	—[b]
179.	1-Phenylphenanthrene	421.66	0.31[c]
180.	1-Methylchrysene	422.87	0.06[c]
181.	7-Methylbenz[a]anthracene	423.14	—[b]
182.	o-Quaterphenyl	423.63	—[a]
183.	2,2′-Binaphthyl	423.91	—[b]
184.	2,(2′-Naphthyl)-benzo[b]thiophene	428.11	0.37[d]
185.	1,3-Dimethyltriphenylene	432.32	—[b]
186.	1,12-Dimethylbenz[a]anthracene	436.82	—[b]
187.	Benzo[j]fluoranthene	440.92	—[b]
188.	Benzo[b]fluoranthene	441.74	0.48[c]
189.	Benzo[k]fluoranthene	442.56	—[b]
190.	7,12-Dimethylbenz[a]anthracene	443.38	0.12[c]
191.	1,6,11-Trimethyltriphenylene	446.24	—[b]
192.	Dinaphtho[1,2-b; 1′,2′-d]furan	450.20	—[b]
193.	Benzo[e]pyrene	450.73	0.17[c]
194.	Dibenzo[c,kl]xanthene	451.57	—[b]
195.	Benzo[a]pyrene	453.44	—[b]
196.	Perylene	456.22	0.29[c]
197.	1,3,6,11-Tetramethyltriphenylene	461.72	—[b]
198.	3-Methylcholanthrene	468.44	—[b]
199.	m-Quaterphenyl	472.81	0.23[c]
200.	Indeno[1,2,3-cd]pyrene	481.87	0.09[c]
201.	Pentacene	486.81	—[b]
202.	p-Quaterphenyl	488.18	0.38[d]
203.	Dibenz[a,c]anthracene	495.01	0.08[c]
204.	Dibenz[a,h]anthracene	495.45	—[a]
205.	Benzol[b]chrysene	497.66	—[a]
206.	Picene	500.00	
207.	Benzo[ghi]perylene	501.32	0.18[e]
208.	Dibenzo[def,mno]chrysene	503.89	—[b]
209.	2,3-Dihydrodibenzo[def,mno]chrysene	503.91	—[b]

[a] One determination.
[b] Two determinations.
[c] Three determinations.
[d] Four determinations.
[e] Five determinations.
[f] Six determinations.
[g] Seven determinations.
[h] Eight determinations.

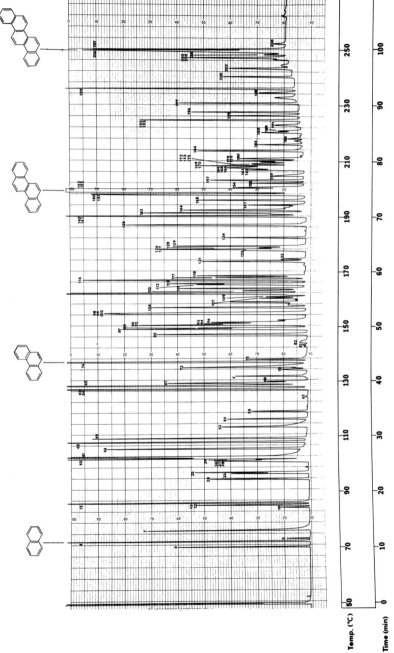

Fig. 7-14. Retention pattern of 208 standard PAC on a 12 m × 0.29 mm i.d. glass capillary column coated with a 0.34 μm film of SE-52 silicone gum. Numbered peaks are detailed in Table 7-2. [Reproduced with permission from M. L. Lee, D. L. Vassilaros, C. M. White, and M. Novotny, *Anal. Chem.* **51**, 768 (1979). Copyright, American Chemical Society.]

TABLE 7-3

Retention Indices from Capillary Columns with Different Stationary
Phase Film Thicknesses[a]

Compound	Normal retention index		PAH index	
	Column A	Column B	Column A	Column B
Naphthalene	1168.74	1166.21	200.00	200.00
Acenaphthylene	1425.03	1413.00	244.65	244.67
Fluorene	1555.87	1549.28	268.14	268.22
Phenanthrene	1744.70	1734.95	300.00	300.00
Anthracene	1754.20	1744.40	301.73	301.76
4H-Cyclopenta[def]phenanthrene	1876.18	1864.28	321.95	322.09
Pyrene	2063.99	2048.56	351.13	351.25
Benzo[a]fluorene	2167.27	2153.04	366.64	366.75
Benzo[e]pyrene	2770.65	2751.03	450.66	450.80
Perylene	2812.49	2815.42	456.12	456.23

[a] According to Lee et al. (73).

adverse effects were minimized. A retention index system for programmed temperature runs was developed (73), using naphthalene (200.00), phenanthrene (300.00), chrysene (400.00), and picene (500.00) as retention standards. The average 95% confidence limits for four measurements on more than 200 compounds were ±0.25 index units (see Table 7-2).

Even column-to-column variations were found to be unimportant when using the above retention measurement approach. Whereas, the retention data measured on two capillary columns with different film thicknesses (Table 7-3) for "normal" retention standards (n-hydrocarbons) show their limited practical utility, the data gathered with the PAH index are much more useful.

The retention of a number of PAH including the four chosen PAH standards is demonstrated in Fig. 7-14. The PAH retention system was found beneficial in a subsequent identification of numerous isomers in coal-derived materials (74). A similar high-precision retention index system has been developed for alkylated pyridines and quinolines (75). Here, pyridine, quinoline, and acridine were used as reference compounds.

IV. DETECTORS AND ANCILLARY TECHNIQUES

A. Quantitative Analysis

Depending on the concentration of PAC in various samples, column inertness and sensitivity of GC detectors may be critical. For nanogram- and picogram-level determinations, adverse column effects (e.g., irreversible adsorption and tailing) must be minimized.

Considering column aspects first, it should be pointed out that different losses of various PAC compounds may result from less-than-adequate column deactivation. Thus, the column alone may be a major source of analytical inaccuracy. An insufficiently deactivated solid support or capillary column inner wall can cause incomplete elution of PAC compounds. This problem could be particularly severe when working with very small samples. Consequently, columns should be periodically checked with standard compounds to determine their sample "thresholds."

TABLE 7-4

FID Response Factors of Selected PAH[a]

Compound	Response factor relative to fluoranthene
Biphenyl	0.751
Fluorene	0.864
9,10-Dihydrophenanthrene	0.827
9,10-Dihydroanthracene	0.803
Phenanthrene	0.920
Anthracene	0.880
Dihydropyrene	0.962
Pyrene	1.067
Fluoranthene	1.000
2-Methylfluoranthene	1.070
3-Methylpyrene	1.142
1-Methylpyrene	1.138
Benz[a]anthracene	1.245
Chrysene	1.239
1-Methylchrysene	1.334
6-Methylchrysene	1.321
Benzo[j]fluoranthene	1.293
Benzo[k]fluoranthene	1.426
Benzo[b]fluoranthene	1.330
Benzo[a]pyrene	1.322
Benzo[e]pyrene	1.331
Perylene	1.332
3-Methylcholanthrene	1.337
Benzo[b]chrysene	1.348
Picene	1.354
Benzo[ghi]perylene	1.356
Anthanthrene	1.350
Coronene	1.483

[a] According to Lao et al. (76).

Although other nonselective detectors were used in earlier PAC analyses, the flame ionization detector is now almost universally employed for this task. Response linearity, sensitivity, and day-to-day quantitative reliability in routine determinations are the most important assets of this detector. Although its response to aromatic solutes is slightly reduced as compared to aliphatic compounds, typical detection limits for PAC should be slightly below 1 ng. Different detector designs and instruments will cause some variations.

Since the detector response is somewhat different for various PAC, determination of response factors (coefficients relating detected peak areas to the solute concentration) has been of interest to many workers. Disagreements among the different literature sources indicate that analysts may be wise to determine their own values for most accurate results. These disagreements could be the result of different column conditions, detector designs, etc. Quite frequently, the assumption of equal response for all mixture components is made.

Response factors for a number of PAH compounds were determined by Lao *et al.* (76). Some typical values (relative to fluoranthene) extracted from their work are presented in Table 7-4. Whereas, these figures show that some variations among individual PAH exist, they are not as large as some previously published data indicate (5, 77). Lao *et al.* (76) have also found that the response factors of methylated PAH are typically 5–7% larger than those of parent compounds, and detector response appears to be decreased by ring saturation.

While semiquantitative data will suffice in many investigations (e.g., when comparing PAC chromatographic profiles under different circumstances), accurate quantification may be of increasing concern in evaluating human exposure to various carcinogens.

There are many general aspects of GC quantification that are beyond the scope of this book. An interested reader may find additional information in a monograph (78).

B. Selective Detectors

The main advantage of selective detectors in chromatography is their enhanced response to compounds with certain structural features. If their selectivity is sufficiently high, such devices completely "ignore" the other coeluting substances in complex mixtures. In addition, many selective detectors are inherently more sensitive than the flame ionization detector. However, less quantitative reliability may often be the price paid for the advantages of selective detectors.

Perhaps the ultimate in detection selectivity is furnished by special mass spectroscopic techniques, such as selective ion monitoring. Their utilization in PAC analysis has already been demonstrated, and pertinent aspects will be included later in Chapter 8. This discussion will be primarily dedicated to three detector types: (1) electron capture detector, (2) detectors specific to PAC containing heteroatoms, and (3) gas-phase (optical) spectroscopic detectors. All of these have been successfully used in PAC analysis.

The parallel utilization of several detectors (both selective and nonselective types) with capillary columns is now feasible. Some aspects of capillary GC and selective detection have been discussed in a recent review article (47).

C. Electron Capture Detector

This detector may well be the most sensitive device presently available to a chemist for the detection of organic compounds (79). Its sensitivity frequently extends from the picogram (10^{-12} g) down to the femtogram (10^{-15} g) range. Although PAC compounds are not among the best electron acceptors, picogram determinations should clearly be feasible.

The electron-capture properties of some PAH compounds were noticed by Lovelock in one of his initial studies (80) on the detector. He hypothesized that the affinity of some organic compounds for low-energy electrons may have a wider biological significance. A subsequent paper (81) dealt more specifically with the possible implication of this physical phenomenon in the carcinogenic properties of certain PAH. Although there are exceptions to the hypothesis (e.g., the detector response is not enhanced by either the presence or the position of a methyl group in obviously carcinogenic PAH) and additional correlations of PAH carcinogenesis have been suggested (as discussed in Chapter 3), it remains an important fact that some PAH capture thermal-energy electrons. Consequently, the electron capture detector can be used as a sensitive and selective device in this field.

As early as 1965, Cantuti et al. (77) recognized the fact that a differing response of the electron capture detector from one PAH compound to another (in contrast to a nearly uniform response of the same substances in the flame ionization detector) could be used as a means of qualitative distinction. Selected response measurements are shown in Table 7-5. It should be pointed out that these data cannot be relied on as absolute, since responses will vary for different detector designs and conditions. Nevertheless, the results are sufficient for the demonstration of the detector selectivity. The enhanced sensitivity of the electron capture detector can be utilized in some cases where there are sample-size limitations. Although the sensitivity of this detector is currently greater than that of any ancillary identification method, the combined information on compound retention and detector response can still be useful. In addition, the electron capture detector can be employed

TABLE 7-5

Detector Weight Responses of Selected PAH with the Flame Ionization and Electron Capture Detectors, Relative to Benzo[mno]fluoranthene[a]

Compound	FID response	ECD response	Ratio
Anthracene	0.75	0.06	20.2
Fluoranthene	0.76	0.09	32.5
Pyrene	0.84	0.40	124.3
Benzo[a]fluorene	1.00	0.02	5.5
3-Methylpyrene	0.97	0.27	69.2
Benzo[mno]fluoranthene	1.00	1.00	250.00
Benz[a]anthracene	0.90	0.87	267.2
Chrysene	0.85	0.00_5	1.5
Benzo[a]pyrene	1.53	2.15	343.5
Benzo[e]pyrene	0.66	0.75	310.0
Benzo[k]fluoranthene	1.01	0.67	180.2
Perylene	1.80	0.01	1.5

[a] According to Cantuti et al. (77).

for the selective detection of PAH in mixtures containing other interfering compounds (e.g., alkanes). The complementary response of the electron capture detector (ECD) to that of the flame ionization detector (FID) was demonstrated with chromatographic profiles of an extract of airborne material (77). Similarly, Carugno and Rossi (5) employed this method to monitor compounds from tobacco-smoke condensate.

More recently, Bjørseth and Eklund (82) measured the ECD/FID response ratios for a number of PAH. They found that many isomers could be differentiated by measurements of these ratios. Figure 7-15 shows a simultaneous FID/ECD chromatogram of a standard mixture of PAH obtained by them.

It should be pointed out that the more recent advances in the preparation of highly inert glass capillary columns will undoubtedly extend the applicability of the electron capture detector in trace PAC analysis. The combination of this detector with capillary columns is feasible either through detector cell miniaturization (83) or the use of a make-up gas at the column outlet (to overcome dead-volume problems).

D. Detectors Selective to PAC Containing Heteroatoms

Among the nitrogen-containing PAC, many compounds are either proven or suspected carcinogens. Although neither the analytical chemistry nor toxicological studies of this compound group are greatly developed, they have received increasing attention lately. Since they are frequently present in relatively small quantities, their sensitive detection is essential.

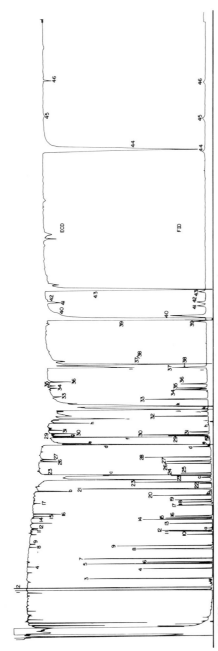

Fig. 7-15. FID/ECD dual-trace capillary gas chromatogram of a standard PAH mixture. Chromatographic conditions: 50 m × 0.35 mm i.d. glass column coated with SE-54, temperature programmed from 100° to 250°C at 3 C/min. Selected peak identifications: (1) 2-methylnaphthalene, (14) phenanthrene, (15) anthracene, (23) fluoranthene, (24) pyrene, (29) benz[a]anthracene, (30) chrysene, (34) benzo[e]pyrene, (35) benzo[a]pyrene, (36) perylene, (45) coronene. [Reproduced with permission from A. Bjørseth and G. Eklund, *J. High Resoln. Chromatogr./Chromatogr. Commun.* **2**, 22 (1979).]

It has been recognized for some time that many nitrogen-containing compounds can be detected by detectors based on the thermionic principle. While the first such detector design of Aue *et al.* (*84*), with the alkali source attached directly to the jet of a flame ionization detector, showed clearly the detection possibilities, the detector arrangement lacked both sufficient selectivity and quantitative reliability. The newer versions of this detector use an externally heated alkali source (e.g., a rubidium silicate bead). The mechanism of response to nitrogen compounds as explained by Kolb and Bischoff (*85*) consists of the formation of cyano radicals in a relatively cool flame and their subsequent interaction with the excited rubidium atoms that form a cloud around the heated source. The liberation of an electron results from such an interaction and the measured detector current increases.

Currently available nitrogen-sensitive thermionic detectors demonstrate adequate analytical performance. For example, the detector investigated by Hartigan *et al.* (*86*) had a sensitivity around 10^{-13} g/sec, a response linearity over several orders of magnitude, and a selectivity factor of 10^3 to 10^4. Although several operational properties of this detector must be adequately controlled, its routine utilization is feasible.

The simultaneous detection of PAH and azaarenes should be possible using a parallel arrangement of the flame ionization and nitrogen-sensitive detectors. Since the nitrogen detector is the more sensitive of the two, complementary profiles of both substance classes can be generated. This was demonstrated with samples of airborne particulates (*86*) and engine oils (*87*). Figure 7-16 shows the chromatographic results obtained from the nitromethane fraction of engine oil samples. While the flame ionization profiles are only a little dissimilar, the differences in the "nitrogen" chromatograms are rather dramatic.

As discussed in Chapter 2, sulfur-containing PAC have been found in a variety of combustion products. Although their environmental importance is not completely clear, sulfur-containing PAC are likely to receive more attention in the future. Although it would be tempting to conclude that the sulfur-sensitive (flame photometric) detector must be applicable in selective detection of such compounds, there are some practical problems that limit its utility.

The flame photometric detector (FPD) originally developed by Brody and Chaney (*88*) has enjoyed some popularity with a number of analysts. However, one of its serious drawbacks is the phenomenon of response "quenching" due to the presence of other (nonsulfur) compounds within the same peak, as pointed out by Perry (*89*). A recent investigation of the composition of coal-derived materials (*74*) fully supports Perry's observations. Although mass-spectral data established the presence of sulfur PAC beyond any doubt, the peaks were mostly undetectable with the flame photometric

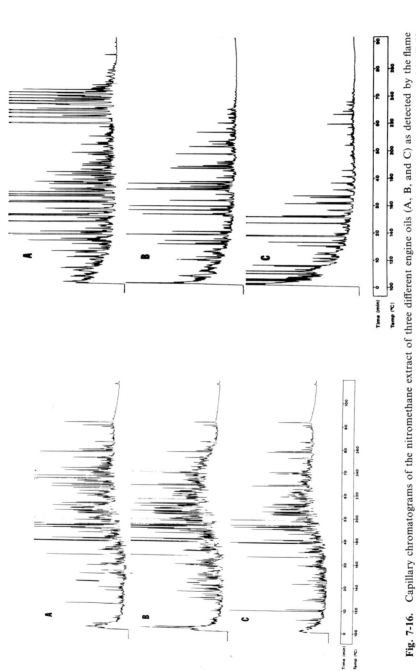

Fig. 7-16. Capillary chromatograms of the nitromethane extract of three different engine oils (A, B, and C) as detected by the flame ionization detector (left) and a nitrogen-sensitive detector (right). Conditions: 22 m × 0.26 mm i.d. glass capillary column coated with SE-52 silicone polymer. [Reproduced with permission from M. L. Lee, K. D. Bartle, and M. Novotny, *Anal. Chem.* **47**, 540 (1975). Copyright, American Chemical Society.]

detector. Of course, when good resolution is achieved with capillary columns, the interference problem becomes less severe. A recent study produced simultaneous FID/FPD chromatograms of several distillation cuts of a coal gasification tar (*90*). The compounds were identified by the further separation of the sulfur fraction from the other compounds and analysis by gas chromatography/mass spectrometry.

Recently developed ionization detectors that offer some potential in PAC analysis are the Hall electrolytic conductivity detector (HECD) and the photoionization detector (PID). The HECD can be used to detect selectively both nitrogen and sulfur-bearing PAC (*91*). As components elute from the column, they are swept into a nickel reaction tube along with an appropriate reaction gas. The compounds are pyrolyzed at temperatures between 700° and 1000°C to form a mixture of conducting and nonconducting reaction products, which are first passed over a scrubber to remove interfering species and then fed into the conductivity cell. The species of interest ionizes in the flowing electrolyte and the change in conductivity is monitored. Under optimum conditions, the selectivity factor for nitrogen over hydrocarbons is $10^6:1$, and for sulfur over hydrocarbons is $10^5:1$. The use of the HECD with capillary columns for the selective detection of sulfur and nitrogen compounds in various samples has recently been reported (*91*).

The photoionization detector (*92*) operates by ionization of eluted compounds by UV light and detection of the ions produced. The device is highly sensitive, especially for PAH, for which the PID/FID response ratio is between 5 and 10. The use of UV lamps with different wavelengths allows some selectivity in detection, since only radiation with energy greater or equal to the ionization potential of the species will produce a signal.

E. Selective Spectroscopic Detectors

Much of what has been said about LC spectroscopic monitors applies also for the gas-phase detectors. While the established universal detector in GC is the flame ionization detector, selectively responding devices are of considerable interest. Although the incentive for the development of spectroscopic detectors has mostly been the insufficient resolution of PAH components in packed-column chromatography, their potential utilization goes far beyond this aspect.

Even when capillary GC can minimize the problem of overlapping peaks, selective detectors could be beneficial in providing reliable quantitation *directly* in complex samples. For example, in order to determine PAC profiles from airborne particulates, the extract must be partitioned, for example between cyclohexane and nitromethane, for the removal of interfering alkanes. If capillary GC were employed together with a detector specific to

PAC compounds of interest, no such solvent partition step would be necessary.

While the sensitivities of several gas-phase detectors used with packed columns appear somewhat lower than expected, much improvement can theoretically be obtained in high-efficiency GC. Firstly, the narrower capillary peaks are known to enhance the signal-to-noise ratio, resulting in smaller detectable amounts (probably, two- to fourfold improvement). Secondly, if the detection cells are designed to match more effectively the volumes of capillary fractions, substantial sensitivity gains are realized.

The qualitative meaning of the detector response should not be overlooked. Just as discussed in Chapter 6, the ratios of absorbances or fluorescence intensities measured at different wavelengths can be highly indicative of the presence of certain compounds, even though this information can hardly be considered sufficient for their positive identification. In addition, approaches to acquiring the whole gas-phase spectra during peak elution have been at least partially successful; with further technological advances, the recording of complete spectral information from each fraction will become increasingly feasible (see Chapter 10 concerning FT-IR).

While gas-phase spectroscopic detectors show a certain potential, a considerable knowledge gap exists concerning vapor-phase spectra. In general, there is much more information available in the literature on PAH spectra obtained in the condensed phase. Such knowledge for the vapor phase is needed for both the reliable function of quantitative selective monitors and identification procedures.

Finally, it should be pointed out that developments of gas-phase spectroscopic detectors should be considered complementary rather than competitive to efforts aimed at higher column efficiencies. Different types of such detectors will now be discussed.

F. Ultraviolet Absorption Detector

UV detection may be desirable in a number of practical applications since it is possible to selectively determine aromatics in the presence of other hydrocarbons. Thus, it is not surprising that UV detectors were explored for the first time more than 15 years ago (93, 94).

While Kaye (93) provided a commercial UV spectrophotometer with a heated flow cell to investigate certain aromatics in GC effluents, Merritt et al. (94) developed a special process instrument based on a GC–UV combination. Both instruments provided reliable analyses and their sensitivities compared quite favorably with the then-available GC detectors (at the microgram level).

In a recent development, Novotny et al. (95) succeeded in coupling

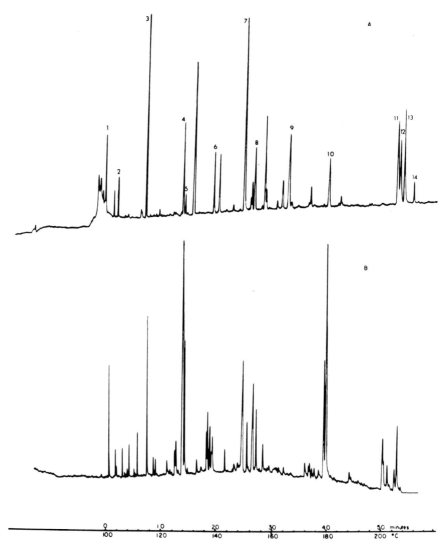

Fig. 7-17. Separation of polycyclic aromatic hydrocarbons on a 0.25 mm i.d. SE-52 glass capillary column and detection with UV detector at 250 nm. (A) Standard mixture containing approximately 30 ng of each of the following: (1) naphthalene; (2) biphenyl; (3) fluorene; (4) phenanthrene; (5) anthracene; (6) 9-methylphenanthrene; (7) fluoranthene; (8) pyrene; (9) 1-methylpyrene; (10) chrysene and triphenylene; (11) benzo[e]pyrene; (13) dibenz[a,c]anthracene; and (14) perylene. (B) 4 μl aliquot of crude coal tar. [Reproduced with permission from F. J. Schwende, M. Novotny, and J. E. Purcell, *Chromatogr. Newsl.* **8**, 1 (1980). Copyright, Perkin-Elmer Corp.]

capillary columns to a variable-wavelength UV detector. With the current cell design (50 μl volume), it is possible to employ wide-bore (0.5–0.7 mm i.d.) glass capillary columns without loss of chromatographic resolution. Alternatively, small-bore (0.25 mm i.d.) capillary columns can also be used if a purge gas is introduced at the column end. While this procedure minimizes the problems of excessive cell and interface volumes, the detection sensitivity is reduced. In spite of this drawback, good results were obtained with PAH samples chromatographed on a short SE-52 coated glass capillary column. The chromatograms obtained of a coal tar sample and PAH standards are shown in Fig. 7-17. The sensitivity was also evaluated, comparing adequately with conventional flame ionization detectors.

Although the selectivity of UV monitors for PAH compounds can be analytically useful, a careful selection of the most suitable wavelength is critical. This is evident from the vapor-phase spectrum (*93*) of naphthalene (Fig. 7-18).

G. Spectrofluorimetric Detectors

GC fluorescence detectors have received considerably more attention than UV detectors in the literature, mainly because fluorescence offers higher sensitivity, combined with a greater degree of selectivity.

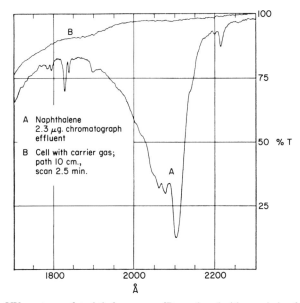

Fig. 7-18. UV spectrum of naphthalene vapor. [Reproduced with permission from W. Kaye, *Anal. Chem.* **34,** 287 (1962). Copyright, American Chemical Society.]

A modified Aminco-Bowman spectrofluorimeter described by Bowman and Beroza (96) can be considered as the forerunner of the presently studied detectors. In their experimental arrangement, the GC effluent was trapped in a suitable solution in which the spectrofluorimetric measurements were subsequently performed. The trapped PAH fractions were easily measured at the nanogram level. However, the cumbersome nature of this detector arrangement forced other investigators to evaluate the merits of direct gas-phase detection. Several reports were published on the subject over the years, but improvements in detection technology are still desirable.

Gas-phase detection with a heated cell was extensively studied by Burchfield *et al.* (97, 98). The construction details pertinent to the detector modification were given. As can be seen from Fig. 7-19, the carrier gas may have an appreciable effect on the detector response, the magnitude of which decreases in the following order: $N_2 > He > CO_2 > H_2$. A general observation from their work (as well as that of other investigators, as discussed later) was a

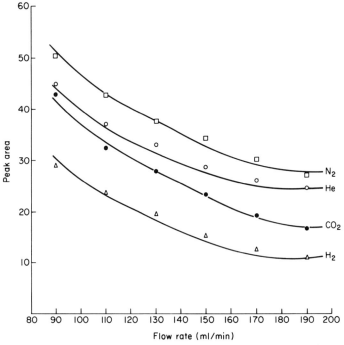

Fig. 7-19.   Effect of carrier-gas type and flow rate on the response of a gas-phase fluorimetric detector to fluorene. $\square = N_2$; $\bigcirc = He$; $\bullet = CO_2$; and $\triangle = H_2$. [Reproduced with permission from H. P. Burchfield, R. J. Wheeler, and J. B. Bernos, *Anal. Chem.* **43,** 1976 (1971). Copyright, American Chemical Society.]

substantial drop in the detector response when going from the solution to the gas phase.

Since fluorimetric detectors belong to the family of concentration-sensitive detectors, the suggestion was made to remove the carrier gas (prior to the detector cell) by means of a suitable enrichment device, similar to GC/MS. When using H_2 as a carrier gas, the palladium separator described by Lovelock et al. (99) would appear most appropriate. For reliable detection by a spectrofluorimetric detector, it is imperative to avoid any presence of oxygen in the detector cell; oxygen was found to quench detector response for several PAH (97).

While the detector response is generally linear with concentration, as expected, some variation appears in the detection limits reported by different authors. In most cases, minimum detectable amounts are of the order of a few nanograms (97, 98, 100–104), although better sensitivities should clearly be feasible. An analysis of the problem is given by Cooney and Winefordner (104), who quantitatively compared several optical systems. The origin of the noise was evaluated for different systems, and the following suggestions for sensitivity improvement were given: (1) use of highly intense stable sources, and (2) reduction of stray light through more efficient monochromation. Under the best conditions, the minimum detectable amount for anthracene was estimated as 0.028 ng. Similarly, Freed and Falkner (102) reasoned that 10^{-14}-g sensitivities should be feasible using optimized gas-phase spectrofluorimeters.

While sensitivity is only one aspect of GC spectrofluorimetric detection, the response selectivity is frequently even more desirable. In complex mixtures where PAC are encountered, the separation of isomers is a problem that can be satisfactorily overcome only through higher column efficiencies. For example, for a conventional stationary phase, the three isomers benzo[e]-pyrene, benzo[a]pyrene, and perylene can only be resolved by a capillary column. On the other hand, fluorimetric detection can alleviate the resolution problem with a packed column if runs are made at two different wavelength λ_{em} settings: benzo[a]pyrene has λ_{em} at 406 nm, whereas, the value for perylene is 440 nm (97). Thus, the three compounds can be measured entirely independently of each other. However, such a solution is hardly universal for all PAC analytical problems.

There are differences between gas-phase detection and measurements carried out in the condensed phase. The decrease of response in the gas phase due to higher temperatures can be explained by the different probability of nonradiative transitions (105). Such a decrease can be very substantial.

Elevated temperatures also cause some alterations in fluorescence spectra which tend to be more diffuse due to free rotational molecular motions and

TABLE 7-6
Gas-Phase Fluorescence Wavelength Maxima for Selected PAH[a]

Compound	Excitation-λ^b	Emission λ^b
Fluorene	272 (300)	330 (321)
Chrysene	262 (264)	386 (381)
Benzo[k]fluoranthene	298 (302)	424 (400)
Benzo[e]pyrene	280 (339)	394 (400)
Benzo[a]pyrene	290 (300)	411 (413)
Perylene	244 (430)	433 (438)

[a] According to Burchfield *et al.* (*98*).
[b] Values in parentheses apply to pentane solutions.

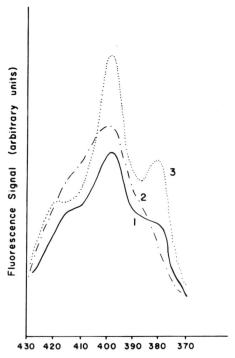

Fig. 7-20. Fluorescence spectra of anthracene: 1, gas phase, $\lambda_{ex} = 367$ nm; 2, gas phase, $\lambda_{ex} = 266$ nm; 3, condensed phase, $\lambda_{ex} = 266$ nm. [Reproduced with permission from R. P. Cooney, T. Vo-Dinh, and J. D. Winefordner, *Anal. Chim. Acta* **89,** 9 (1977). Copyright, Elsevier Sci. Publ. Co.]

to Doppler and collisional broadening processes (*103*). There are also spectral shifts involved, and according to Cooney *et al.* (*103*), a typical blue shift of approximately 5 nm is encountered when going from the liquid to the gas phase. Therefore, well-specified temperature *and* excitation energy are needed to characterize fully a particular measurement. Some spectral values for selected PAH are given in Table 7-6 (*98*). According to one report (*106*), higher sensitivities may be feasible when operating at reduced pressure.

Just as with LC analyses discussed in Chapter 6, it would frequently be desirable to acquire a complete fluorescence spectrum for identification purposes. While Freed and Faulkner (*102*) investigated the merits of rapid spectral scanning (up to 150 nm/sec) in 1972, more emphasis has lately been placed on imaging detectors. Since the subject was already discussed in general terms in Chapter 6, only the pertinent points of gas-phase detection will be mentioned here. Firstly, the important questions relate to the diffuseness of gas-phase spectra and spectral shifts already mentioned. The magnitude of such phenomena vary somewhat among different PAH, as Fig. 7-20 shows (*103*).

Subsequently, one should consider the potential contribution of such measurements to the overall identification process and relate it to the price of imaging detectors. A block diagram of one such device (a silicon intensifier target detector) as used by Cooney *et al.* (*103*) is shown in Fig. 7-21. Although this approach brings considerable more complexity to the analysis, technological improvements in both optics and computers are likely to make imaging detectors more attractive. While the sensitivity is only slightly lower here than with ordinary detection (*103*), possibilities of spectral subtraction are an additional advantage.

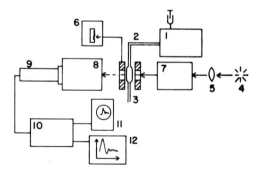

Fig. 7-21. Block diagram of the GC–SIT (SIT = silicon intensifier target) imaging detector according to Cooney *et al.* (*103*). Key: 1, gas chromatograph; 2, transfer line; 3, flow cell; 4, light source; 5, optics; 6, heating device; 7, excitation monochromator; 8, emission monochromator; 9, SIT image converter; 10, optical multichannel analyzer; 11, oscilloscope; and 12, strip-chart recorder. [Reproduced with permission from R. P. Cooney, T. Vo-Dinh, and J. D. Winefordner, *Anal. Chim. Acta* **89**, 9 (1977). Copyright, Elsevier Sci. Publ. Co.]

V. SUPERCRITICAL-FLUID CHROMATOGRAPHY

Through adjustment of pressure and temperature, many mobile phases can be brought to their critical points. While the conventional carrier gases such as nitrogen or argon require extremely high pressures (around 2,000 atm), critical conditions are quite easily achieved for carbon dioxide, ammonia, nitrous oxide, and several other gases. Similarly, many liquids (such as volatile organic solvents) can be brought to similar conditions. Critical data for several representative substances are given in Table 7-7.

The fine manipulation of pressures and/or temperatures in the vicinity of the critical point of a mobile phase affects the retention of sample molecules. As shown many years ago by Desty *et al.* (*107*), solute retention is slightly affected even with conventional gases and fairly low pressures. The solute retention decreases with pressure according to

$$\log k = A + \frac{\bar{P}}{2.303RT}(2B_{1,2} - v_2^0) \tag{3}$$

where k = capacity ratio, $\bar{P}$ = average column pressure, R = gas constant, T = absolute temperature, $B_{1,2}$ = second virial coefficient, A = constant, and v_2^0 = partial molar volume of solute in the liquid phase.

While molecular interactions between solute and mobile-phase molecules are obviously present in liquid chromatography, supercritical fluid–solute interactions produce certain advantages. The chromatographic theory (*108*) indicates that the mobile-phase viscosity and the solute diffusivity in the mobile phase are the two most important parameters involved in the separation process. Specifically, the mobile-phase velocity is related to its viscosity, which limits the ultimate column length. Simultaneously, the mobile-phase solute diffusivity is crucial in radial diffusion and efficient mass transfer processes. The two regions of particular chromatographic interest

TABLE 7-7

Supercritical Fluids of Interest for PAH Separations

Mobile phase	Critical pressure (atm)	Critical temperature (°C)	Corresponding density (g/ml)
n-Pentane	33.3	196.6	0.232
Dichlorotetrafluoroethane	35.5	146.7	0.582
Isopropanol	47.0	253.3	0.273
Carbon dioxide	72.9	31.3	0.448
Sulfur hexafluoride	37.1	45.6	0.752

are the so-called "dense-gas" and "supercritical-fluid" (*109*) conditions. Importantly, viscosities here are typically 100 times lower, while the solute diffusion coefficients are 10–200 times larger (*110*).

Both the initial successful results of Klesper *et al.* (*111*) in migrating porphyrines with supercritical chlorofluorocarbons and the theoretical predictions of Giddings (*12*) unmistakably pointed at the potential of this area of chromatographic separations. However, the later achievements in this field were overshadowed by the rapid advances in HPLC and discouraging experimental difficulties with supercritical fluids. Nonetheless, the potential still exists to be explored.

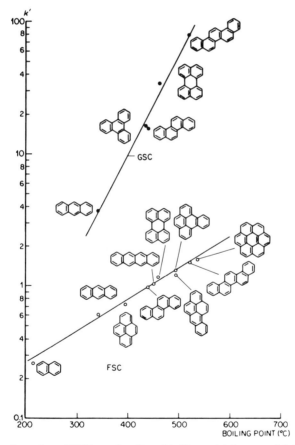

Fig. 7-22. Retention of PAH as a function of boiling point for gas-solid chromatography (GSC) at 460°C with helium carrier gas, and fluid–solid chromatography (FSC) at 245°C. Adsorbent, untreated alumina. [Reproduced with permission from S. T. Sie and G. W. A Rijnders, *Anal. Chim. Acta* **38**, 31 (1967). Copyright, Elsevier Sci. Publ. Co.]

Because of the ease of using the UV detector in supercritical-fluid chromatography, PAC were frequently used as model substances for studying different mobile-phase conditions and various column types. Figure 7-22 compares the retention of a series of PAH under the conditions of ordinary gas-solid chromatography (at 460°C) and supercritical-fluid–solid chromatography (at 245°C) using isopropanol (112). A wider molecular-weight range of compounds can generally be chromatographed in supercritical-fluid chromatography and at lower temperatures than required for gas chromatography.

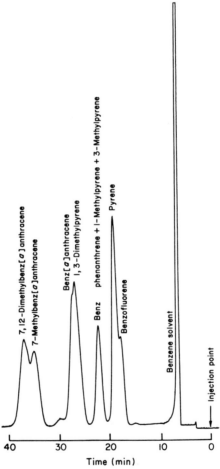

Fig. 7-23. Separation of standard PAH by supercritical-fluid chromatography on a reversed-phase packed column, using CO_2 as the mobile phase. [Reproduced with permission from T. H. Gouw and R. E. Jentoft, *Adv. Chromatogr.* **13**, 2 (1975). Copyright, Marcel Dekker, Inc.]

While supercritical-fluid chromatography has demonstrated some interesting possibilities in the separation of PAH (*113*) (Fig. 7-23), its potential is far from being developed (*114*). Efficiency studies indicate (*115*) that column pressure drops present the most serious difficulties. Chromatographic columns with high average pressures, but with very low pressure gradients, will undoubtedly be favored. The earlier work on supercritical-fluid chromatography, summarized in the review articles by Gouw and Jentoft (*110, 114*), should be consulted for more detailed information on the many aspects of this chromatographic technique.

REFERENCES

1. W. J. A. VandenHeuvel, C. C. Sweeley, and E. C. Horning, *J. Am. Chem. Soc.* **82,** 3481 (1960).

2. L. S. Ettre, *in* "Open Tubular Columns in Gas Chromatography," p. 47. Plenum, New York, 1965.

3. J. Oro', J. Han, and A. Zlatkis, *Anal. Chem.* **39,** 27 (1967).

4. A. Liberti, G. P. Cartoni, and V. Cantuti, *J. Chromatogr.* **15,** 141 (1964).

5. N. Carugno, and S. Rossi, *J. Gas Chromatogr.* **5,** 103 (1967).

6. M. Novotny, M. L. Lee, and K. D. Bartle, *J. Chromatogr. Sci.* **12,** 606 (1974).

7. M. L. Lee, M. Novotny, and K. D. Bartle, *Anal. Chem.* **48,** 405 (1976).

8. M. L. Lee, M. Novotny, and K. D. Bartle, *Anal. Chem.* **48,** 1566 (1976).

9. S. T. Sie, J. P. A. Bleumer, and G. W. A. Rijnders, *in* "Gas Chromatography 1968" (C. L. A. Harbourn, ed.), p. 235. The Institute of Petroleum, London, 1969.

10. J. H. Purnell, *Nature (London)* **184,** 2009 (1959).

11. B. Karger, *Anal. Chem.* **39,** 24A (1967).

12. J. C. Giddings, *in* "Gas Chromatography 1964" (A. Goldup, ed.), p. 2. The Institute of Petroleum, London, 1965.

13. F. Dupire and G. Botquin, *Anal. Chim. Acta* **18,** 282 (1958).

14. F. Dupire, *Fresenius Z. Anal. Chem.* **170,** 317 (1959).

15. B. J. Gudzinowicz and W. R. Smith, *Anal. Chem.* **32,** 1767 (1960).

16. L. J. Wood, *J. Appl. Chem. Biotechnol.* **11,** 130 (1961).

17. W. Lijinsky, I. Domsky, G. Mason, H. Y. Ramahi, and T. Safavi, *Anal. Chem.* **35,** 952 (1963).

18. R. E. Schaad, *Chromatogr. Rev.* **13,** 61 (1970).

19. L. S. Lysyuk and A. N. Korol, *Chromatographia* **10,** 712 (1977).

20. M. Novotny, R. Segura, and A. Zlatkis, *Anal. Chem.* **44,** 9 (1972).

21. T. B. Gavrilova, M. Krejci, H. Dubsky, and J. Janak, *Collect. Czech. Chem. Commun.* **29,** 2753 (1964).

22. K. Bhatia, *Anal. Chem.* **43,** 609 (1971).

23. W. A. Aue, C. R. Hastings, and K. O. Gerhardt, *J. Chromatogr.* **99,** 45 (1974).

24. F. W. Karasek, D. W. Denney, K. W. Chan, and R. E. Clement, *Anal. Chem.* **50,** 82 (1978).

25. P. W. Solomon, *Anal. Chem.* **36,** 476 (1964).

26. O. T. Chortyk, W. S. Schlotzhauer, and R. L. Stedman, *J. Gas Chromatogr.* **3,** 394 (1965).

27. B. H. Gump, *J. Chromatogr. Sci.* **7,** 755 (1969).

28. A. V. Kiselev, *Adv. Chromatogr.* **4,** 113 (1967).

29. A. V. Kiselev, J. Janak, K. Tesarik, and F. Onuska, *J. Chromatogr.* **34,** 81 (1968).

30. A. Zane, *J. Chromatogr.* **38**, 130 (1968).
31. J. Frycka, *J. Chromatogr.* **65**, 341 (1972).
32. J. Frycka, *J. Chromatogr.* **65**, 432 (1972).
33. C. Vidal-Madjar, J. Ganansia, and G. Guiochon, *in* "Gas Chromatography 1970" (R. Stock, ed.), p. 20. The Institute of Petroleum, London, 1971.
34. F. Bruner, P. Ciccioli, G. Bertoni, and A. Liberti, *J. Chromatogr. Sci.* **12**, 758 (1974).
35. C. Vidal-Madjar, S. Bekassy, M. F. Gonnord, P. Arpino, and G. Guiochon, *Anal. Chem.* **49**, 768 (1977).
36. L. C. Chow and D. E. Martire, *J. Phys. Chem.* **75**, 2005 (1971).
37. H. Kelker and E. von Schivizhoffen, *Adv. Chromatogr.* **6**, 247 (1968).
38. G. M. Janini, K. Johnston, and W. L. Zielinski, Jr., *Anal. Chem.* **47**, 670 (1975).
39. S. Wasik and S. Chesler, *J. Chromatogr.* **122**, 451 (1976).
40. G. M. Janini, G. M. Muschik, and W. L. Zielinski, Jr., *Anal. Chem.* **48**, 809 (1976).
41. G. M. Janini, G. M. Muschik, J. A. Schroer, and W. L. Zielinski, Jr., *Anal. Chem.* **48**, 1879 (1976).
42. H. Pailer and V. Hlozek, *J. Chromatogr.* **128**, 163 (1976).
43. D. H. Desty and A. Goldup, *in* "Gas Chromatography 1960" (R. P. W. Scott, ed.), p. 162. Butterworths, London, 1960.
44. D. H. Desty, *Adv. Chromatogr.* **1**, 199 (1965).
45. M. Novotny, *Anal. Chem.* **50**, 16A (1978).
46. M. Novotny, submitted for publication.
47. M. Novotny, *in* "Advances in Analytical Chemistry (D. F. S. Natusch, ed.), in press.
48. W. Jennings, "Gas Chromatography with Glass Capillary Columns." Academic Press, New York, 1980).
49. J. R. Wilmhurst, *J. Chromatogr.* **17**, 50 (1965).
50. E. Proksch, *Fresenius Z. Anal. Chem.* **223**, 23 (1966).
51. T. H. Gouw, I. M. Whittemore, and R. E. Jentoft, *Anal. Chem.* **42**, 1394 (1970).
52. B. W. Wright and M. L. Lee, *J. High Resoln. Chromatog./Chromatog. Commun.* **3**, 352 (1980).
53. G. Schomburg, R. Dielmann, H. Borwitzky, and H. Husmann, *J. Chromatogr.* **167**, 337 (1978).
54. M. L. Lee and B. W. Wright, *J. Chromatogr. Sci.* **18**, 345 (1980).
55. L. Blomberg and T. Wannman, *J. Chromatogr.* **168**, 81 (1979).
56. L. Blomberg and T. Wannman, *J. Chromatogr.* **186**, 159 (1980).
57. K. Grob and G. Grob, *J. Chromatogr. Sci.* **7**, 584 (1969).
58. K. Grob and K. Grob, Jr., *J. Chromatogr.* **94**, 53 (1974).
59. F. J. Yang, A. C. Brown III, and S. P. Cram, *J. Chromatogr.* **158**, 91 (1978).
60. G. Schomburg, H. Behlau, R. Dielmann, F. Weeke, and H. Husmann, *J. Chromatogr.* **142**, 87 (1977).
61. M. Novotny and R. Farlow, *J. Chromatogr.* **103**, 1 (1975).
62. M. L. Lee, Doctoral Thesis, Indiana University, Bloomington, 1975.
63. G. Schomburg, H. Husmann, and F. Weeke, *J. Chromatogr.* **112**, 205 (1975).
64. R. Dandeneau, P. Bente, T. Rooney, and R. Hiskes, *Am. Lab.* **11**, 61 (1979).
65. S. Nygren, *J. Chromatogr.* **142**, 109 (1977).
66. E. sz. Kováts, *Helv. Chim. Acta* **41**, 1915 (1958).
67. G. Schomburg, *Adv. Chromatogr.* **6**, 211 (1968).
68. F. Riedo, D. Fritz, G. Tarján, and E. sz. Kováts, *J. Chromatogr.* **126**, 63 (1976).
69. K. Tesarik and M. Novotny, *Chromatographia* **2**, 384 (1969).
70. A. J. Solo and S. W. Pelletier, *Anal. Chem.* **35**, 1584 (1963).
71. L. Soják and J. A. Rijks, *J. Chromatogr.* **119**, 505 (1976).
72. L. Soják, J. Janák, and J. A. Rijks, *J. Chromatogr.* **142**, 177 (1977).
73. M. L. Lee, D. L. Vassilaros, C. M. White, and M. Novotny, *Anal. Chem.* **51**, 768 (1979).

74. R. V. Schultz, J. W. Jorgenson, M. P. Maskarinec, M. Novotny, and L. J. Todd, *Fuel* **58**, 783 (1979).
75. M. Novotny, R. Kump, F. Merli, and L. J. Todd, *Anal. Chem.* **52**, 401 (1980).
76. R. C. Lao, R. S. Thomas, H. Oja, and L. Dubois, *Anal. Chem.* **45**, 908 (1973).
77. V. Cantuti, G. P. Cartoni, A. Liberti, and A. G. Torri, *J. Chromatogr.* **17**, 60 (1965).
78. J. Novák, "Quantitative Analysis by Gas Chromatography." Dekker, New York, 1975.
79. J. E. Lovelock, *J. Chromatogr.* **99**, 3 (1974).
80. J. E. Lovelock, *Nature (London)* **189**, 729 (1961).
81. J. E. Lovelock, A. Zlatkis, and R. S. Becker, *Nature (London)* **193**, 540 (1962).
82. A. Bjørseth and G. Eklund, *J. High Resoln. Chromatog./Chromatog. Commun.* **2**, 22 (1979).
83. K. Grob, *Chromatographia* **8**, 423 (1975).
84. W. A. Aue, C. W. Gehrke, R. C. Tindle, D. L. Stalling, and C. D. Ruyle, *J. Gas Chromatogr.* **5**, 381 (1967).
85. B. Kolb and J. Bischoff, *J. Chromatogr. Sci.* **12**, 625 (1974).
86. M. J. Hartigan, J. E. Purcell, M. Novotny, M. L. McConnell, and M. L. Lee, *J. Chromatogr.* **99**, 339 (1974).
87. M. L. Lee, K. D. Bartle, and M. Novotny, *Anal. Chem.* **47**, 540 (1975).
88. S. S. Brody and J. E. Chaney, *J. Gas Chromatogr.* **4**, 42 (1966).
89. S. G. Perry and F. W. G. Carter, *in* "Gas Chromatography 1970" (R. Stock, ed.), p. 381. The Institute of Petroleum, London, 1971.
90. M. L. Lee, C. Willey, R. N. Castle, and C. M. White, *in* "Polynuclear Aromatic Hydrocarbons: Chemistry and Biological Effects" (A. Bjørseth and A. J. Dennis, eds.), p. 59. Battele Press, Columbus, Ohio, 1980.
91. V. F. Cox and R. J. Anderson, *Pittsburgh Conf. Anal. Chem. Appl. Spectrosc.*, Atlantic City, New Jersey, 1980.
92. J. N. Driscoll, *Am. Lab.* **9**, 71 (1976).
93. W. Kaye, *Anal. Chem.* **34**, 287 (1962).
94. J. Merritt, F. Comendant, S. T. Abrams, and V. N. Smith, *Anal. Chem.* **35**, 1461 (1963).
95. M. Novotny, F. J. Schwende, M. J. Hartigan, and J. E. Purcell, *Anal. Chem.* **52**, 736 (1980).
96. M. Bowman and M. Beroza, *Anal. Chem.* **40**, 535 (1968).
97. H. P. Burchfield, R. J. Wheeler, and J. B. Bernos, *Anal. Chem.* **43**, 1976 (1971).
98. H. P. Burchfield, E. E. Green, R. J. Wheeler, and S. M. Billedeau, *J. Chromatogr.* **99**, 697 (1974).
99. J. E. Lovelock, K. W. Charlton, and P. G. Simmonds, *Anal. Chem.* **41**, 1048 (1969).
100. J. W. Robinson and J. P. Goodbread, *Anal. Chim. Acta* **66**, 239 (1973).
101. M. M. Cooke, M. F. Guyer, G. M. Semeniuk, and E. Sawicki, *Anal. Lett.* **8**, 511 (1975).
102. P. J. Freed and L. R. Faulkner, *Anal. Chem.* **44**, 1194 (1972).
103. R. P. Cooney, T. Vo-Dinh, and J. D. Winefordner, *Anal. Chim. Acta* **89**, 9 (1977).
104. R. P. Cooney and J. D. Winefordner, *Anal. Chem.* **49**, 1057 (1977).
105. N. A. Borisevich and V. A. Tolkachev, *Bull. Acad. Sci. USSR, Phys. Ser.* **27**, 566 (1960).
106. G. W. Robinson, *J. Chem. Phys.* **47**, 1967 (1967).
107. D. H. Destry, A. Goldup, G. R. Luckhurst, and W. T. Swanton, *in* "Gas Chromatography 1962" (M. van Swaay, ed.), p. 67. Butterworths, Washington, 1962.
108. J. C. Giddings, *Anal. Chem.* **36**, 1890 (1964).
109. J. C. Giddings, M. N. Myers, L. McLaren, and R. A. Keller, *Science* **162**, 67 (1968).
110. T. H. Gouw and R. E. Jentoft, *J. Chromatogr.* **68**, 303 (1972).
111. E. Klesper, A. H. Corwin, and D. A. Turner, *J. Org. Chem.* **27**, 700 (1962).
112. S. T. Sie and G. W. A. Rijnders, *Anal. Chim. Acta* **38**, 31 (1967).
113. R. E. Jentoft and T. H. Gouw, *J. Chromatogr. Sci.* **8**, 138 (1970).
114. T. H. Gouw and R. E. Jentoft, *Adv. Chromatogr.* **13**, 2 (1975).
115. M. Novotny, W. Bertsch, and A. Zlatkis, *J. Chromatogr.* **61**, 17 (1971).

8

Mass Spectrometry

I. INTRODUCTION

In the last decade, mass spectrometry has gained wide acceptance for the analysis of PAC. Although the technique is generally less sensitive than UV absorption and fluorescence methods (Chapter 9), it does provide much meaningful information about a complex mixture of PAC because of the characteristically simple mass spectra obtained from these compounds. Furthermore, new rapid scan techniques coupled with high-resolution chromatography have, so far, greatly surpassed any other method or combination of methods used for PAC sample analysis. High-resolution chromatographic retention data with complementary mass spectral information can often provide sufficient information for unambiguous identifications of numerous components. New ionization methods which differentiate between isomeric PAC promise increased diagnostic power in the future.

II. ELECTRON IMPACT

A. Electron-Impact Fragmentation

The electron-impact mass spectra of PAC are well-characterized as being quite simple, mainly consisting of an intense molecular ion and small ions due to the loss of one to four hydrogen atoms. These $(M - 1)^+$, $(M - 2)^+$, $(M - 3)^+$, and $(M - 4)^+$ ions range from 0 to 50% abundance of the parent ion. These higher abundances are particularly apparent in the spectra of higher molecular weight PAH. The $(M + 1)^+$ ion is always present and is due mainly to the ^{13}C isotope abundance. Doubly charged molecular ions are quite common and are usually near 20% of the abundance of the

242

molecular ion. Ions due to the expulsion of C_2H_2 are present, but in very low abundances. The general fragmentation pathway for most PAH is given in Eq. (1) (*1*).

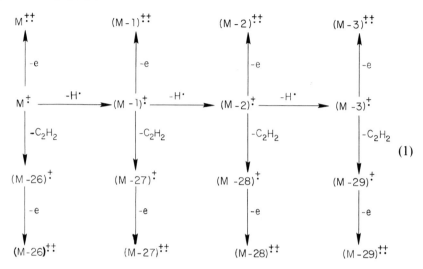

$$(1)$$

Alkylated PAH demonstrate the normal $(M - 15)^+$, $(M - 29)^+$, etc. fragmentation pattern for alkyl chains, even in the case of methyl-substituted compounds. The $(M - 15)^+$ ion is lower in abundance for methyl-substituted compounds than for PAH with longer alkyl sidechains because of the favorable loss of a proton followed by ring expansion to form the tropylium ion. Figure 8-1 compares the mass spectra for naphthalene, 2-methylnaphthalene, and 2-ethylnaphthalene. (The mass spectra in this section were taken from references *1–4*.) In the fragmentation of 2-methylnaphthalene [Eq. (2)], elimination of a proton is followed by ring expansion

$$(2)$$

m/e 142   *m/e* 141   *m/e* 115

to the benzotropylium ion which then can eliminate acetylene to give the indenyl ion. The stability of the tropylium ion is again demonstrated by the loss of a methyl group in the spectrum of 2-ethylnaphthalene followed by ring expansion, leading to the benzotropylium ion as the most abundant ion in the spectrum.

In most cases, differentiation of PAH isomers by electron-impact mass spectra alone cannot be done. Figure 8-2 gives the mass spectra for the

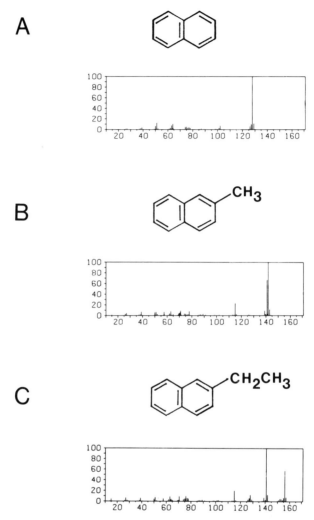

Fig. 8-1. Electron-impact mass spectra of naphthalene (A), 2-methylnaphthalene (B), and 2-ethylnaphthalene (C).

isomeric three-ring compounds phenanthrene and anthracene. Although slight differences are seen, they are not significant enough for positive identification within the experimental variations normally encountered when analyzing real samples. Even in the cases of isomers with very different structures, such as fluoranthene and pyrene, the mass spectra are most often indistinguishable (see Fig. 8-3). Figure 8-4 compares the mass spectra of

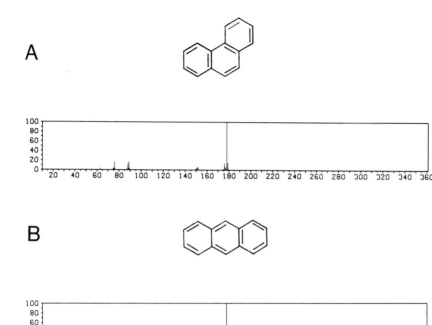

Fig. 8-2. Electron-impact mass spectra of phenanthrene (A) and anthracene (B).

the four-ring isomers triphenylene, chrysene, benz[*a*]anthracene, naphthacene, and benzo[*c*]phenanthrene. All mass spectra are essentially identical except for benzo[*c*]phenanthrene. In this case, steric interaction between the protons on the 1 and 12 carbons facilitate the loss of these two protons with the subsequent formation of the benzo[*ghi*]fluoranthene ion [see Eq. (3)]. Similarly, steric interactions are probably responsible for increased

$$\text{(3)}$$

m/e 228   m/2 226

intensities of the $(M - 1)^+$, $(M - 2)^+$, and $(M - 3)^+$ ions for 4-methylphenanthrene, 1-methyltriphenylene, 12-methylbenz[*a*]anthracene, 4-methylchrysene, and 5-methylchrysene as compared with their respective isomers.

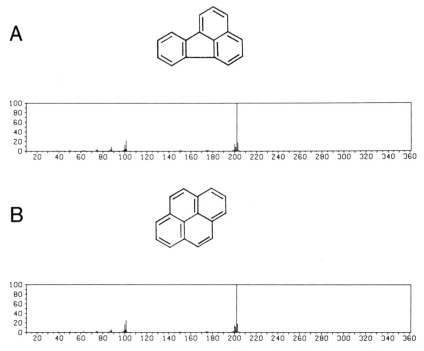

Fig. 8-3. Electron-impact mass spectra of fluoranthene (A) and pyrene (B).

The mass spectra for hydroaromatic compounds such as fluorene and acenaphthene (Fig. 8-5) show the ease of removal of protons from saturated carbons under electron-impact conditions to give abundant $(M - 1)^+$ ions for both compounds and an abundant $(M - 2)^+$ ion for acenaphthene.

The substitution of a heteroatom in the ring makes little difference in the appearance of the mass spectra. Figure 8-6 gives the mass spectra for the three-ring isomers phenanthridine, benzo[h]quinoline, and acridine. Again, the position of nitrogen substitution in the ring makes essentially no difference in the observed spectra. The same results are found in the mass spectra of sulfur-containing PAC (Fig. 8-7). In these spectra, the ^{34}S isotope peaks are present and the $(M + 2)^+$ ion represents the ^{34}S molecular ion. In comparing the mass spectra for dinaphtho[2,1-b:1′,2′-d]thiophene and dinaphtho-[1,2-b: 1′,2′-d]thiophene (Fig. 8-8), large abundances of the $(M - 1)^+$ and $(M - 2)^+$ ions are seen in the former. This is due probably to the formation of the perylo[1,12-bcd]thiophene ion [Eq. (4)] by the same mechanism that

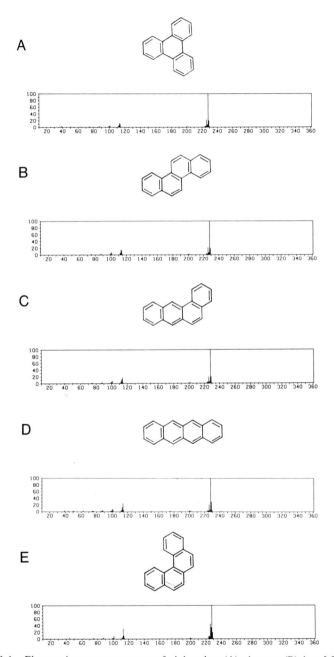

Fig. 8-4. Electron-impact mass spectra of triphenylene (A), chrysene (B), benz[*a*]anthracene (C), naphthacene (D), and benzo[*c*]phenanthrene (E).

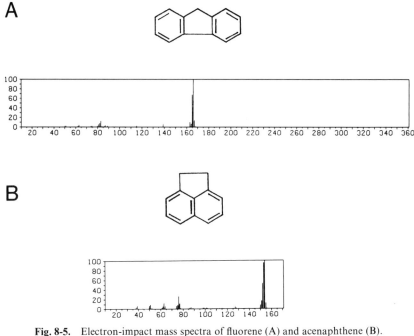

Fig. 8-5. Electron-impact mass spectra of fluorene (A) and acenaphthene (B).

was observed previously [Eq. (3)].

$$m/e\ 284 \xrightarrow{-2\mathrm{H}\cdot} m/e\ 282 \tag{4}$$

A discussion of the mass spectra of PAH would not be complete without a word concerning the benzopyrene/perylene/benzofluoranthene isomers. The mass spectra of these compounds are given in Fig. 8-9. Again, they are indistinguishable by electron-impact mass spectrometry alone. The same results are seen in the mass spectra of the isomeric dibenzopyrenes (5).

B. Group-Type Analysis

The capacity of the early commercial mass spectrometers to give information on the carbon–hydrogen ratios and molecular sizes of hydro-

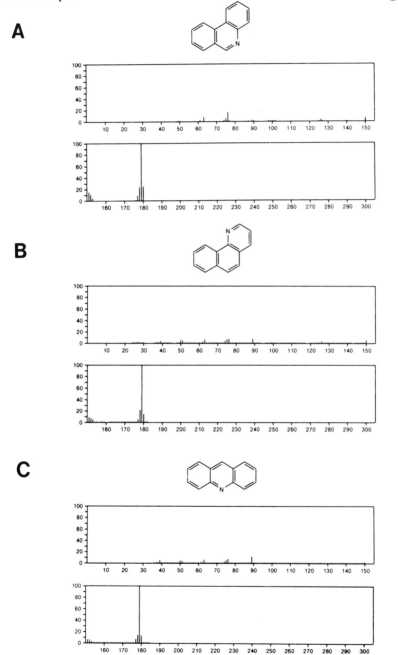

Fig. 8-6.   Electron-impact mass spectra of phenanthridine (A), benzo[*h*]quinoline (B), and acridine (C).

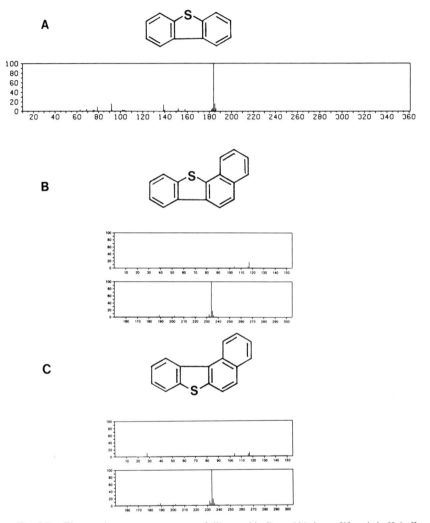

Fig. 8-7. Electron-impact mass spectra of dibenzothiophene (A), benzo[*b*]naphtho[2,1-*d*]-thiophene (B), and benzo[*b*]naphtho[1,2-*d*]thiophene (C).

carbon mixtures led to the widespread use of the technique in the early 1950s for the analysis of petroleum fractions. In 1951, Brown (*6*) showed that compound types could be reliably determined in gasoline by mass spectrometric analysis. After studying the fragmentation patterns of over 170 different compounds, he found that ions which appear at m/e ratios 43, 57, 71, 85, and 99 were generally most abundant in paraffins. Similarily,

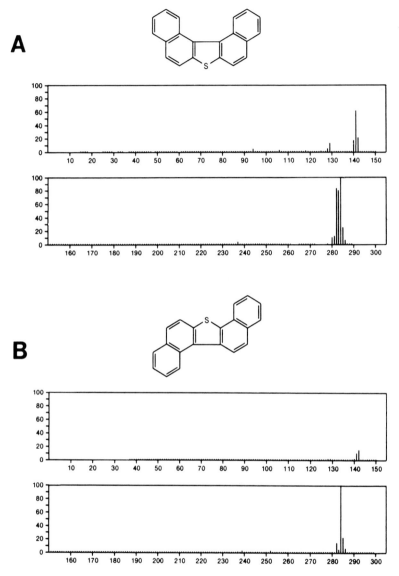

Fig. 8-8. Electron-impact mass spectra of dinaphtho[2,1-*b*:1′,2′-*d*]thiophene (A) and dinaphtho[1,2-*b*:1′,2′-*d*]thiophene (B).

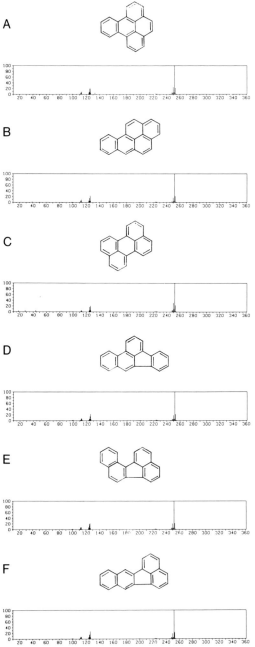

Fig. 8-9.   Electron-impact mass spectra of benzo[*e*]pyrene (A), benzo[*a*]pyrene (B), perylene (C) benz[*e*]acephenanthrylene (D), benzo[*j*]fluoranthene (E), and, benzo[*k*]fluoranthene (F).

relatively large peaks at 41, 55, 69, 83, and 97 appeared to be characteristic of cycloparaffins and monoolefins, whereas ions at masses 67, 68, 81, 82, 95, and 96 are contributed generally by cycloolefins, diolefins, and acetylenes. Aromatic fragments were found at m/e values of 77, 78, 79, 91, 92, 105, 106, 119, 120, 133, and 134. When the intensities of the peaks in each of the four groups were summed, the resultant numbers were used to describe the complex mixture in terms of only four component types.

Brown's (6) work outlined the necessary steps in developing a group-type analysis.

1. The selection of the "analytical" peak or peaks is the first step. Qualitative analysis must precede quantitative analysis. Mass spectrometry requires that some feature or combination of features shown by the complete spectrum be unique to each of the components in the mixture. Only after careful study of the mass spectrum of the mixture of interest and selected standard compounds can group types and their corresponding analytical peaks be selected.

2. The peak heights from the sample spectrum then must be adjusted with the peak response per unit of sample material. This response factor is usually referred to as the sensitivity of the given compound. Different compounds have different sensitivities.

3. Overlapping contributions to analytical peaks from different components in the mixture must be removed. Most mixtures are too complicated to expect to find a single peak unique to each given component. The most common method of doing this is to set up a series of linear simultaneous equations relating peak heights with sensitivities for selected peaks and then solving by computer.

Unlike the gasoline samples analyzed by Brown, most higher molecular weight mixtures contain such a large number of components that a single spectrum cannot be interpreted in terms of compound types. By the late 1950s, however, several mass spectrometer group-type methods were published for determining separately the compositions of saturate and aromatic fractions (7, 8). During this period, high-resolution mass spectrometers were not available, and analyses were done at low resolution. In 1956, Hastings *et al.* (8) described the development of a method for the compound-type analysis of the aromatic fractions of virgin gas oils. Twelve compound types were selected as being most important in these samples, and the characteristic masses chosen for each of three types are given in Table 8-1. Summations were made of the most intense $(M - 1)^+$ peaks for each type. The calibration data are given in Table 8-2, and the final tabulation of the heavy gas oil composition is given in Table 8-3.

Robinson and Cook (9) have described a similar mass spectrometric

TABLE 8-1

Characteristic Mass Sums for Major Compound Types in
Aromatic Fraction of Virgin Gas Oils

Compound type	Mass series
Benzenes	$\Sigma 91 = 91 + 105 + 119 + 133$
Indanes and/or tetralins	$\Sigma 117 = 117 + 131 + 145 + 159$
Dinaphthenebenzenes	$\Sigma 129 = 129 + 143 + 157$
Naphthalenes and trinaphthenebenzenes	$\Sigma 141 = 141 + 155 + 169$
Acenaphthenes and other naphthenenaphthalenes	$\Sigma 167 = 167 + 181 + 195 + 209$
Acenaphthylenes and dinaphthenenaphthalenes	$\Sigma 165 = 165 + 179 + 193 + 207$
Phenanthrenes, anthracenes, and trinaphthenenaphthalenes	$\Sigma 191 = 191 + 205 + 219$
Pyrenes	$\Sigma 215 = 215 + 229 + 213$
Chrysenes	$\Sigma 241 = 241 + 255 + 269$
Benzothiophenes	$\Sigma 147 = 147 + 161 + 175$
Dibenzothiophenes and/or naphthothiophenes	$\Sigma 197 = 197 + 211 + 225$
Naphthobenzothiophenes	$\Sigma 247 = 247 + 261 + 275$

procedure for determining up to 21 compound types in petroleum aromatic fractions. Sums of both the M^+ and $(M - 1)^+$ series were used. A comparison with results obtained from low-voltage mass spectral analysis is included and appears to be quite consistent. An extension of this work has been published (*10*).

Swansiger *et al.* (*11*) extended the knowledge gained from group-type analysis of petroleum products to the analysis of coal liquids. It was found that the petroleum methods did not contain the proper calibration standards to analyze accurately coal liquids. The short-chain alkyl substitution of coal liquids led to the development of a new analytical matrix. The characteristic ions for the group sums in the matrix are listed in Table 8-4. The choices of M^+ and $(M - 1)^+$ ions for the group sums were made because of their strong intensity and inherent resolution. Table 8-5 gives the results of an analysis of a coal liquid aromatic fraction from a Big Horn subbituminous coal.

In the 1960s, high-resolution mass spectrometry was introduced for group-type analysis of petroleum (*12*). The methods were more rapid in that preliminary silica gel separations were not necessary. Ultrahigh-resolution mass spectrometry of petroleum and coal products was introduced in the mid 1970s (*13*). Table 8-6 illustrates the advantages in using high-resolution mass spectrometry. Without the high resolving power, many isomers could

TABLE 8-2

Calibration Data for Analysis of Aromatic Fraction of Virgin Gas Oils by Mass Spectrometry

Mass series	Benzenes	Indanes	Dinaphthene-benzenes	Naphthalenes	Ace-naphthenes	Ace-naphthylenes	Phenanthrenes	Pyrenes	Chrysenes	Benzothiophenes	Dibenzothiophenes	Naphthobenzothiophenes	Diagonal
Σ91	1.0000	0.0500	0.0500	0.0140	0.0140	0.0140	0.0140	0.0200	0.0200	0.0540	0.0110	0.0230	3,500
Σ117	0.0500	1.0000	0.0500	0.0320	0.0100	0.0100	0.0100	0.0200	0.0200	0.0300	0.0080	0.0230	3,100
Σ129	0.0500	0.0500	1.0000	0.0670	0.0170	0.0170	0.0110	0.0330	0.0330	0.0610	0.0110	0.0330	1,800
Σ141	0.0420	0.0420	0.0420	1.0000	0.0230	0.0230	0.0050	0.0120	0.0120	0.0230	0.0050	0.0120	4,300
Σ167	0.0610	0.0610	0.0610	0.0610	1.0000	0.0610	0.0180	0.0180	0.0120	0.0610	0.0180	0.0120	1,650
Σ165	0.0880	0.0880	0.0880	0.0880	0.0880	1.0000	0.0880	0.0260	0.0260	0.0880	0.0880	0.0260	1,130
Σ191	0.0670	0.1330	0.1780	0.2220	0.3780	0.3560	1.0000	0.0440	0.0440	0.2670	0.2220	0.0890	450
Σ215	0.1860	0.1860	0.1860	0.2330	0.2330	0.2330	0.4650	1.0000	0.4650	0.2330	0.2790	0.2330	430
Σ241	0.1540	0.1150	0.1150	0.0770	0.0770	0.0770	0.1350	0.0960	1.0000	0.0770	0.1150	0.1920	520
Σ147	0.1600	0.1440	0.1440	0.0800	0.0400	0.0400	0.0160	0.0240	0.0240	1.0000	0.0080	0.0240	1,250
Σ197	0.1040	0.1040	0.1040	0.1300	0.1560	0.1560	0.1300	0.0390	0.0390	0.1430	1.0000	0.0390	770
Σ247	0.0360	0.0360	0.0360	0.1400	0.1950	0.1950	0.2440	0.1220	0.1220	0.1460	0.1950	1.0000	410

TABLE 8-3

Mass Spectral Analysis of a Heavy Gas Oil before and after Desulfurization

Stock compound type (%)	Feed (as is)	Gas oil product from run (feed basis)	
		A	B
Saturates[a]	52.8	46.2	47.8
Benzenes	6.4	7.8	7.2
Indanes	3.2	3.7	3.4
Dinaphthenebenzenes	3.8	3.0	2.6
Naphthalenes	0.9	0.8	0.6
Acenaphthenes	3.7	3.6	3.2
Acenaphthylenes	5.5	5.3	4.7
Phenanthrenes	3.2	3.0	3.0
Pyrenes	0.8	1.5	1.3
Chrysenes	2.0	0.9	0.6
Benzothiophenes	3.7	2.1	1.3
Dibenzothiophenes	6.1	2.4	0.8
Naphthobenzothiophenes	1.4	0.3	0.1
Sulfides[b]	6.5	0.0	0.0
% S (actual)	1.68	0.46	0.20
% S (calculated from analysis)	1.64	0.44	0.21

[a] Determined by silica gel.
[b] Determined by iodine complex.

not be separated and individually determined. Peters and Bendoraitis (*14*) have described a high-resolution (15,000–25,000) mass spectrographic method for the analysis of nitrogen- and oxygen-containing material derived from petroleum. Figure 8-10 shows a photograph of the oscilloscope display of a quartet at m/e 279 from their work. The exact mass, formula, relative intensity, and mass measurement error in millimass units for each peak are given.

C. Low-Voltage Electron-Impact Mass Spectrometry

Although much work is continuing in the normal electron-impact mass spectrometry of PAC mixtures, the use of low-ionizing-voltage mass spectrometry has become widespread for the analysis of complex samples. The original description of the method and many of its advantages and difficulties was given by Field and Hastings (*15*) and later by Lumpkin (*16*). The mass

TABLE 8-4

Characteristic Ions for Group Sums in Matrix

Type	Characteristic ions (MW)	Z no.
Benzopyrenes	251, 252, 265, 266, 279, 280	−28
Chrysenes	227, 228, 241, 242, 255, 256	−24
Decahydropyrenes	211, 212, 225, 226, 239, 240	−12
Hexahydropyrenes	193, 194, 207, 208, 221, 222, 235, 236,	−16
Tetrahydrofluoranthenes	205, 206, 219, 220, 233, 234	−18
Pyrenes/fluoranthenes	201, 202, 215, 216, 229, 230, 243, 244, 257, 258, 271, 272	−22
Dihydropyrenes	189, 190, 203, 204, 217, 218, 231, 232, 245, 246, 259, 260	−20
Octahydrophenanthrenes	185, 186, 199, 200, 213, 214	−10
Hexahydrophenanthrenes	183, 184, 197, 198	−16
Tetrahydrophenanthrenes	181, 182, 195, 196, 209, 210, 223, 224	−14
Phenanthrenes	177, 178, 191, 192	−18
Fluorenes/dihydrophenanthrenes	165, 166, 179, 180	−16
Acenaphthenes/biphenyls/dibenzofuran	153, 154, 167, 168	−14
Tetralins/benzofuran	103, 104, 117, 118, 131, 132, 145, 146	−8
Tetrahydroacenaphthenes	129, 130, 157, 158, 171, 172	−10
Naphthalenes	128, 141, 142, 155, 156, 169, 170	−12
Benzenes	77, 78, 79, 91, 92, 105, 106, 119, 120	−6

spectral peaks, both parent and fragments, of mixture components are often obscured by other fragment peaks in the normal spectra produced at 70 eV. When the mass spectrometer is operated so that the bombarding electrons have only sufficient energy to form the molecular ion and insufficient energy to cause carbon–hydrogen or carbon–carbon bond cleavage, only parent ions are observed in the spectra. If the ionizing voltage is selected at a relatively low value, only the molecular ions of compounds having an ionization potential at or below the selected voltage are formed. Since most of the saturated hydrocarbon types have ionization potentials in the 10 to 13 eV

TABLE 8-5

Total Aromatic Fraction of Coal Liquid

Z no.	Compound type	Group sum	Partial volume	Volume % (based on 100)
−28	Benzopyrenes	219.3	565.9	2.0
−24	Chrysenes	408.7	1092.9	3.9
−12	Decahydropyrenes	447.2	2214.9	7.9
−16	Hexahydropyrenes	976.7	4250.9	15.3
−18	Tetrahydrofluoranthenes	571.5	1271.9	4.6
−22	Pyrenes/fluoranthenes	2349.1	4490.4	16.1
−20	Dihydropyrenes	1618.2	2863.7	10.3
−10	Octahydrophenanthrenes	875.3	2661.6	9.6
−16	Hexahydrophenanthrenes	394.4	−51.9	0.0
−14	Tetrahydrophenanthrenes	1298.8	5106.5	18.3
−18	Phenanthrenes	1434.9	1729.2	6.2
−16	Fluorenes/dihydrophenanthrenes	1743.5	−240.7	0.0
−14	Acenaphthenes/biphenyls	1090.9	−333.3	0.0
−8	Tetralins	895.4	1374.5	4.9
−10	Tetrahydroacenaphthenes	882.3	197.5	0.7
−12	Naphthalenes	1295.2	−398.3	0.0
−6	Benzenes	588.4	44.9	0.2
	Totals	17,089.8	27,864.8	100.0

range, and the double-bond-containing molecules have ionization potentials below about 10 eV, it is possible to select an ionizing voltage which will give a spectrum composed almost exclusively of the parent peaks of double-bond-containing molecules. A major advantage of this technique is that the absence of fragment peaks makes unnecessary the use of simultaneous equations to correct for fragment ion contributions to peak intensities. The major disadvantages are the relatively low sensitivity and reproducibility of the method.

Mass sensitivity data must still be calculated when using the low-voltage technique. Lumpkin (16) measured the sensitivities of a number of aromatic compounds at an ionizing voltage of 7.2 eV. As can be seen from Fig. 8-11, the sensitivities increase rapidly with size of the parent compound. The effect of alkyl substitution on sensitivity can also be seen. The sensitivity values increase initially, reach maxima at C_4-substituted benzenes and C_2-substituted naphthalenes, and then decrease regularly. The unavailability of standard compounds has limited the determination of sensitivities for a wide range of compounds. The work by Lumpkin and Aczel (17) allowed the prediction of sensitivities for many aromatic compounds which

TABLE 8-6

Mass Doublets and Resolving Power (RP)

Doublet	ΔMass	Example	Max mass at RP		
			10 K	80 K	150 K
H_{12}–C	0.094	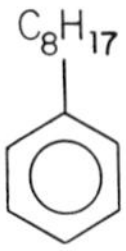 C_8H_{17} $C_{14}H_{22}$ 190.1721 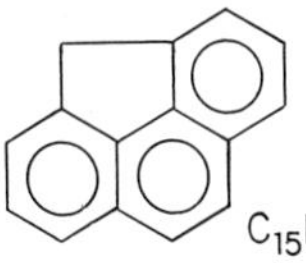 / $C_{15}H_{10}$ 190.0782	940	7520	14000
C_2H_8–S	0.091	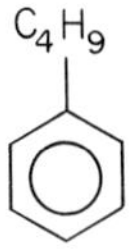 C_4H_9 $C_{10}H_{14}$ 134.1095 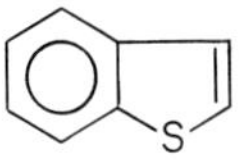 / C_8H_6S 134.0190	910	7280	13500
CH_4–O	0.0364	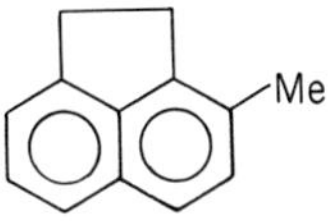 Me $C_{13}H_{12}$ 168.0939 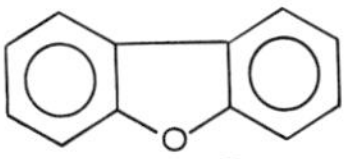 / $C_{12}H_8O$ 168.0575	364	2910	5460
CH_2–N	0.0126	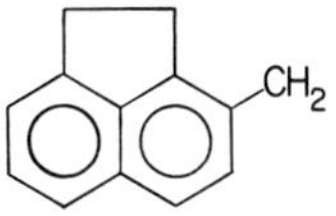 CH_2 $C_{13}H_{11}$ 167.0861 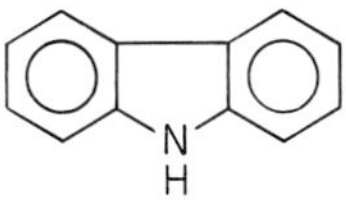 / $C_{12}H_9N$ 167.0735	126	1010	1890
SH_4–C_3	0.0034	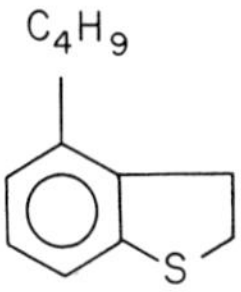 C_4H_9 $C_{12}H_{14}S$ 190.0816 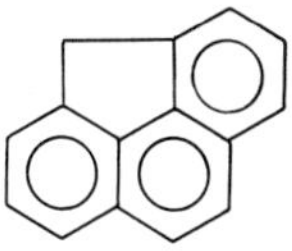 / $C_{15}H_{10}$ 190.0782	34	270	500

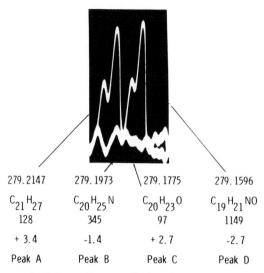

279.2147	279.1973	279.1775	279.1596
$C_{21}H_{27}$	$C_{20}H_{25}N$	$C_{20}H_{23}O$	$C_{19}H_{21}NO$
128	345	97	1149
+ 3.4	-1.4	+ 2.7	-2.7
Peak A	Peak B	Peak C	Peak D

Fig. 8-10. Photograph of the oscilloscope display of a quartet at m/e 279, including the exact mass, formula, relative intensity, and mass measurement error in millimass units for each peak. [Reproduced with permission from A. W. Peters and J. G. Bendoraitis, *Anal. Chem.* **48**, 968 (1976). Copyright, American Chemical Society.]

were not available from any source, and the decrease in sensitivity upon alkylation was measured for many compound types. Shultz *et al.* (*18*), reported that the sensitivities of hydroaromatic compounds could be predicted from the sensitivity of the corresponding aromatic ring compound, and Schiller (*19*) synthesized and measured the sensitivities of a number of methylated PAH. The potential of the combination of high-resolution and low-voltage techniques was recognized by Lumpkin (*20*). The peak-matching mass measurement technique (*21*) applied by him is extremely lengthy when applied to very complex mixtures, and is not applicable to the determination of minor and trace components. Johnson and Aczel (*22*) described a method for making mass measurements directly from the recorded mass spectrum which greatly reduced the time of analysis and permitted the identification of trace components. A portion of a typical spectrum obtained under these conditions is shown in Fig. 8-12.

The computer system necessary to handle the enormous number of data generated by high-resolution low-voltage analysis was described by Aczel *et al.* (*23*). The use of computer systems for data acquisition is necessary for the rapid calculation of formulas and intensities in a format suitable for further calculations and quantitative analysis. A novel system of reference standards has been used (*23*) for mass calibration, because none of the

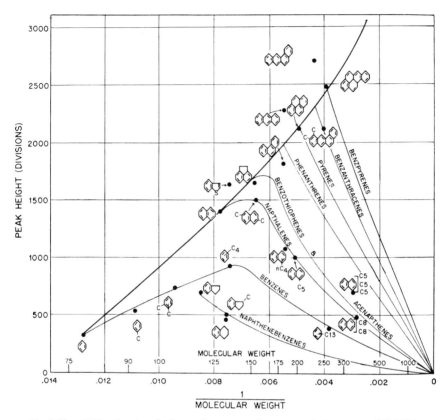

Fig. 8-11. Calibration data for low-voltage aromatic compounds; benzene = 325 divisions. [Reproduced with permission from H. E. Lumpkin, *Anal. Chem.* **30**, 321 (1958). Copyright, American Chemical Society.]

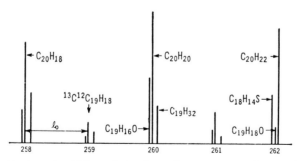

Fig. 8-12. Portion of a high-resolution low-voltage spectrum of a petroleum mixture. [Reproduced with permission from B. H. Johnson and T. Aczel, *Anal. Chem.* **39**, 682 (1967). Copyright, American Chemical Society.]

conventional compounds used for this purpose (perfluorokerosene, hepta-cosaperfluorobutylamine) yield peaks at low ionization voltages. Table 8-7 lists the reference standards. A blend of these compounds is suitable as a standard for hydrocarbon samples, since they yield masses with considerable negative mass defects and characteristic isotopes, and can easily be recognized either by examination of the printed output, or automatically by the computer.

The recent interest in nitrogen and oxygen heterocycles in petroleum has led to the use of mass-spectral group-type analysis for their detection (24–27).

Although group-type mass-spectral analysis has been used most extensively in the analysis of petroleum, the techniques have recently been extended to the study of coal and coal-derived products (11, 13, 28–32) and the analysis of environmental pollutants (33–46). The typical analytical approach is to introduce an aliquot of the PAC sample into the mass spectrometer through the direct introduction probe system and slowly vaporize

TABLE 8-7

Reference Blend for Precise Mass Measurement at Low Voltages

	Peaks used			
Component	1	2	3	4
Pyrrole	67.042197			
Fluorobenzene	96.037525			
Chlorobenzene	112.007976	114.005026		
Chlorofluorobenzene	129.998554	131.995604		
Dichlorobenzene	145.969005	147.966055	149.963105	
Bromobenzene	155.957513	157.955543		
Chloronaphthalene	162.023625	164.020675		
Trichlorobenzene	179.930033	181.927083	183.924133	
Chlorobromobenzene	189.918543		193.912643	
Bromonaphthalene	205.973162	207.971192		
Tetrachlorothiophene	219.847485	221.844396	223.841586	225.838636
Iodochlorobenzene	237.904812	239.901862		
Iodonaphthalene	253.959432			
Perfluoronaphthalene	271.987218			
Perfluoroxylene	285.984022			
Dibromotetrafluorobenzene	305.830389	307.828419	309.826449	
Perfluorodiphenyl	333.984022			
Perfluoroacetophenone	361.978930			
Diiodotetrafluorobenzene	401.802929			
Tetrabromomonofluorophenol	425.672731	427.670761	429.668791	
Octafluorodibromodiphenyl	453.823998	455.822028	457.820058	
Hexabromobenzene	549.506400	551.504430	553.502460	

the sample with increasing temperature while the data is being collected. In the case of high-resolution mass spectrometry, a number of timed exposures on a photographic plate is made. These plates are developed, read on a comparator, and the exact masses with intensities are recorded. Plots can then be constructed that display the relative abundance of each parent compound and its alkyl derivatives as a function of carbon number. These are referred to as *alkyl homolog plots*.

With a low-resolution instrument which is interfaced to a computer, data from repetitive scans can be stored and later summed to give the same type of results. The high-resolution mass spectrometric procedure gives the most reliable data because it can discriminate between m/e ratios that differ slightly in mass in the third or fourth decimal place. Therefore, impure PAC fractions can be analyzed without significant interferences from unrelated mass fragments. On the other hand, a high abundance of any ion can saturate the spectral line on the photographic plate, resulting in incorrect comparisons. In using low-resolution instruments, the PAC fractions must

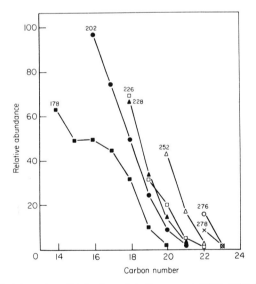

Fig. 8-13. Alkyl homolog distribution plots for several PAH series in coal combustion products. The lines are labeled by the molecular weight of the unsubstituted species. Example isomers are 178 ($Z = -18$) phenanthrene, 202 ($Z = -22$) pyrene, 226 ($Z = -26$) cyclopenta-[cd]pyrene, 228 ($Z = -24$) chrysene, 252 ($Z = -28$) benzo[a]pyrene, 276 ($Z = -32$) indeno-[1,2,3-cd]pyrene, and 278 ($Z = -30$) dibenzanthracene. The value of Z is derived from the general formula C_nH_{2n+Z}. [Reproduced with permission from M. L. Lee, G. P. Prado, J. B. Howard, and R. A. Hites, *Biomed. Mass Spectrom.* **4**, 182 (1977). Copyright, Heyden and Son, Ltd.]

be carefully purified of all other organic compounds and separated into each group of homologs (parent compound plus alkyl substituents) before mass spectral analysis. Again, it must be emphasized that the molecular ion abundances are not a linear function of compound abundances; therefore, the comparisons are only semiquantitative indications of the relative abundances of parent PAC and their alkyl-substituted derivatives.

In recent studies by Blumer *et al.* (*38–40*), alkyl homolog plots were constructed for PAH isolated from different sediments, soils, and fossil fuels. Their analyses confirmed the presence in soils and sediments of homologs extending to at least seven alkyl carbons, with the unsubstituted polycyclic hydrocarbons usually being the most abundant. On the other hand, PAH homologs from fossil fuels displayed the highest abundance in the three or four carbon alkyl-substituted PAH. Using a similar procedure, Hites *et al.* (*41, 42*) found a similar trend in PAH profiles from sediments. Lee *et al.* (*43*) found that the alkyl homolog plots from air particulate matter in Indianapolis (a high coal-burning area) resemble plots from coal combustion products while those from air particulate matter in Boston (a low coal-burning area) resembled plots from PAH collected from a kerosene turbulent diffusion burner. Figure 8-13 shows representative alkyl homolog plots for several PAH series in coal combustion products. The comparison of the pyrene-type homolog series for various combustion products as compared to air particulate matter is shown in Fig. 8-14.

Blumer *et al.* (*39*) have reported the homolog distribution of nitrogen-containing PAC in recent marine sediments. These azaarenes range from three- to eight-membered rings, with homologs containing up to eight alkyl carbons.

D. Gas Chromatography/Mass Spectrometry

The group-type analysis of PAC mixtures has provided a great amount of information concerning their general compositions. However, the exact identification and quantification of individual constituents is most desirable. The only method that can even approach this task today is combined gas chromatograhy/mass spectrometry. As discussed in Chapter 7, the use of high-resolution capillary gas chromatographic columns has greatly increased during the last several years. The maximum resolution of mixture components before mass-spectral analysis is of utmost importance in providing unambiguous identifications of individual compounds. This is especially true in the case of PAC because conventional mass spectra of many isomers are identical. The mass spectrometer is therefore unable to provide the correct structure by itself and, therefore, prior separation by high-resolution capillary columns is essential.

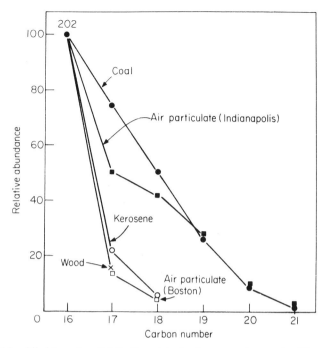

Fig. 8-14. Alkyl homolog distribution plots for the pyrene-type series ($Z = -22$) in the combustion products of coal, wood, and kerosene, and in air particulate matter from Indianapolis and Boston. The abundance of the parent compound in each series was normalized to 100. [Reproduced with permission from M. L. Lee, G. P. Prado, J. B. Howard, and R. A. Hites, *Biomed. Mass Spectrom.* **4**, 182 (1977). Copyright, Heyden and Son, Ltd.]

The gas chromatographic peaks obtained from capillary columns are often extremely narrow, sometimes only a few seconds. This means that the scan times of the mass spectrometer must be short in order to obtain several spectra per peak and without having too much distortion of the mass spectra as a result of the changing concentration along the peak elution profile. Modern GC/MS/computer systems have been built that easily handle four mass spectra per second. The quadrupole mass spectrometer is particularly well-suited for this purpose. The claimed advantages of this instrument are (a) its initial relatively low cost, (b) its straightforward maintenance and repair, and (c) its extremely high-speed linear mass scan, simplifying system control, data logging, and spectra interpretation (47). The disadvantages of the quadrupole are its low mass resolution and low mass sensitivity at higher masses.

With the short cycle times for obtaining the mass spectra, the number of spectra taken during the gas chromatographic run can easily run into thousands. At four mass spectra per second, in 45 min the number of mass spectra is already more than 10,000. These figures emphasize the need for computerized data acquisition and reduction. In modern instruments, final results are produced at high speed and with higher accuracy than by manual data processing. Computerized data reduction also allows for the subtraction of spectral background, thus preventing misinterpretation or confusion of mass spectra with spurious ion interferences contributed by GC column bleeding or other impurities.

The power of the combined GC/MS system for the analysis of PAC is demonstrated by the numerous applications found in the literature. Since 1972, GC/MS instrumentation has been used to separate and identify PAC in air particulate matter (*48–63*), combustion effluents (*41–43, 50, 53, 64–71*), coal tar (*50, 53, 72–74*), aqueous industrial effluents (*69, 75*), fossil fuels (*76–80*), tobacco- and marijuana-smoke condensate (*81–84*), carbon blacks (*85*), forest fire smoke (*86*), and in sediments (*41, 42, 45, 87*). A large number of these studies made use of the resolving power of capillary column gas chromatography prior to mass spectrometry (*48, 51, 54, 57, 59, 63, 69, 71–75, 77, 80, 83, 85, 87*). Undoubtedly, this technique will find even more widespread use in the future.

The quantitative analysis by GC/MS of individual components in a complex mixture has always been a problem of great concern. Other substances with identical retention times interfere, giving values too high or too low. In 1968 (*88*), mass fragmentography was introduced as a means to avoid many of these problems. The first of several different approaches is single ion monitoring (SIM). In this method the mass spectrometer is focused on one m/e value characteristic of the compound under consideration. In this mode of operation, the mass spectrometer is nothing but a detector, but a sensitive and selective one. SIM sensitivity is higher than the regular scan mode sensitivity because the mass spectrometer monitors only a few m/e values instead of scanning a wide atomic mass unit range. In addition, the change of signal as a function of time is slow, allowing amplifiers with a small bandwidth. This improves the signal-to-noise ratio. The amount necessary for quantitative measurement is in the low picogram range for most compounds.

In another method of operation, the mass spectrometer jumps sequentially over a number of preset m/e values, characteristic for the substance under investigation (*89*). Mass chromatograms are obtained for each m/e value chosen. This method, in which more than one ion is monitored, is called multiple ion detection (MID). Sensitivity is increased also in this mode because of the longer times spent at each chosen mass.

If the GC/MS system is coupled to a computer, mass chromatograms of any desired masses can be generated. During the gas chromatographic run,

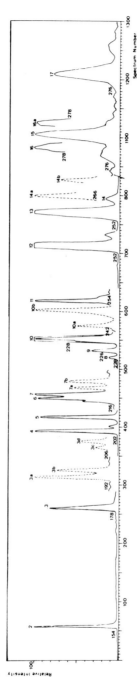

Fig. 8-15. Mass chromatograms for polyaromatic hydrocarbons. Peaks: 2 = biphenyl; 3 = phenanthrene, anthracene; 3a = methylphenanthrene; 3b = methylanthracene; 3c = ethylphenanthrene; 3d = ethylanthracene; 4 = fluoranthene; 5 = pyrene; 6 = benzo[*a*]fluorene; 7 = benzo[*c*]fluorene; 7a = methylfluoranthene; 7b = methylpyrene; 8 = benzo[*c*]phenanthrene; 9 = benzo[*ghi*]fluoranthene; 10 = benz[*a*]anthracene, chrysene; 10a = methylbenz[*a*]anthracene; 10b = methylchrysene; 11 = 2,2′-binaphthyl; 12 = benzo[*k*]fluoranthene, benzo[*b*]fluoranthene; 13 = benzo[*a*]pyrene, benzo[*e*]pyrene; 14 = perylene; 14a = methylbenzo[*k*/*b*]fluoranthene; 14b = methylbenzo[*a*/*e*]pyrene; 15 = benzo[*b*]-chrysene, indeno[1,2,3-*cd*]pyrene; 16 = dibenzanthracenes; 16a = picene, benzo[*c*]tetraphene; 17 = benzo[*ghi*]perylene, ananthrene. [Reproduced with permission from W. Cautreels and K. Van Cauwenberghe, *J. Chromatogr.* **131**, 253 (1977). Copyright, Elsevier Sci. Publ. Co.].

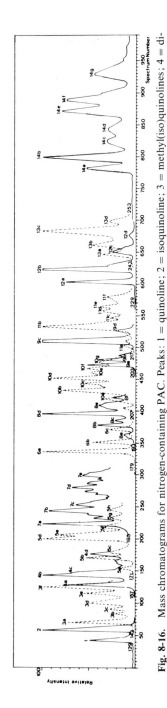

Fig. 8-16. Mass chromatograms for nitrogen-containing PAC. Peaks: 1 = quinoline; 2 = isoquinoline; 3 = methyl(iso)quinoline; 4 = dimethyl(iso)quinolines; 5 = trimethyl(iso)quinolines; 6 = acridine, phenanthridine, and benzo(iso)quinolines; 7 = tetramethyl(iso)quinolines; 8 = methylacridines, methylphenanthridines, and methylbenzo(iso)quinolines; 9 = azafluoranthenes and azapyrenes; 10 = dimethylacridines, dimethylphenanthridines, and dimethylbenzo(iso)quinolines; 11 = azabenzofluorenes, methylazapyrenes, and methylazafluoranthenes; 12 = azabenz[*a*]anthracenes, azachrysenes, and dibenzo(iso)quinolines; 13 = methylbenzacridines, methylbenzophenanthridines, and methyldibenzo-(iso)quinolines; 14 = azabenzopyrenes and azabenzofluoranthenes. [Reproduced with permission from W. Cautreels and K. Van Cauwenberghe, *J. Chromatogr.* **131**, 353 (1977). Copyright, Elsevier Sci. Publ. Co.]

the mass spectrometer scans periodically while the mass spectra are stored in the computer. After the run is over, the computer can then be asked for any desired mass chromatogram. This method is not as sensitive as the SIM method because the instrument is continually scanning. A series of mass chromatograms generated from the analysis of air particulate matter is shown in Figs. 8-15 and 8-16 (55). In most cases the molecular ion was chosen as the characteristic mass.

III. CHEMICAL IONIZATION AND NEGATIVE IONS

Conventional chemical ionization mass spectrometry, using methane as the reagent gas, produces mass spectra of PAC that appear quite similar to those produced by electron impact. Two major differences are that the most abundant ion is the $(M + 1)^+$ or protonated molecular ion, and the second most abundant ion is the $(M + 29)^+$ ion. Because of the proton affinities of PAC, the major ionization process is protonation of the aromatic compound by the reagent gas, which produces the $(M + 1)^+$ ion. Of secondary importance is the addition of an ethyl group to the aromatic system to produce the $(M + 29)^+$ ion. Munson and Field (90) have discussed the

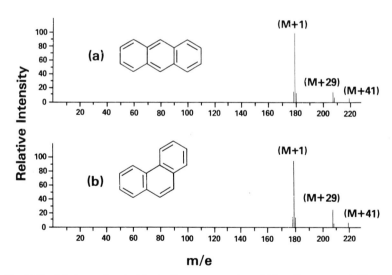

Fig. 8-17. Methane chemical ionization mass spectra of (a) anthracene and (b) phenanthrene. [Reproduced from M. L. Lee, D. L. Vassilaros, W. S. Pipkin, and W. L. Sorensen, Trace Organic Analysis: A New Frontier in Analytical Chemistry, NBS Spec. Publ. 519, U.S. Gov. Printing office, Washington, D. C., 1979.]

chemical ionization of a number of alkylbenzenes and two alkylnaphthalenes, but little has been published on larger ring systems. Recent studies (63, 91, 92), however, do indicate that the mass spectra of different isomers appear in most cases to be identical. Figure 8-17 compares the methane chemical ionization mass spectra of anthracene and phenanthrene.

The use of a mixed charge exchange–chemical ionization reagent gas for mass spectrometry of PAH that differentiates between isomers has recently been reported (63, 91, 93). The theory behind the use of the mixed reagent gas was discussed by Arsenault (94) and later by Beggs (95). When a charge exchange–chemical ionization reagent gas such as an argon and methane mixture is used under chemical ionization conditions, a number of reactions occur. Argon is first ionized to Ar^+ by the electron beam [Eq. (5)].

$$Ar + e^- \rightarrow Ar^+ + 2e^- \tag{5}$$

Charge transfer then occurs between the Ar^+ and PAH molecule to give an abundant molecular ion as in electron impact [Eq. (6)]. Ionization of methane

$$M + Ar^+ \rightarrow M^+ + Ar \tag{6}$$

also occurs by the electron beam [Eq. (7)] or by the Ar^+ ion [Eq. (8)]. Further

$$CH_4 + e^- \rightarrow CH_4^+ + 2e^- \tag{7}$$

$$CH_4 + Ar^+ \rightarrow CH_4^+ + Ar \tag{8}$$

reaction with another methane molecule gives rise to the reactant ion CH_5^+ [Eq. (9)] which then reacts with the PAH molecule to produce an abundant

$$CH_4^+ + CH_4 \rightarrow CH_5^+ + CH_3 \tag{9}$$

$(M + 1)^+$ ion [Eq. (10)]. The relative rates of the charge exchange [Eq. (6)] and proton exchange [Eq. (10)] reactions, and hence the ratio of the abun-

$$M + CH_5^+ \rightarrow MH^+ + CH_4 \tag{10}$$

dance of the $(M + 1)^+$ ion to the abundance of the M^+ ion, will vary according to the proton affinity and/or ionization potential of each PAH. Since the ionization potentials of PAH isomers are dependent on the specific structure of the molecule, the argon–methane reagent can produce quite different spectra for different isomers. This is seen by comparison of the argon–methane chemical ionization (CI) spectra for anthracene and phenanthrene (Fig. 8-18).

In applications involving GC/MS, selected ions can be monitored for increased sensitivity. Since the molecular and protonated molecular ions are the important ions in mixed charge exchange–chemical ionization, they can

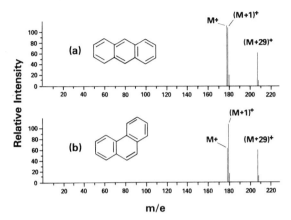

Fig. 8-18. Argon–methane chemical ionization mass spectra of (*a*) anthracene and (*b*) phenanthrene. [Reproduced from M. L. Lee, D. L. Vassilaros, W. S. Pipkin, and W. L. Sorensen, Trace Organic Analysis: A New Frontier in Analytical Chemistry, NBS Spec. Publ. 519, U.S. Gov. Printing Office, Washington, D.C., 1979.]

be simultaneously monitored. Figure 8-19 shows the selected ion plots (or mass chromatograms) for the molecular and protonated molecular ions of the methylanthracene/methylphenanthrene region in air particulate matter. The ratios are characteristic for the methylphenanthrenes. However, the position of alkylation cannot be determined by this method.

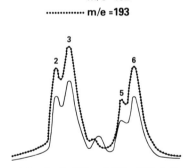

Fig. 8-19. Selected ion plots of *m/e* 192 and 193 of Utah County air particulate matter. Peak numbers refer to compounds identified as (2) 3-methylphenanthrene; (3) 2-methylphenanthrene; (5) 9-methylphenanthrene; and (6) 1-methylphenanthrene. [Reproduced from M. L. Lee, D. L. Vassilaros, W. S. Pipkin, and W. L. Sorensen, Trace Organic Analysis: A New Frontier in Analytical Chemistry, NBS Spec. Publ. 519, U.S. Gov. Printing Office, Washington, D.C., 1979.]

It has been determined (91) that the $(M + 1)/M$ ratio has a high positive correlation with ionization potential ($r = 0.877$, $P \ll 0.01$). This trend results from the fact that as the ionization potential increases, charge transfer processes will be less effective for electron extraction while at the same time protonation becomes more favorable. Since many standard PAH are not available for comparison, structures of unknown compounds may be predicted by calculating the ionization potentials of suspected compounds from molecular orbital theory and comparing them with the mixed charge exchange–chemical ionization mass spectra.

There has been one report that the use of pure methane at low ion source pressures produces spectra for PAH similar to those obtained with argon–methane (96).

Hunt et al. (97–100) have recently described the use of oxygen as reagent in both positive and negative chemical ionization mass spectrometry of PAH. Electron bombardment of oxygen at 1 torr pressure yields O_2^+ and O^+ in the positive ion mode [Eq. (11)], and O_2^- and O^- in the negative ion mode [Eq. (12)]. Addition of hydrogen (10%) to the reagent gas eliminates O^- [Eq. (13)].

$$O_2 + e^- \rightarrow O^+ + O_2^+ \tag{11}$$

$$O_2 + e^- \rightarrow O^- + O_2^- \tag{12}$$

$$O^- + H_2 \rightarrow H_2O + e^- \tag{13}$$

The elimination of O^- is necessary, since the more highly exothermic reactions of O^- with PAH provide little information unique to that particular system. The reaction of O_2^+ with aromatic systems like pyrene [Eq. (14)]

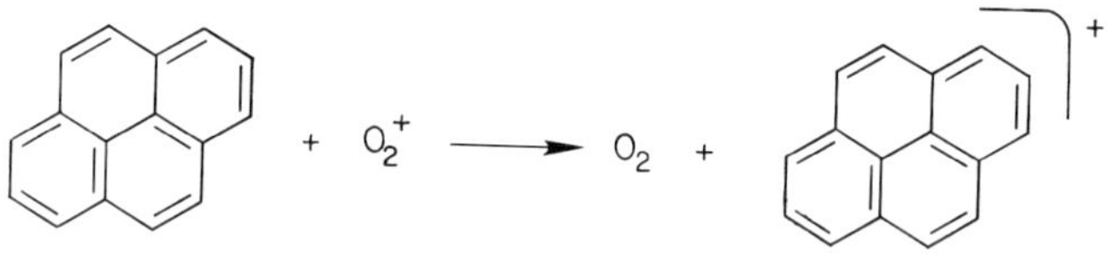

$$\tag{14}$$

yields spectra containing a single ion, M^+ (see Fig. 8-20). Thus ionization by O_2^+ is the CI equivalent of the low-voltage electron impact technique, except that the CI method does not sacrifice sensitivity. In the negative ion mode, CI (O_2/H_2) spectra show an abundant $(M + 15)^-$ ion formed by either capture of a thermal electron followed by reaction with neutral oxygen [Eq. (15)], or by direct reaction with O_2^- [Eq. (16)].

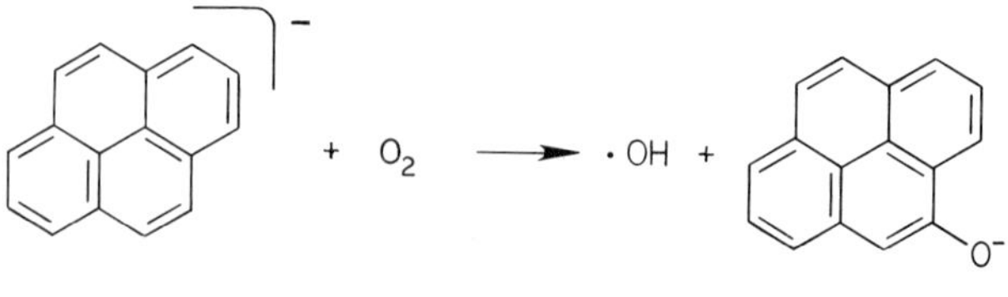

$$\tag{15}$$

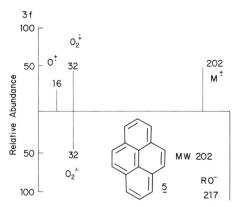

Fig. 8-20. Pulsed positive–negative ion chemical-ionization (PPNICI) mass spectrum of pyrene. [Reproduced with permission from D. F. Hunt, G. C. Stafford, Jr., F. W. Crow, and J. W. Russell, *Anal. Chem.* **48**, 2098 (1976). Copyright, American Chemical Society.]

$$+ \ O_2^- \longrightarrow \cdot OH \ + \qquad\qquad \tag{16}$$

Simultaneous recording of positive- and negative-ion CI mass spectra can be accomplished by pulsing the polarity of the ion source potential and the focusing lens potential (*98*). Under these conditions, packets of positive and negative ions are ejected from the ion source in rapid succession. Using a quadrupole mass filter, ions of identical m/e but different polarity traverse the quadrupole field and exit the rods at the same point. Detection of ions is accomplished simultaneously by two continuous diode multipliers operating with first dynode potentials of opposite polarity. The result is that positive and negative ions are recorded simultaneously as deflections in opposite direction on a conventional light-beam oscillograph (Fig. 8-20).

An example of the differentiation between isomeric PAH using pulsed positive-ion/negative-ion chemical ionization can be seen by comparing the spectra obtained for benzo[*ghi*]perylene and indeno[1,2,3-*cd*]pyrene (Fig. 8-21). Both isomers exhibit a single ion corresponding to M^+ in the positive ion mode, but in the negative mode, benzo[*ghi*]perylene shows ions corresponding to M^- and $(M + 15)^-$ in a 1:1 ratio while indeno[1,2,3-*cd*]pyrene has a much more abundant M^- ion. The presence of the five-membered ring in the indenopyrene facilitates formation of the stable negative ion. Similar results were obtained in the comparison between fluoranthene and pyrene when using O_2 as reagent and CO as the O^- scavenger (*100*).

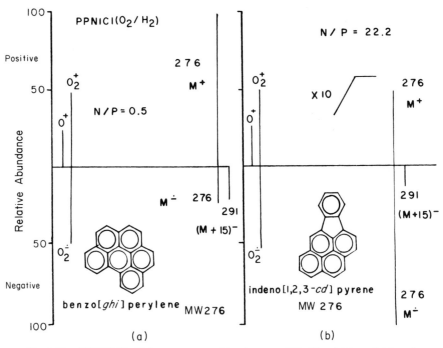

Fig. 8-21. PPNICI (O_2/H_2) mass spectra of two isomers of $C_{22}H_{12}$: (*a*) benzo[*ghi*]perylene and (*b*) indeno[1,2,3-*cd*]pyrene. The intensity of reagent ions is 50 to 100 times greater than as shown. [Reproduced with permission from D. F. Hunt and S. K. Sèthi, *in* "High Performance Mass Spectrometry: Chemical Applications" (M. L. Gross, ed.), p. 150. ACS Symp. Ser. 70, Am. Chem. Soc., Washington, D.C., 1978.]

Negative-ion CI (O_2/H_2) spectra of aromatic sulfides (*98*) such as dibenzo-thiophene (Fig. 8-22) show ions corresponding to M^- and $(M + O_2)^-$. The $(M + O_2)^-$ ion is due most likely to the formation of a sulfone type structure.

The use of pulsed positive- and negative-ion chemical ionization mass spectrometry as described above allows one to determine the molecular weight of the sample from the positive-ion spectrum and to determine the structural identity from the negative-ion spectrum. Furthermore, the electron-capturing characteristics of PAH facilitate the identification and quantification at subpicogram levels of many of these compounds.

Hunt *et al.* (*100*) have also shown that careful selection of reagent gases based on proton affinities can provide a means for differentiating PAH isomers. The use of C_2H_5OD as a reagent gas can produce different spectra for phenanthrene and anthracene. The $C_2H_5OD_2^+$ reagent ion donates a D^+ ion to anthracene, giving $(M + D)^+$ as the predominant ion, while exchange

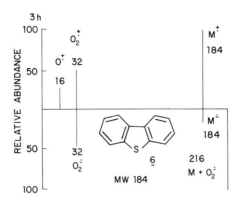

Fig. 8-22. PPNICI mass spectra of dibenzothiophene. [Reproduced with permission from D. F. Hunt, G. C. Stafford, Jr., F. W. Crow, and J. W. Russell, *Anal. Chem.* **48**, 2098 (1976). Copyright, American Chemical Society.]

of H for D on phenanthrene produces $(M + D)^+$, d_1-$(M + D)^+$,..., d_{10}-$(M + D)^+$ ions. In contrast to this, when ND_3 is used as a reagent, the ND_4^+ ion reacts with anthracene to produce not only $(M + D)^+$, but ions corresponding to exchange as well: d_1-$(M + D)^+$ and d_2-$(M + D)^+$. However, D^+ is not transferred to phenanthrene, since the proton affinity of ND_3 is greater than that of phenanthrene.

Another method (*100*) for differentiating isomers is illustrated with fluorene (MW = 166) and dihydroanthracene (MW = 180). Substitution of a methyl group on fluorene would make these compounds isomeric. The methane CI spectra of these two compounds yield $(M + H)^+$ ions in the positive mode and $(M - H)^-$ ions in the negative mode. The addition of $\sim 10\%$ nitromethane to the methane reagent gas provides a source of protons for neutralization of the anions as they are formed. However, since the proton affinity of $-CH_2NO_2$ lies between those of the two product anions, a proton will be transferred from nitromethane only to the dihydroanthracene anion and not the fluorene anion. Thus, the positive and negative CI (CH_4/CH_3NO_2) spectra of a mixture of fluorene and dihydroanthracene show $(M + H)^+$ ions for both species and an $(M - H)^-$ ion only for fluorene.

The formation of negative ions by interaction of electrons and sample molecules can occur by three different mechanisms (*101*):

(a) resonance capture

$$AB + e^- \rightarrow AB^- \tag{17}$$

TABLE 8-8

Negative-Ion Mass Spectra of Aromatic Hydrocarbons

Compound	Formula	Base peak (m/e)	Other peaks (m/e)
Benzene	C_6H_6	77 (M − 1)	75
Naphthalene	$C_{10}H_8$	127 (M − 1)	126, 125
Fluorene	$C_{13}H_{10}$	165 (M − 1)	166
Anthracene	$C_{14}H_{10}$	178 (M)	177
Phenanthrene	$C_{14}H_{10}$	177 (M − 1)	178, 176, 175
Pyrene	$C_{16}H_{10}$	202 (M)	201
Fluoranthene	$C_{16}H_{10}$	202 (M)	201, 200
Naphthacene	$C_{18}H_{12}$	228 (M)	227
Benzo[a]pyrene	$C_{20}H_{12}$	252 (M)	251
3-Methylcholanthrene	$C_{21}H_{16}$	266 (M)	265
Coronene	$C_{24}H_{12}$	300 (M)	299

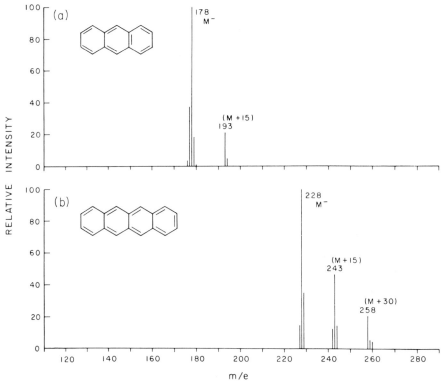

Fig. 8-23. Negative ion spectrum of (a) anthracene and of (b) naphthacene. [Reproduced with permission from S. Safe and O. Hutzinger, "Mass Spectrometry of Pesticides and Pollutants," p. 13. CRC Press, Cleveland, Ohio, 1973.]

(b) dissociative resonance capture

$$AB + e^- \rightarrow A + B^- \tag{18}$$

(c) ion-pair production

$$AB + e^- \rightarrow A^+ + B^- + e^- \tag{19}$$

These processes show strong dependence on electron energy (98). Resonance capture involves electrons with energies near 0 eV, dissociative electron capture is observed with electrons in the energy range 0–15 eV, and ion-pair production usually requires electron energies above 10 eV.

Von Ardenne (102) employed a low-pressure (10^{-2} torr) argon discharge to increase the formation of negative ions upon electron impact. A plasma containing positive argon ions and a large population of low-energy electrons was generated by this Duoplasmatron ion source. Ionization of the sample occurs predominantly by the resonance capture and dissociative resonance capture mechanisms, and the resulting spectra of PAH contained ions with

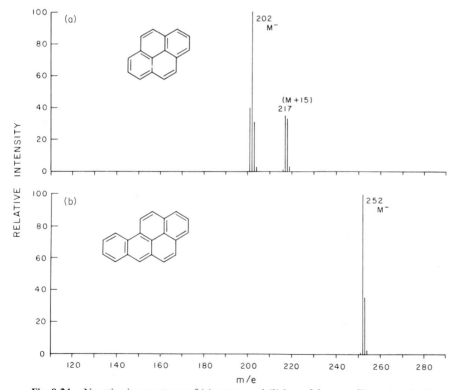

Fig. 8-24. Negative ion spectrum of (a) pyrene and (b) benzo[a]pyrene. [Reproduced with permission from S. Safe and O. Hutzinger, "Mass Spectrometry of Pesticides and Pollutants," p. 13. CRC Press, Cleveland, Ohio, 1973.]

masses near the molecular ion, such as M^- (resonance capture), $(M - 1)^-$ (loss of hydrogen), and $(M + 15)^-$. The latter two species were formed by reaction with oxygen (O^-) present in the ion source. Table 8-8 lists the most important ions in the negative ion mass spectra of a number of PAH. Several representative spectra are given in Figs. 8-23 and 8-24. Negative-ion mass spectrometry of PAH has not been generally accepted as a very useful tool because of the low yield of negative ions usually obtained, and a corresponding decrease in sensitivity. Even though the Duoplasmatron ion source has solved these problems to some extent, it has not enjoyed widespread utilization because of the high cost and extensive nature of the modifications required to attach the source to commercially available mass spectrometers.

IV. METASTABLE IONS AND COLLISION SPECTROSCOPY

If the average lifetime of an ion (m_1^+), which was formed in the ion source, is greater than 5×10^{-6} s, the ion becomes accelerated, separated, and recorded as m_1^+. On the other hand, if the average lifetime is less than 5×10^{-6} s, it fragments before acceleration to give a new ion of mass m_2^+ [Eq. (20)] where $(m_1 - m_2)$ represents a neutral fragment. An ion that de-

$$m_1^+ \rightarrow m_2^+ + (m_1 - m_2) \tag{20}$$

composes in the field-free region (the zone between the source and the analyzer) appears in the mass spectrum at m*, which is given by the relationship shown in Eq. (21).

$$m^* = m_2^2/m_1 \tag{21}$$

Such a decomposition is called a "metastable transition," and the peak corresponding to m* is the metastable peak. These metastable peaks are easily recognized in a mass spectrum because they appear as broad peaks that do not usually appear at whole mass numbers. Metastable peaks are useful in the study of fragmentation patterns, because of the information gained from the unimolecular decompositions of singly and multiply charged ions.

A relatively new technique which is growing rapidly and is related to the processes involved in metastable ions is collision spectroscopy. By introducing a collision gas in the field-free region, intermolecular processes such as charge exchange and collision-induced fragmentations can be studied. The collision-induced dissociation can be considered as involving two distinct steps, collisional excitation [Eq. (22)] followed by unimolecular dissociation [Eq. (23)].

$$m_1^+ + N \rightarrow m_1^{+*} + N \tag{22}$$

$$m_1^{+*} \rightarrow m_2^+ + (m_1 - m_2) \tag{23}$$

The analytical power of observations of spontaneous (103) and collisionally induced (104) fragmentations of ions in field-free regions of sector mass spectrometers has been well documented. Such ions possess lower internal energy, on the average, than do ions fragmenting on the time scale appropriate to the ion source, such that their fragmentation reactions are often more sensitive to the structure of the neutral precursor than are the latter.

Studies of collision-induced dissociation (CID) can be made using a variety of spectrometer configurations (104). Although some useful data can be obtained using only a single mass or energy analyzer, relatively little work has been done this way because of the possibility of overlap of peaks due to different processes. The study of CID in a single-focusing-sector mass spectrometer involves the use of the field-free region as the collision chamber. With a magnetic sector, a single scan records all the ions leaving the field-free region including the collision-induced ions, metastable ions, and the stable ions.

The use of a single electric sector for direct translational energy analysis has the advantages that the energy scale is more readily calibrated, hysteresis effects are minimized, the field strength is more readily controlled, and interference from signals due to stable ions is avoided (104).

The use of instruments with two analyzers allows a single reaction to be isolated for study. Instruments employing dual mass analyzers (105, 106) with an intermediate collision chamber allow selection of the reactant ion by mass in the first stage of the instrument and mass analysis of the product ions in the second stage. Except for this case, most double analyzer instruments have an electric and a magnetic sector, and it is possible to scan in various ways so as to observe the products of metastable or collision-induced decomposition without interference from the stable ions that make up the normal mass spectrum (107).

One reported study of the fragmentation processes of aromatic hydrocarbons in the first field-free region employed a mass spectrometer with fields arranged in the order accelerating voltage (V), electric sector voltage (E), and magnetic field (B), and used a detector located behind the energy resolving slit located between the sectors (108). Scanning of the electric sector voltage gave a fingerprint of all fragmentations without prior mass separation. This is called an ion-kinetic-energy (IKE) scan because the ions are separated on the basis of their kinetic energies. Figure 8-25 shows the high energy peaks in the IKE spectrum of anthracene. Subsequent mass

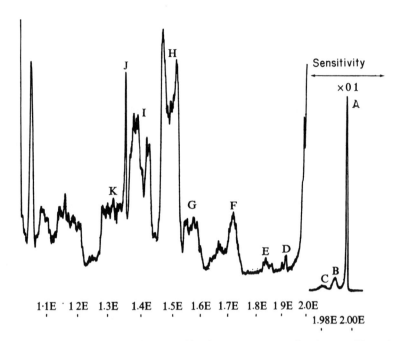

Fig. 8-25. High-energy peaks in the ion kinetic energy spectrum of anthracene. [Reproduced with permission from J. H. Beynon, R. M. Caprioli, W. E. Baitinger, and J. W. Amy, *Org. Mass. Spectrom.* **3,** 455 (1970). Copyright, Heyden and Son, Ltd.]

analysis of peak A in the IKE spectrum is shown in Figure 8-26. Both the IKE spectra and mass spectra can be used to give detailed fingerprints of organic compounds. PAH that have been studied using this technique include naphthalene, 2-methylnaphthalene, biphenyl, anthracene, dimethylnaphthalene, benzo[*e*]pyrene, and benzo[*k*]fluoranthene isomers (*110*).

Shushan *et al.* (*111*) recently described a method in which both B and E are scanned so that B/E is held at the constant value required to transmit stable parent ions of preselected m/e ratio; in this way fragment ions formed in the first field-free region, from the preselected parent ions, are successively transmitted. Figure 8-27 shows the complete daughter-ion spectrum of the molecular ion of chrysene with no collision gas. The daughter-ion spectra of the four isomers chrysene, triphenylene, benz[*a*]anthracene, and naphthacene showed considerable differences in the relative abundances of the M^+, $(M - H)^+$, and $(M - 2H)^+$ ions, the latter two being the products of fragmentation of M^+ ions in the first field-free region. Plots of the

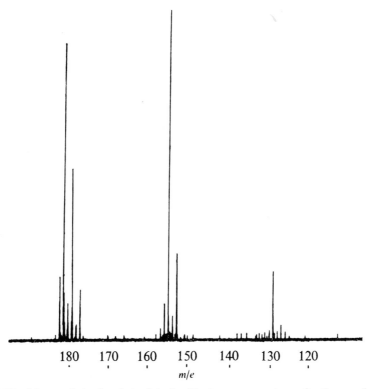

Fig. 8-26. Mass analysis of peak A of the ion kinetic energy spectrum of anthracene (see Fig. 8-25). [Reproduced with permission from J. H. Beynon, R. M. Caprioli, W. E. Baitinger, and J. W. Amy, *Org. Mass Spectrom.* **3**, 455 (1970). Copyright, Heyden and Son, Ltd.]

$(M - H)/M$ and $(M - 2H)/(M - H)$ intensity ratios against the number of benzo interactions per molecule are given in Fig. 8-28. As is readily seen, the structural assignments of these isomers can easily be made using this technique. It was found in this study that even though the use of a collision gas increased appreciably the percentage of fragmentation occurring in the first field-free region, the resultant intensity ratios of the collisionally induced reactions were much closer than were the corresponding ratios for the unimolecular case. In addition, the ratios were very sensitive to variations in collision gas pressure, which was difficult to control precisely.

Another widely used configuration is that in which the second stage of the instrument is an energy rather than a mass analyzer (*112*). With the arrangement of fields V, B, E, fragmentations between the sectors can be observed.

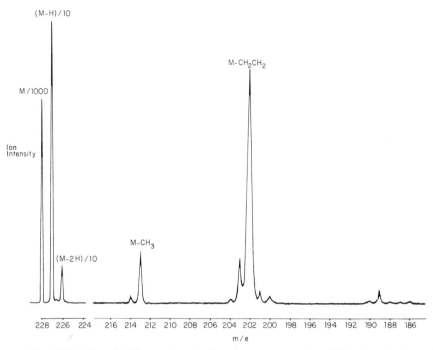

Fig. 8-27. First field-free region daughter-ion spectrum (no collisional activation) of molecular ion ($m/e = 228$) of chrysene; B/E linked scan, source temperature 210°C, nominal electron energy 70 eV, direct probe inlet. [Reproduced with permission from B. Shushan, S. H. Safe, and R. K. Boyd, *Anal. Chem.* **51**, 156 (1979). Copyright, American Chemical Society.]

The initial magnetic field is set to select the m_1^+ ions, and the fragmentation products of these ions can then be plotted in a single spectrum by varying the electric sector voltage. This is called a MIKE scan, for mass-analyzed ion kinetic energy. An analytical application of such a scan has been termed DADI (*113*), for direct analysis of daughter ions.

The number of analytical applications of metastable-ion and collision spectroscopy have multiplied rapidly in the last few years (*103, 104, 114*), but the potential uses of these techniques have been little explored. The full usefulness of these methods—i.e., in the analysis of PAH—has yet to be determined.

V. FIELD IONIZATION AND FIELD DESORPTION

Field ionization (FI) and field desorption (FD) mass spectrometry produce virtually fragment-ion-free mass spectra and should be ideally suited

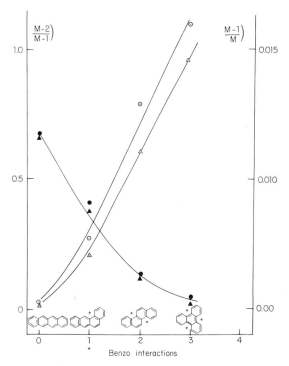

Fig. 8-28. Correlation of relative peak-height ratios from B/E linked scan spectra of PAH isomers with number of hydrogen benzo interactions per molecule. Circles, ion source 210°C; triangles, 160°C. [Reproduced with permission from B. Shushan, S. H. Safe, and R. K. Boyd, *Anal. Chem.* **51**, 156 (1979). Copyright, American Chemical Society.]

to the group-type quantitative analysis of PAH mixtures. FI differs from FD in that the sample is introduced as a vapor and may yield much lower sensitivity, especially for compounds of low vapor pressure. Nevertheless, both techniques provide virtually single molecular ion spectra for compounds like PAH, which makes them attractive techniques for the determination of molecular weight distributions in mixtures. Although these techniques have not enjoyed very widespread use, the recent determination of FI relative sensitivities for a number of PAH (*115*) and improvements in quantitative aspects of FD (*116, 117*) have helped. The application of FI and FD mass spectrometry in the analysis of the aromatic fractions of fossil fuels has been described in several recent studies (*115–121*). Figure 8-29 shows the FI mass spectrum of an aromatic hydrocarbon fraction from the Athabasca Tar Sands (*118*).

An interesting application of FD mass spectrometry was recently reported

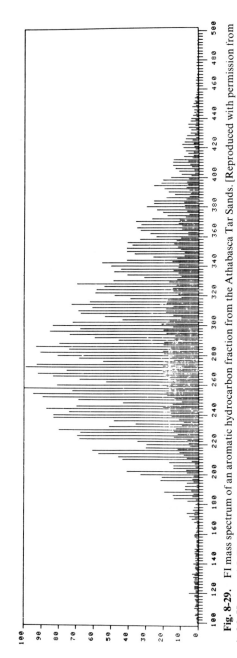

Fig. 8-29. FI mass spectrum of an aromatic hydrocarbon fraction from the Athabasca Tar Sands. [Reproduced with permission from A. M. Hogg and J. D. Payzant, *Int. J. Mass Spectrom. Ion Phys.* **27**, 291 (1978). Copyright, Elsevier Sci. Publ. Co.]

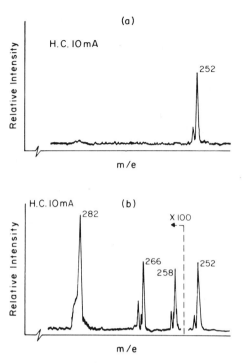

Fig. 8-30. Field desorption mass spectra: (*a*) benzo[*a*]pyrene (*m/e* 252), nonirradiated; (*b*) benzo[*a*]pyrene, irradiated 40 min. [Reproduced with permission from D. F. Barofsky and E.J. Baum, *J. Am. Chem. Soc.* **98**, 8286 (1976). Copyright, American Chemical Society.]

(*122*) in which the photooxidation of PAH adsorbed on carbon was studied. Mass spectra of PAH irradiated directly on carbon microneedle emitters were discussed. Figure 8-30 shows the field desorption mass spectra for both nonirradiated and irradiated benzo[*a*]pyrene. In addition to the parent ion peak at *m/e* 252, the mass spectrum of irradiated benzo[*a*]pyrene exhibits mass peaks at *m/e* 258, 266, and 282. The mass peak at *m/e* 282 corresponds to a dione. The mass peaks at *m/e* 258 and 266 were not identified, but the absence of contaminant or artifact peaks in the reference spectrum (nonirradiated benzo[*a*]pyrene) suggests that they represent photoproducts.

VI. PHOTOIONIZATION

Photoionization (PI) techniques yield pure molecular ion spectra for aromatics, but the low-voltage electron impact technique is preferable because of the experimental difficulties of PI. A description of the method for analysis of mixtures including the analysis of aromatic hydrocarbons is given by Severin (*123*).

The mass spectroscopic method chosen for the analysis of PAH is dependent on the type of results desired. For group-type analysis, low- and high-voltage electron impact, field ionization, field desorption, photoionization and sometimes chemical ionization may be used. By far the most universal approach is low-voltage electron impact. The availability of instrumentation, sensitivity data, and reproducibility are the main reasons for this.

On the other hand, if structural identification of individual mixture components is desired, the combination of gas chromatography/mass spectrometry is the method of choice. Because of the high numbers of structural isomers in complex PAH mixtures, separation before mass spectral identification is essential. Even then, conventional mass spectrometry cannot, in most cases, differentiate between isomers. The relatively new techniques of chemical-ionization, metastable-ion, and collision-induced mass spectrometry have the greatest potential for solving these problems.

REFERENCES

1. S. Safe and O. Hutzinger, "Mass Spectrometry of Pesticides and Pollutants," p. 77. CRC Press, Cleveland, Ohio, 1973.

2. E. Stenhagen, S. Abrahamsson, and F. W. McLafferty, "Registry of Mass Spectra Data," Vols. 1–4, Wiley, New York, 1974.

3. S. R. Heller and G. W. A. Milne, "EPA/NIH Mass Spectral Data Base," Vols. 1–4, U.S. Gov. Printing Office, Washington, D.C., 1978.

4. "Eight Peak Index of Mass Spectra," Vols. 1–4, Mass Spectrometry Data Centre, Aldermaston, Reading, United Kingdon, 1974.

5. R. S. Thomas and R. C. Lao, *Adv. Mass Spectrom.* **7,** 1709 (1978).

6. R. A. Brown, *Anal. Chem.* **23,** 430 (1951).

7. H. E. Lumpkin and B. H. Johnson, *Anal. Chem.* **26,** 1719 (1954).

8. S. H. Hastings, B. H. Johnson, and H. E. Lumpkin, *Anal. Chem.* **28,** 1243 (1956).

9. C. J. Robinson and G. L. Cook, *Anal. Chem.* **41,** 1548 (1969).

10. C. J. Robinson, *Anal. Chem.* **43,** 1425 (1971).

11. J. T. Swansiger, F. E. Dickson, and H. T. Best, *Anal. Chem.* **46,** 730 (1974).

12. E. J. Gallegos, J. W. Green, L. P. Lindeman, R. L. LeTourneau, and R. M. Teeter, *Anal. Chem.* **39,** 1833 (1967).

13. H. E. Lumpkin and T. Aczel, *in* "High Performance Mass Spectrometry: Chemical Applications" (M. L. Gross, ed.), p. 261. ACS Symp. Ser. 70, Am. Chem. Soc., Washington, D.C., 1978.

14. A. W. Peters and J. G. Bendoraitis, *Anal. Chem.* **48,** 968 (1976).

15. F. H. Field and S. H. Hastings, *Anal. Chem.* **28,** 1248 (1956).

16. H. E. Lumpkin, *Anal. Chem.* **30,** 321 (1958).

17. H. E. Lumpkin and T. Aczel, *Anal. Chem.* **36,** 181 (1964).

18. J. L. Shultz, A. G. Sharkey, Jr., and R. A. Brown, *Anal. Chem.* **44,** 1486 (1972).

19. J. E. Schiller, *Anal. Chem.* **49,** 1260 (1977).

20. H. E. Lumpkin, *Anal. Chem.* **36**, 2399 (1964).
21. K. S. Quisenberry, T. T. Scolman, and A. O. Nier, *Phys. Rev.* **102**, 1071 (1956).
22. B. H. Johnson and T. Aczel, *Anal. Chem.* **39**, 682 (1967).
23. T. Aczel, D. E. Allan, J. H. Harding, and E. A. Knipp, *Anal. Chem.* **42**, 341 (1970).
24. L. R. Snyder, *Anal. Chem.* **41**, 314 (1969).
25. L. R. Snyder, *Anal. Chem.* **41**, 1084 (1969).
26. L. R. Snyder, B. E. Buell, and H. E. Howard, *Anal. Chem.* **40**, 1303 (1968).
27. J. R. McKay, J. H. Weber, and D. R. Latham, *Anal. Chem.* **48**, 891 (1976).
28. H. W. Holden and J. C. Robb, *Fuel* **39**, 39 (1960).
29. T. Kessler, R. Raymond, and A. G. Sharkey, Jr., *Fuel* **48**, 179 (1969).
30. J. L. Shultz, R. A. Friedel, and A. G. Sharkey, Jr., Mass Spectrometric Analyses of Coal-Tar Distillates and Residues, U.S. Dep. Interior, Bureau of Mines, Washington, D.C., 1967.
31. A. G. Sharkey, Jr., J. L. Shultz, and R. A. Friedel, *Fuel* **41**, 359 (1962).
32. A. G. Sharkey, Jr., *in* "Carcinogenesis—A Comprehensive Survey" (R. I. Freudenthal and P. W. Jones, eds.), Vol. I, p. 341. Raven, New York, 1976.
33. A. G. Sharkey, Jr., J. L. Shultz, T. Kessler, and R. A. Friedel, *Res. Dev.* **20**, 30 (1969).
34. A. G. Sharkey, Jr., J. L. Shultz, T. Kessler, and R. A. Friedel, *Proc. Int. Clean Air Congr. 2nd, 1971*, p. 539.
35. D. Schuetzle, A. L. Crittenden, and R. J. Charlson, *J. Air Pollut. Control Assoc.* **23**, 704 (1973).
36. J. L. Shultz, A. G. Sharkey, Jr., and R. A. Friedel, *Biomed. Mass Spectrom.* **1**, 137 (1974).
37. D. Schuetzle, *Biomed. Mass Spectrom.* **2**, 288 (1975).
38. W. Giger and M. Blumer, *Anal. Chem.* **46**, 1663 (1974).
39. M. Blumer and W. W. Youngblood, *Science* **188**, 53 (1975).
40. W. W. Youngblood and M. Blumer, *Geochim. Cosmochim. Acta* **39**, 1303 (1975).
41. R. A. Hites and W. G. Biemann, *Adv. Chem. Ser.* **147**, 188 (1975).
42. A. Hase and R. A. Hites, *Geochim. Cosmochim. Acta* **40**, 1141 (1976).
43. M. L. Lee, G. P. Prado, J. B. Howard, and R. A. Hites, *Biomed. Mass Spectrom.* **4**, 182 (1977).
44. M. Blumer, T. Dorsey, and J. Sass, *Science* **195**, 283 (1977).
45. R. E. Laflamme and R. A. Hites, *Geochim. Cosmochim. Acta* **42**, 289 (1978).
46. J. L. Stauffer, P. L. Levins, and J. E. Oberholtzer, *in* "Carcinogenesis—A Comprehensive Survey" (P. W. Jones and R. I. Freudenthal, eds.), Vol. 3, p. 89. Raven, New York, 1978.
47. R. C. Lao, H. Oja, R. S. Thomas, and J. L. Monkman, *Sci. Total Environ.* **2**, 223 (1973).
48. G. Grimmer and H. Bohnke, *Fresenius Z. Anal. Chem.* **261**, 310 (1972).
49. R. C. Lao, R. S. Thomas, H. Oja, and L. Dubois, *Anal. Chem.* **45**, 908 (1973).
50. R. C. Lao, R. S. Thomas, and J. L. Monkman, *J. Chromatogr.* **112**, 618 (1975).
51. G. Ketseridis and J. Hahn, *Fresenius Z. Anal. Chem.* **273**, 257 (1975).
52. W. Cautreels and K. Van Cauwenberghe, *Atmos. Environ.* **10**, 447 (1976).
53. R. C. Lao, R. S. Thomas, and J. L. Monkman, *in* "Carcinogenesis—A Comprehensive Survey" (R. I. Freudenthal and P. W. Jones, eds.), Vol. 1, p. 271. Raven, New York, 1976.
54. M. L. Lee, M. Novotny, and K. D. Bartle, *Anal. Chem.* **48**, 1566 (1976).
55. W. Cautreels and K. Van Cauwenberghe, *J. Chromatogr.* **131**, 253 (1977).
56. M. W. Dong, D. C. Locke, and D. Hoffmann, *Environ. Sci. Technol.* **11**, 612 (1977).
57. A. Bjørseth and G. Lunde, *J. Am. Ind. Hyg. Assoc.* **38**, 224 (1977).
58. W. Cautreels and K. Van Cauwenberghe, *Atmos. Environ.* **12**, 1133 (1978).
59. W. Giger and C. Schaffner, *Anal. Chem.* **50**, 243 (1978).
60. L. Van Vaeck and K. Van Cauwenberghe, *Atmos. Environ.* **12**, 2229 (1978).
61. F. W. Karasek, D. W. Denney, K. W. Chan, and R. E. Clement, *Anal. Chem.* **50**, 82 (1978).

62. M. Dong, I. Schmeltz, E. LaVoie, and D. Hoffmann, in "Carcinogenesis—A Comprehensive Survey" (P. W. Jones and R. I. Freudenthal, eds.), Vol. 3, p. 97. Raven, New York, 1978.

63. M. L. Lee, D. L. Vassilaros, W. S. Pipkin, and W. L. Sorensen, in Trace Organic Analysis: A New Frontier in Analytical Chemistry, NBS Spec. Publ. 519, p. 731. U.S. Government Printing Office, Washington, D.C., 1979.

64. F. W. Karasek, R. J. Smythe, and R. J. Laub, J. Chromatogr. **101**, 125 (1974).

65. A. Hase, P. H. Lin, and R. A. Hites, in "Carcinogenesis—A Comprehensive Survey" (R. I. Freudenthal and P. W. Jones, eds.), Vol. 1, p. 435. Raven, New York, 1976.

66. P. E. Strup, R. D. Giammar, T. B. Stanford, and P. W. Jones, in "Carcinogenesis—A Comprehensive Survey" (R. I. Freudenthal and P. W. Jones, eds.), Vol. 1, p. 241. Raven, New York, 1976.

67. G. Broddin, L. Van Vaeck, and K. Van Cauwenberghe, Atmos. Environ. **11**, 1061 (1977).

68. H. Tausch and G. Stehlik, Chromatographia **10**, 350 (1977).

69. P. W. Jones, Proc. Anal. Div. Chem. Soc. **15**, 158 (1978).

70. J. C. Liao and R. F. Browner, Anal. Chem. **50**, 1683 (1978).

71. L. Sucre, W. Jennings, G. L. Fisher, O. G. Raabe, and J. Olechno, in Trace Organic Analysis: A New Frontier in Analytical Chemistry, NBS Spec. Publ. 519, p. 109. U.S. Govt. Printing Office, Washington, D.C., 1979.

72. H. Borwitzky, D. Henneberg, G. Schomburg, H. D. Sauerland, and M. Zander, Erdoel Kohle, Erdgas, Petrochem. **30**, 370 (1977).

73. H. Borwitzky, G. Schomburg, H. D. Sauerland, and M. Zander, Erdoel. Kohle, Erdgas, Petrochem. **31**, 371 (1978).

74. H. Borwitzky and G. Schomburg, J. Chromatogr. **170**, 99 (1979).

75. P. E. Strup, J. E. Wilkinson, and P. W. Jones, in "Carcinogenesis—A Comprehensive Survey" (P. W. Jones and R. I. Freudenthal, eds.), Vol. 3, p. 131. Raven, New York, 1978.

76. G. Grimmer and H. Bohnke, Chromatographia **9**, 30 (1976).

77. W. Bertsch, E. Anderson, and G. Holzer, J. Chromatogr. **126**, 213 (1976).

78. D. L. Fishel and T. F. Longo, Jr., Adv. Mass Spectrom. **7**, 1323 (1978).

79. M. R. Guerin, J. L. Epler, W. H. Griest, B. R. Clark, and R. K. Rao, in "Carcinogenesis—A Comprehensive Survey" (P. W. Jones and R. I. Freudenthal, eds.), Vol. 3, p. 21. Raven, New York, 1978.

80. C. M. White, A. G. Sharkey, Jr., M. L. Lee, and D. L. Vassilaros, in "Polynuclear Aromatic Hydrocarbons" (P. W. Jones and P. Leber, eds.), p. 261. Ann Arbor Science, Ann Arbor, Michigan, 1979.

81. M. E. Snook, R. F. Severson, H. C. Higman, R. F. Arrendale, and O. T. Chortyk, Beitr. Tabakforsch. **8**, 250 (1975).

82. R. F. Severson, M. E. Snook, H. C. Higman, O. T. Chortyk, and F. J. Akin, in "Carcinogenesis—A Comprehensive Survey" (R. I. Freudenthal and P. W. Jones, eds.), Vol. 1, p. 253. Raven, New York, 1976.

83. M. L. Lee, M. Novotny, and K. D. Bartle, Anal. Chem. **48**, 405 (1976).

84. M. E. Snook, in "Carcinogenesis—A Comprehensive Survey" (P. W. Jones and R. I. Freudenthal, eds.), Vol. 3, p. 203. Raven, New York, 1978.

85. M. L. Lee and R. A. Hites, Anal. Chem. **48**, 1890 (1976).

86. C. K. McMahon and S. N. Tsoukalas, in "Carcinogenesis—A Comprehensive Survey" (P. W. Jones and R. I. Freudenthal, eds.), Vol. 3, p. 61. Raven, New York, 1978.

87. R. H. Bieri, M. K. Cueman, C. L. Smith, and C.-W. Su, Int. J. Environ. Anal. Chem. **5**, 293 (1978).

88. G.-G. Hammar, B. Holmstedt, and R. Ryhage, Anal. Biochem. **25**, 532 (1968).

89. J. M. Strong, A. J. Atkinson, and R. J. Ferguson, Anal. Chem. **47**, 1720 (1975).

90. M. S. B. Munson and F. H. Field, *J. Am. Chem. Soc.* **89,** 1047 (1967).

91. M. L. Lee and R. A. Hites, *J. Am. Chem. Soc.* **99,** 2008 (1977).

92. J. S. Warner, *Anal. Chem.* **48,** 578 (1976).

93. R. A. Hites and G. R. Dubay, *in* "Carcinogenesis—A Comprehensive Survey" (P. W. Jones and R. I. Freudenthal, eds.), Vol. 3, p. 85. Raven, New York, 1978.

94. G. P. Arsenault, *J. Am. Chem. Soc.* **94,** 8241 (1972).

95. D. P. Beggs, Hewlett-Packard Appl. Note No. 176-8. Avondale, Pennsylvania.

96. T. R. Henderson and R. L. Hanson, *Abstracts 26th Annu. Conf. Mass Spectrom. Allied Topics, St. Louis, Missouri,* 1978, p. 610.

97. D. F. Hunt, C. N. McEwen, and T. M. Harvey, *Anal. Chem.* **47,** 1730 (1975).

98. D. F. Hunt, G. C. Stafford, Jr., F. W. Crow, and J. W. Russell, *Anal. Chem.* **48,** 2098 (1976).

99. D. F. Hunt and S. K. Sethi, *in* "High Performance Mass Spectrometry: Chemical Applications" (M. L. Gross, ed.), p. 150. ACS Symp. Ser. 70, Am. Chem. Soc., Washington, D.C., 1978.

100. D. F. Hunt, P. J. Gale, and S. K. Sethi, *Abstracts 26th Annu. Conf. Mass Spectrom. Allied Topics, St. Louis, Missouri, May, 1978,* p. 151.

101. S. Safe and O. Hutzinger, "Mass Spectrometry of Pesticides and Pollutants," p. 13. CRC Press, Cleveland, Ohio, 1973.

102. M. von Ardenne, K. Steinfelder, and R. Tummler, "Electronenanlagerungs—Massenspektrographie Organischer Substanzen." Springer-Verlag, Berlin and New York, 1971.

103. R. G. Cooks, J. H. Beynon, R. M. Caprioli, and G. R. Lester, "Metastable Ions." Elsevier, Amsterdam, 1973.

104. R. G. Cooks, ed., "Collision Spectroscopy" Plenum, New York, 1978.

105. S. E. Kupriyanov and A. A. Perov, *Russ. J. Phys. Chem. (Engl. Transl.)* **39,** 871 (1965).

106. A. Giardini-Guidoni, R. Plantania, and F. Zucchi, *Int. J. Mass Spectrom. Ion Phys.* **13,** 453 (1974).

107. R. K. Boyd and J. H. Beynon, *Org. Mass Spectrom.* **12,** 163 (1977).

108. J. H. Beynon, R. M. Caprioli, W. E. Baitinger, and J. W. Amy, *Org. Mass Spectrom.* **3,** 455 (1970).

109. T. Ast, J. H. Beynon, and R. G. Cooks, *Org. Mass Spectrom.* **6,** 749 (1972).

110. R. C. Lao, R. S. Thomas, and J. L. Monkman, *Adv. Mass Spectrom.* **6,** 129 (1974).

111. B. Shushan, S. H. Safe, and R. K. Boyd, *Anal. Chem.* **51,** 156 (1979).

112. J. H. Beynon, R. G. Cooks, J. W. Amy, W. E. Baitinger, and T. Y. Ridley, *Anal. Chem.* **45,** 1023A (1973).

113. V. P. Schlunegger, *Angew. Chem. Int. Ed. Engl.* **14,** 679 (1975).

114. J. H. Beynon and R. K. Boyd, *Adv. Mass Spectrom.* **7,** 1115 (1978).

115. S. E. Scheppele, P. L. Grizzle, G. J. Greenwood, T. D. Marriott, and N. B. Perreira, *Anal. Chem.* **48,** 2105 (1976).

116. D. F. Barofsky, E. Barofsky, and R. Held-Aigner, *Adv. Mass Spectrom.* **7,** 109 (1978).

117. S. Pfeifer, H. D. Beckey, and H.-R. Schulten, *Fresenius Z. Anal. Chem.* **284,** 193 (1977).

118. A. M. Hogg and J. D. Payzant, *Int. J. Mass Spectrom. Ion Phys.* **27,** 291 (1978).

119. S. E. Scheppele, G. J. Greenwood, P. L. Grizzle, T. D. Marriott, C. S. Hsu, N. B. Perreira, P. A. Benzon, K. N. Detwiler, and G. M. Stewart, *Prepr. Div. Pet. Chem., Am. Chem. Soc.* **22,** 665 (1977).

120. M. Anbar and G. A. St. John, *Fuel* **57,** 105 (1978).

121. G. A. St. John, S. E. Buttrill, and M. Anbar, *in* "Organic Chemistry of Coal" (J. W. Larsen, ed.), p. 223. ACS Symp. Ser. 71, Am. Chem. Soc., Washington, D.C., 1978.

122. D. F. Barofsky and E. J. Baum, *J. Am. Chem. Soc.* **98,** 8286 (1976).

123. D. Severin, *Compend. Dtsch. Ges. Mineraloelwiss. Kohlechem.* **76–77,** 949 (1976).

9

Ultraviolet Absorption
and Luminescence Spectroscopy

I. INTRODUCTION

Irradiation of a PAC molecule in the ground state with light of an appropriate wavelength in the ultraviolet (UV) or visible regions generally results in absorption as π-electrons are promoted to higher π-electronic energy states which are usually also vibrationally excited (Fig. 1-1). Fluorescence (the emission of light with longer wavelength than that of the exciting radiation as the electron returns to the ground state from the lowest vibrational level of an excited electronic state) is also common for PAC. However, other possibilities for energy transfer exist, in particular, radiationless transitions from singlet to triplet states, so-called intersystem crossing. The subsequent decay of triplet states with emission of radiation is termed phosphorescence and has been observed for many PAC.

A definitive review published in 1970 described the routine method of analysis for PAH by column chromatography on alumina, followed by UV absorptiometry (1). A volume of 10^4 m^3 of air, yielding 100 mg of organic airborne particulates, were required for an analysis which took 2–3 days. Although variations on this UV method are still applied, the much greater sensitivity of fluorescence/phosphorescence techniques (in certain cases similar to that of radiochemical methods) (2) has made luminescence analysis one of the most widely applied procedures for the identification and estimation of environmental PAC. The use of intense light sources such as lasers and detection by photon-counting now makes analysis below the part-per-trillion level possible.

If measured in liquid solution, as is most common, many UV and luminescence spectra, although specific, are broad and featureless. Elaborate separation methods are thus often required if PAC are to be unambiguously identified by conventional spectrophotometry. Luminescence offers at least two advantages over the simple UV-absorption spectrum: first, greater sensitivity, and second, in that four spectra are available (excitation and emission of fluorescence and phosphorescence), although overlap is possible. A corollary of the necessity for separation is the now widespread use of UV and fluorimetric detectors in chromatography, especially HPLC, as was discussed in Chapter 6.

The overlap of generally broad spectra can make quantitation by spectrophotometry very difficult if the sample contains more than two or three individual PAC. Various methods for selective excitation of fluorescing components are under continuing study, and attention has also been focused on a number of other approaches to the analysis of mixtures by luminescence. Thus, samples in frozen solutions and isolated in matrices not only yield well-resolved fine structure, permitting unambiguous identification, but may also be analyzed with fewer of the constraints which intermolecular energy transfers impose on relations between intensity and concentration. Selective quenching, however, can be a valuable aid in analysis. Newer instrumental methods for the analysis of mixtures by luminescence, such as modulation, derivative techniques, and time resolution are also being devised, and the field is one of very rapid expansion.

II. ULTRAVIOLET ABSORPTION

The molar absorptivities of PAC in the UV are high (10^4–10^5). For certain PAC, the energies of the absorption bands are characteristic enough for the concentration of individual hydrocarbons to be determined from measurements of one or more peaks. Thus, Fig. 9-1 illustrates the correspondence between the UV absorption spectra of single PAH recovered from a paraffin wax sample (3) and those of pure compounds, and demonstrates the degree of specificity possible.

For mixtures of unknown composition, separated by column (4–6), thin–layer (7, 8), paper (9), gas–liquid (10–12), and liquid chromatography (13), identification and quantitation of components is more difficult, and the precision of measurement is likely to be poor unless the composition is simple. In Fig. 9-2, the overlap of the UV spectra of important carcinogens in column chromatographic fractions (4) from air particulates is clear. Analysis by high-resolution GC (Chapter 7) has since shown that such fractions are even more complex than previously suspected. The high content of alkyl derivatives of PAH with the same ring system make the UV spectra of such mixtures liable to errors in interpretation (14), since bathochromic

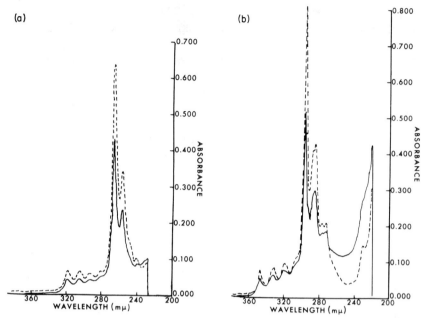

Fig. 9-1. (a) Ultraviolet absorption spectra of chrysene recovered from a 100-g wax sample at a level of 0.1 ppm. (Solid line = recovery curve; broken line = reference curve.) (b) Ultraviolet absorption spectra of dibenz[a,h]anthracene recovered from a 100-g wax sample at a level of 0.1 ppm. (Solid line = recovery curve; broken line = reference curve.) [Reproduced with permission from J. W. Howard and E. O. Haenni, *J. Assoc. Offic. Agric. Chem.* **46**, 933 (1963). Copyright The Association of Official Analytical Chemists, Inc.]

shifts of absorption maxima occur on substitution (*15*). Changes of up to 8 nm in wavelength may be observed in the spectra of pyrene fractions containing methylpyrenes (*4*). Alkyl derivatives of phenanthrene (*16*), benz[a]anthracene (*17*), benzo[c]phenanthrene (*18*), chrysene (*4, 19*), triphenylene (*20*), and fluoranthene (*4*) show similar effects. Interference from other, unsubstituted, PAH can also make analysis difficult. Benzo[a]pyrene, benzo[k]fluoranthene, and benzo[ghi]perylene are difficult to separate by column chromatography (Fig. 9-2); the UV spectra of benzo[a]pyrene and benzo[ghi]perylene are practically identical, and while benzo[a]pyrene has an absorption maximum at 402 nm, absent in the benzo[ghi]perylene spectrum, a peak at 401 nm is also found in the spectrum of benzo[k]fluoranthene (*21*).

Many of the quantitative data in the literature for benzo[a]pyrene in environmental samples are thus probably erroneous (*4, 21*), especially those obtained before 1960. In general, the conditions summarized by Moore *et al.*

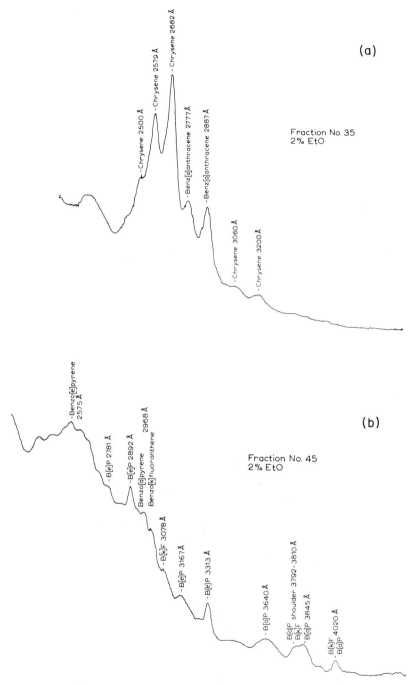

Fig. 9-2. Ultraviolet absorption spectra of column chromatographic fractions of air-particulate PAH. (a) Benz[a]anthracene and chrysene, fraction 35. (b) Benzo[a]pyrene, benzo-[e]pyrene, and benzo[k]fluoranthene, fraction 45. [Reproduced with permission from G. E. Moore, R. S. Thomas, and J. L. Monkman, *J. Chromatogr.* **26,** 456 (1960). Copyright, Elsevier Scientific Publishing Co.]

(4) for the identification by UV spectrophotometry of a given compound in a mixture are seldom met: (a) the band wavelengths of the unknown must be identical with the bands of known standard solutions either recorded in the laboratory or from literature compilations (15, 22–25); (b) the shapes and ratios of the various peaks in the unknown and standard spectra must be comparable; and (c) each of the bands displayed by the standard should be

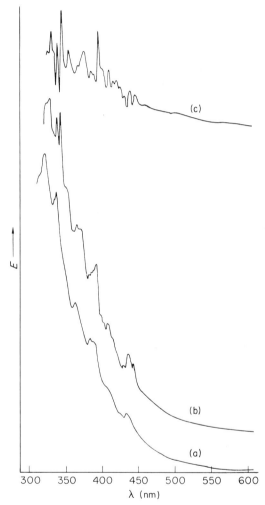

Fig. 9-3. UV/visible spectrum of coal-tar pitch in ethanol: (a) at room temperature; (b) at −178°C; (c) first derivative of (b). [Reproduced with permission from K. D. Bartle, G. Collin, J. W. Stadelhofer, and M. Zander, *J. Chem. Technol.* **29**, 531 (1979). Copyright, Soc. Chem. Ind.]

identifiable in the spectrum of the unknown sample. Since the "background" constituents of environmental PAC samples, including some aliphatic hydrocarbons, absorb in the region where most PAC show structure (240–300 nm), the presence of all bands is often obscured unless prior separation is very efficient (4).

The reported detection limit of UV spectrophotometry is of the order of micrograms [insufficiently sensitive for applications in water analysis as compared to the previous extensive use in air pollution studies (26)]. Present-day applications of UV spectrophotometry alone in PAC analysis are few; in general, the method is applied in combination with luminescence methods or in the continuous detection of PAH eluted from HPLC columns (Chapter 6).

A number of approaches have not, however, been fully used in realizing the potential of UV absorption spectroscopy; for example, the close relationship between the UV spectra of benzologs resulting from the annellation principle (27) (Section II, Chapter 1) has not been exploited. Little use has been made so far of more sophisticated UV methods in environmental analysis. Low-temperature (Fig. 9-3b) and derivative (Fig. 9-3c) absorption spectra markedly increase the information gained relative to conventional spectra recorded at room temperature (Fig. 9-3a). Thus, differentiation of UV absorption and luminescence spectra emphasizes secondary spectral features such as flat maxima, slopes, and shoulders of the peaks which comprise the more generally observed "zeroth-derivative" spectrum (28) arising from the overlapping of lines considerably broadened by collisional processes in a solution. However, a prototype instrument has been developed (29) for the monitoring of PAC vapors in the field from their second-derivative UV absorption spectra. The incident wavelength is modulated by a few nanometers, and the second harmonic of the modulation frequency is detected. The intensity of this signal is proportional to the concentration of the PAC. Second-derivative absorption is independent of sample opacity, so that it is unnecessary to remove particulate matter prior to the analysis. The instrument may also be used in the analysis of PAC in solution, e.g., to monitor wastewater on-line.

III. FLUORESCENCE

Methods of analysis of mixtures of PAC using luminescence provide greater specificity and sensitivity than absorption spectrophotometry. Thus, while benzo[a]pyrene and benzo[ghi]perylene have similar absorption spectra, and therefore similar fluorescence excitation spectra, they are readily distinguished by their fluorescence emission spectra (Fig. 9-4) (30, 31). In

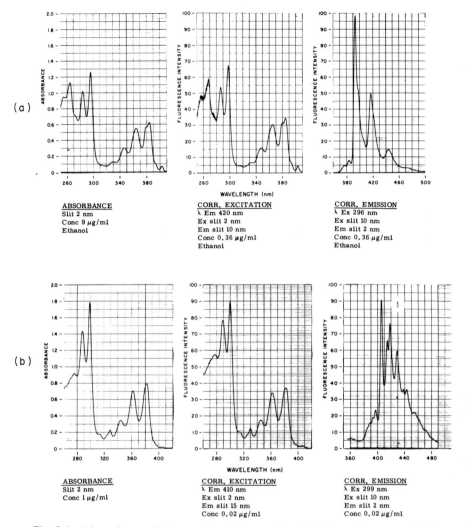

Fig. 9-4. Absorption, excitation, and emission spectra of (a) benzo[*a*]pyrene, and (b) benzo[*ghi*]perylene. [Reproduced with permission from T. J. Porro, *J. Assoc. Offic. Anal. Chem.* **56**, 607 (1973). Copyright, The Association of Official Analytical Chemists, Inc.]

any case, the excitation spectrum often offers advantages over the absorption spectrum and may be recorded in the presence of absorbing impurities; fewer compounds fluoresce than absorb light. The sensitivity of luminescence methods is frequently from 10 to 10^3 times as great as that of absorption methods (*32*).

Sawicki reviewed the early (up to 1969) applications of fluorescence in the analysis of polluted air (33). Comparisons were made of the complexity of the various combined separation/photometric methods, the sensitivity, precision, and accuracy of photometric procedures, along with the importance of interference from other compounds including impurities in standards. The optimum conditions for the application of absorption, fluorescence, and phosphorescence methods have also been reviewed by Winefordner (34).

Numerous methods for the determination by fluorescence techniques of single components in mixtures containing interfering components are available (35), and many of these refer to benzo[a]pyrene, in the presence of benzo[k]fluoranthene (36), and in various materials as diverse as cigarette-smoke condensate (37), cotton seeds (38), and human lungs (39). Sawicki has compared the methods for the determination of benzo[a]pyrene in air particulates (26). The same group reported in detail on analytical methods (along with data on the reproducibility, accuracy, and precision) for the determination of the amounts of total PAC (40), benzo[a]pyrene (39–42), and benzo[k]fluoranthene (41) in environmental samples by chromatography on alumina followed by spectrophotometry and fluorimetry. Detection limits of between 3 ng and 200 ng (benzo[a]pyrene) were quoted.

A similar method, employing column chromatography followed by fluorimetry, has been recommended by IUPAC as an analytical method for use in occupational hygiene to determine benzo[a]pyrene and benzo[k]-fluoranthene in airborne particulates (43). Fractions containing these PAH are bulked, and the emission at 410 nm is measured with excitation at 309 and 385 nm (Fig. 9-5). The emission for excitation at 309 nm is essentially due to benzo[k]fluoranthene only, so that the concentration of this compound can be determined directly; the benzo[a]pyrene concentration is then obtained from the intensity of the emission for excitation at 385 nm after correction for the contribution of benzo[k]fluoranthene.

Thin-layer chromatography continues to be prominent as a separation method to precede the use of fluorescence (44). A method due to Borneff (45) for the determination of the levels of certain PAH in drinking water, the subject of World Health Organization (WHO) recommendations, is widely applied. The PAH are extracted from the water with an organic solvent, and separated by two-dimensional TLC, after a clean-up procedure. Each PAC is identified on the plate by its position and fluorescence color under UV light (Fig. 9-6). Quantitation is then by fluorimetry: either semi-quantitatively by visual densitometric comparison with color intensities from reference plates, or by elution of each spot individually and measurement of fluorescence emission intensities at characteristic wavelengths for comparison with a calibration curve. The wavelengths for the PAH, specified by WHO, are listed in Table 9-1.

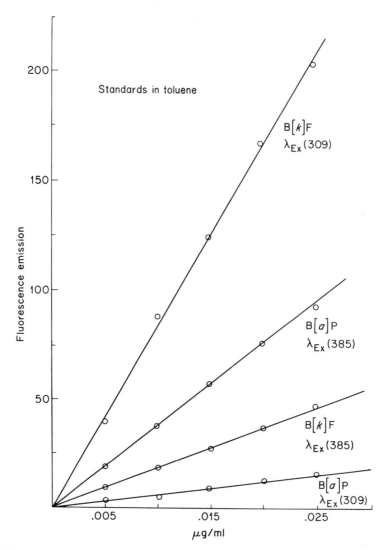

Fig. 9-5. Calibration curves for fluorimetric determination of benzo[*a*]pyrene and benzo-[*k*]fluoranthene in toluene. [Reproduced with permission from W. Galley, ed., "Environmental Pollutants—Selected Analytical Methods" (Scope 6), p. 277. Butterworths, London, 1975.

PAH may also be microsublimed from TLC plates before fluorimetry (*46*), or fluorescence-emission spectra may be obtained directly from TLC spots (*47*). Light from an appropriate source (Fig. 9-7) passes through the excitation monochromator and is focused on the plate as a narrow line. The reflected light is viewed by the analyzer monochromator. The plate is

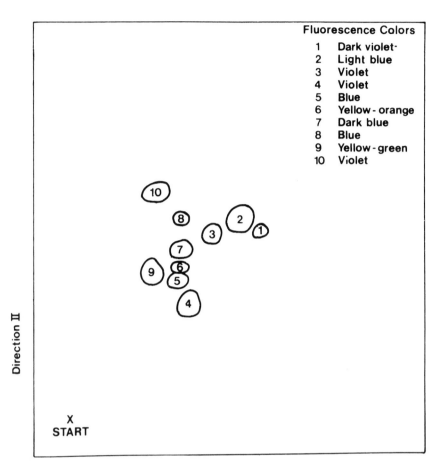

Fig. 9-6. Two-dimensional thin-layer chromatogram of a PAH mixture: 1, pyrene; 2, fluoranthene; 3, benz[a]anthracene; 4, benzo[a]pyrene; 5, benzo[b]fluoranthene; 6, benzo[j]-fluoranthene; 7, benzo[k]fluoranthene; 8, perylene, 9, indeno[1,2,3-cd]pyrene; and 10, benzo-[ghi]perylene. [Reproduced with permission from R. M. Harrison, R. Perry and R. A. Wellings, *Water Res.* **9,** 331 (1975). Copyright, Pergamon Press, Inc.]

driven in x or y directions by a digital motor, and the signal corresponding to each spot is integrated (47). Optimal excitation and emission wavelengths for this procedure are listed in Table 9-1, and detection limits from 1 to 4 ng with standard deviations of 7–14% have been claimed (47). Similar detection limits (0.1–2 ng) were described by Kaschani and Reiter (48) in rapid TLC/fluorimetric analysis of the "marker" PAH benzo[a]pyrene, benzo[b]-fluoranthene, anthanthrene, indeno[1,2,3-cd]pyrene, benzo[k]fluoranthene,

TABLE 9-1

Recommended Fluorescence Excitation and Emission Wavelengths (nm) in
the Analysis of PAH Separated by TLC Procedures

Compound	Borneff (45)[a] Emission	Tomingas et al. (47) Excitation	Tomingas et al. (47) Emission	Katz et al. (56) Excitation	Katz et al. (56) Emission
Fluoranthene	465	350	445		
Pyrene		325	398	332	392
Benz[a]anthracene		280	410		
Chrysene		308	405		
Benzo[b]fluoranthene	454	340	425	301	424
Benzo[k]fluoranthene	434	300	432	305	402
Benzo[a]pyrene	431	355	410	381	402
Benzo[e]pyrene		325	412	329	389
Perylene		377	440	407	438
Dibenz[a,h]anthracene		290	420	299	392
Dibenzo[b,def] chrysene				312	451
Dibenzo[def,mno] chrysene				302	429
Naphtho[1,2,3,4-def] chrysene				304	396
Benzo[ghi]perylene	420	375	425	382	419
Benzo[rst]pentaphene				393	432
Indeno[1,2,3-cd]pyrene	505				
Coronene		298	430		

[a] Excitation at 365 nm.

and perylene in vehicle exhaust condensates, mineral oil products, and waste water. Benzo[a]pyrene was detected at a lower limit of 1–2 ppm in shale oil by fluorimetry on the chromatoplate after separation (49, 50). During an investigation of the efficiency of a process for the fluorimetric analysis of environmental PAH separated on TLC plates, detection limits down to 1 ng were again claimed (51). On the other hand, on the basis of tests with ^{14}C-labeled compounds, de Wiest et al. (52) have shown how the reproducibility of procedures based on TLC/fluorimetry is poor for quantities below 1 μg. Care is clearly necessary in interpreting the results of fluorescence measurements on TLC plates. To this end, Hurtubise et al. have used fluoranthene as a model compound in obtaining theoretical fluorescence densitometric calibration curves from equations governing the scattering of light from particles. Comparisons were then made with experimental fluorescence reflectance and transmission on alumina and silica plates (53).

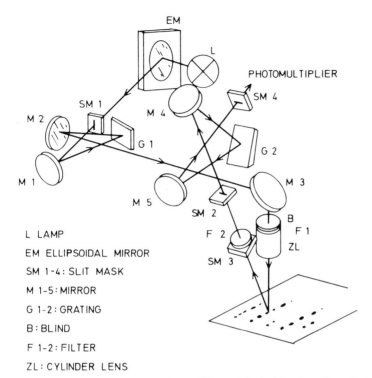

EM

L

PHOTOMULTIPLIER

SM 4

SM 1

M 2

M 4

G 1

G 2

M 1

M 3

M 5

SM 2

B

L LAMP

F 2

F 1

EM ELLIPSOIDAL MIRROR

ZL

SM 1-4: SLIT MASK

SM 3

M 1-5: MIRROR

G 1-2: GRATING

B : BLIND

F 1-2: FILTER

ZL: CYLINDER LENS

Fig. 9-7. UV/vis-chromatogram analyzer of Farrand Optical Co., Inc., New York, optical schematic. [Reproduced with permission from R. Tomingas, G. Voltmer and R. Bednarik, *Sci. Total Environ.* **7,** 261 (1977). Elsevier Scientific Publishing Co.]

Reflecting the recent shift in emphasis to the higher molecular weight PAC (*44, 54*), Pierce and Katz resolved 12 penta- and hexacyclic arenes from atmospheric aerosols by a two-stage TLC process and made determinations by fluorimetry (*54–56*). The perimeters of the fluorescent areas were scribed, adsorbent containing the compound of interest was scraped from the plate, and the PAH were extracted with dichloromethane (from alumina–silica gel and magnesium hydroxide) or diethyl ether (from cellulose). Excitation and emission spectra of hexane solutions of each PAH were recorded at room temperature with a xenon-arc lamp source and grating monochromator (Table 9-1). The superimposed spectra (e.g., Fig. 9-8) of standard PAH and of PAH from air particulates were significantly similar. This method has also been applied to determine both the distribution of PAH in relation to particle size (*55*) (Chapter 4) and the concentration levels of PAH over corresponding three-month periods for different cities in Ontario (*56*).

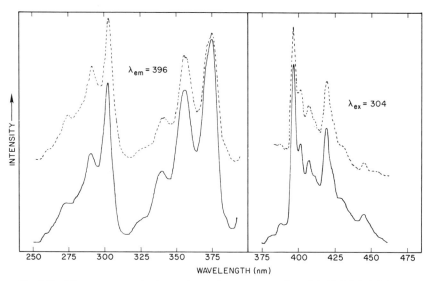

Fig. 9-8. Fluorescence excitation and emission spectrum of naphtho[1,2,3,4-*def*]chrysene. Solid line = model compound, broken line = extract from benzene soluble fraction of air particulates. [Reproduced with permission from: R. C. Pierce and M. Katz, *Anal. Chem.* **47**, 1743 (1975). Copyright, American Chemical Society.]

The use of fluorescence as a detector for HPLC provides additional flexibility to complement UV-absorption detection, and was discussed in Chapter 6. Tandem use of UV-absorption and fluorescence spectra of aromatic constituents can allow the identification of oil spills without prior separations of the PAH fraction in a "fingerprint" method (*57, 58*). Thus, Levy used a simple measurement of UV absorbance at 256 nm as an indicator of oil contamination in water (*59*), and also used the ratio of absorbances at 228 and 256 nm to identify petroleum products in samples from the marine environment (*60*). Hellman (*61*) similarly used differences in the relative absorbances at 313, 360, 365, and 465 nm of PAH from different sources, and was thus able to classify their origins (as biogenic, exhaust gas, petroleum, or pyrogenic) from these ratios and also by separating them on TLC plates and recording fluorescence as a function of retention index. Klimisch and Szonn (*62*) related the (integrated) intensities in fluorescence excitation and emission spectra (presumed a measure of total PAC concentration) to the biological activity of fractions of cigarette-smoke condensate.

A simple ratio method was used by Thurston and Knight to identify crude oils (*63*). From excitation at 340 nm, the intensity of the emission at 386 nm is divided by that at 440 nm for different concentrations. The full emission spectra for excitation at 254 and 290 nm yield more characteristic

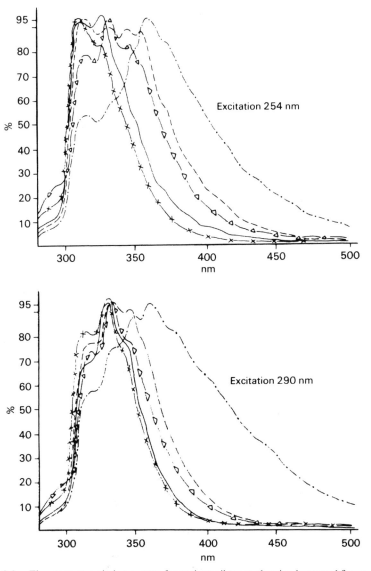

Fig. 9-9. Fluorescence emission spectra for various oil types, showing improved fingerprint obtained with excitation at (a) 254 nm rather than (b) 290 nm (1 : 10⁴ in cyclohexane): (——) No. 2 fuel oil; (—·—) No. 4 fuel oil; (—x—) gas oil; (– – –) diesel oil; and (—△—) lube oil. [Reproduced with permission from A. P. Bentz, *Anal. Chem.* **48**, 454A (1976). Copyright, American Chemical Soc.]

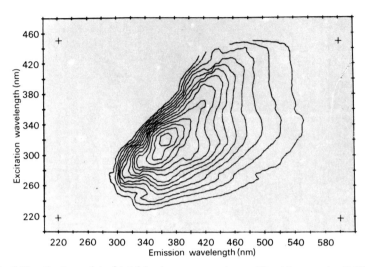

Fig. 9-10. Contour plot of total luminescence spectrum of Louisiana crude oil (100 ppm in cyclohexane). Bandwidth: 10 nm. Contour interval: 50 nm. [Reproduced with permission from A. P. Bentz, *Anal. Chem.* **48**, 454A (1976). Copyright, American Chemical Soc.]

fingerprints (Fig. 9-9), and this approach is widely applied by the U.S. Coast Guard to identify spilled oil (*58*). More elaborate procedures are possible: thus Freegarde *et al.* (*64*) recorded the emission spectra of samples at many different excitation wavelengths to yield a characteristic contour map of equal intensities or fluorogram (*65*). The automated version of this procedure is "total luminescence" (see Section VII), and this is being applied to the identification of crude oils (*58, 65*) (e.g., Fig. 9-10) along with the new technique of synchronous excitation fluorescence (*66*).

IV. PHOSPHORESCENCE

The selectivity of phosphorimetry is greater than that of spectrofluorimetry although its range of application is not as great; fewer compounds have measurable phosphorescence than have fluorescence, but the spectra (e.g., Fig. 9-11), especially of PAC, may be more characteristic (*67*). The limits of detection of PAC in phosphorimetry and fluorimetry (Table 9-2) are of the same order of magnitude. The wider slits which may be used in phosphorimetry compensate for the generally lower quantum yield of phosphorescence compared with fluorescence. Phosphorescence (triplet state) lifetimes (10 μs to 10 s) are much longer than excited singlet state life-

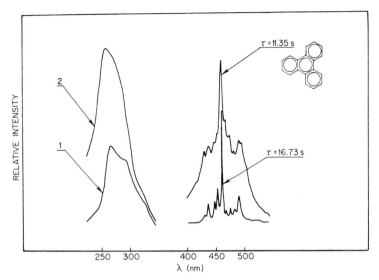

Fig. 9-11. Phosphorescence excitation (*left*) and emission (*right*) spectra of triphenylene standard (1) and triphenylene fraction (2) from GPC of aromatic concentrate from a white petroleum product. Excitation at 280 nm, emission 461 nm, τ = mean lifetime. [Reproduced with permission from M. Popl, M. Stejskal, and J. Mostecky, *Anal. Chem.* **47,** 1947 (1975). Copyright, American Chemical Society.]

times of fluorescence (100 ps to 500 ns), so that this parameter can be more easily used in analysis.

The phosphorimetric experimental method is more complex because nonradiative interaction with other molecules in solution is favored by the relatively long lifetimes of the triplet states of PAC; phosphorescence is hence generally observed not at room temperature (except in the special circumstances described below) but in solvent glasses at low temperature. The cryogenic limitation coupled with the poorer precision and accuracy [1% for UV absorption and fluorimetry, 10% for phosphorimetry, although the last figure may be improved by a rotating sample cell (*68*) (Fig. 9-12)] have led to only sparing use of phosphorimetry in the analysis of environmental PAC (*69*). Nonetheless, Hood and Winefordner studied the fluorescence and phosphorescence emission wavelengths, phosphorescence decay times, and limits of detection of a large number of PAC. They concluded that, if concentrations are available at which both techniques give analytically useful signals, it is convenient to measure both fluorescence and phosphorescence (*70*); by combining the two techniques, an analytical range over four decades results, and most carcinogens can be detected as ethanolic

TABLE 9-2

Detection Limits (ng/ml) for Determination of PAH by Luminescence[a]

Compound	Fluorescence at 25°C		Fluorescence at -196°C	Shpol'skii fluorescence at -196°C		Phosphorescence at -196°C
	Deuterium lamp	Laser excitation				
Naphthalene	0.03 (155)	1 × 10⁻³ (157)				700 (104)
Anthracene	0.03 (155)	9 × 10⁻³ (157)			10 (139)	
9,10-Dimethylanthracene						
Phenanthrene				50 (110)	10 (139)	30 (175)
Benz[a]anthracene			30 (70)	50 (110)	10 (139)	50 (70)
7,12-Dimethylbenz[a]anthracene				100 (110)		
Chrysene	50 (175)		100 (175)	100 (110)	10 (139)	100 (175)
Triphenylene	50 (175)		100 (175) 300 (70)	200 (110)	10 (139)	2 (175) 3 (70)
Benzo[b]fluorene			10 (70)	50 (110)		400 (70)
Pyrene		0.5 × 10⁻³ (157)	2 (70)	100 (110)	10 (139)	400 (70)
3-Methylpyrene				200 (110)		
Perylene			2 (70)	50 (110)	1 (139)	
Benzo[a]pyrene	100 (175)		500 (175) 3 (70)	5 (110) 0.1 (122)	0.1 (139)	100 (175) 2000 (70)
Benzo[e]pyrene	0.10 (155)		30 (70)	70 (110)	10 (139)	20 (70)
Dibenz[a,c]anthracene			7 (70)	10 (139)		90 (70)
Dibenz[a,h]anthracene	50 (175)		20 (175) 8 (70)	100 (110)	10 (139)	100 (175) 20 (70)
Benzo[ghi]perylene			5 (70)	1 (139)		90 (70)
Dibenzo[a,e]pyrene			7 (70)	80 (110)		600 (70)
Dibenzo[b,def]chrysene			3 (70)	40 (110)		
Benzo[rst]pentaphene			50 (70)	100 (110)	1 (139)	500 (70)
Dibenzo[def,p]chrysene			70 (70)	80 (110)		
Indeno[1,2,3-cd]pyrene				300 (110)		
Coronene	20 (175)		100 (175) 4 (70)	100 (110)		20 (175) 4 (70)

[a] Numbers in parentheses are reference numbers.

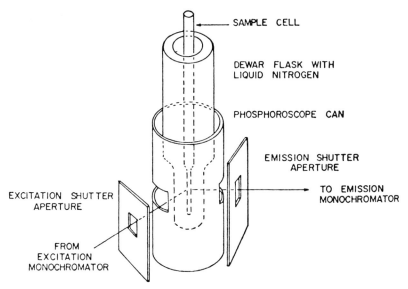

SAMPLE CELL

DEWAR FLASK WITH
LIQUID NITROGEN

PHOSPHOROSCOPE CAN

EMISSION SHUTTER
APERTURE

EXCITATION SHUTTER
APERTURE

TO EMISSION
MONOCHROMATOR

FROM
EXCITATION
MONOCHROMATOR

Fig. 9-12. Schematic rotating-can phosphoroscope. [Reproduced with permission from P. F. Lott and R. J. Hurtubise, *J. Chem. Ed.* **51**, A315 (1974). Copyright, American Chemical Society.]

extracts from TLC plates at the 0.1 μg level. Phosphorimetry is most useful where the mixture contains strongly fluorescent but weakly phosphorescent interfering species. For example, perylene interferes with the fluorimetric determination of dibenzo[*a*,*i*]pyrene but gives negligible interference in phosphorimetry (*70*).

Zander (*67*) has reviewed a number of elegant applications of phosphorimetry in the determination of impurities in pure hydrocarbons, and in the analysis of coal-tar fractions. The method has also found some use in the area of petroleum products; Popl *et al.* (*71*) separated PAH from white petroleum by column chromatography and GLC and made identifications from both phosphorescence spectra (Fig. 9-11) and lifetimes, although the latter are generally similar for carcinogens. Applications of phosphorimetry in air pollution work were pioneered by Sawicki (*72–74*), who detected PAH on thin-layer chromatograms by their phosphorescence in liquid nitrogen (*72*). The same group demonstrated the acquisition of phosphorescence spectra from spots cut from glass-fiber and paper chromatograms of air-particulate PAC (*74*).

Phosphorescence emission may be observed at room temperature from salts of ionic compounds adsorbed on silica gel, alumina, and even filter paper (*75*). The effect, ascribed to increased molecular rigidity through adsorption, which is presumed to reduce quenching (i.e., collisional de-activation), has been developed as a sensitive spectroscopic method of analysis (76–78). Winefordner has shown (77) that sodium iodide and silver nitrate induce room-temperature phosphorescence emission from a number of PAH and allow detection at the nanogram level for phenanthrene and pyrene. Strong RTP can be induced from nitrogen heterocycles adsorbed on silica gel chromatoplates containing a polymeric binder (78). These approaches may increase the application of phosphorescence in the analysis of environmental PAC.

V. SELECTIVE QUENCHING IN LUMINESCENCE ANALYSIS

As discussed in Sections III and V, the luminescence characteristics of PAC are markedly dependent on the environment. Thus, reduction of luminescence intensity, or quenching, occurs if an electronically excited molecule gives up its energy to a solvent or other adjacent molecule. This property may be used to increase the selectivity of luminescence in various ways, known collectively as *quenchofluorimetry*.

Henrich and Güsten have shown (79) how selective quenching by oxygen can enable fluorimetric determination of PAH in difficult-to-separate mixtures such as benz[a]anthracene/chrysene/triphenylene and benzo[ghi]-perylene/anthanthrene. A more commonly applied approach uses the property of a substituted atom or group such as chlorine, bromine, or nitro in the solvent to enhance phosphorescence at the expense of fluorescence by making intersystem crossing from excited singlet to phosphorescent triplet states (Fig. 1-1) easier through an increase in spin–orbit coupling (the "heavy-atom" effect) (80). Zander (81) found that the addition of 10% by volume of iodomethane to his standard solvent system of diethyl ether, isopentane, and ethanol (5:5:2 by volume) gave significantly improved detection limits e.g., from 2×10^{-6} g/ml to 5×10^{-7} g/ml for the determination of fluoranthene by phosphorescence. The fluorescence of this compound is correspondingly quenched, as is that of numerous other PAH (82), e.g., naphthalene, anthracene, naphthacene, chrysene, and benzo[a]pyrene.

However, certain PAH are exceptions: dibenzo[b,def]chrysene shows only slight quenching in the presence of iodomethane (81), and perylene, 3-methyl-perylene, and dibenzo[a,f]perylene show no change in intensity (82, 83). It is thus possible to determine quantitatively small concentrations (0.1–2%) of perylene in the presence of large excesses of benzo[e]pyrene and naphthacene by fluorimetry. This application may be important in view of the suggested use of perylene as a geochemical marker (84, 85). On the other hand, no *internal* heavy-atom effect operates for perylene: the quantum yields and decay times of bromoperylenes are nearly identical to those of the parent (86).

The reduction in fluorescence emission intensity by quenching is expressed by the Stern–Volmer equation

$$f_0/f_q = KQ + 1 \ldots \tag{1}$$

where f_0 and f_q are the fluorescence yields in absence and presence of quencher respectively, K is the Stern–Volmer coefficient, and Q is the concentration of quencher. For efficient quenching

$$K = k_q \tau_0 \ldots \tag{2}$$

where k_q is the (diffusion-controlled) rate constant for the quenching process, and τ_0 is the unquenched lifetime (87). Quenching effects therefore depend on fluorescence lifetimes which vary extensively and therefore account, at least in part, for the observations shown above. Graphs of f_0/f_q against Q (Stern–Volmer plots) can be interpreted to yield qualitative information about the composition of complex mixtures and allow quantitative analysis of simpler mixtures (88).

Other heavy-atom perturbers of fluorescence which have been investigated include iodoethane in ethanol solutions of PAH in which two kinds of behavior are observed (89): (a) fluorescence is quenched and phosphorescence enhanced relative to ethanol alone; or (b) fluorescence is enhanced and phosphorescence quenched according to perturber concentration. The first effect is typical of benzo[a]- and benzo[b]fluorene, benz[a]anthracene, naphthacene, benzo[a]pyrene, and dibenz[a,c]anthracene, and it increases with increasing iodoethane concentration. However, the limits of detection of these compounds by phosphorescence decrease accordingly. Naphthalene, acenaphthene, anthracene, phenanthrene, and triphenylene exhibit the second effect. A variety of selective luminescence methods are thus possible for environmental PAH mixtures using heavy-atom solvents (90).

Since a heavy atom may perturb the vibrational structure of the phosphorescence spectrum of a PAC, especially naphthalene homologs and heteroaromatics, different spectra may be observed in inert and heavy-atom solvents. Thus, the band at 430 nm in the phosphorescence spectrum of triphenylene is characteristically strengthened in a heavy-atom solvent (91).

Zander has shown how silver nitrate selectively enhances the phosphorescence of azaaromatics relative to that of PAH because of the strong electron-donor properties of the nitrogen-containing compounds (92). This effect should have applications in environmental analysis by luminescence.

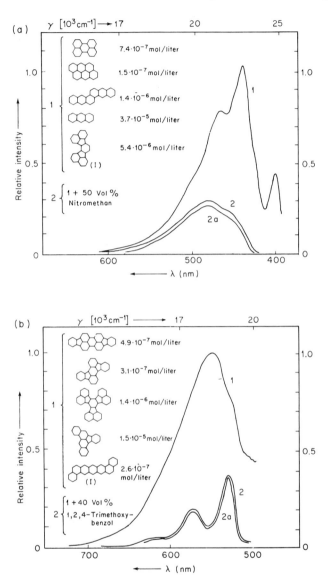

Fig. 9-13. (a) Fluorescence spectra in acetonitrile with excitation at 383 nm of (1) a mixture of mainly alternant PAH; (2) same mixture in acetonitrile containing 50% v/v nitromethane; and (2a) spectrum of pure acenaphtho[1,2-j]fluoranthene. (b) Fluorescence spectra in acetonitrile with excitation at 430 nm of (1) a mixture of mainly nonalternant PAH; (2) same mixture in acetonitrile containing 40% v/v 1,2,4-trimethoxybenzene; and (2a) spectrum of pure dibenzo-[a,l]pentacene. [Reproduced with permission from U. Breymann, N. Dreeskamp, E. Koch, and M. Zander, *Fresenius Z. Anal. Chem.* **293,** 208 (1978). Copyright, Springer-Verlag.]

Another useful application of the heavy-atom effect in the observation of room-temperature phosphorescence of PAC in presence of silver and sodium nitrates has already been noted in Section IV.

Selective quenching of the fluorescence of PAH with only six-membered rings by electron acceptors such as nitromethane was noted by Sawicki (93); the fluorescence of PAH containing the fluoranthene skeleton was said not to be quenched. Dreeskamp et al. (94) have shown a few exceptions to this "rule"; benzo[b]- and benzo[k]fluoranthene showed significant quenching, while even fluoranthene was itself slightly quenched. However, this phenomenon is still general enough (95) to allow the identification of non-alternant PAH after separation—e.g., on thin-layer plates, and in mixtures with alternant PAH (Fig. 9-13a). Complementary quenchofluorimetry with electron donors such as 1,2,4-trimethoxybenzene, moreover, suppresses the fluorescence of nonalternants and allows recognition of alternants in mixtures (Fig. 9-13b) (96).

VI. LOW-TEMPERATURE LUMINESCENCE

The broadened peaks of the fluorescence spectra of PAH recorded in liquid solution (e.g., Fig. 9-4) may inhibit analysis (97); moreover, although some substituted PAH have markedly different spectra from the parent [e.g., methylfluoranthenes (98)], such effects may be small (e.g., methylbenz-[a]anthracenes and methylnaphtho[2,1-a]fluorenes) (99) and may lead to confusion between parent and derivative. Urban atmospheres contain relatively large amounts of alkylated PAH (100). An attractive alternative approach is to make use of the Shpol'skii effect (101) [first observed in 1952 (102) and confirmed in 1954 (103)], which involves the resolution of characteristic vibrational fine structure in the luminescence emission spectra of PAH (e.g., compare Figs. 9-4 and 9-14, and see Fig. 9-15) in frozen solutions in certain n-alkane, alcohol, and ether solvents.

The usefulness of low-temperature luminescence for the determination of PAH is well known, particularly in the enhancement of sensitivity which results from the increase in fluorescent intensity and narrowing of lines from PAH in frozen-solution glasses (70, 104) (Table 9-2). Highly resolved low-temperature fluorescence spectra may be obtained with a double-monochromator spectrofluorimeter (70). However, the origin of quasilinear Shpol'skii spectra is thought to be the strict orientation of PAH molecules separated by large distances when embedded in a crystalline solvent lattice like an orientated gas molecule. Vibrational and rotational energy are reduced, interaction between the solute molecules is prevented, and compounds having liquid-solution spectra with half-bandwidths of several nm

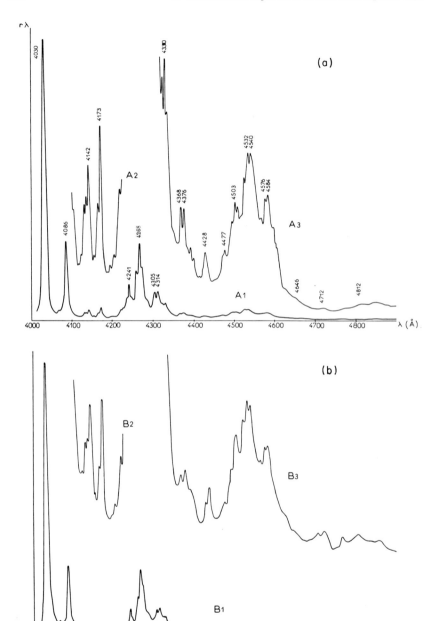

Fig. 9-14. Shpol'skii spectra of benzo[a]pyrene recorded with 320–390 nm excitation for solutions in n-octane at −190°C: (a) authentic sample; (b) sample from cigarette-smoke condensate. [Reproduced with permission from B. Muel and G. Lacroix, *Bull. Soc. Chim. Fr.*, 2139 (1960). Copyright, Masson et Cie.]

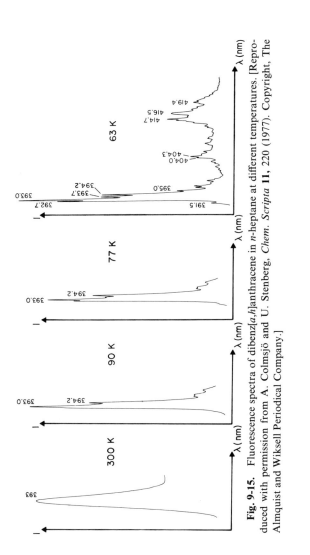

Fig. 9-15. Fluorescence spectra of dibenz[a,h]anthracene in n-heptane at different temperatures. [Reproduced with permission from A. Colmsjö and U. Stenberg, *Chem. Scripta* **11**, 220 (1977). Copyright, The Almquist and Wiksell Periodical Company.]

have frozen-solution half-bandwidths of one or two nm, further split by the Shpol'skii effect to extremely sharp lines with half-bandwidths of only a few hundredths of a nanometer. Further sharpening of lines is possible with laser excitation.

For Shpol'skii luminescence to be observed, the dimensions of the solute molecules must be approximately equal to those of the solvent molecules (2); only then can correct embedding occur. Pfister (105) claimed fine distinctions between alkane solvents, one of which could generally be selected to yield extremely narrow lines for a given PAH. This "lock and key" rule was examined by Dekkers (106) for several n-alkane polycrystals, but no exact solvent specificity was detected; thus quasiline spectra of anthracene are not restricted to n-heptane, nor of naphthacene to n-nonane. Shpol'skii spectra are fairly general for PAH if the correct concentration and cooling rate are employed. Freezing must be rapid or many of the embedded molecules are frozen in their lowest vibrational states with loss of many quasilines; in addition, formation of microcrystalline aggregates of solute may occur with consequent quantitative imprecision from excimer fluorescence, quenching, and sensitization (107). n-Alkanes of suitable molecular dimensions are the most commonly used solvents (108–110), although spectra are also readily observed in tetrahydrofuran (111). The physics of the ordering of molecules within the solvent matrix has been reviewed by Shpol'skii and Bolotnikova (112).

The sharpness of the Shpol'skii spectra depends on the temperature: at $-196°C$, the spectrum of pyrene in n-hexane contains about 60 lines, whereas at $-269°C$ the same solution shows more than 220 lines (113). Although the majority of the many quasilinear luminescence emission spectra of PAH so far reported (101, 108–111) have been obtained near $-196°C$, even a modest temperature reduction to $-210°C$ with a cell cooled by nitrogen at its freezing temperature allows a marked improvement in resolution (114) (Fig. 9-15).

A mercury-vapor lamp has usually been employed as source, and a glass or quartz Dewar tube as sample cell. Causey et al. (115) have compared a number of excitation sources (including a helium–cadmium laser), sample cells, and detection systems and concluded that a xenon arc or mercury lamp with a specially designed copper cryostat cell and dc integration of the luminescence signals constitute a simple and reliable system. The monochromator must, of course, have adequate resolving power (greater than that generally available with commercial fluorimeters), and the light-gathering power of the instrument must be high (115). Detection limits between 0.1 ng/ml (benzo[a]pyrene) and 300 ng/ml (indeno[1,2,3-cd]pyrene) have been quoted (110) (Table 9-2).

Shpol'skii spectra can provide very useful fingerprints of specific compounds, and a variety of qualitative analyses have been reported. Thus, line-rich spectra were obtained by Colmsjö and Stenberg (*116*) for PAH from a number of environmental sources such as automobile exhaust gas. The PAH were first separated by HPLC or microsublimed from thin-layer plates and then frozen in *n*-pentane or *n*-hexane. As many as 50 peaks corresponded in intensity and wavelength with the reference Shpol'skii spectrum for each PAH, demonstrating the high degree of specificity possible. Although the compounds identified were all within the compass of other techniques, it was suggested that compounds difficult to elute from a GC column should be especially amenable to Shpol'skii spectroscopy (*116*).

Drake *et al.* (*117*) demonstrated how such spectra allowed 12 PAC containing between three and 10 rings to be identified in extracts of coal and coal-tar pitch (Fig. 9-16), including ovalene (previously only postulated as a constituent of coal). In these materials, only species resistant to deactivation and quenching are detected because of the formation of inhomogeneous micellar aggregates (see below). This effect favors the detection of benzo[*a*]-pyrene at levels down to 0.1 ppm, and many applications of the quasilinear luminescence method have been applied to this compound. In contrast, benzo[*a*]pyrene quenches the spectrum of benzo[*e*]pyrene (*116*) so the latter isomers must be separated before Shpol'skii analysis.

Parker and Hatchard (*118*) distinguished benzo[*a*]pyrene from its photochemical reaction products, and this carcinogen has also been determined in atmospheric particulates by direct measurement (*119*) of the intensity of the emission at −194°C. Dikun (*120*) determined benzo[*a*]pyrene by Shpol'skii fluorescence using benzo[*ghi*]perylene as internal standard, and a similar method was reported by Personov, using coronene for the same purpose (*121*). The standard addition procedure to minimize the effect of quenching was preferred by Muel and Lacroix (*122*) for the determination of benzo[*a*]-pyrene in cigarette smoke, water, and various alcoholic drinks from the 403.0 nm emission line in *n*-octane at −190°C (Fig. 9-14); a relative precision of 10% was claimed. Jäger (*123, 124*) also determined benzo[*a*]pyrene by this method in a fraction of gasoline engine exhaust, as well as detecting a number of other hydrocarbons.

Combined internal-standard and standard-addition methods enabled Dikun (*125*) to determine benzo[*a*]pyrene in smoked fish. A Shpol'skii fluorescence method due to Khesina for the determination of benzo[*a*]pyrene and other PAC (*126, 127*) with a claimed relative error of less than 2% by standard addition and internal standardization has been applied to complex mixtures including aviation-engine soots (*128, 129*).

Quantitative determinations by quasilinear luminescence of benzo[*a*]-pyrene, mono- and dibenzanthracenes, and benzo[*ghi*]perylene in PAH separated from waste water have also been reported (*130*). The presence

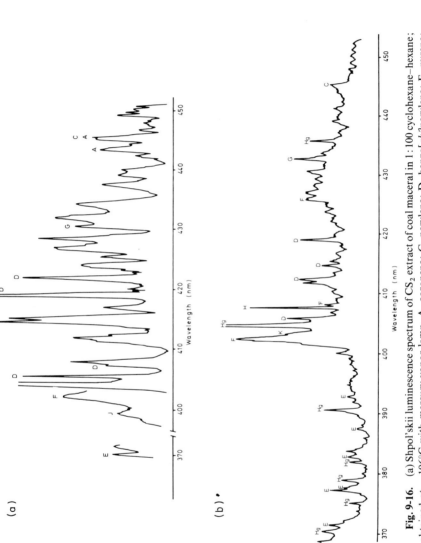

Fig. 9-16. (a) Shpol'skii luminescence spectrum of CS$_2$ extract of coal maceral in 1 : 100 cyclohexane–hexane; obtained at −196°C with mercury-vapor lamp. A, coronene; C, perylene; D, benzo[*ghi*]perylene; E, pyrene; F, benzo[*a*]pyrene; G, dibenzo[*a,l*]pyrene; H, chrysene; J, 9,10-dimethylanthracene; and K, benzo[*b*]fluoranthene. (b) Shpol'skii luminescence spectrum of cyclohexane pitch extract in hexane obtained at −196°C with mercury-vapor lamp (Hg denotes mercury lines). [Reproduced with permission from J. A. G. Drake, D. W. Jones, B. S. Causey, and G. F. Kirkbright, *Fuel* **57**, 663 (1978). Copyright, IPC Science and Tech. Press.]

has been shown, and determinations made of benzo[*a*]pyrene, perylene, and benzo[*ghi*]perylene in geological specimens by this method (*131*). Numerous PAC with between three and seven rings have also been identified in smoked fish, air particulates, soil, snow, and industrial oils from quasilinear spectra at low temperatures (*132, 133*).

In spite of these applications, the criticism has been made (*115, 134*) that the routine analytical utility of the Shpol'skii effect may be limited for a number of reasons: (a) difficulties with choice of solvent (*107*); (b) the bandwidths of quasilines may be dependent both on freezing rate (*135, 136*) and on solute concentration (*137*); and (c) the intensities of quasilines may

TABLE 9-3

Conditions for the Quantitative Analysis of PAH in Mixtures by Shpol'skii Fluorescence[a]

Compound	Wavelength (nm)		Standard	Maximum range of linearity of intensity/ concentration graph (10^{-6} g/ml)	Reproducibility of determination (% standard deviations of 10 determinations)
	Analyte line	Standard line			
Anthracene	286.7	419.5	Benzo[*ghi*]perylene	1	3.7
Phenanthrene	461.6	537.0	Benzo[*e*]pyrene	1	6.3
Benz[*a*]-anthracene	384.6	419.5	Benzo[*ghi*]perylene	1	9.3
Chrysene	498.6	537.0	Benzo[*e*]pyrene	1	5.9
Triphenylene	461.8	537.0	Benzo[*e*]pyrene	1	8.5
Pyrene	382.4	419.5	Benzo[*ghi*]perylene	1	4.7
Perylene	451.1	419.5	Benzo[*ghi*]perylene	3	6.0
Benzo[*a*]-pyrene	403.0	419.5	Benzo[*ghi*]perylene	0.1	6.6
Benzo[*e*]-pyrene	537.0	461.8	Triphenylene	1	2.2
Dibenz[*a,c*]-anthracene	386.2	419.5	Benzo[*ghi*]perylene	1	5.4
Dibenz[*a,h*]-anthracene	394.1	419.5	Benzo[*ghi*]perylene	1	3.5
Benzo[*ghi*]-perylene	419.5	408.5	Benzo[*a*]pyrene	2	6.7
Benzo[*rst*]-pentaphene	431.7	419.5	Benzo[*ghi*]perylene	2	4.4
Coronene	444.6	419.5	Benzo[*ghi*]perylene	1	4.2
Fluoranthene	534.0	537.00	Benzo[*e*]pyrene	2	4.3

[a] From Gaevaya and Khesina (*139*).

be influenced by freezing rate (*135, 136*), through the formation of micro-crystals (*107*) of solute and consequent energy transfer within the matrix. Hence they may have an irregular and irreproducible concentration de-pendence (*137, 138*). Few examples of the published applications (*122*) have quoted reproducibility, accuracy, or precision (*115*), and the linear working range of concentration is restricted (*121, 126, 127, 134*).

Two detailed studies have been made of the applicability of Shpol'skii luminescence in the analysis of PAH. Gaevaya and Khesina investigated the conditions necessary for the quantitative determination of 15 PAH (Table 9-3) in multicomponent solutions by using the effect (*139*). A com-

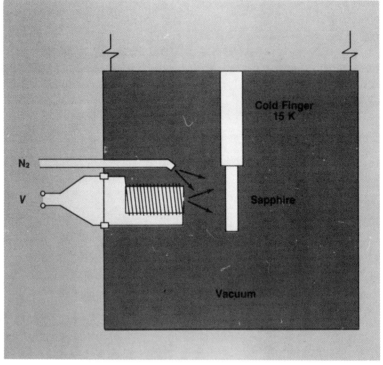

Fig. 9-17. Schematic diagram of cryostat head design used for matrix isolation of poly-cyclic aromatic compounds. Samples were placed in Knudsen cell (wrapped with heating wire maintained at voltage V), mixed with N_2 and deposited on a window (in this case, sapphire for fluorescence spectrometry). Head is maintained at low temperature by closed-cycle (helium) refrigerator and evacuated to pressure of about 10^{-5} torr. [Reproduced with permission from E. L. Wehry and G. Mamantov, *Anal. Chem.* **51**, 643A (1979). Copyright, American Chemical Society.]

bination of addition and internal standard methods was necessary, and standard deviations of 2–10% were found (Table 9-3) for sensitivities of 10^{-8}–10^{-10} g/ml. Kirkbright and de Lima (110) first demonstrated that unambiguous qualitative identification of PAH at very low concentration (Table 9-2) is possible. A careful investigation of the conditions necessary for the quantitative analysis of a mixture of dibenzopyrenes also showed that a combined standard-addition/internal-standard calibration procedure was necessary. Although internal standardization improves precision by decreasing random errors, an "inner filter" effect may occur because of overlap of the excitation spectrum with that of the analyte, with consequent reduction in emission intensity and increase in detection limit (110).

The limitations of Shpol'skii luminescence led Wehry et al. (134, 140) to investigate matrix isolation (MI) fluorescence in the analysis of PAH. Hydrocarbon vapors from a Knudsen cell (141) are mixed in an evacuable cryostat head with a large excess of nitrogen and condensed on a cold (−262°– −258°C) sapphire surface (Fig. 9-17). Detection limits of the order of 10^{-11} g and linear quantitative working curves over five decades were observed for the fluorescence emission spectra of the resulting matrices. MI fluorimetry is substantially free from interference by intramolecular energy transfer or inner-filter effects, even for samples containing microgram concentrations of several closely related PAH (134). Thus, the working curve for chrysene in a four-component mixture is virtually the same as that of pure chrysene, while linear calibration curves were also obtained for 4-methylchrysene in the presence of a 50-fold excess of three of its isomers in addition to benz[a]-anthracene, and the benzo[b]fluorene internal standard (142).

Bandwidths in nitrogen and argon matrices were generally greater than in Shpol'skii matrices, so that MI in such matrices is less useful for the identification of PAH in mixtures, although MI experiments with benz[a]anthracene in n-hexane or n-heptane as the matrix gave quasilinear spectra similar to those in conventional Shpol'skii fluorescence if annealing procedures were carried out (143). Thus, much greater spectral resolution can be obtained in vapor-deposited n-heptane matrices, and selective dye laser excitation of individual compounds in complex mixtures is then possible. Moreover, the sensitivity is greater by between a factor of 10 and 10^3, while the linear dynamic range for the fluorescence of PAH in vapor-deposited alkane matrices is comparable to, and may exceed that observed for nitrogen (144).

In a discussion of methods of overcoming the selectivity limitations of fluorimetry in liquid solution imposed by solute–solvent interactions, Brown et al. (145) have recommended fluorescence-line narrowing spectroscopy in organic glasses at −269°C, rather than in crystalline Shpol'skii matrices. In such glasses, narrow laser-line excitation yields sharp line spectra, and the problems of solubility, concentration gradients, and light

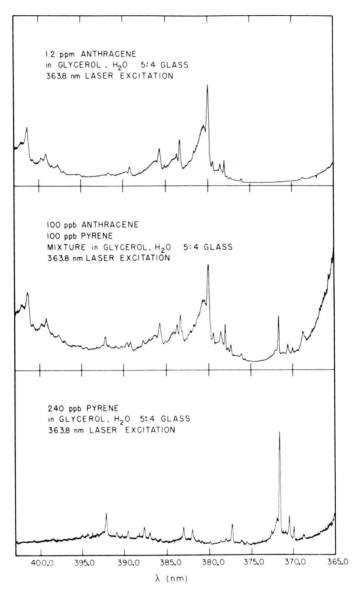

Fig. 9-18. The resolution of a mixture of two PAH through fluorescence-line narrowing spectroscopy. All spectra were measured at $-269°C$ and excited by $\lambda_{ex} = 363.8$ nm. [Reproduced with permission from J. C. Brown, M. Edelson, and G. J. Small, *Anal. Chem.* **50,** 1394 (1978). Copyright, American Chemical Society.]

scattering are minimized. A glass of 1 : 1 glycerol–water was found to be ideal, forming easily at −269°C without cracking, and allowing analysis of contaminated water. Sharp line spectra in such glasses have been obtained for a number of PAH (Fig. 9-18). Selective excitation provided by a tunable dye laser should allow the analysis of complex mixtures by this method.

Other current developments in low-temperature luminescence analysis of

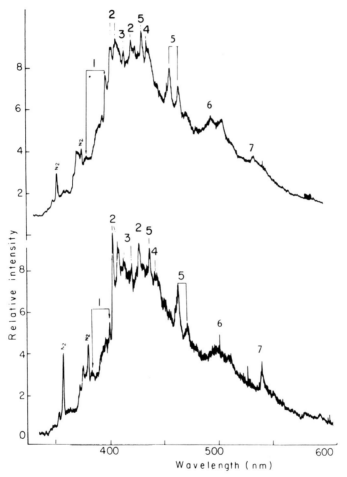

Fig. 9-19. X-ray excited optical luminescence spectra of PAH isolated from coal samples obtained from Iowa (upper) and Illinois (lower): 1, benz[a]anthracene; 2, benzo[a]pyrene; 3, benzo[ghi]perylene; 4, dibenzo[fg,op]naphthacene; 5, perylene; 6, phenanthrene; and 7, benzo-[e]pyrene. [Reproduced with permission from C. S. Woo, A. P. D'Silva, V. E. Fassel, and G. Oestreich, *Environ. Sci. Technol.* **12,** 173 (1978). Copyright, American Chemical Society.]

PAH include the observation of X-ray excited optical sharp line luminescence of PAH in frozen solution in cyclohexane, and better, in *n*-heptane,
which may result in the population of electronic levels not accessible to UV
excitation, and hence yield spectral features not normally observed, especially in the near- and vacuum-UV spectral ranges (*146*). Optical interaction ("cross talk") between exciting and emitted luminescence radiation
is eliminated by X-ray excitation. The technique allows profiles to be
obtained for the complex mixture of PAH extracted from coal, from the
by-products of coal conversion, and from shale oil and fuel oil; ppm levels
in the Shpol'skii solvent *n*-heptane were detected (*147, 148*), and phenanthrene, perylene, benz[*a*]anthracene, benzo[*a*]- and benzo[*e*]pyrene, benzo-
[*ghi*]perylene, and dibenzo[*fg,op*]naphthacene were all shown to be present
in Iowa and Illinois coals (Fig. 9-19).

A further potential application of X-ray excitation is the substantial
phosphorescence observed (*146*) even when UV excitation results in negligible emission. Qualitative and quantitative analyses have, in fact, already
been shown with the aid of certain phosphorescence spectra in *n*-alkane
solvents at −196°C which show the Shpol'skii effect (*139, 149, 150*).

VII. NEW LUMINESCENCE METHODS

Recent approaches to the luminescence analysis of PAC have been aimed
at increasing the sensitivity and selectivity of the methods through instrumental techniques; new excitation sources and dispersive systems, more
sophisticated detectors, and data-handling systems based on computer
methods are all under active study (*151–154*).

The requirement of the routine determination of PAH in water has
prompted the use of high-intensity excitation sources such as special deuterium lamps (*155, 156*) and pulsed lasers (*157*). The former have high
output below 300 nm and allow direct determination of PAH at the
0.1 μg/liter level (*155*) (Table 9-2). Since lasers are high-power sources of
coherent light easily focused on small sample volumes, and, if pulsed, allow
discrimination between emission and scattering, their use results in sensitive
and selective detection in luminescence. The fluorescence of several PAH
induced by a nitrogen-pumped dye laser continuously tunable from 258 to
750 nm with peak power of the order of kW and repetition rates up to 50 Hz
led to detection limits in the sub-ppt range. All the compounds studied
exhibited a linear dependence of fluorescence intensity on concentration,
sometimes extending over six decades (*157*) (Fig. 9-20).

In many cases, the time resolution inherent in the use of pulsed lasers is
advantageous in improving sensitivity, and this method has been shown to

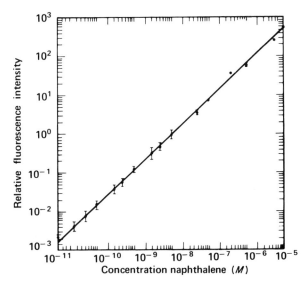

Fig. 9-20. Naphthalene fluorescence intensity vs. concentration with laser excitation. [Reproduced with permission from J. H. Richardson and M. E. Ando, *Anal. Chem.* **49**, 955 (1977). Copyright, American Chemical Society.]

be a selective means of detecting PAH in very dilute aqueous solutions by fluorimetry (*157*) and of resolving overlap of spectral features in matrix isolation spectra of PAC mixtures (*158*). For example, while benzo[*a*]pyrene and benzo[*k*]fluoranthene cannot readily be distinguished from each other by steady-state MI fluorimetry, their fluorescence decay times in N_2 matrices are sufficiently different (78 ns for BaP and 13 ns for BkF) to produce excellent temporal resolution of emission spectra for these compounds. Correspondingly, pulsed-source time-resolved phosphorimetry shows improved sensitivity and selectivity compared with continuously operated source methods and can, of course, be applied to PAC (*159*).

Computer methods allow luminescence spectra to be both time-resolved and component-resolved (*160*). Thus, phosphorescence spectra are recorded as a family of decay curves so that the spectra of binary mixtures may be kinetically resolved to yield the individual spectra and lifetimes of the individual compounds.

Both wavelength modulation, digital computation, or electronic differentiation may be used to obtain first- or second-derivative luminescence spectra (*161*). An analog computer may be used to produce higher order derivative spectra (*162*).

Although random noise is enhanced, systematic errors from light-source drift, background and stray light, etc., are eliminated, and much more

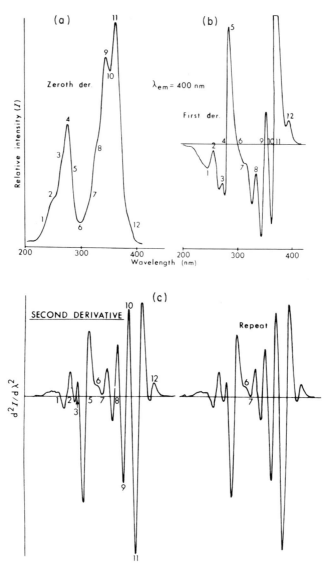

Fig. 9-21. Normal (a), first-derivative and (b), second-derivative (c) fluorescence excitation spectra of 1 ppm benzo[e]pyrene in methanol–water. [Reproduced with permission from G. L. Green and T. C. O'Haver, *Anal. Chem.* **46,** 2191 (1974). Copyright, American Chemical Society.]

information is contained in the derivative spectrum than in the "zeroth-order" spectrum, as is illustrated in Fig. 9-21. Quantitative measurements on a mixture of anthracene and pyrene showed a reduced effect for excess anthracene concentration, since baseline corrections are unnecessary (161) (Fig. 9-22).

Selective modulation fluorescence excitation and/or emission spectra can be recorded for either component of a mixture of two fluorophors, the spectra of which overlap too severely for conventional wavelength selection to be effective (163). The technique involves wavelength-modulating one monochromator, scanning the other, and detecting the resulting photodetector system with a lock-in amplifier. The principle is illustrated in Fig. 9-23a; if A and B are the excitation spectra of two (hypothetical) molecules in a mixture, modulation of the excitation wavelength over the indicated interval at a modulation frequency of F Hz modulates the fluorescence intensity of B at the same frequency, while the fluorescence intensity of A will be modulated at $2F$ Hz. Wavelength modulation across a minimum (Fig. 9-23b) also produces an intensity modulation at twice the wavelength modulation frequency. Benz[a]anthracene may be determined in mixtures with chrysene and vice versa by this method, in spite of the considerable overlap which is

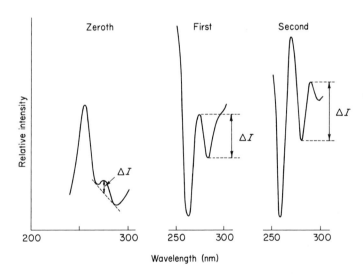

Fig. 9-22. (a) Normal, (b) first-derivative, and (c) second-derivative fluorescence excitation spectra of a mixture of 440 ppb pyrene and 360 ppb anthracene in isopropanol, illustrating the normal (zeroth derivative) base-line measurement and the first- and second-derivative (peak-to-peak) measurements of the pyrene band at 270 nm. The emission wavelength is 385 nm optimized for pyrene. [Reproduced with permission from G. L. Green and T. C. O'Haver, *Anal. Chem.* **46**, 2191 (1974). Copyright, American Chemical Society.]

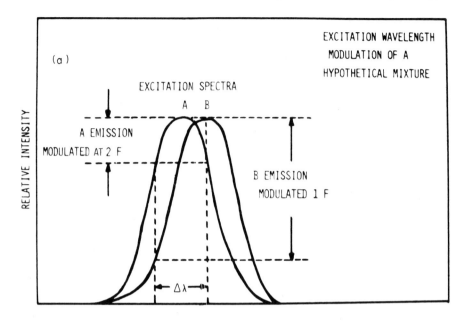

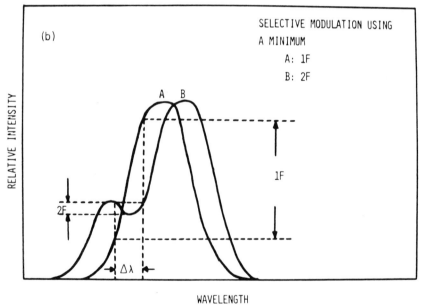

Fig. 9-23. The principle of selective modulation. [Reproduced with permission from T. C. O'Haver and W. M. Parks, *Anal. Chem.* **46**, 1886 (1974). Copyright, American Chemical Society.]

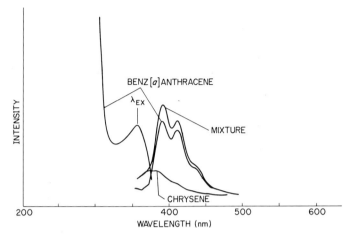

Fig. 9-24. Selective excitation of 0.27 ppm benz[*a*]anthracene in the presence of 1.0 ppm chrysene in cyclohexane at 25°C. [Reproduced with permission from T. C. O'Haver and W. M. Parks, *Anal. Chem.* **46**, 1886 (1974). Copyright, American Chemical Society.]

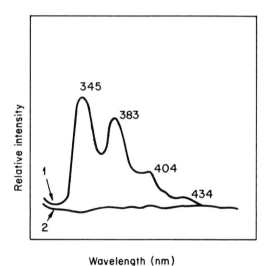

Wavelength (nm)

Fig. 9-25. Curve 1: Selectively modulated fluorescence emission spectrum of a mixture of chrysene and benz[*a*]anthracene from air particulates; excitation modulation at the absorption maxima of chrysene. Curve 2: Selectively modulated (i.e., nulled) fluorescence emission spectrum of a standard chrysene sample under conditions as in curve 1. [Reproduced with permission from M. A. Fox and S. W. Staley, *Anal. Chem.* **48**, 992 (1976). Copyright, American Chemical Society.]

observed even with excitation wavelength optimized (Fig. 9-24). Modulation of the excitation monochromator in the region of the appropriate absorption maximum nulls the signals of the interfering hydrocarbon almost completely (Fig. 9-25). Modulation fluorescence spectroscopy was used by Fox and Staley to demonstrate that the compound eluted with benzo[e]pyrene in the HPLC of air particulate PAH is benzo[k]fluoranthene (164).

The practical limitations of the application of conventional luminescence methods with fixed excitation or emission wavelength to mixtures of PAH

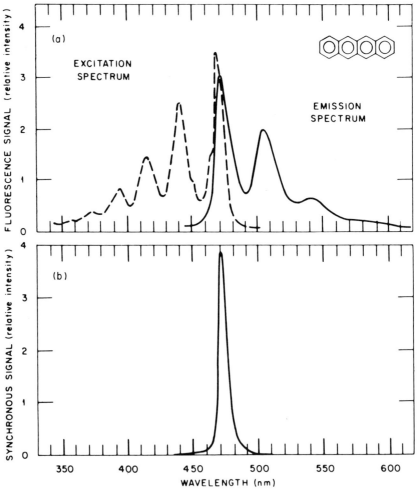

Fig. 9-26. (a) Fluorescence emission and excitation spectra of naphthacene. (b) Synchronous fluorescence signal of naphthacene. [Reproduced with permission from: T. Vo-Dinh, *Anal. Chem.* **50**, 396 (1978). Copyright, American Chemical Society.]

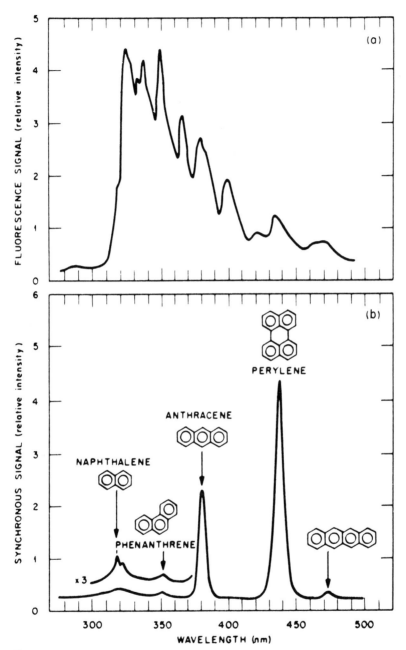

Fig. 9-27. (a) Conventional fluorescence spectrum of a mixture of naphthalene, phenanthrene, anthracene, perylene, and naphthacene. (b) Synchronous spectrum of the mixture. [Reproduced with permission from T. Vo-Dinh, *Anal. Chem.* **50**, 396 (1978). Copyright, American Chemical Society.]

with superimposed spectra led Vo-Dinh (165) to investigate synchronous
luminescence spectroscopy (166). Both excitation and emission wavelengths
are scanned while keeping a constant interval $\Delta\lambda$ between them. The principle
is illustrated in Fig. 9-26a. Here, the conventional excitation and emission
spectra of naphthacene in ethanol are compared. With excitation at 462 nm,
the fluorescence emission spectrum covers the range of 460 to 600 nm, while
the excitation spectrum monitored at 507 nm ranges from 350 to 480 nm.
The small wavelength difference of 3 nm between the O—O band peaks in
the emission and excitation spectra is called the Stokes shift. With a wavelength interval $\Delta\lambda = 3$ nm matching the Stokes shift, the synchronous
spectrum of naphthacene, shown in Fig. 9-26b, now consists simply of one
single peak located at 473 nm. For a mixture (e.g., Fig. 9-27b), the synchronous spectrum (using $\Delta\lambda$ as 3 nm, close to most Stokes shifts) consists
of single bands with halfwidths of 10–15 nm appearing at the O—O band
position (347 nm for phenanthrene, 381 nm for anthracene, and 440 nm for
perylene) rather than a range of bands over 200 nm in the conventional
fluorescence spectrum (Fig. 9-27a).

Vo-Dinh et al. (167, 168) applied synchronous excitation to characterize
the naphthalene derivatives in waste water from a coal conversion process
and in an interesting development of room-temperature phosphorescence
(RTP) (Section IV) to multicomponent analysis of PAH adsorbed on filter
paper. The presence of pyrene in Synthoil has been shown by this method
(168). RTP excitation and emission wavelengths and suitable values of $\Delta\lambda$
for a number of PAC are summarized in Table 9-4. Figure 9-28, the RTP

TABLE 9-4

**RTP Excitation and Emission Bands for Various PAC and $\Delta\lambda$ Values for Their
Synchronous RTP Analysis**[a]

Compounds	Excitation peaks (nm)	Emission peaks (nm)	Experimental optimal $\Delta\lambda$ values (nm)
Acridine	360	643	280
Chrysene	325	515	190
Fluorene	305	430	125
Naphthalene	275	472	197
Phenanthrene	250	460, 508	207
Pyrene	350	600	250
Quinoline	315	465	150
Benzo[e]pyrene	335	543	208
Benzo[a]pyrene	392	690	298
Dibenz[a,h]anthracene	301	555	254
Dibenz[a,c]anthracene	296	570	274

[a] From Vo-Dinh (167, 168).

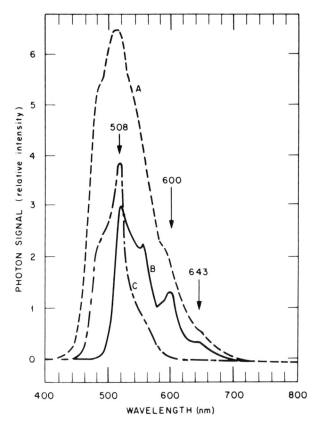

Fig. 9-28. RTP spectra of sample containing 17 ng of pyrene, 10 ng of phenanthrene, and 26 ng of acridine: (A) conventional RTP spectrum with fixed λ_{ex} (295 nm); (B) Synchronous RTP spectrum with $\Delta\lambda = 256$ nm; (C) Synchronous spectrum with $\Delta\lambda = 207$ nm. [Reproduced with permission from T. Vo-Dinh, R. B. Gammage, A. R. Hawthorne, and J. H. Thorngate, *Environ. Sci. Technol.* **12,** 1297 (1978). Copyright, American Chemical Society.]

spectrum of a mixture of pyrene, phenanthrene, and acridine, shows how synchronous excitation with $\Delta\lambda = 207$ nm reveals the presence of the phenanthrene emission peak at 508 nm in which the phosphorescences of pyrene and acridine, expected at 600 and 643 nm, are strongly reduced (*167*).

Wakeham (*85*) has also applied synchronously excited fluorescence emission spectroscopy to the determination of perylene in contaminated and uncontaminated environmental samples. Lloyd (*169*) showed how the technique can provide "fingerprints" in forensic science and enable mixtures separated by chromatography to be analyzed (*169*), particularly if partial heavy-atom quenching is simultaneously employed (*88*) (cf. Section V).

Computerized fluorimetry promises to be an important method for the analysis of fluorescent samples containing multiple components. Warner *et al.* (*170*) have proposed a systematic procedure in which the "emission–excitation matrix" **M** is measured (*171*), the elements of which M_{ij}, represent the fluorescent intensity measured at wavelength λ_j for excitation at λ_i. A row of **M** represents a fluorescence emission spectrum, and a column an excitation spectrum. A computer method is then used to determine the number of independent components contributing to the spectra, and if only two are found to be present, to derive the individual spectra.

While the method is slow if a conventional fluorimeter is used, even if computer controlled to eleviate tedium, the recently developed technique of videofluorimetry allows simultaneous acquisition of up to 241 spectra, each with 241 excitation wavelengths, in only 16.7 msec (*172*).

The concept of "total luminescence" follows. This is defined as the function $I(\lambda_{ex}, \lambda_{em})$, where I is the intensity, and λ_{ex} and λ_{em} represent all possible excitation and monitoring wavelengths. If a diode array-imaging device

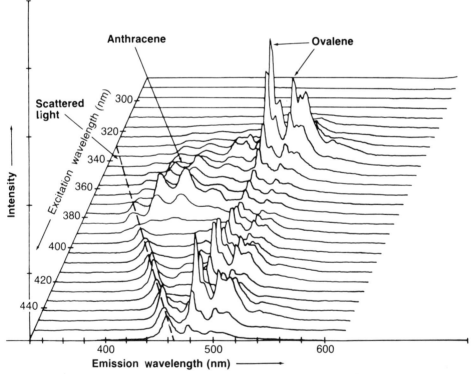

Fig. 9-29. Total luminescence spectrum of anthracene and ovalene. [Reproduced with permission from Y. Talmi, D. C. Baker, J. R. Jadamec, and W. A. Saner, *Anal. Chem.* **50**, 936A (1978). Copyright, American Chemical Society.]

(*173*) is placed at the focal plane of the spectrometer emission polychromator, rapid scanning of the image allows acquisition of all the $I(\lambda_{ex}, \lambda_{em})$ data and display either as contours (Fig. 9-10) or isometric projections (Fig. 9-29). Emission decay times may also be measured. This method clearly has great potential in the characterization of mixtures of PAH eluted from HPLC and GC columns (Chapters 6 and 7).

The use of parallel optoelectronic image detectors (OID) in luminescence analysis allows extension of many of the new techniques discussed above. The parallel multichannel OID allows the entire spectrum to be recorded simultaneously, as compared with the single channel available in the photomultiplier detector of conventional spectrofluorimeters. Parallel detection results in a considerable reduction in analysis time, so that the method is readily applicable to transient spectrofluorimetry, as in HPLC and GC detection, or for photochemically labile compounds. The signal-to-noise ratio is markedly improved if a silicon-intensifier target vidicon is used as detector, and multiple scans are accumulated by a linked multichannel analyzer in conjunction with a digital memory; the system is also particularly suitable for use with pulsed-laser excitation (*174*).

Spectra obtained with the aid of OID may be subjected to a variety of computational treatments such as smoothing, sensitivity correction, differentiation, source compensation, and constant-energy detection. Laser-induced phosphorescence decay times may also be measured, and OID with diagonal scans across the spectral matrix can yield information identical to that obtained by synchronous luminescence (with multiple scanning).

REFERENCES

1. E. Sawicki, *Crit. Rev. Anal. Chem.* **1**, 275 (1970).
2. C. A. Parker, "Photoluminescence of Solutions." Elsevier, Amsterdam, 1968.
3. J. W. Howard and E. O. Haenni, J. Assoc. Off. Agric. Chem. **46**, 933 (1963).
4. G. E. Moore, R. S. Thomas, and J. L. Monkman, *J. Chromatogr.* **26**, 456 (1967).
5. M. Popl, V. Dolansky, and J. Mostecky, *J. Chromatogr.* **59**, 329 (1971).
6. R. Gladen, *Chromatographia* **5**, 236 (1972).
7. H. Thielemann, *Mikrochim. Acta*, 838 (1971).
8. L. E. Stromberg and G. Widmark, *J. Chromatogr.* **47**, 27 (1970).
9. G. Grimmer and A. Hildebrandt, *J. Chromatogr.* **20**, 89 (1965).
10. T. D. Searl, F. J. Cassidy, W. H. King, and R. A. Brown, *Anal. Chem.* **42**, 954 (1970).
11. R. A. Greinke and I. C. Lewis, *Anal. Chem.* **47**, 2151 (1975).
12. G. Grimmer, M. Böhnke, and A. Glaser, *Erdoel. Kohle, Erdgas, Petrochem.* **30**, 411 (1977).
13. M. Dong, D. C. Locke, and E. Ferrand, *Anal. Chem.* **48**, 368 (1976).
14. W. Giger and M. Blumer, *Anal. Chem.* **46**, 1663 (1974).
15. E. Clar, "Polycyclic Hydrocarbons," Vols. 1 and 2. Academic Press, New York, 1964.
16. M. Gelus, J. M. Bonnier, and P. Traynard, *C.R. Hebd. Seances Acad. Sci.* **255**, 2576 (1962).

17. C. Haig and C. A. Coulson, *Tetrahedron* **19**, 527 (1963).

18. M. S. Newman, R. G. Mentzer, and G. Slomp, *J. Am. Chem. Soc.* **85**, 4018 (1963).

19. D. Cagniant, *Bull. Soc. Chim. Fr.*, 2325 (1966).

20. K. D. Bartle, H. Heaney, D. W. Jones, and P. Lees, *Tetrahedron* **21**, 3289 (1965).

21. L. Dubois, A. Zdrojewski, and J. L. Monkman, *Mikrochim. Acta* 834 (1967).

22. R. A. Friedel and M. Orchin, "Ultraviolet Spectra of Aromatic Compounds." Wiley, New York, 1951.

23. K. Hirayama, "Handbook of Ultraviolet and Visible Absorption Spectra of Organic Compounds." Plenum, New York, 1967.

24. H. E. Ungnade, ed., "Organic Electronic Spectral Data," Vol. 2, 1953–1955, 1960; O. H. Wheeler and L. A. Kaplan (eds.), Vol. 3, 1956–1957, 1966; J. P. Phillips and F. C. Nachod (eds.), Vol. 4, 1958–1959, 1963; J. P. Phillips, R. E. Lyle, and P. R. Jones (eds.), Vol. 5, 1960–1961, 1969; J. P. Phillips, L. D. Freedman, and J. C. Craig (eds.), Vol. 6, 1962–1963, 1970; J. P. Phillips, J. C. Dacons, and R. G. Rice (eds.), Vol. 7, 1964–1965, 1971; J. P. Phillips, H. Feuer, and B. S. Thyagarajan (eds.), Vols. 8, 9, 10, 1966, 1967, and 1968. 1972, 1973, and 1974; J. P. Phillips, H. Feuer, P. M. Laughton, and B. S. Thyagarajan (eds.), Vol. 11, 1969, 1975; J. P. Phillips, H. Feuer, and B. S. Thyagarajan (eds.), Vol. 12, 1970, 1976; J. P. Phillips, D. Bates, H. Feuer, and B. S. Thyagarajan (eds.), Vol. 13, 1971, 1977.

25. "DMS UV Atlas of Organic Compounds." Butterworths, London, 1966.

26. E. Sawicki, T. W. Stanley, W. C. Elbert, J. Meeker, and S. McPherson, *Atmos. Environ.* **1**, 131 (1967).

27. E. Clar, "The Aromatic Sextet," Wiley, New York, 1972.

28. A. R. Hawthorne and J. H. Thorngate, *Appl. Opt.* **17**, 724 (1978).

29. A. R. Hawthorne, J. H. Thorngate, R. B. Gammage, and T. Vo-Dinh, *in* "Polynuclear Aromatic Hydrocarbons" (P. W. Jones and P. Leber, eds.), p. 299. Ann Arbor Sci. Publ., Ann Arbor, 1979.

30. T. J. Porro, *J. Assoc. Off. Anal. Chem.* **56**, 607 (1973).

31. I. D. Berlman, "Handbook of Fluorescence Spectra of Aromatic Molecules." Academic Press, New York, 1965.

32. G. G. Guilbault, "Practical Fluorescence: Theory, Methods and Techniques." Dekker, New York, 1973.

33. E. Sawicki, *Talanta* **16**, 1231 (1969).

34. J. D. Winefordner, W. J. McCarthy, and P. A. St. John, *J. Chem. Ed.* **44**, 80, 136, 215 (1967); **45**, 98 (1968).

35. O. Hutzinger, S. Safe, and M. Zander, *Analabs Res. Notes* **13**, (3) 13 (1973).

36. L. Dubois, A. Zdrojewski, and J. L. Monkman, *Mikrochim. Acta*, 903 (1967).

37. E. T. Oakley, L. F. Johnson, and H. M. Stahr, *Tob. Sci.* **16**, 19 (1970).

38. V. P. Rzhekhin, *Tr. Vses. Naukchno. Issled. Inst. Zhirov*, **27**, 46 (1970); *Chem. Abstr.* **76**, 81979 (1970).

39. H. Weisz, *Zentralbl. Bakteriol. Parasitenk. Infektionskr. Hyg.*, *Abt. 1: Orig. Reihe B* **155**, 78 (1971).

40. E. Sawicki, R. C. Corey, A. E. Dooley, J. B. Gisclard, J. L. Monkman, R. E. Neligan, and L. A. Ripperton, *Health Lab. Sci.* **7**, 31 (1970).

41. E. Sawicki, R. C. Corey, A. E. Dooley, J. B. Gisclard, J. L. Monkman, R. E. Neligan, and L. A. Ripperton, *Health Lab. Sci.* **7**, 56 (1970).

42. E. Sawicki, R. C. Corey, A. E. Dooley, J. B. Gisclard, J. L. Monkman, R. E. Neligan, and L. A. Ripperton, *Health Lab.* **7**, 60 (1970).

43. W. Galley, ed., "Environmental Pollutants—Selected Analytical Methods (Scope 6)." p. 277. Butterworths, London, 1975.

44. J. F. McKay and D. R. Latham, *Anal. Chem.* **45**, 1050 (1973).
45. J. Borneff, "Fate of Pollutants in the Air and Water Environment" (I. H. Suffet, ed.), Part 2, p. 393. Wiley, New York, 1977.
46. A. Colmsjö and V. Stenberg, *J. Chromatogr.* **169**, 205 (1979).
47. R. Tomingas, G. Voltmer, and R. Bednarik, *Sci. Total Environ.* **7**, 261 (1977).
48. D. T. Kaschani and R. Reiter, *Fresenius Z. Anal. Chem.* **292**, 141 (1978).
49. R. J. Hurtubise, J. F. Schabron, J. D. Feaster, and D. H. Therkildson, *Anal. Chim. Acta* **89**, 377 (1977).
50. R. J. Hurtubise, G. T. Skar, and R. Poulson, *Anal. Chim. Acta* **97**, 13 (1978).
51. H. Kunte, *Arch. Hyg. Bakteriol.* **151**, 193 (1967).
52. F. de Wiest, D. Rondia and H. Della Fiorentina, *J. Chromatogr.* **104**, 399 (1975).
53. R. J. Hurtubise, *Anal. Chem.* **49**, 2160 (1977).
54. R. C. Pierce and M. Katz, *Anal. Chem.* **47**, 1743 (1975).
55. M. Katz and R. C. Pierce, *in* "Carcinogenesis—A Comprehensive Survey" (R. I. Freudenthal and P. W. Jones, eds.), Vol. 1, p. 413. Raven, New York, 1976.
56. M. Katz, T. Sakuma, and A. Ho, *Environ. Sci. Technol.* **12**, 909 (1978).
57. E. R. Adlard, *J. Inst. Petrol.* **58**, 63 (1973).
58. A. P. Bentz, *Anal. Chem.* **48**, 454A (1976).
59. E. M. Levy, *Water Res.* **5**, 723 (1971).
60. E. M. Levy, *Water Res.* **6**, 57 (1972).
61. H. Hellmann, *Fresenius Z. Anal. Chem.* **281**, 125 (1976); *Fresenius Z. Anal. Chem.* **272**, 30 (1974).
62. H. J. Klimisch and W. Szonn, *Fresenius Z. Anal. Chem.* **265**, 7 (1972).
63. A. D. Thurston and R. W. Knight, *Environ. Sci. Technol.* **5**, 64 (1971).
64. M. Freegarde, C. G. Hatchard, and C. A. Parker, *Lab. Pract.* **20**, 35 (1971).
65. J. H. Rho and J. L. Stuart, *Anal. Chem.* **50**, 620 (1978).
66. P. John and I. Soutar, *Anal. Chem.* **48**, 520 (1976).
67. M. Zander, "Phosphorimetry." Academic Press, New York, 1968.
68. M. C. Hollifield and J. D. Winefordner, *Anal. Chem.* **40**, 1759 (1968).
69. R. Zweidinger and J. D. Winefordner, *Anal. Chem.* **42**, 639 (1970).
70. L. V. S. Hood and J. D. Winefordner, *Anal. Chim. Acta* **42**, 199 (1968).
71. M. Popl, M. Stejskal, and J. Mostecky, *Anal. Chem.* **47**, 1947 (1975).
72. E. Sawicki and H. Johnson, *Microchem. J.* **8**, 85 (1964).
73. E. Sawicki, *Chemist-Analyst* **53**, 88 (1964).
74. E. Sawicki and J. D. Pfaff, *Anal. Chim. Acta* **32**, 521 (1964).
75. E. M. Schulman and C. Walling, *Science* **178**, 53 (1972).
76. S. L. Wellons, R. A. Paynter, and J. D. Winefordner, *Spectrochim. Acta* **30A**, 2133 (1974); *Anal. Chem.* **46**, 736 (1974).
77. T. Vo-Dinh, F. Lue Yen, and J. D. Winefordner, *Anal. Chem.* **48**, 1186 (1976); *Talanta* **24**, 146 (1977).
78. C. D. Ford and R. J. Hurtubise, *Anal. Chem.* **52**, 656 (1980).
79. G. Heinrich and H. Güsten, *Fresenius Z. Anal. Chem.* **278**, 257 (1976).
80. W. J. McCarthy, *in* "Spectrochemical Methods of Analysis" (J. D. Winefordner, ed.), Chapter 8. Wiley (Interscience), New York, 1971.
81. M. Zander, *Fresenius Z. Anal. Chem.* **226**, 251 (1967); *Erdoel. Kohle, Erdgas, Petrochem.* **22**, 81 (1969).
82. M. Zander, *Fresenius Z. Anal. Chem.* **263**, 19 (1973).
83. M. Zander, *Fresenius Z. Anal. Chem.* **229**, 352 (1967).
84. R. E. Laflamme and R. A. Hites, *Geochim. Cosmochim. Acta* **42**, 298 (1978).
85. S. G. Wakeham, *Environ. Sci. Technol.* **11**, 272 (1977).

86. H. Dreeskamp, E. Koch, and M. Zander, *Chem. Phys. Lett.* **31,** 251 (1978).

87. J. B. Birks, "Photophysics of Aromatic Molecules," Chapter 4. Wiley (Interscience), New York, 1970.

88. J. B. F. Lloyd, *Analyst (London)* **99,** 729 (1974).

89. L. V. S. Hood and J. D. Winefordner, *Anal. Chem.* **38,** 1922 (1968).

90. M. Zander, *Int. J. Environ. Anal. Chem.* **3,** 29 (1973).

91. M. Zander, *Fresenius Z. Anal. Chem.* **226,** 251 (1967).

92. M. Zander, *Z. Naturforsch.* **33a,** 998 (1978).

93. E. Sawicki, T. W. Stanley, and W. C. Elbert, *Talanta* **11,** 1433 (1964).

94. H. Dreeskamp, E. Koch, and M. Zander, *Z. Naturforsch.* **30a,** 1311 (1975).

95. M. Zander, U. Breymann, H. Dreeskamp, and E. Koch, *Z. Naturforsch.* **32a,** 1561 (1977).

96. U. Breymann, H. Dreeskamp, E. Koch, and M. Zander, *Fresenius Z. Anal. Chem.* **293,** 208 (1978).

97. B. L. Van Duuren, *Chem. Rev.* **63,** 325 (1963).

98. B. L. Van Duuren, *Anal. Chem.* **32,** 1436 (1960).

99. R. Schoental and E. J. Y. Scott, *J. Chem. Soc.*, 1683 (1949).

100. E. Sawicki, T. W. Stanley, S. McPherson, and M. Morgan, *Talanta* **13,** 619 (1966).

101. E. V. Shpol'skii, *Soviet Phys. Usp.* **2,** 378 (1959); **3,** 372 (1960); **5,** 522 (1962); **6,** 252, 411 (1963).

102. E. V. Shpol'skii, A. A. Ilina, and L. A. Klimova, *Dokl. Akad. Nauk. SSSR* **87,** 935 (1952).

103. E. J. Bowen and B. Brocklehurst, *J. Chem. Soc.*, 3875 (1954); *Trans. Faraday Soc.* **51,** 774 (1955).

104. S. P. McGlyn, B. T. Neely, and C. Neely, *Anal. Chim. Acta* **28,** 472 (1963).

105. C. Pfister, *Chim. Phys.* **2,** 171 (1973).

106. J. J. Dekkers, G. P. Hoorneg, G. Visser, C. Maclean, and N. H. Velthoorst, *Chem. Phys. Lett.* **47,** 457 (1977).

107. R. J. McDonald and B. K. Selinger, *Aust. J. Chem.* **24,** 249 (1971).

108. D. M. Grebenshchikov, N. A. Koyrizhnykh, and S. A. Kozlov, *Opt. Spectrosc.* **31,** 392 (1971).

109. A. Moissan-Pellois, R. Collorec, and J. Ripoche, *C. R. Hebd. Seances Acad. Sci. Ser. B* **269,** 1305 (1969).

110. G. F. Kirkbright and C. G. deLima, *Analyst (London)* **99,** 338 (1974).

111. G. F. Kirkbright and C. G. deLima, *Chem. Phys. Lett.* **37,** 165 (1976).

112. E. V. Shpol'skii and T. N. Bolotnikova, *Pure Appl. Chem.* **37,** 183 (1974).

113. L. A. Klimova, *Opt. Spectrosc.* **15,** 185 (1963).

114. A. Colmsjö and U. Stenberg, *Chem. Scr.* **11,** 220 (1977).

115. B. S. Causey, G. F. Kirkbright, and C. G. deLima, *Analyst (London)* **101,** 367 (1976).

116. A. Colmsjö and U. Stenberg, *Anal. Chem.* **51,** 145 (1979); *in* "Polynuclear Aromatic Hydrocarbons" (P. W. Jones and P. Leber, eds.), p. 121. Ann Arbor Sci. Publ., Ann Arbor, Michigan, 1979.

117. J. A. G. Drake, D. W. Jones, B. S. Causey, and G. F. Kirkbright, *Fuel* **57,** 663 (1978).

118. C. A. Parker and C. G. Hatchard, *Photochem. Photobiol.* **5,** 699 (1966).

119. M. J. Eichhoff and N. Köhler, *Fresenius Z. Anal. Chem.* **197,** 272 (1963).

120. P. O. Dikum, *Vop. Onkol.* **7,** 42 (1961); *Chem. Abstr.* **57,** 658 (1961).

121. R. I. Personov, *Zh. Anal. Khim.* **17,** 506 (1962).

122. B. Muel and G. Lacroix, *Bull. Soc. Chim. Fr.*, 2139 (1960).

123. J. Jäger, *Chem. Listy* **60,** 1184 (1966).

124. J. Jäger, *Atmos. Environ.* **2,** 293 (1968).

125. P. P. Dikun, N. D. Krasnistkaya, N. D. Gorelova, and I. A. Kalinina, *J. Appl. Spectrosc.* **8,** 254 (1968).

126. G. E. Danil'tseva and A. Ya. Khesina, *J. Appl. Spectrosc.* **5**, 196 (1966).
127. G. E. Fedoseeva and A. Ya Khesina, *J. Appl. Spectrosc.* **9**, 838 (1968).
128. L. M. Shabad and G. A. Smirnov, *Atmos. Environ.* **6**, 153 (1972).
129. I. L. Varshavskii, L. M. Shabad, A. Ya. Khesina, S. S. Khitrovo, V. G. Chalabov, and A. I. Pakhol'nik, *J. Appl. Spectrosc.* **2**, 68 (1965).
130. M. Stepanova, R. I. I'lina and Y. K. Shaposhnikov, *Zh. Anal. Khim.* **27**, 1201 (1972).
131. V. N. Florovskaya, T. A. Teplitskaya, and R. I. Personov, *Geochem. Int.* **3**, 419 (1966), and references therein.
132. P. P. Dikun, *J. Appl. Spectrosc.* **6**, 130 (1967).
133. F. I. Gurov and Yu. V. Novikov, *Gig. Sanit.* **36**, 409 (1971).
134. R. C. Stroupe, P. Tokousbalides, R. B. Dickinson, Jr., E. R. Wehry, and G. Mamantov, *Anal. Chem.* **49**, 701 (1977).
135. D. M. Grebenshchikov, N. A. Kovizhnykh, and S. A. Kozlov, *Opt. Spectrosc.* **37**, 155 (1974).
136. G. L. LeBel and J. D. Laposa, *J. Mol. Spectrosc.* **41**, 249 (1972).
137. E. V. Shpol'skii, L. A. Klimova, G. N. Nersesova, and V. I. Glyadkovskii, *Opt. Spectrosc.* **24**, 25 (1968).
138. T. N. Bolotnikova and T. M. Naumova, *Opt. Spectrosc.* **24**, 253 (1968).
139. T. Ya. Gaevaya and A. Ya. Khesina, *Z. Anal. Khim.* **29**, 2225 (1974).
140. E. L. Wehry and G. Mamantov, *Anal. Chem.* **51**, 643A (1979).
141. G. Mamantov, E. L. Wehry, R. R. Kemmerer, and E. R. Hinton, *Anal. Chem.* **49**, 86 (1977).
142. P. Tokousbalides, E. R. Hinton, R. B. Dickinson, Jr., P. V. Bilotta, E. R. Wehry, and G. Mamantov, *Anal. Chem.* **50**, 1189 (1978).
143. P. Tokousbalides, E. L. Wehry, and G. Mamantov, *J. Phys. Chem.* **81**, 1769 (1977).
144. J. R. Maple, E. L. Wehry, and G. Mamantov, *Anal. Chem.* **52**, 920 (1980).
145. J. C. Brown, M. Edelson, and G. J. Small, *Anal. Chem.* **50**, 1394 (1978).
146. A. P. D'Silva, G. J. Oestreich, and V. E. Fassell, *Anal. Chem.* **48**, 915 (1976).
147. C. P. Woo, A. P. D'Silva, V. E. Fassel, and G. Oestreich, *Environ. Sci. Technol.* **12**, 173 (1978).
148. C. S. Woo, A. P. D'Silva, and V. E. Fassell, *Anal. Chem.* **52**, 159 (1980).
149. V. D. Tuan and U. P. Wild, *J. Lumin.* **6**, 296 (1973).
150. L. A. Klimova, A. I. Oglobina, and V. I. Glyadkovskii, *Opt. Spectrosc.* **30**, 384 (1971).
151. C. M. O'Donnell and T. N. Solle, *Anal. Chem.* **48**, 175R (1976); **50**, 189R (1978).
152. K. R. Naqvi, A. R. Holzwarth, and U. P. Wild, *Appl. Spectrosc. Rev.* **12**, 131 (1977).
153. M. Soutif, *Pure Appl. Chem.* **48**, 99 (1976).
154. E. L. Wehry, ed., "Modern Fluorescence Spectroscopy," Vols. 1 and 2. Plenum Press, New York, 1976.
155. F. P. Schwarz and S. P. Wasik, *Anal. Chem.* **48**, 524 (1976).
156. B. S. Das and G. H. Thomas, *Anal. Chem.* **50**, 967 (1978).
157. J. H. Richardson and M. E. Ando, *Anal. Chem.* **49**, 955 (1977); *Proc. Mater. Res. Symp. Trace Org. Analysis,* 9th Gaithersburg, Maryland, 1978, p. 723. National Bureau of Standards Spec. Publ. 519, Washington, D.C., 1979.
158. R. B. Dickinson, Jr., and E. L. Wehry, *Anal. Chem.* **51**, 778 (1979).
159. G. D. Boutillier and J. D. Winefordner, *Anal. Chem.* **51**, 1384 (1979).
160. R. W. Marshall and T. L. Miller, *Anal. Chem.* **47**, 256 (1975).
161. G. L. Green and T. C. O'Haver, *Anal. Chem.* **46**, 2191 (1974).
162. G. Talsky, L. Mayring, and H. Kreuzer, *Angew. Chem. Int. Ed.* **17**, 785 (1978).
163. T. C. O'Haver and W. M. Parks, *Anal. Chem.* **46**, 1886 (1974).
164. M. A. Fox and S. W. Staley, *Anal. Chem.* **48**, 992 (1976).
165. T. Vo-Dinh, *Anal. Chem.* **50**, 396 (1978).

166. J. B. F. Lloyd, *Nature* (*London*) **231,** 64 (1971).

167. T. Vo-Dinh, R. B. Gammage, A. R. Hawthorne, and J. H. Thorngate, *Environ. Sci. Technol.* **12,** 1297 (1978).

168. T. Vo-Dinh, R. B. Gammage, and A. R. Hawthorne, *in* "Polynuclear Aromatic Hydrocarbons" (P. W. Jones and P. Leber, eds.), p. 111. Ann Arbor Sci. Publ., Ann Arbor, 1979.

169. J. B. F. Lloyd, *J. Forensic Sci.* **11,** 153, 235 (1971).

170. I. M. Warner, G. D. Christian, E. R. Davidson, and J. B. Callis, *Anal. Chem.* **49,** 564 (1977).

171. G. Weber, *Nature* (*London*) **190,** 27 (1961).

172. I. M. Warner, J. B. Callis, E. R. Davidson, M. Gouterman, and G. D. Christian, *Anal. Lett.* **8,** 665 (1975).

173. D. W. Johnson, J. B. Callis, and G. D. Christian, *Anal. Chem.* **49,** 575 (1974).

174. Y. Talmi, D. C. Baker, J. R. Jadamec, and W. A. Saner, *Anal. Chem.* **50,** 936A (1978).

175. M. D. Sauerland and M. Zander, *Erdoel. Kohle, Erdgas, Petrochem.* **19,** 502 (1966).

176. K. D. Bartle, G. Collin, J. W. Stadelhofer, and M. Zander, *J. Chem. Technol.* **29,** 531 (1979).

10

Nuclear Magnetic Resonance and Infrared Spectroscopy

I. INTRODUCTION

Among the variety of available spectroscopies, nuclear magnetic resonance (NMR) and infrared (IR) are probably the most important in the identification of organic molecules. For different reasons, however, neither has yet had much impact in the area of PAC analysis, but current developments in instrumentation are likely to lead to an increasing future use.

The NMR spectra of PAC are valuable fingerprints, and NMR is also distinctive in sensing aspects of molecular symmetry. ¹H NMR allows positions of substitution to be determined either through characteristic spin-spin coupling patterns of remaining aromatic protons (*1*) or from the chemical shifts (*2, 3*) and/or benzylic (*4*) couplings of proton-bearing substituents. Reference compounds need not always be available since *a priori* interpretation of the ¹H-NMR spectrum is often possible for compounds of this type (*1*). The uses of ¹H NMR for identifying polycyclic aromatic molecules have been widespread in organic chemistry, and have been reviewed (*5*). This wide application has not yet been paralleled for environmental PAC because of the poor sensitivity of NMR when compared with other physical methods. The sample requirement of ¹H NMR, large in comparison with the capacity of chromatographic techniques used in separation processes, has until recently limited the applications of this powerful identification tool. Only with the advent of rapid spectrum accumulation through pulse Fourier-transform (FT) methods (*6*), which use multichannel excitation to allow many scans to be recorded and averaged in a comparatively short time, have signal-to-noise (S/N) ratios become adequate for the analysis of the dilute solutions available. Similar considerations apply to the use of ¹³C NMR, except that the low isotopic abundance and small magnetogyric ratio of this nucleus impose a further reduction in sensitivity (by a factor of 10^4) as compared with ^1H NMR.

The IR spectra of PAC generally consist (7, 8) of absorptions arising from C—H stretching vibrations near 3000–3100 cm^{-1}, C—C skeletal vibrations at 1500–1600 cm^{-1}, and out-of-plane C—H vibrations between 650 and 900 cm^{-1}. The latter are the most characteristic, since from one to four adjacent hydrogen atoms on the same ring yield strong absorptions in different regions, and in favorable cases the out-of-plane C—H vibrations may be used to locate the position of substitution (7, 9). However, the two chief disadvantages of conventional dispersion IR spectroscopy, (a) the absence of unique features in the spectra and (b) the lack of proportionality between band strengths and concentration, have restricted this method from the analysis of PAC in the environment (10). Again, however, the introduction of the FT technique has, in conjunction with matrix-isolation (MI) spectrometry, begun to allow analysis by IR in this context. IR spectroscopy does, of course, allow useful information to be gained about the structure of unknown heterocyclic PAC compounds, and FT-IR is a potentially important detection method for HPLC and GC (see Chapters 6 and 7).

However, because the interactions of solute molecules both with each other and with the solvent are minimized, the IR spectra of matrix-isolated species at cryogenic temperatures consist of very sharp vibrational lines (11, 12). It is hence possible to resolve features separated by 1 cm^{-1}, to make measurements with a precision of ± 0.1 cm^{-1}, and to construct linear Beer's law plots. Very characteristic fingerprint IR spectra of PAC are thus available by the MI method (13, 14) (see Chapter 9), but the limitations of the energy of the source of a conventional spectrometer do not allow an adequate S/N ratio to be attained for the now dilute analyte.

Multiplex IR radically improves the S/N ratio compared with dispersion IR spectroscopy for measurements made in the same time by allowing concurrent viewing of all the resolution elements of the spectrum. Repetition of the (rapid) spectrum acquisition allows averaging of the accumulated signals. A further increase in S/N arises from the higher optical throughput of the FT-IR spectrometer. A Michelson interferometer is employed to convert the optical signals to frequencies which can be read and distinguished by IR detectors. There is also an improvement in the frequency accuracy of FT-IR because of the laser reference of the interferometer (12).

FT techniques may thus be used to allow applications of both NMR and IR in the analysis of PAC. In FT-NMR, overlapping signals from the free induction decay which follow a pulse of radiofrequency radiation are decoded into separate sine and cosine frequency components, which is conversion from the time domain to the frequency domain. This capability of measuring all the frequencies in the NMR spectrum at one time, followed by

FT decoding, results in the multiplex advantage over measuring frequencies singly in conventional NMR. Analogous decoding converts a multiplex IR spectrum obtained from an interferometer into the conventional spectrum— a graph of absorbance against wavenumber.

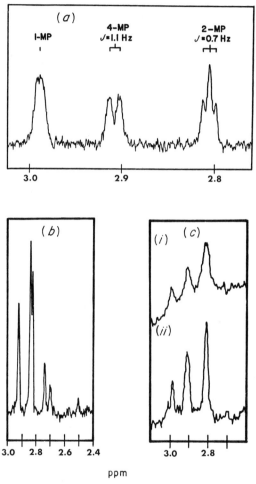

Fig. 10-1. (*a*) Methyl region of continuous-wave ¹H-NMR 60-MHz spectrum of equimolar mixture of methylpyrenes. (*b*) Methyl region of continuous wave ¹H-NMR 60-MHz spectrum of monomethyl pyrene fraction of amber petrolatum. (*c*) Methyl regions of continuous-wave ¹H-NMR 60-MHz spectra of (*i*) 220 μg of the monomethyl pyrene fraction of asphalt (microcell and time averaging) and (*ii*) 20:38:42 mixture of 1-, 4-, and 2-methylpyrenes under similar conditions. [Reproduced with permission from L. K. Keefer, L. Wallcave, J. Loo, and R. S. Peterson, *Anal. Chem.* **43**, 1411 (1971). Copyright, American Chemical Society.]

II. CONTINUOUS-WAVE ^{1}H NMR

Conventional single-scan NMR may confirm the presence of a given class of compounds, e.g., alkyl-substituted derivatives in the PAC fraction of shale oil (15), and may also be applied to the identification of specific individual compounds. Thus, Burham et al. (16) isolated acenaphthylene from contaminated well water and confirmed its identity, in part, by a comparison of the continuous wave ^{1}H-NMR spectrum with that of an authentic sample. Partially interpreted ^{1}H-NMR spectra were also used in the identification of sulfur isosteres of pyrene, benzo[a]pyrene, and benzo[ghi]perylene present in reference samples of these compounds (17). Keefer et al. (18) carefully investigated the methods of identification of methyl-substituted PAH from environmental samples by continuous-wave ^{1}H NMR. They studied in particular the use of the chemical shifts of CH_3 (and their different concentration dependences), and of the benzylic coupling (see Fig. 10-1a), confirmed by spin-decoupling experiments, between CH_3 and ring protons. The identities of 4-methylpyrene (1.1 Hz doublet) and 2-methylpyrene (0.7 Hz triplet) in the monomethyl pyrene fraction of amber petrolatum (Fig. 10-1b) were confirmed from their chemical shifts; the dimethylpyrene fraction from the same material showed no resonances in regions characteristic of CH_2CH_3 groups so that ethyl derivatives could be excluded (18), a conclusion not possible from mass spectrometric data.

A time-averaging method and use of a microcell as sample tube also allowed Keefer et al. (18) to analyze quantitatively mixtures with 30–40 μg of individual components, e.g., 220 μg of the methylpyrene fraction of asphalt (Fig. 10-1c); a similar approach showed that the monomethyl triphenylene fraction of petrolatum is mainly composed of the 2-isomer.

III. PULSE FOURIER-TRANSFORM NMR

Pulse Fourier-transform ^{1}H NMR can be applied with much more facility to small samples. Thus, Bartle et al. (19) were able to identify by FT ^{1}H NMR as little as 20 μg of individual PAH in mixtures generally totalling less than 1 mg separated from atmospheric dust and from the condensates of tobacco and marijuana smoke. Continuous-wave spectra of the fractions revealed little except in very favorable instances where high molecular symmetry resulted in intense signals. For example, the single-scan spectrum of a high-molecular-weight (about 300) fraction from air-pollutant PAH consisted (19) only of a diffuse band in the aromatic region but with a sharp singlet at 8.95 ppm, showing the entire spectrum of coronene (20). However, pulse FT spectra at 90 MHz of other fractions allowed identification of both parent

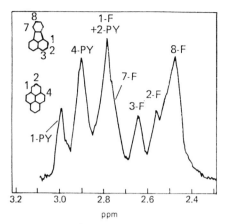

Fig. 10-2. Methyl region of FT ^{1}H-NMR 90-MHz spectrum of fluoranthene/pyrene fraction (0.9 mg) of tobacco-smoke condensate. [Reproduced with permission from K. D. Bartle, M. L. Lee, and M. Novotny, *Analyst* (*London*) **102**, 731 (1977). Copyright, Wiley and Sons, Inc.]

PAH and their methyl derivatives (Table 10-1). Thus, the spectra of all the fluoranthene–pyrene fractions showed a number of other peaks near 2.5–3.0 ppm, in addition to those of CH_3 of the three methylpyrenes indicating the presence of all five methylfluoranthenes (Fig. 10-2). This is in agreement with the GC/MS results (*21*) which suggested eight methylated compounds with molecular weight 216. Bioassays of the methylfluoranthenes have shown (*22*) tumor-initiating and carcinogenic activity for 2- and 3-methylfluoranthenes (see Appendix 5).

The examples discussed above and in Chapter 11 show how FT ^{1}H NMR can thus yield valuable information concerning the identity of derivatives of PAH with up to four rings (Table 10-1). For higher molecular weight series, the parent hydrocarbons may be identified from characteristic singlets or from fingerprint fine structure. Thus Clarke reported (*23*) the identification of metabolites of benzo[*a*]pyrene from the FT ^{1}H-NMR signals of aromatic protons. Comparisons were made among spectrometers working at different frequencies, but only at 220 MHz was chemical-shift assignment possible for all protons (cf. Refs. *24* and *25*). Nonetheless, fingerprint spectra at lower frequencies still facilitated not only the identification of metabolites but the estimation of the degree of purity at various stages of HPLC clean-up of, for example, 9-hydroxybenzo[*a*]pyrene (*23*). While benzopyrenes and benzo-fluoranthenes were recognized in this way from 90-MHz FT spectra of high-molecular-weight fractions of environmental PAH (*19*), strong signals in the methyl region could not be found because of overlap of resonances of the

TABLE 10-1

¹H-NMR Data for Alkyl Derivatives of Polynuclear Aromatic Hydrocarbons in Air Particulates and Tobacco- and Marijuana-Smoke Condensates[a]

Fraction type	Compound	¹H-NMR chemical shift (ppm)			
		Literature	Air particulates	Tobacco smoke	Marijuana smoke
Anthracene/phenanthrene	1-methylanthracene	2.82 (*31, 32, 33*)	2.77	2.82	2.82
	2-methylanthracene	2.54 (*31, 32, 33*)	2.56	2.53	2.55
	9-methylanthracene	3.11[a] (*32*)	—	3.09	—
	1-methylphenanthrene	2.74 (*30, 31, 32, 33*)	2.77	2.72	2.75
	2-methylphenanthrene	2.54 (*30, 31, 32, 33*)	2.56	2.53	2.55
	3-methylphenanthrene	2.62 (*30, 31, 32, 33*)	2.64	2.60	2.62
	4-methylphenanthrene	3.12 (*30, 32*)	—	3.09	—
	9-methylphenanthrene	2.72[a] (*30, 31, 32, 33*)	2.73	2.71	2.74
Fluoranthene/pyrene	1-methylfluoranthene	2.79 (*34*)	2.80	2.79	2.75
	2-methylfluoranthene	2.56 (*19*)	2.54	2.54	2.55
	3-methylfluoranthene	2.70[a] (*34*)	2.64	2.65	2.62
	7-methylfluoranthene	2.73 (*19*)	2.76	2.76	2.72
	8-methylfluoranthene	2.43 (*34*)	2.47	2.44	2.43

	1-methylpyrene	2.96 (*31, 32*)	3.00	2.98	2.98
	2-methylpyrene	2.80 (*31, 32*)	2.80	2.79	2.75
	4-methylpyrene	2.89[a] (*31, 32*)	2.91	2.90	2.89
	benzo[*a*]fluorene	4.02 (*35*)	—	4.04	4.04
	benzo[*b*]fluorene	4.06 (*35*)	—	4.07	4.08
	benzo[*c*]fluorene	3.87 (*35*)	—	3.85	3.85
Benz[*a*]anthracene/chrysene	2-methylbenz[*a*]anthracene	2.65 (*18, 33*)	2.60	2.60	2.62
	3-methylbenz[*a*]anthracene	2.56 (*18, 33*)	2.60	2.60	2.55
	4-methylbenz[*a*]anthracene	2.75 (*18, 33*)	2.77	2.70	2.77
	9-methylbenz[*a*]anthracene	2.59 (*18, 33*)	2.60	2.60	2.55
	10-methylbenz[*a*]anthracene	2.59 (*18, 33*)	2.60	2.60	2.55
	1-methylchrysene	2.74 (*19*)	2.68	2.70	2.70
	2-methylchrysene	2.55 (*19*)	2.60	2.60	2.55
	3-methylchrysene	2.63 (*19*)	2.60	2.60	2.55
	5-methylchrysene	3.20[a] (*36*)	—	—	3.17
	6-methylchrysene	2.85[a] (*36*)	2.83	2.82	2.85

[a] Doublet—signal split by benzylic coupling.

345

large number of possible methyl derivatives (e.g., 18 methylbenzopyrenes and 47 methylbenzofluoranthenes). Only spectra at higher frequencies coupled with more efficient chromatographic separation will allow the application of ^{1}H NMR to those materials.

The simplicity of proton-decoupled ^{13}C-NMR spectra make this a promising technique when sufficient sample is available. Signals from four proton-bearing carbons and three quaternary carbons in the spectrum of a sulfur-containing PAC with molecular formula $C_{14}H_8S$ isolated from commercial pyrene showed it to be phenanthro[4,5-*bcd*]thiophene (*25*).

IV. FOURIER-TRANSFORM IR

Wehry *et al.* (*13*, *14*, *26–28*) have conducted detailed investigations of the applicability of MI FT-IR in the qualitative and quantitative analysis of PAH. Samples in nitrogen matrices on a CsI surface at −258°C were pro-

TABLE 10-2

Strong and Medium MI-IR Absorption Bands of Polycyclic Aromatic Compounds (Resolution 2 cm^{-1}) (*26*)

Compounds	IR bands (cm^{-1})
Triphenylene	1502, 1438, 746[a]
Chrysene	1365, 1234, (819, 816 doublet), 765[a]
Benz[*a*]anthracene	1504, 1365, 886, 809, 785, 752[a]
Pyrene	1436, 1185, 846[a], 715
Anthracene	881, (733, 730 doublet)[a], 603
Phenanthrene	1504, 1462, 870, 818, 740.5[a], 617
Carbazole	1496, 1464.5, 1454, 1397.5, 1338[a], 1328, 1241, 754, 729.5, 618, 569
Naphthacene	900.5[a], (747, 744.5 doublet), 551.5
Fluoranthene	1461, 1455.5, 1444, 1138, 830, (782, 778 doublet)[a], 750, 620
Fluorene	(1483.5, 1480 doublet), (1456.6, 1451.5 doublet), 1408, 1301, 1190, 1005, 958, 746[a], 696, 622
Acenaphthene	(1431, 1422 doublet), 1374, 1017.5, 843, 792[a], 750
Perylene	1498, 1385, 818[a], (776, 774 doublet)
Benzo[*b*]fluorene	957, 872, 775[a], 764, 728, 570, 475
Benzo[*a*]fluorene	1223, 1020, (828, 826 doublet), (763, 760.5 doublet)[a]
1,3,5-Triphenylbenzene	1580, 1558, 1503, 1417, 1080, 1032, 884, 758.5[a], 702, 632, 623, (614, 611 doublet), 509
Benzo[*a*]pyrene	1184, (888, 886 doublet), 855, 843, 831, 765[a], 747, 694
Dibenz[*a,h*]anthracene	1513, 1458, 892, (813, 809 doublet)[a], (750, 745 doublet), 667, 649, 529

[a] Strongest MI band.

duced by mixing the vapors of PAH effusing from a Knudsen cell with a large excess of nitrogen gas (cf. Chapter 9, Section VII and Fig. 9-17). Analyses of moderately complex mixtures (up to about 10 components) are made possible by the line-rich spectra obtainable at 2 cm^{-1} resolution. The useful absorption bands for seventeen compounds are summarized in Table 10-2. All have at least one distinctive major absorption band in the 700–900 cm^{-1} (C—H bending deformation) region.

Figure 10-3 illustrates the application of MI FT-IR to a four-component mixture; the PAH can be identified readily in the presence of carbazole, the IR spectrum of which is very complex (26). Routine measurements on mixtures of PAH at the μg level are possible (Fig. 10-4). The standard deviations of absorbances in repeated experiments were reduced by annealing the matrix (26). Beer's law plots (e.g., Fig. 10-5) are linear over the working concentration range (27). Like FT-NMR, the method has obvious potential, and it has already been used to characterize coal liquids and shale oils (28); an application to an important PAH mixture—the methylchrysenes—is discussed in Chapter 11.

The recording "on-the-fly" of FT-IR spectra in the gas phase of compounds eluted from a GC column allows a "chemigram" to be determined.

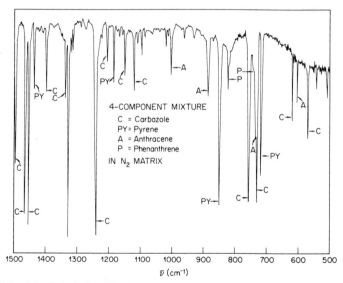

Fig. 10-3. Matrix isolation FT-IR spectrum of a four-component mixture co-deposited with nitrogen, 100 scans at 2 cm^{-1} resolution. [Reproduced with permission from E. L. Wehry, G. Mamantov, R. R. Kemmerer, M. O. Brotherton, and R. C. Stroupe, *in* "Carcinogenesis— A Comprehensive Survey" (R. I. Freudenthal and P. W. Jones, eds.), Vol. 1, p. 299. Raven, New York, 1976.]

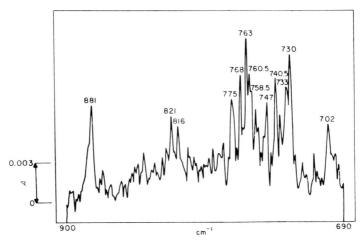

Fig. 10-4. Matrix isolation FT-IR spectrum of a five-component mixture consisting of 7 μg benzo[a]fluorene (763, 760.5 cm^{-1}), 7 μg benzo[b]fluorene (775 cm^{-1}), 7 μg triphenylbenzene (758.5, 702 cm^{-1}), 10 μg anthracene (733, 730 cm^{-1}), and 10 μg phenanthrene (740.5 cm^{-1}) in nitrogen (range of nitrogen/PAH mole ratio: 1.46 × 10^5 to 3.56 × 10^5). Resolution: 2.0 cm^{-1}. The sample was annealed for 30 min at −248°C. [Reproduced with permission from G. Mamantov, E. L. Wehry, R. R. Kemmerer, and E. R. Hinton, *Anal. Chem.* **49,** 86 (1977). Copyright, American Chemical Society.]

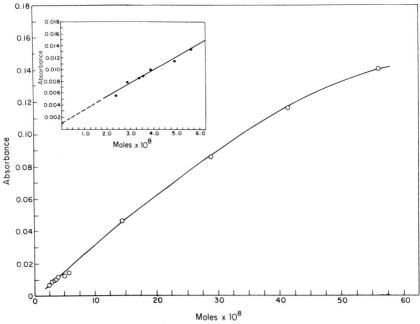

Fig. 10-5. Matrix isolation FT-IR Beer's law plot for the 733-cm^{-1} band of anthracene (range of nitrogen/anthracene mole ratio: 1.46 × 10^4 to 2.05 × 10^6). Resolution: 2.0 cm^{-1}. [Reproduced with permission from: G. Mamantov, E. L. Wehry, R. R. Kemmerer, and E. R. Hinton, *Anal. Chem.* **49,** 86 (1977). Copyright, American Chemical Society.]

Thus, Erickson *et al.* (*29*) graphed the absorbance in four IR spectral windows, corresponding to different functional group frequencies, against time as the components of a fraction of diesel-exhaust particulates emerged from a glass capillary column. The series of nitrogen heterocycles compatible with GC/MS results were hence shown to be absent, and it was concluded that the compounds present, in fact, contained carbonyl groups. The complete IR spectrum of the major component decoded from interferograms by Fourier transformation was consistent with a C_2-alkyl-fluorene-9-one (*29*).

REFERENCES

1. K. D. Bartle, *J. Assoc. Off. Anal. Chem.* **55**, 1101 (1972).
2. W. Carruthers, H. N. M. Stewart, P. G. Hansell, and K. M. Kelly, *J. Chem. Soc. C*, 2067 (1967).
3. F. F. Yew and B. J. Mair, *Anal. Chem.* **36**, 843 (1964).
4. K. D. Bartle, D. W. Jones, and R. S. Matthews, *Rev. Pure Appl. Chem.* **19**, 191 (1969).
5. K. D. Bartle and D. W. Jones, *Adv. Org. Chem.* **8**, 317 (1972).
6. K. Mullen and P. S. Pregosin, "Fourier Transform NMR Techniques, A Practical Approach." Academic Press, New York, 1976.
7. C. G. Cannon and G. B. M. Sutherland, *Spectrochim. Acta* **4**, 373 (1951).
8. N. Fusan and M. L. Josien, *J. Am. Chem. Soc.* **78**, 3049 (1956).
9. C. N. R. Rao, "Chemical Applications of IR Spectroscopy," p. 182. Academic Press, New York, 1963.
10. National Research Council, "Particulate Polycyclic Organic Matter." Nat. Acad. Sci., Washington, D.C., 1972.
11. H. E. Hallam and G. F. Scrimshaw, *in* "Vibrational Spectroscopy of Trapped Species" (H. E. Hallam, ed.), p. 11. Wiley, New York, 1973.
12. D. W. Green and G. T. Reedy, *in* "Fourier Transform Infrared Spectroscopy: Applications to Chemical Systems" (J. R. Ferraro and L. J. Basile, eds.), Vol. 1, p. 1. Academic Press, New York, 1978.
13. E. L. Wehry, G. Mamantov, R. R. Kemmerer, H. O. Brotherton, and R. C. Stroupe, *in* "Carcinogenesis—A Comprehensive Survey" (R. I. Freudenthal and P. W. Jones, eds.), Vol. 1, p. 299. Raven, New York, 1976.
14. E. L. Wehry and G. Mamantov, *Anal. Chem.* **51**, 643A (1979).
15. J. J. Schmidt-Collerus, F. Bonomo, K. Gala, and L. Leffler, *in* "Science and Technology of Oil Shale" (T. F. Yen, ed.), p. 115. Ann Arbor Sci. Publ., Ann Arbor, Michigan, 1976.
16. A. K. Burham, G. V. Calder, J. S. Fritz, G. A. Junk, H. J. Svec, and R. Willis, *Anal. Chem.* **44**, 139 (1972).
17. W. Karcher, R. Depaus, J. van Eijk, and J. Jacob, *in* "Polynuclear Aromatic Hydrocarbons" (P. W. Jones and P. Leber, eds.), p. 341. Ann Arbor Sci. Publ., Ann Arbor, Michigan, 1979.
18. L. K. Keefer, L. Wallcave, J. Loo, and R. S. Peterson, *Anal. Chem.* **43**, 1411 (1971).
19. K. D. Bartle, M. L. Lee, and M. Novotny, *Analyst* (*London*) **102**, 731 (1977).
20. R. H. Martin, N. Defay, F. Geerts-Evrad, and S. Delavarenne, *Tetrahedron* **20**, 1073 (1964).
21. M. L. Lee, M. Novotny, and K. D. Bartle, *Anal. Chem.* **48**, 405 (1976).
22. D. Hoffmann, G. Rathkamp, S. Nesnow, and E. L. Wynder, *J. Natl. Cancer Inst.* **49**, 1165 (1972).
23. P. A. Clarke, *in* "Carcinogenesis—A Comprehensive Survey" (R. I. Freudenthal and P. W. Jones, eds.), Vol. 1, p. 311. Raven, New York, 1976.

24. C. W. Haigh and R. B. Mallion, *J. Mol. Spectrosc.* **29,** 478 (1969).

25. K. D. Bartle, D. W. Jones, and R. S. Matthews, *Spectrochim. Acta* **25A,** 1603 (1969).

26. G. Mamantov, E. L. Wehry, R. R. Kemmerer, and E. R. Hinton, *Anal. Chem.* **49,** 86 (1977).

27. E. L. Wehry, G. Mamantov, R. R. Kemmerer, R. C. Stroupe, P. T. Tokousbalides, E. R. Hinton, D. M. Hembree, R. B. Dickinson, Jr., A. A. Garrison, P. V. Bilotta, and R. R. Gore, *in* "Carcinogenesis—A Comprehensive Survey" (P. W. Jones and R. I. Freudenthal, eds.), Vol. 3, p. 193. Raven, New York, 1978.

28. G. Mamantov, E. L. Wehry, R. R. Kemmerer, R. C. Stroupe, E. R. Hinton, and G. Goldstein, *Adv. Chem. Ser.* **170,** 99 (1978).

29. M. D. Erickson, D. L. Newton, E. D. Pellizzari, and K. B. Tomer, *J. Chromatogr. Sci.* **17,** 449 (1979).

30. K. D. Bartle and J. A. S. Smith, *Spectrochim. Acta* **23A,** 1689 (1967).

31. I. C. Lewis, *J. Phys. Chem.* **70,** 1667 (1966).

32. A. Cornu, J. Ulrich, and K. Persaud, *Chim. Anal. (Paris)* **47,** 357 (1965).

33. P. Durand, J. Parello, N. P. Buu-Hoi, and L. Alais, *Bull. Soc. Chim. Fr.,* 2438 (1963).

34. E. Clar, A. Mullen, and U. Sanigok, *Tetrahedron* **25,** 5639 (1969).

35. D. W. Jones, R. S. Matthews, and K. D. Bartle, *Spectrochim. Acta* **28A,** 2053 (1972).

36. D. Cagniant, *Bull. Soc. Chim. Fr.* 2325 (1966).

11

Approaches to Problem Solving in PAC Analysis

I. INTRODUCTION

The scope and limitations of the various analytical methods for PAC have been outlined in the previous chapters. Clearly, none of the procedures is universally applicable. Depending on the nature of the sample and the information required, a variety of methods may be employed, either alone or in combination.

The use of certain techniques is limited by a number of factors such as volatility, molecular weight, presence of heteroatoms, isomerism of compounds, or sample size. Major components may need to be identified (e.g., in coal-derived liquids), or the compounds of interest may be present in trace amounts (e.g., in water). Sensitivity, resolution, precision, time of analysis, and the simplicity of the methods must all be taken into account. Sample enrichment is needed where trace constituents are to be identified. In the characterization of complex mixtures and the identification of individual components often present at low concentration, a combination of chromatographic and spectroscopic techniques is required. Some methods overlap in their ranges of application, e.g., capillary GC and HPLC in the determination of common PAH with two to six rings. In the former method, the problem of interfering compounds is overcome by high chromatographic efficiency, but in the latter, the solution is sought through detector sensitivity, e.g., variable wavelength UV or fluorimetry. GC/MS with selective ion detection and HPLC with fluorescence detection compete with regard to sensitivity.

In this book, we do not pretend to present a unique method for every PAC analysis, but rather we aim to acquaint the reader with the range of methodologies currently available. However, in this final chapter we describe two examples which illustrate the strengths and weakness of many of the procedures we have discussed.

II. PRIORITY POLLUTANTS IN CONTAMINATED WATER

As discussed in Chapter 4, in 1971 the WHO set an upper limit on the total concentration of six PAC in water for domestic use. In 1976 the U.S. Environmental Protection Agency (EPA) proposed a further list of compounds as priority pollutants in effluent waters.

Rapid methods are thus required for detecting and determining the concentrations of PAC of environmental concern present in water. Two approaches have been suggested: (a) direct analysis at the ppm level of the sample without preconcentration, usually making use of the sensitivity and

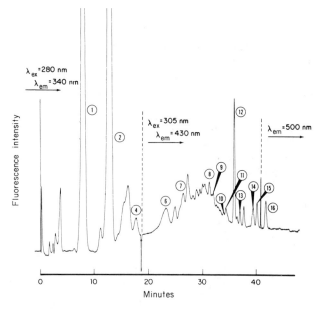

Fig. 11-1. Reversed-phase HPLC chromatogram of the extract of a plastics manufacturing effluent water. Chromatographic conditions: column Perkin-Elmer HC-ODS; mobile phase acetonitrile in water with gradient elution at 0.5 ml/min. Detection was by fluorimetry with excitation and emission wavelengths shown. Peak identification: (1) naphthalene, (2) acenaphthene, (4) phenanthrene, (5) anthracene, (6) fluoranthene, (7) pyrene, (8) benz[a]anthracene, (9) chrysene, (10) benzo[e]pyrene, (11) benzo[b]fluoranthene, (12) benzo[k]fluoranthene, (13) benzo[a]pyrene, (14) dibenz[a,h]anthracene, (15) benzo[ghi]perylene, (16) indeno[1,2,3-cd]-pyrene. [Reproduced with permission from K. Ogan, E. Katz, and W. Slavin, *Anal. Chem.* **51**, 1315 (1979). Copyright, American Chemical Society.]

selectivity of fluorescence; or (b) concentration of PAH from the water, followed by single-step analysis of a concentrate, using HPLC with fluorescence detection.

The direct approach is exemplified by detection of individual PAH in heavily polluted water contaminated by industrial sources, using fluorimetry with a deuterium lamp and photon counting (1) (c.f. Chapter 9). Detection levels of 0.03 μg/liter (naphthalene) to 0.17 μg/liter (fluoranthene) are typical. For the levels found in domestic water, however, greater sensitivity is necessary. Limits of detection below parts per trillion (i.e., reduced by two orders of magnitude) are made possible by fluorescence analysis with pulsed laser excitation (2) (Chapter 9). Levels of 0.5 $\times$ 10^{-3} μg/liter (pyrene) to 8.9 $\times$ 10^{-3} μg/liter (naphthalene) may now be detected directly in aqueous samples. Sub-ppt detection is also possible for PAC by fluorescence line-narrowing spectroscopy of glasses in which aqueous samples are mixed with glycerol or glycerol–dimethyl sulfoxide, frozen to $-269°$C (Chapter 9), and excited with a tunable dye laser (3).

Simplified analysis of polluted water for PAC may be carried out by HPLC of a concentrate obtained by use of a porous polymer or an HPLC precolumn with C_{18} bonded phase, or more simply by liquid-liquid extraction into an organic solvent (Chapter 4). For example, Ogan et al. (4) described how dichloromethane extracts of contaminated waters were injected, without clean-up, onto a reversed-phase HPLC column packed with 10-μm C_{18} bonded phase. Elution was by acetonitrile in water with a concentration gradient. Fluorescence with different λ_{ex} and λ_{em} in the early and later parts of the chromatogram allowed detection of 15 of the EPA priority-pollutant PAH (Fig. 11-1). The limits of detection were of the order of pg, corresponding to μg/liter in the original water sample.

Both approaches thus employ fluorimetry to overcome any interference from other compounds present, but it is noteworthy that the HPLC procedure does not have the limitations which oxygen quenching or sample turbidity may impose on the direct fluorescence method.

III. ANALYSIS OF METHYLCHRYSENES

While methyl derivatives of PAH are relatively minor constituents of environmental mixtures (5,6), the enhanced carcinogenicity of certain alkylated derivatives relative to the parent makes the analysis of such mixtures of particular interest. For example, 2-, 3-, 4-, and 6-methylchrysenes are all moderately carcinogenic, and the 5-methyl isomer, as a more potent tumor initiator than benz[a]pyrene, and with comparable activity as a complete carcinogen, is one of the most dangerous PAH yet reported (7).

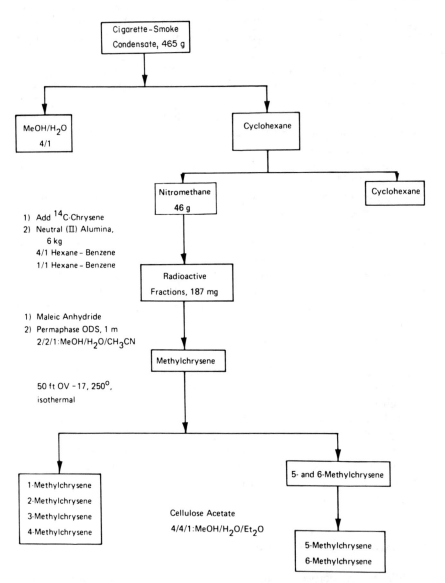

Fig. 11-2. The isolation of methylchrysenes from tobacco-smoke condensate. [Reproduced with permission from K. D. Brunnemann and D. Hoffmann *in* "Polynuclear Aromatic Hydrocarbons: Chemistry, Metabolism, and Carcinogenesis" (R. I. Freudenthal and P. W. Jones, eds.), p. 283. Raven Press, New York, 1976.]

A number of different approaches to the analysis of mixtures containing methylchrysenes and their congeners have been made. As conventionally isolated, the methylchrysene fraction usually contains methylbenz[a]-anthracenes which interfere because of their similar retention behavior in column chromatography, HPLC, and GC. Moreover, the final resolution of the methylchrysenes themselves is a particularly difficult problem unless specialized spectroscopic or chromatographic techniques are applied.

Hoffmann et al. (8, 9) separated the methylchrysene fraction from cigarette smoke by the scheme outlined in Fig. 11-2. The condensate was solvent partitioned to yield a PAH fraction in nitromethane which was separated, after the addition of 5,6-^{14}C-chrysene, by column chromatography on neutral alumina. The chrysene fraction, identified by its radioactivity, was first reacted with maleic anhydride and then enriched by reversed-phase HPLC. The Diels–Alder benz[a]anthracene/maleic anhydride adducts were thus easily separated. The HPLC peaks corresponding to the methylchrysenes were collected and analyzed by GLC on a long (15.2 m) packed column with 3% OV-17 stationary phase. The 1-, 2-, 3-, and 4-isomers could be separated (Fig. 11-3), but 5- and 6-methylchrysenes were eluted together and were collected and then separated by preparative paper chromatography on acetylated cellulose; UV spectroscopy (Fig. 11-4) was used for quantification in the last step. The conclusions from this procedure were that methylchrysenes are present in total at a level corresponding to half the concentration of the parent (Table 11-1).

A similar approach was made by Goeckner and Griest (10) to the analysis of a coal-liquefaction product for methylchrysenes. PAH were extracted

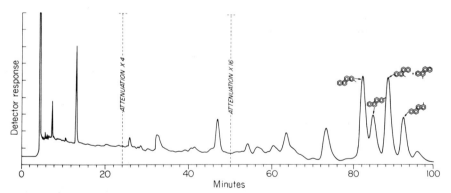

Fig. 11-3. Gas chromatogram of methylchrysene fraction from tobacco-smoke condensate: column 15.2 m × 3 mm, 3% OV-17 on 60–80 Gas Chrom Q, isothermal at 240°C. [Reproduced with permission from S. S. Hecht, W. E. Bondinell, and D. Hoffmann, *J. Natl. Cancer Inst.* **53**, 1121 (1974).]

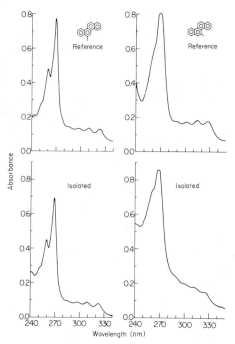

Fig. 11-4. UV spectra of 5- and 6-methylchrysenes. Upper, reference samples; lower, isolated from tobacco-smoke condensate. [Reproduced with permission from S. S. Hecht, W. E. Bondinell, and D. Hoffmann, *J. Natl. Cancer Inst.* **53**, 1121 (1974).]

TABLE 11-1

Methylchrysenes in Cigarette Smoke and a Coal-Liquefaction Product

Isomer	Cigarette smoke[a] (ng per cigarette)	Coal-liquefaction[b] product (ppm)
1-Methylchrysene	3.0	c
2-Methylchrysene	1.2	102
3-Methylchrysene	6.1	106
4-Methylchrysene	0	19[d]
5-Methylchrysene	0.6	
6-Methylchrysene	7.2	64
Chrysene	36.5	98

[a] From S. S. Hecht, *et al.* (8).
[b] From N. A. Goeckner and W. M. Griest (10).
[c] Analysis not possible—interference from alkyltriphenylene.
[d] Maximum upper limit.

with dimethyl sulfoxide and precipitated by adding aqueous NaCl so that an evaporation step involving nitromethane was avoided. Preliminary column chromatography on Florisil cleaned up the sample before the alumina chromatography, and this method was also used to remove the maleic anhydride adducts of benz[*a*]anthracenes. Gas chromatography on a 6-m 3.2% OV-22 column finally allowed only incomplete resolution of the major isomers present: 2-, 3-, and 6-methylchrysenes, together 270% of the chrysene content (Table 11-1). A major PAH contaminant, possibly a methyltriphenylene [2-methyltriphenylene, present (*11*) in petroleum?] interfered with the 1-methylchrysene peak.

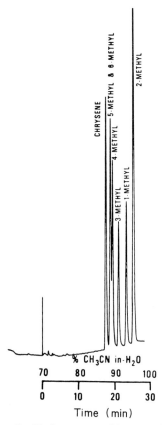

Fig. 11-5. Reversed-phase liquid chromatographic separation of chrysene and methylchrysenes. Column: Vydac 201 TP Reversed-phase. Detection: UV absorbance at 256 nm. Linear solvent gradient from 80–100% acetonitrile in water at 1% min^{-1} and 1 ml/min. [Reproduced with permission from S. A. Wise, W. J. Bonnett, and W. E. May, *in* "Polynuclear Aromatic Hydrocarbons: Chemistry and Biological Effects" (A. Bjørseth and A. J. Dennis, eds.), p. 791. Battelle Press, Columbus, Ohio, 1980.]

Packed column GC on Dexsil 300 of enriched gel-permeation chromatography fractions of tobacco smoke also allows only partial resolution of
methylchrysenes (*12*); as many as four components are present in each peak.
Even very high resolution reversed-phase LC with gradient elution does not
separate 5- from 6-methylchrysene, and allows only partial resolution of the
4-isomer (Fig. 11.5) (*13*). High-resolution (capillary column) GC seems
necessary. Detailed retention data were compiled (*14*) for all the methylchrysenes, and all but one of the methylbenz[*a*]anthracenes (Table 7-2). For
short but highly efficient glass columns coated with films of SE-52 methyl-
phenyl silicone, compounds with retention index differing by one unit are
better than 50% resolved since the peak width at half-height averages 0.8
index unit. However, even though this phase is well documented as being
especially suitable for PAH (*15–17*), on such columns resolution of the
methylchrysenes is incomplete and there is considerable overlap with
methylbenz[*a*]anthracenes (Table 7-2).

A complement to high resolution chromatography is the application of
newer spectroscopic methods. Fourier transform ^{1}H NMR was applied by
Bartle *et al.* (*15, 16, 18*) to PAH fractions from atmospheric dust, tobacco
smoke, and marijuana smoke which were separated in a sequence of solvent
partition, column chromatography, and HPLC (Fig. 4-18), and shown by
UV and fluorescence to be mixtures of methyl derivatives of chrysenes and

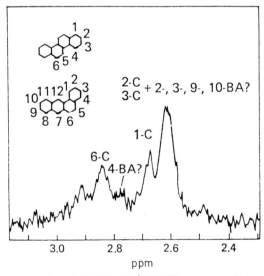

Fig. 11-6. Methyl region of 90-MHz FT ^{1}H-NMR spectrum of chrysene fraction of air-
pollutant PAH. [Reproduced with permission from K. D. Bartle, M. L. Lee, and M. Novotny,
Analyst (London) **102**, 731 (1977).]

benz[a]anthracenes (15, 16). The presence of four methylchrysenes was inferred from the methyl region of the 90-MHz ^{1}H FT-NMR spectrum of air-pollutant PAC (Fig. 11-6). No resonances for the sterically hindered 4- and 5-methylchrysenes (expected at 3.12 and 3.20 ppm respectively) were detected. The remaining methyl derivative may be either 4-methylbenz[a]-anthracene (small peak near 2.77 ppm) or 2-, 3-, 9-, or 10-methylbenz[a]-anthracene (overlapping the strong peak centered at 2.60 ppm). Similar spectra were observed for the tobacco- and marijuana-smoke benz[a]-anthracene–chrysene fractions, except that a number of dimethyl and trimethyl derivatives were also present. 1-, 2-, 3-, and 6-Methylchrysene again contribute to the ^{1}H-NMR spectrum, but a small peak at 3.17 ppm in the marijuana fraction is assigned to CH_3 of 5-methylchrysene.

Such information concerning the identity of the methylchrysenes present in the environment is a vital part of understanding the carcinogenic action of air-pollutant and other PAH. While methylbenz[a]anthracenes can be identified only tentatively by ^{1}H NMR, the absence of the strongly carcinogenic 7- and 12-methyl isomers from all the materials studied by Bartle et al. may be strongly inferred.

While the FT technique is required in NMR to overcome the limitations of the small samples available, the multiplex advantage allows the FT

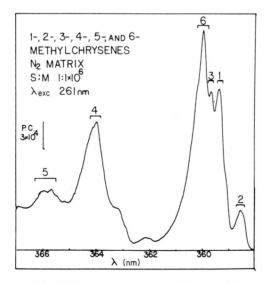

Fig. 11-7. Portion of the MI fluorescence spectrum of the methylchrysenes (200 ng of each). [Reproduced with permission from P. Tokousbalides, E. R. Hinton, Jr., R. B. Dickinson, Jr., P. V. Bilotta, E. L. Wehry, and G. Mamantov, *Anal. Chem.* **50,** 1189 (1978). Copyright, American Chemical Society.]

infrared spectra of dilute matrix-isolated (MI) PAH to be recorded with consequent resolution of extra fine structure. Narrow lines are also resolved in the MI fluorescence spectra. Both approaches were applied by Tokousbalides et al. to the methylchrysenes problem (19). The fluorescence spectra in a N_2 matrix at $-261°C$ are most specific and have the narrowest lines between 360 and 370 nm. Five of the six isomers can be resolved (Fig. 11-7) by excitation at 261 nm, but by choice of a more appropriate, specific, excitation wavelength for each isomer (Table 11-2), any one methylchrysene may be recognized in the presence of the others. By contrast, only 6-methylchrysene could be resolved in the (Shpol'skii) quasilinear fluorescence spectra of methylchrysenes in frozen solutions in n-heptane, presumably because the optimum solvent, concentration, and freezing rate differ for each isomer.

Each methylchrysene exhibits at least two intense features in the 700–800 cm^{-1} region of the MI FT-IR spectrum (Table 11-2). In a mixture of all six isomers, any of 1-, 2-, 3-, and 5-isomer may be unambiguously identified (Table 11-2 and Fig. 11-8), but the 4- and 6-isomers do not have unique bands. However, MI fluorescence and FT-IR are complementary since the 3-isomer, the most difficult to detect in a mixture of all six isomers by fluorescence (Fig. 11-6), can be readily identified in the MI FT-IR spectrum from the doublet at 839, 835.6 cm^{-1} (Table 11-2). Similarly, 4- and 6-methylchrysenes cannot be distinguished from the other isomers by FT-IR but are easily identified by MI fluorimetry.

While MI spectroscopy thus allows any methylchrysene to be identified in the presence of its isomers, benz[a]anthracenes need not be removed since

TABLE 11-2

MI Fluorescence and FT-IR Spectroscopy of Methylchrysenes[a]

Isomer	Optimum excitation wavelength (nm)	IR absorption bands[b] 700–900 cm^{-1} (cm^{-1})
1	305	822, (797,795.5 doublet), 774[c]
2	300	880, 818, 805, 754.5[c]
3	302	(865.5,862 doublet), (839,836.5 doublet), 827.5, 774, 754.5[c]
4	308	828[c], (800,798 doublet), 767.5, 756[c]
5	261	(892,883 doublet), (864,867 doublet), 824, 795.5, 765[c], 748
6	304	823.5, 764.5[c]

[a] From P. Tokousbalides et al. (19).
[b] Measured at $-268°C$ for nitrogen matrix. Sample: matrix ratio $1:10^5$.
[c] Indicates strongest peak.

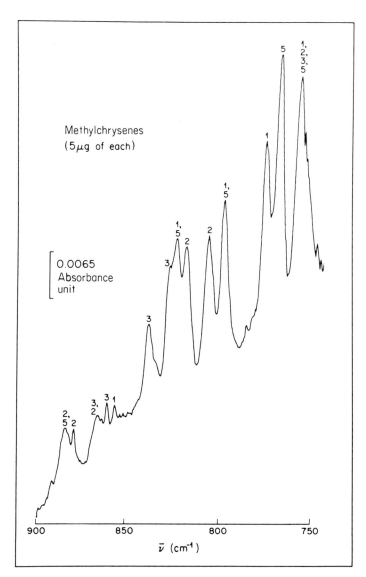

Fig. 11-8. Matrix-isolation FT-IR spectrum of a four-component methylchrysene mixture (1-, 2-, 3-, and 5-isomers; 5 μg of each) at −258°C in a nitrogen matrix. [Reproduced with permission from P. Tokousbalides, E. R. Hinton, Jr., R. B. Dickinson, Jr., P. V. Bilotta, E. L. Wehry, and G. Mamantov, *Anal. Chem.* **50,** 1189 (1978). Copyright, American Chemical Society.]

their MI fluorescence and FT-IR spectra do not interfere. Moreover, quantitative analysis is also easily possible by this technique. Beer's law plots for the MI fluorescence and FT-IR intensities of methylchrysenes, related to an internal standard, are linear over a wide range, even in the presence of large excesses of other interfering PAH.

REFERENCES

1. F. P. Schwarz and S. P. Wasik, *Anal. Chem.* **48**, 526 (1976).
2. J. H. Richardson and M. E. Ando, *Anal. Chem.* **49**, 955 (1977).
3. J. C. Brown, J. A. Duncanson, Jr., and G. J. Small, *Anal. Chem.* **52**, 1711 (1980).
4. K. Ogan, E. Katz, and W. Slavin, *Anal. Chem.* **51**, 1315 (1979).
5. D. Hoffmann and E. L. Wynder, *in* "Air Pollution" (A. C. Stern, ed.), Vol. 2, p. 361. Academic Press, New York, 1976.
6. National Academy of Sciences, Particulate Polycyclic Organic Matter. Rep. Comm. Biol. Effects of Atmospheric Pollutants, Washington, D.C., 1972.
7. S. S. Hecht, M. Loy, R. Mazzarese, and D. Hoffmann *in* "Polycyclic Hydrocarbons and Cancer" (H. V. Gelboin and P. O. P. Ts'o, eds.), Vol. 1, p. 119. Academic Press, New York, 1976.
8. S. S. Hecht, W. E. Bondinell, and D. Hoffmann, *J. Natl. Cancer Inst.* **53**, 1121 (1974).
9. K. D. Brunnemann and D. Hoffmann, *in* "Polynuclear Aromatic Hydrocarbons: Chemistry, Metabolism, and Carcinogenesis" (R. I. Freudenthal and P. W. Jones, eds.), p. 283. Raven, New York, 1976.
10. N. A. Goeckner and W. H. Griest, *Sci. Total Environ.* **8**, 187 (1977).
11. L. K. Keefer, L. Wallcave, J. Loo, and R. S. Peterson, *Anal. Chem.* **43**, 1411 (1971).
12. R. F. Severson, M. E. Snook, H. C. Higman, R. F. Arrendale, and O. T. Chortyk, *Beitr. Tabakforsch.* **8**, 250 (1976).
13. S. A. Wise, W. J. Bonnett, and W. E. May, *in* "Polynuclear Aromatic Hydrocarbons: Chemistry and Biological Effects" (A. Bjørseth and A. J. Dennis, eds.), p. 791. Battelle Press, Columbus, Ohio, 1980.
14. M. L. Lee, D. L. Vassilaros, C. M. White, and M. Novotny, *Anal. Chem.* **51**, 768 (1979).
15. M. L. Lee, M. Novotny, and K. D. Bartle, *Anal. Chem.* **48**, 405 (1976).
16. M. L. Lee, M. Novotny, and K. D. Bartle, *Anal. Chem.* **48**, 1566 (1976).
17. M. L. Lee, G. P. Prado, J. B. Howard, and R. A. Hites, *Biomed. Mass Spectrom.* **4**, 182 (1977).
18. K. D. Bartle, M. L. Lee, and M. Novotny, *Analyst (London)* **102**, 731 (1977).
19. P. Tokousbalides, E. R. Hinton, Jr., R. B. Dickinson, Jr., P. V. Bilotta, E. L. Wehry, and G. Mamantov, *Anal. Chem.* **50**, 1189 (1978).

1. Polycyclic Aromatic Hydrocarbons:
Names, Formulas, Structures, and Numbering

Name	Molecular formula (weight)	Structure
Indene (abbreviated prefix indeno)	C_9H_8 (116)	
Naphthalene (abbreviated prefix naphtho)	$C_{10}H_8$ (128)	
Acenaphthylene (abbreviated prefix acenaphtho)	$C_{12}H_8$ (152)	
Acenaphthene	$C_{12}H_{10}$ (154)	
Fluorene	$C_{13}H_{10}$ (166)	
Phenalene (abbreviated prefix phenaleno)	$C_{13}H_{10}$ (166)	
Anthracene (abbreviated prefix anthra)	$C_{14}H_{10}$ (178)	
Phenanthrene (abbreviated prefix phenanthro)	$C_{14}H_{10}$ (178)	
4H-Cyclopenta[def]phenanthrene	$C_{15}H_{10}$ (190)	

(Continued)

Appendix 1 (*Continued*)

Name	Molecular formula (weight)	Structure
Aceanthrylene	$C_{16}H_{10}$ (202)	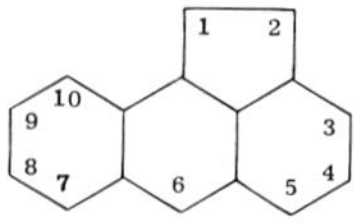
Acephenanthrylene	$C_{16}H_{10}$ (202)	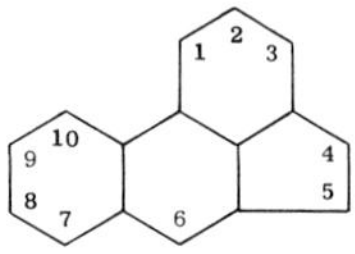
Fluoranthene	$C_{16}H_{10}$ (202)	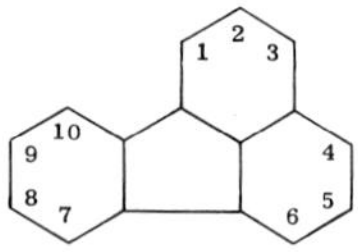
Pyrene (abbreviated prefix pyrano)	$C_{16}H_{10}$ (202)	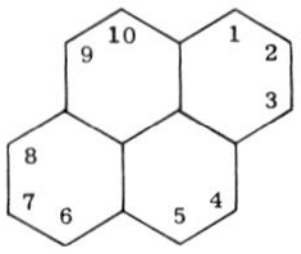
11H-Benzo[a]fluorene	$C_{17}H_{12}$ (216)	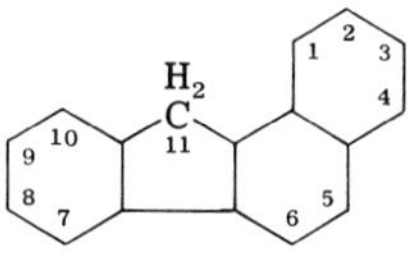
11H-Benzo[b]fluorene	$C_{17}H_{12}$ (216)	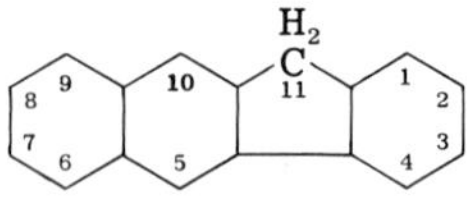
7H-Benzo[c]fluorene	$C_{17}H_{12}$ (216)	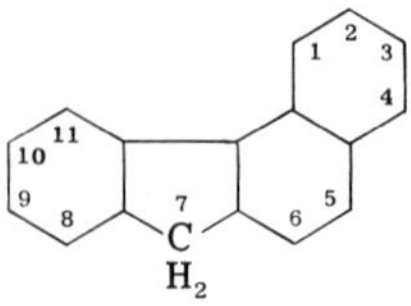
1H-Benz[de]anthracene	$C_{17}H_{12}$ (216)	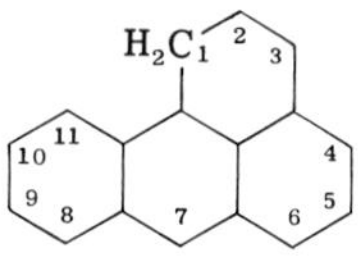

(*Continued*)

Appendix 1 (*Continued*)

Name	Molecular formula (weight)	Structure
Benzo[*ghi*]fluoranthene	$C_{18}H_{10}$ (226)	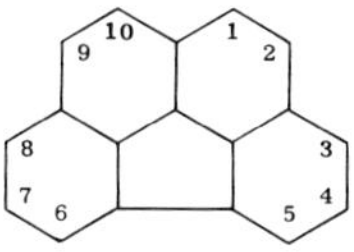
Cyclopenta[*cd*]pyrene	$C_{18}H_{10}$ (226)	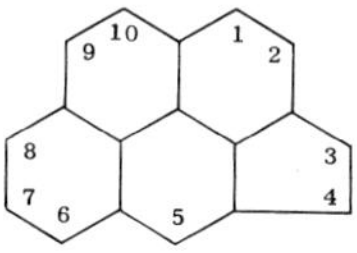
Benz[*a*]anthracene	$C_{18}H_{12}$ (228)	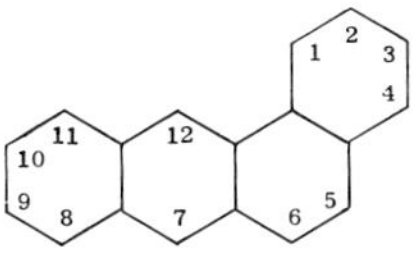
Chrysene (abbreviated prefix chryseno)	$C_{18}H_{12}$ (228)	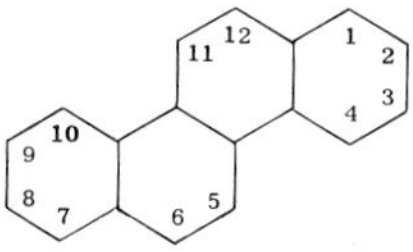
Naphthacene (abbreviated prefix naphthaceno)	$C_{18}H_{12}$ (228)	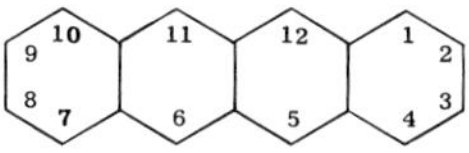
Benzo[*c*]phenanthrene	$C_{18}H_{12}$ (228)	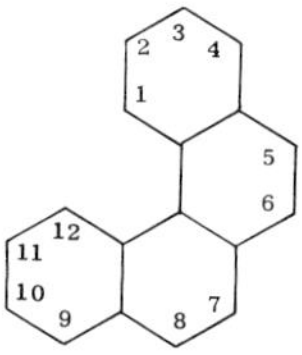
Triphenylene (abbreviated prefix triphenyleno)	$C_{18}H_{12}$ (228)	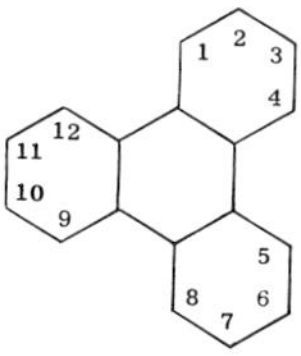
11*H*-Benz[*bc*]aceanthrylene	$C_{19}H_{12}$ (240)	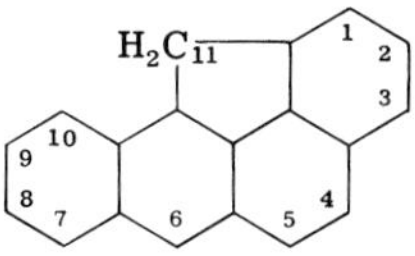

(*Continued*)

Appendix 1 (*Continued*)

Name	Molecular formula (weight)	Structure
4*H*-Cyclopenta[*def*]chrysene	$C_{19}H_{12}$ (240)	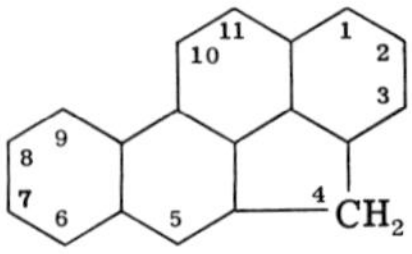
4*H*-Cyclopenta[*def*]triphenylene	$C_{19}H_{12}$ (240)	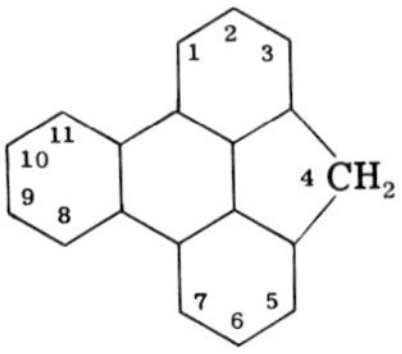
Perylene (abbreviated prefix perylo)	$C_{20}H_{12}$ (252)	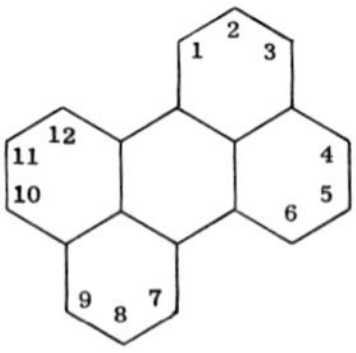
Benzo[*a*]pyrene	$C_{20}H_{12}$ (252)	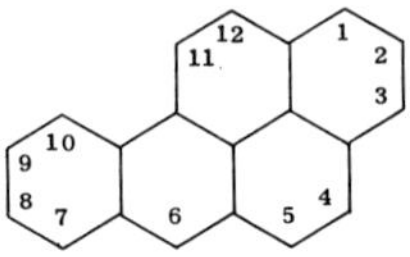
Benzo[*e*]pyrene	$C_{20}H_{12}$ (252)	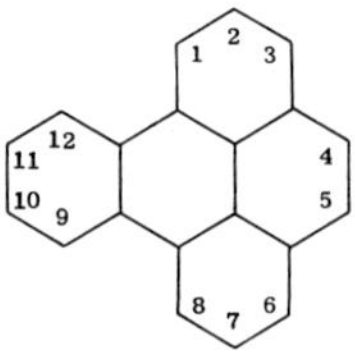
Benz[*a*]aceanthrylene	$C_{20}H_{12}$ (252)	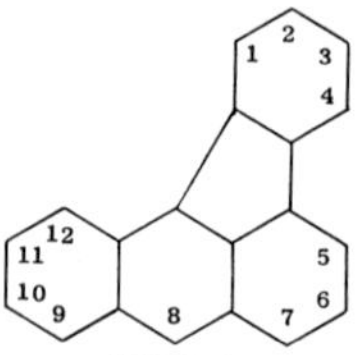
Benz[*d*]aceanthrylene	$C_{20}H_{12}$ (252)	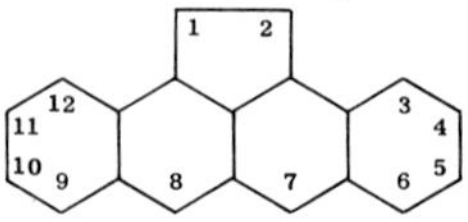

(*Continued*)

Appendix 1 (*Continued*)

Name	Molecular formula (weight)	Structure
Benz[*e*]aceanthrylene	$C_{20}H_{12}$ (252)	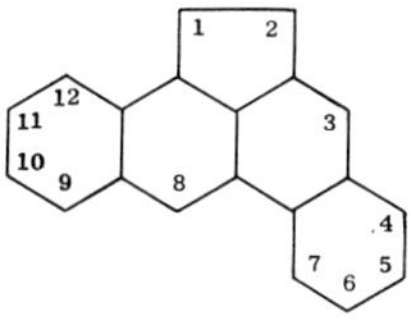
Benz[*l*]acephenanthrylene	$C_{20}H_{12}$ (252)	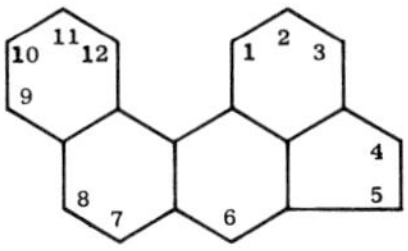
Benz[*j*]aceanthrylene	$C_{20}H_{12}$ (252)	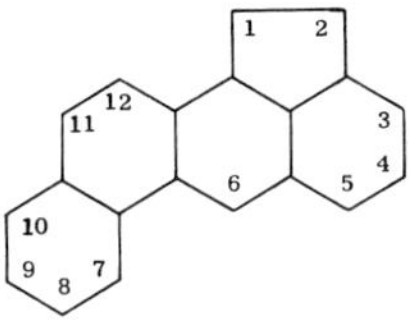
Cyclopenta[*de*]naphthacene	$C_{20}H_{12}$ (252)	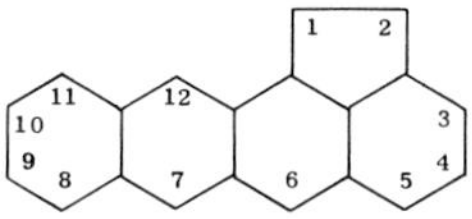
Benz[*l*]aceanthrylene	$C_{20}H_{12}$ (252)	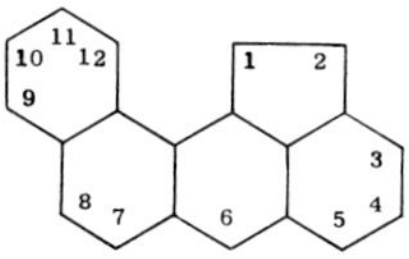
Benz[*a*]acephenanthrylene	$C_{20}H_{12}$ (252)	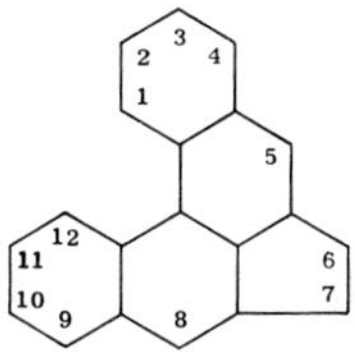
Benz[*e*]acephenanthrylene	$C_{20}H_{12}$ (252)	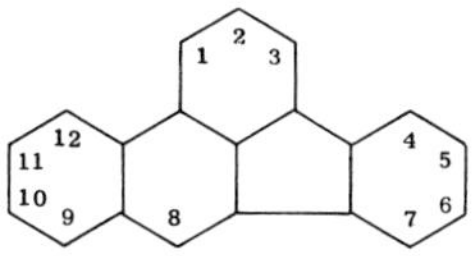

(*Continued*)

Appendix 1 (*Continued*)

Name	Molecular formula (weight)	Structure
Benz[*k*]acephenanthrylene	$C_{20}H_{12}$ (252)	
Benzo[*j*]fluoranthene	$C_{20}H_{12}$ (252)	
Benzo[*k*]fluoranthene	$C_{20}H_{12}$ (252)	
Cyclopenta[*hi*]chrysene	$C_{20}H_{12}$ (252)	
4*H*-Benzo[*def*]cyclopenta[*mno*]-chrysene	$C_{21}H_{12}$ (264)	
11*H*-Cyclopenta[*ghi*]perylene	$C_{21}H_{12}$ (264)	
11*H*-Indeno[2,1,7-*cde*]pyrene	$C_{21}H_{12}$ (264)	

Appendix 1 (*Continued*)

Name	Molecular formula (weight)	Structure
13*H*-Dibenzo[*a*,*g*]fluorene	$C_{21}H_{14}$ (266)	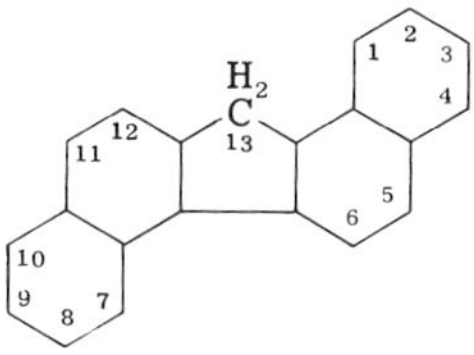
13*H*-Dibenzo[*a*,*h*]fluorene	$C_{21}H_{14}$ (266)	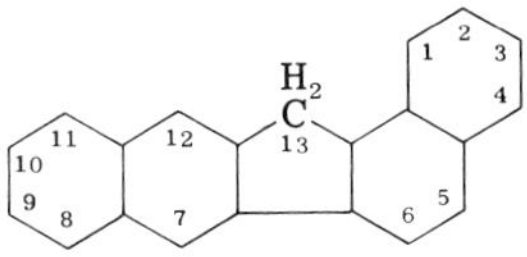
13*H*-Dibenzo[*a*,*i*]fluorene	$C_{21}H_{14}$ (266)	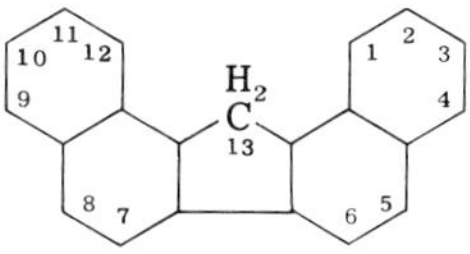
7*H*-Dibenzo[*b*,*g*]fluorene	$C_{21}H_{14}$ (266)	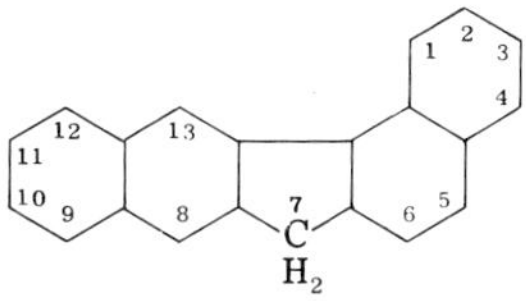
8*H*-Indeno[2,1-*b*]phenanthrene	$C_{21}H_{14}$ (266)	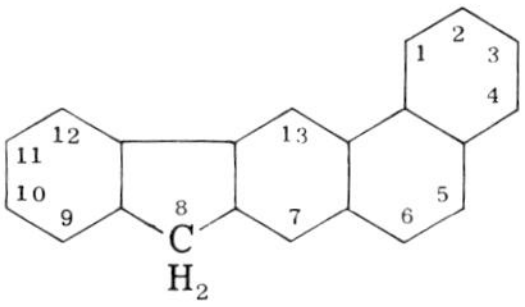
9*H*-Indeno[2,1-*c*]phenanthrene	$C_{21}H_{14}$ (266)	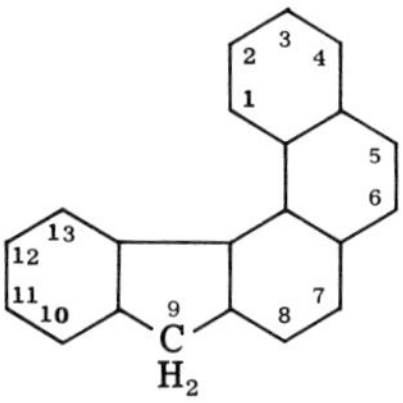
12*H*-Dibenzo[*b*,*h*]fluorene	$C_{21}H_{14}$ (266)	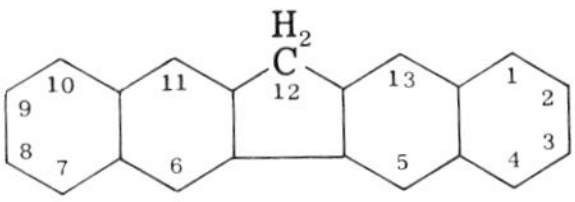

(*Continued*)

Appendix 1 (*Continued*)

Name	Molecular formula (weight)	Structure
7*H*-Dibenzo[*c*,*g*]fluorene	$C_{21}H_{14}$ (266)	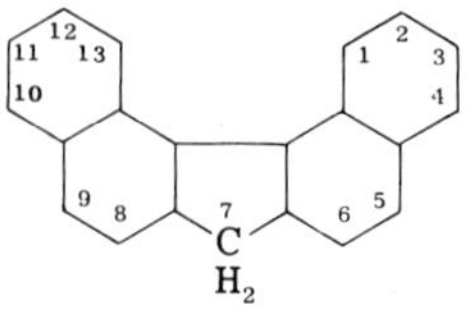
8*H*-Indeno[1,2-*a*]anthracene	$C_{21}H_{14}$ (266)	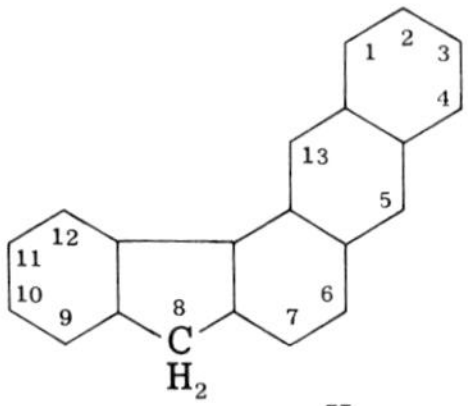
13*H*-Indeno[1,2-*b*]anthracene	$C_{21}H_{14}$ (266)	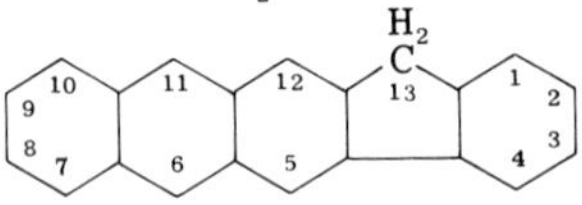
13*H*-Indeno[2,1-*a*]anthracene	$C_{21}H_{14}$ (266)	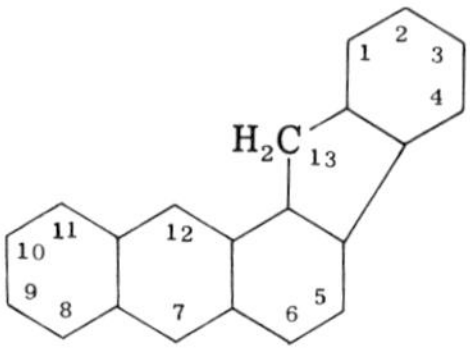
12*H*-Indeno[1,2-*b*]phenanthrene	$C_{21}H_{14}$ (266)	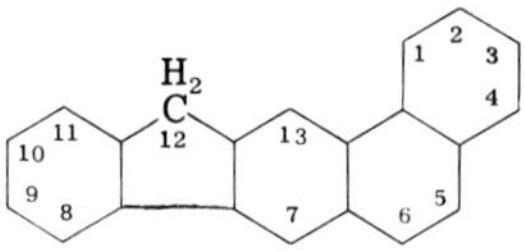
13*H*-Indeno[1,2-*c*]phenanthrene	$C_{21}H_{14}$ (266)	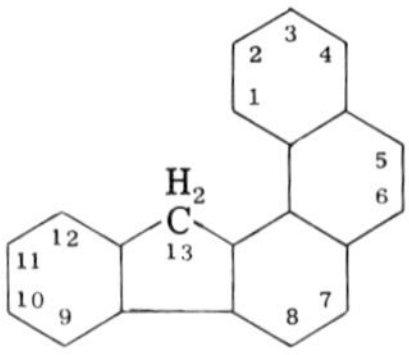
13*H*-Indeno[1,2-*l*]phenanthrene	$C_{21}H_{14}$ (266)	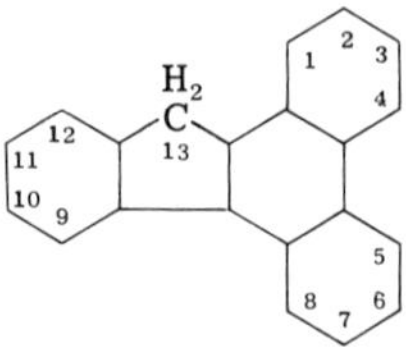

(*Continued*)

Appendix 1 (*Continued*)

Name	Molecular formula (weight)	Structure
11*H*-Indeno[2,1-*a*]phenanthrene	$C_{21}H_{14}$ (266)	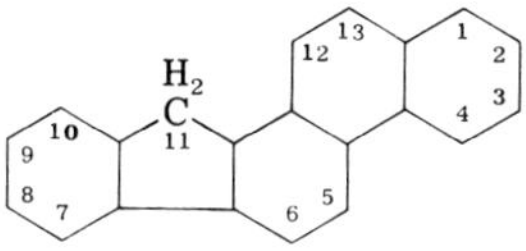
Benzo[*def*]cyclopenta[*qr*]chrysene	$C_{22}H_{12}$ (276)	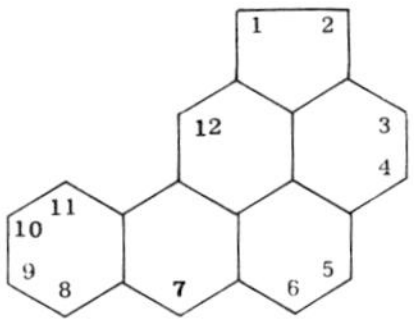
Indeno[1,2,3-*cd*]pyrene	$C_{22}H_{12}$ (276)	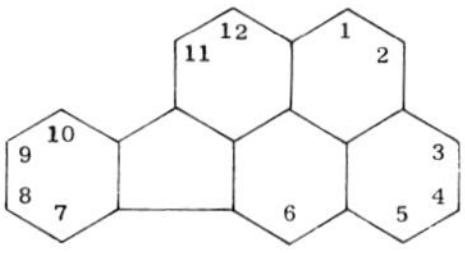
Indeno[1,7-*ab*]pyrene	$C_{22}H_{12}$ (276)	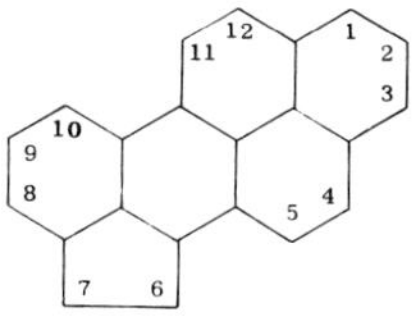
Indeno[1,7,6,5-*cdef*]chrysene	$C_{22}H_{12}$ (276)	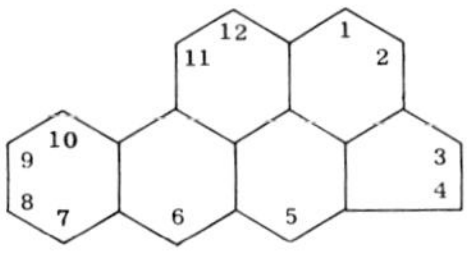
Indeno[5,6,7,1-*defg*]chrysene	$C_{22}H_{12}$ (276)	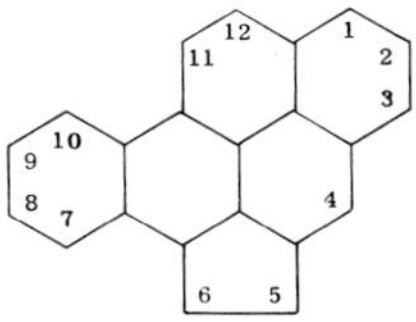
Benzo[*e*]cyclopenta[*jk*]pyrene	$C_{22}H_{12}$ (276)	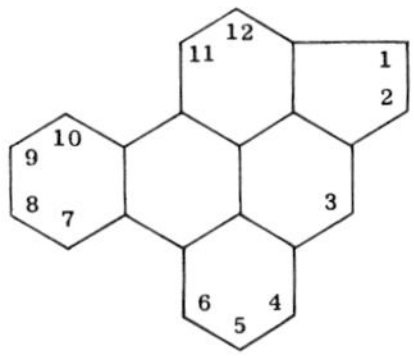

(*Continued*)

Appendix 1 (*Continued*)

Name	Molecular formula (weight)	Structure
Cyclopenta[*cd*]perylene	$C_{22}H_{12}$ (276)	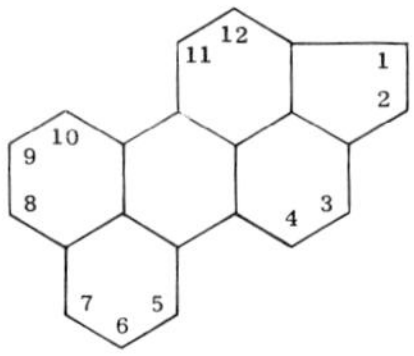
Dibenzo[*def,mno*]chrysene (Anthanthrene)	$C_{22}H_{12}$ (276)	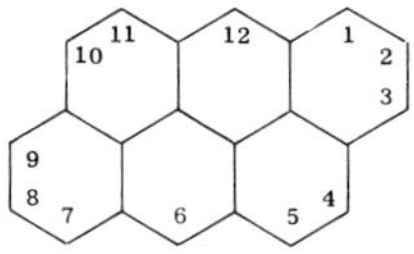
Benzo[*ghi*]perylene	$C_{22}H_{12}$ (276)	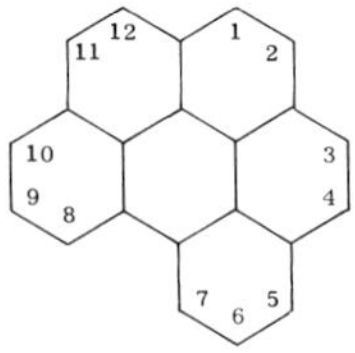
Benzo[*g*]chrysene	$C_{22}H_{14}$ (278)	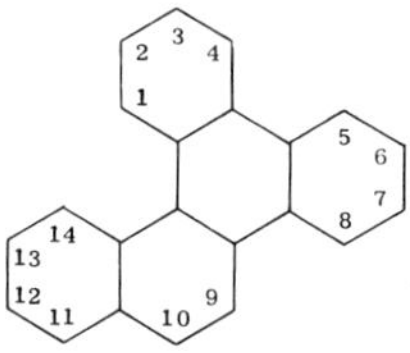
Benzo[*a*]naphthacene	$C_{22}H_{14}$ (278)	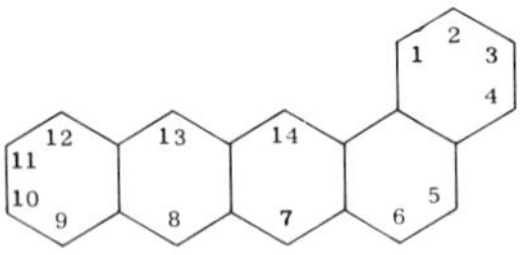
Pentacene	$C_{22}H_{14}$ (278)	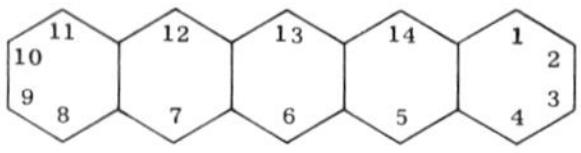
Pentaphene	$C_{22}H_{14}$ (278)	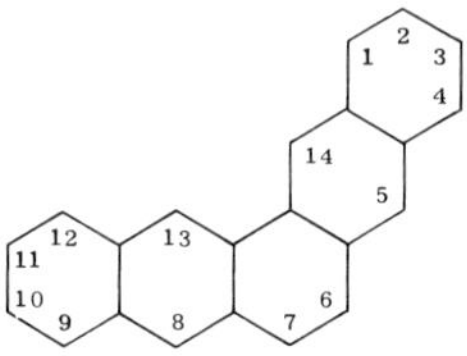

(*Continued*)

Appendix 1 (*Continued*)

Name	Molecular formula (weight)	Structure
Dibenzo[*b,g*]phenanthrene	$C_{22}H_{14}$ (278)	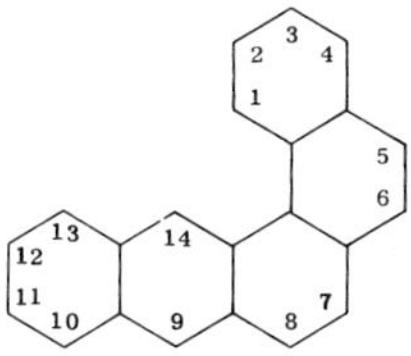
Dibenzo[*c,g*]phenanthrene	$C_{22}H_{14}$ (278)	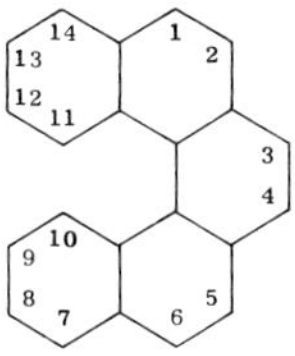
Picene	$C_{22}H_{14}$ (278)	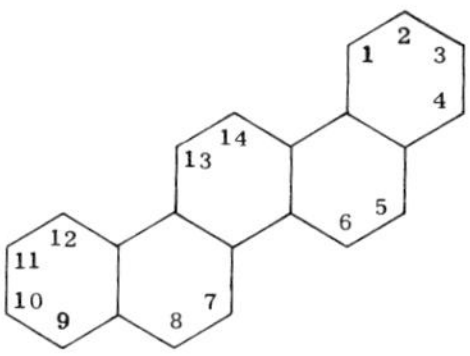
Dibenz[*a,c*]anthracene	$C_{22}H_{14}$ (278)	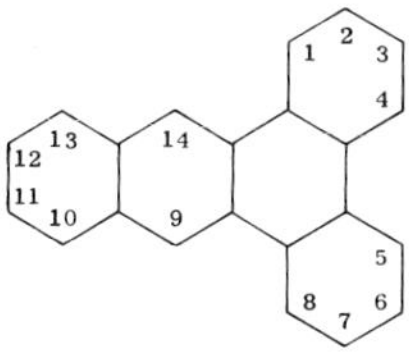
Dibenz[*a,h*]anthracene	$C_{22}H_{14}$ (278)	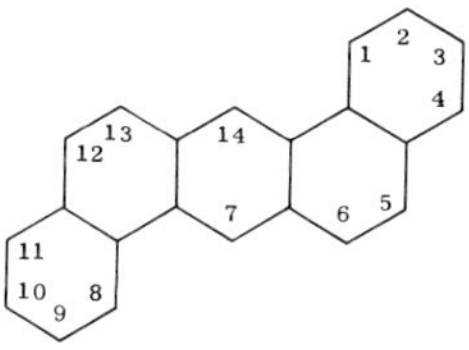
Dibenz[*a,j*]anthracene	$C_{22}H_{14}$ (278)	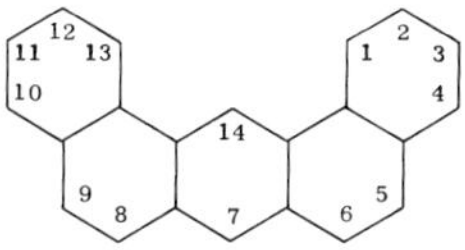

(*Continued*)

Appendix 1 (*Continued*)

Name	Molecular formula (weight)	Structure
Benzo[*b*]chrysene	$C_{22}H_{14}$ (278)	
Benzo[*c*]chrysene	$C_{22}H_{14}$ (278)	
5*H*-Benzo[*b*]cyclopenta[*def*]-chrysene	$C_{23}H_{14}$ (290)	
4*H*-Benzo[*b*]cyclopenta[*mno*]-chrysene	$C_{23}H_{14}$ (290)	
4*H*-Benzo[*c*]cyclopenta[*mno*]-chrysene	$C_{23}H_{14}$ (290)	
8*H*-Benzo[*g*]cyclopenta[*mno*]-chrysene	$C_{23}H_{14}$ (290)	
13*H*-Benzo[*b*]cyclopenta[*def*]-triphenylene	$C_{23}H_{14}$ (290)	

(*Continued*)

Appendix 1 (*Continued*)

Name	Molecular formula (weight)	Structure
13*H*-Cyclopenta[*rst*]pentaphene	$C_{23}H_{14}$ (290)	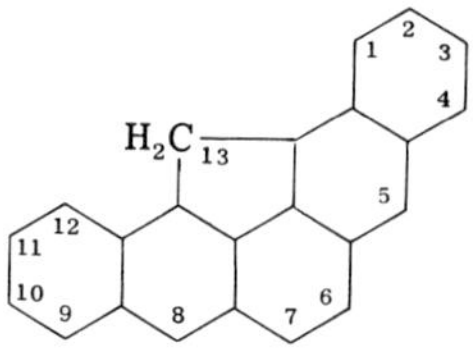
13*H*-Cyclopenta[*pqr*]picene	$C_{23}H_{14}$ (290)	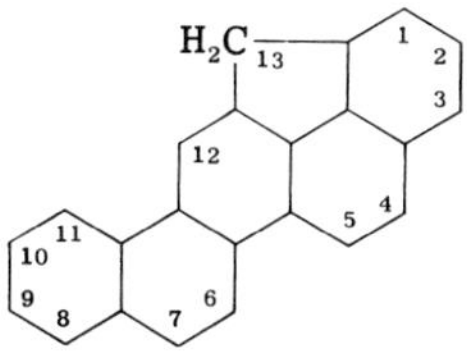
6*H*-Cyclopenta[*ghi*]picene	$C_{23}H_{14}$ (290)	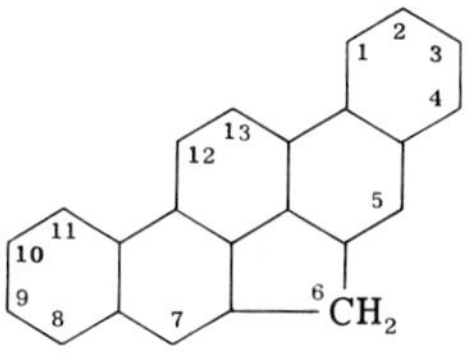
13*H*-Dibenz[*bc,j*]aceanthrylene	$C_{23}H_{14}$ (290)	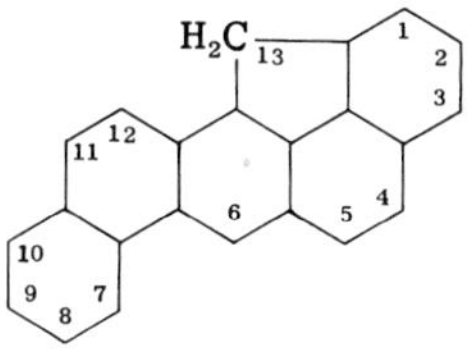
13*H*-Dibenz[*bc,l*]aceanthrylene	$C_{23}H_{14}$ (290)	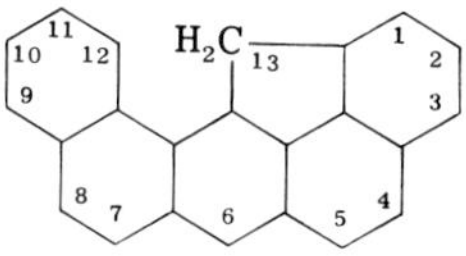
13*H*-Indeno[2,1,7-*qra*]-naphthacene	$C_{23}H_{14}$ (290)	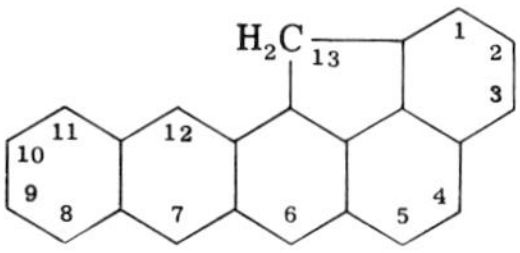

(*Continued*)

Appendix 1 (*Continued*)

Name	Molecular formula (weight)	Structure
4*H*-Indeno[7,1,2-*ghi*]chrysene	$C_{23}H_{14}$ (290)	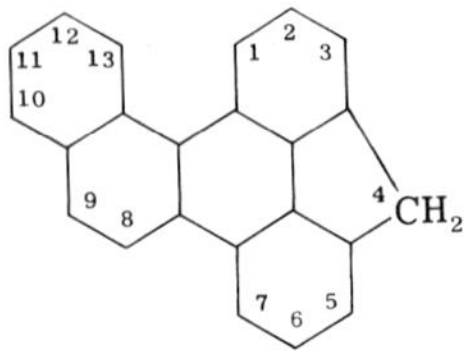
7*H*-Indeno[1,2-*a*]pyrene	$C_{23}H_{14}$ (290)	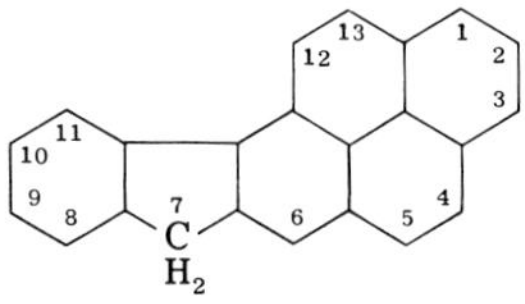
11*H*-Indeno[2,1-*a*]pyrene	$C_{23}H_{14}$ (290)	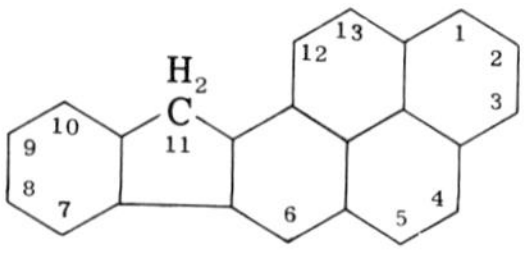
9*H*-Indeno[1,2-*e*]pyrene	$C_{23}H_{14}$ (290)	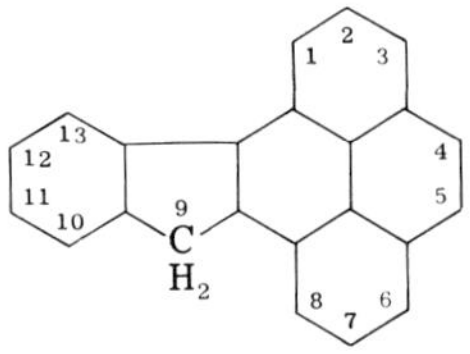
Coronene	$C_{24}H_{12}$ (300)	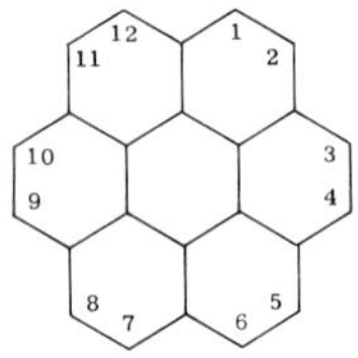
Benzo[*b*]perylene	$C_{24}H_{14}$ (302)	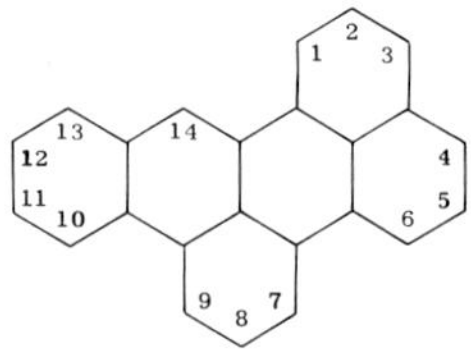

(*Continued*)

Appendix 1 (*Continued*)

Name	Molecular formula (weight)	Structure
Benzo[*pqr*]picene	$C_{24}H_{14}$ (302)	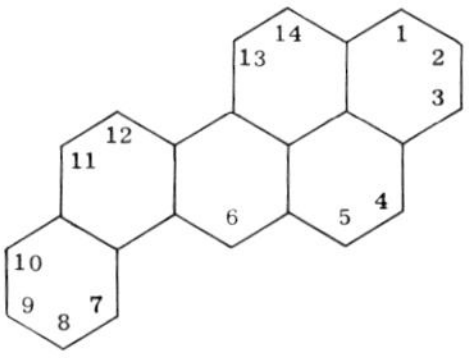
Naphtho[1,2-*e*]pyrene	$C_{24}H_{14}$ (302)	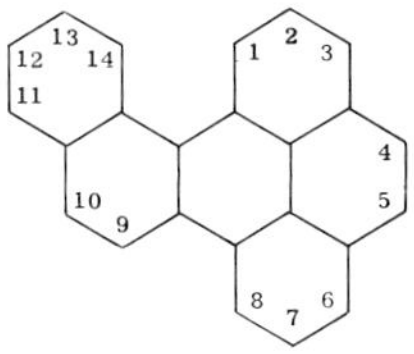
Dibenzo[*b,def*]chrysene	$C_{24}H_{14}$ (302)	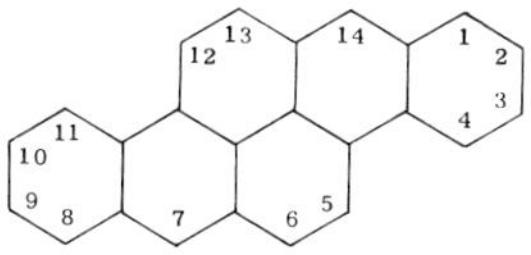
Dibenzo[*c,mno*]chrysene	$C_{24}H_{14}$ (302)	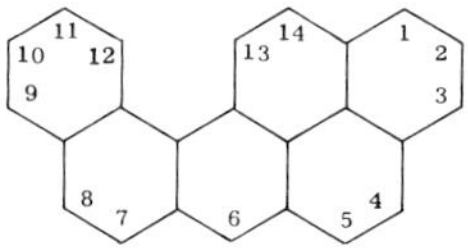
Dibenzo[*def,p*]chrysene	$C_{24}H_{14}$ (302)	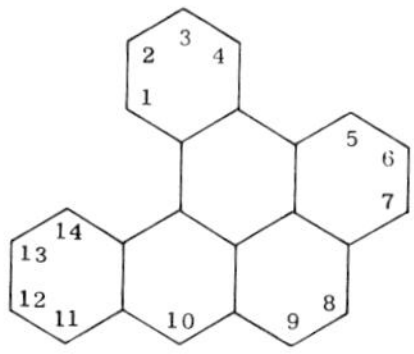
Naphtho[1,2,3,4-*def*]chrysene	$C_{24}H_{14}$ (302)	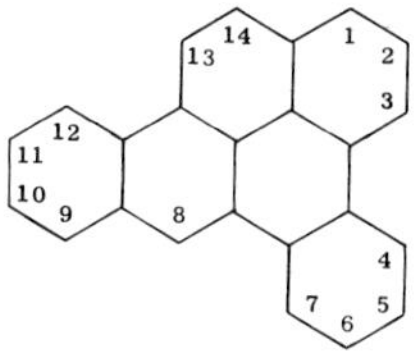

(*Continued*)

Appendix 1 (*Continued*)

Name	Molecular formula (weight)	Structure
Dibenzo[*de,mn*]naphthacene	C$_{24}$H$_{14}$ (302)	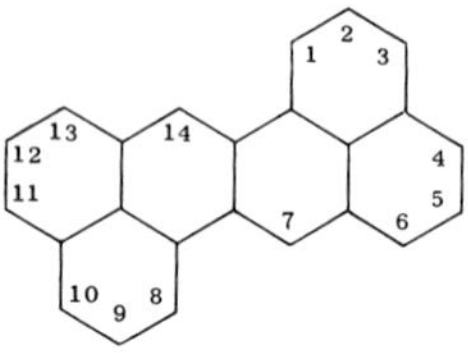
Dibenzo[*de,qr*]naphthacene	C$_{24}$H$_{14}$ (302)	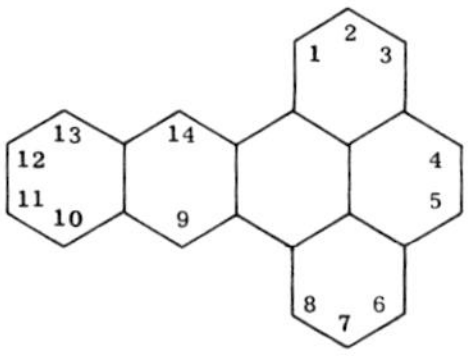
Naphtho[2,1,8-*qra*]naphthacene	C$_{24}$H$_{14}$ (302)	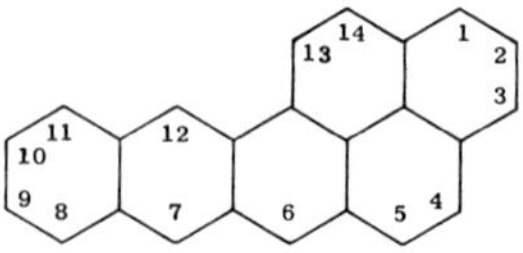
Dibenzo[*fg,op*]naphthacene	C$_{24}$H$_{14}$ (302)	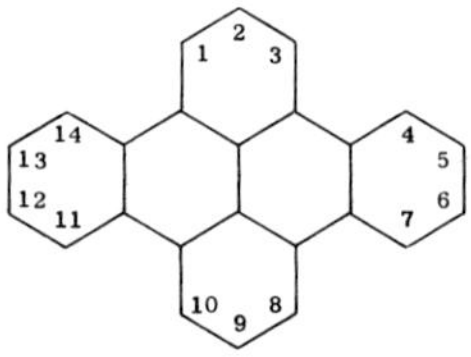
Benzo[*rst*]pentaphene	C$_{24}$H$_{14}$ (302)	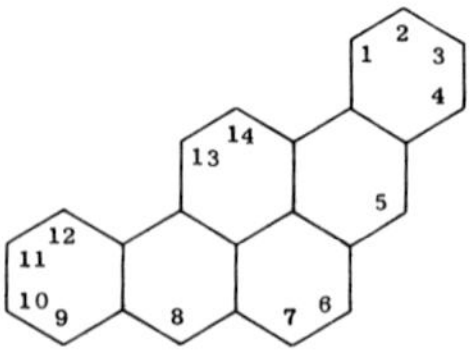
Benzo[*a*]perylene	C$_{24}$H$_{14}$ (302)	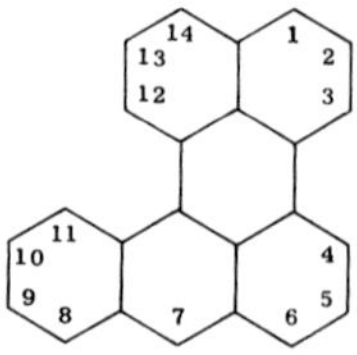

(*Continued*)

Appendix 1 (*Continued*)

Name	Molecular formula (weight)	Structure
Naphtho[2,3-*j*]fluoranthene	C$_{24}$H$_{14}$ (302)	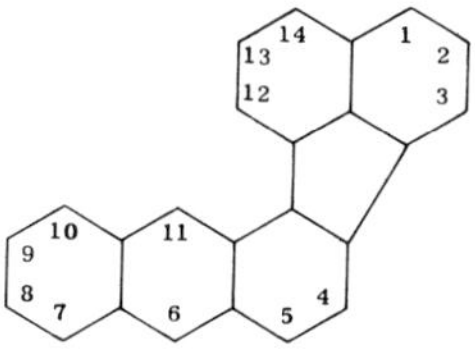
Naphtho[2,3-*k*]fluoranthene	C$_{24}$H$_{14}$ (302)	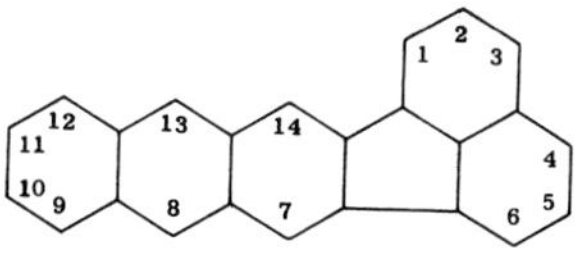
Benzo[*b*]cyclopenta[*hi*]chrysene	C$_{24}$H$_{14}$ (302)	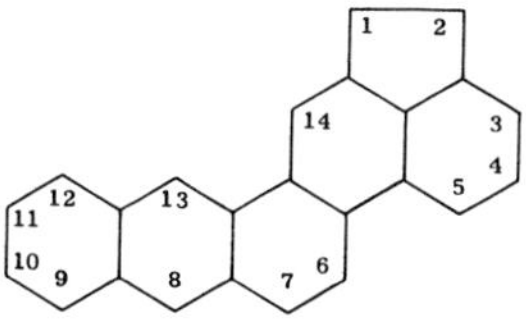
Benzo[*c*]cyclopenta[*hi*]chrysene	C$_{24}$H$_{14}$ (302)	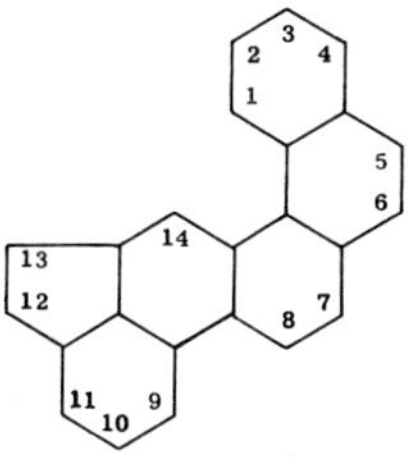
Benzo[*b*]cyclopenta[*qr*]chrysene	C$_{24}$H$_{14}$ (302)	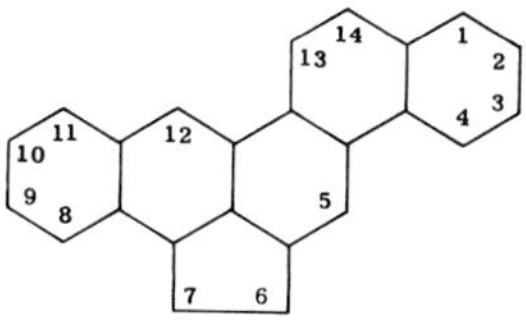
Benzo[*c*]cyclopenta[*qr*]chrysene	C$_{24}$H$_{14}$ (302)	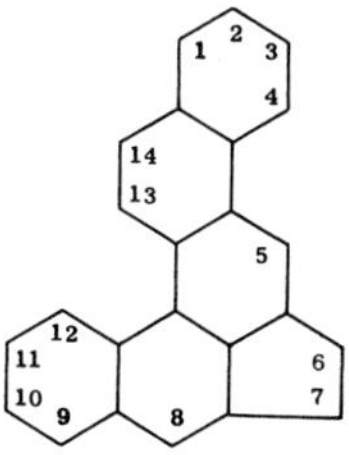

(*Continued*)

Appendix 1 (*Continued*)

Name	Molecular formula (weight)	Structure
Benzo[a]cyclopenta[de]-naphthacene	$C_{24}H_{14}$ (302)	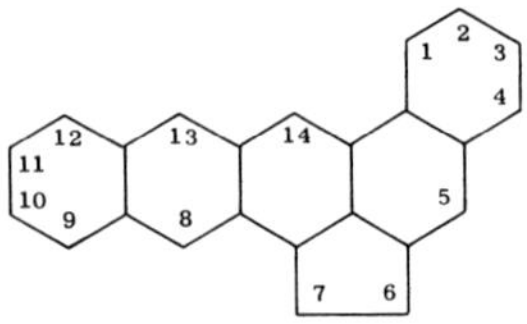
Benzo[a]cyclopenta[fg]-naphthacene	$C_{24}H_{14}$ (302)	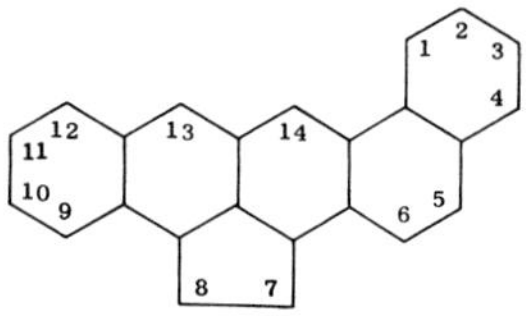
Benzo[a]cyclopenta[hi]-naphthacene	$C_{24}H_{14}$ (302)	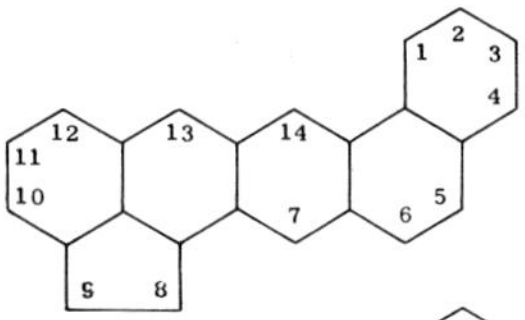
Benzo[a]cyclopenta[mn]-naphthacene	$C_{24}H_{14}$ (302)	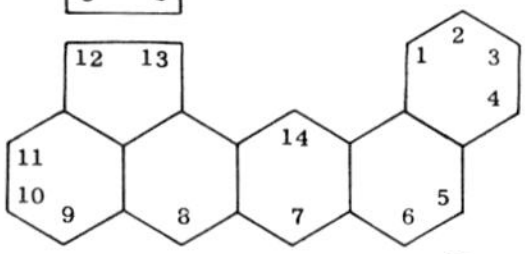
Benzo[a]cyclopenta[op]-naphthacene	$C_{24}H_{14}$ (302)	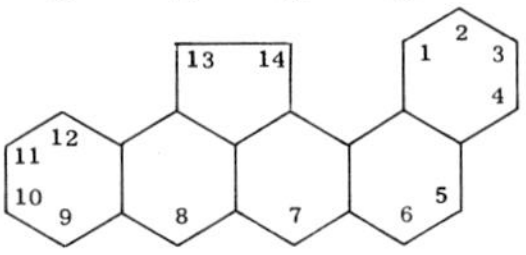
Cyclopenta[de]pentacene	$C_{24}H_{14}$ (302)	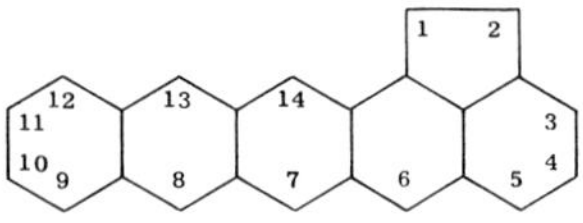
Cyclopenta[fg]pentacene	$C_{24}H_{14}$ (302)	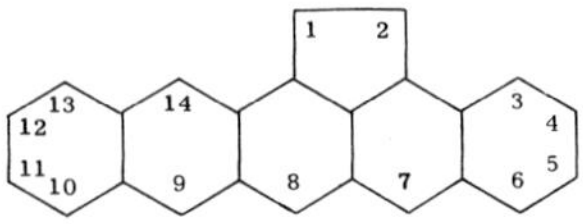
Cyclopenta[de]pentaphene	$C_{24}H_{14}$ (302)	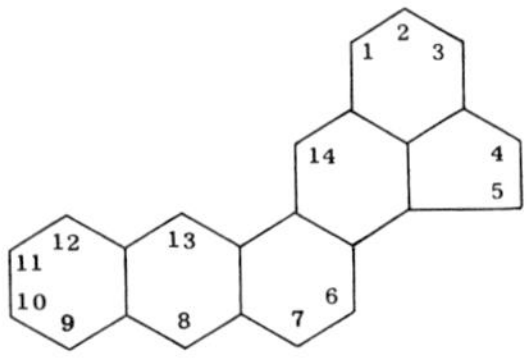

(*Continued*)

Appendix 1 (*Continued*)

Name	Molecular formula (weight)	Structure
Cyclopenta[*fg*]pentaphene	C$_{24}$H$_{14}$ (302)	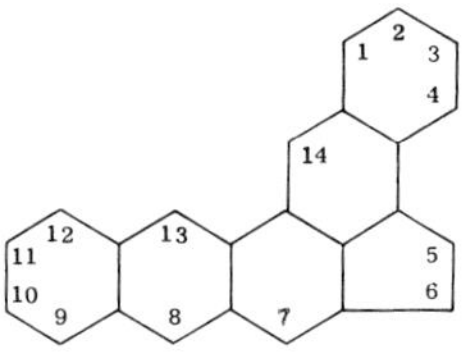
Cyclopenta[*pq*]pentaphene	C$_{24}$H$_{14}$ (302)	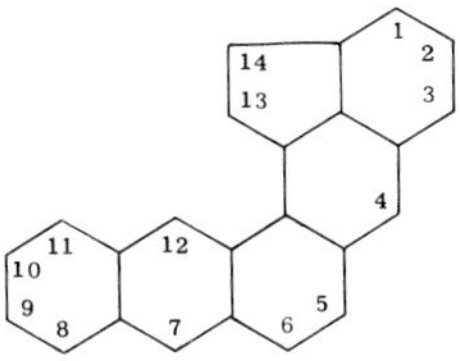
Dibenz[*a,e*]aceanthrylene	C$_{24}$H$_{14}$ (302)	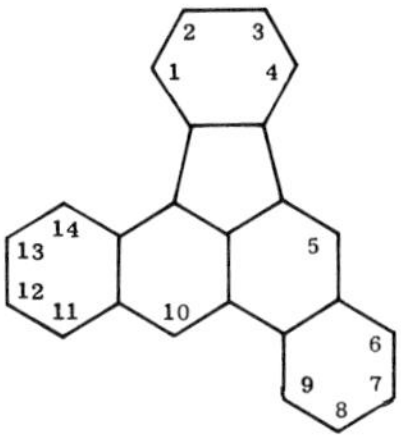
Dibenz[*a,j*]aceanthrylene	C$_{24}$H$_{14}$ (302)	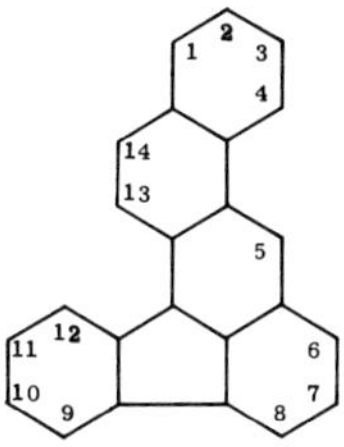
Dibenz[*a,l*]aceanthrylene	C$_{24}$H$_{14}$ (302)	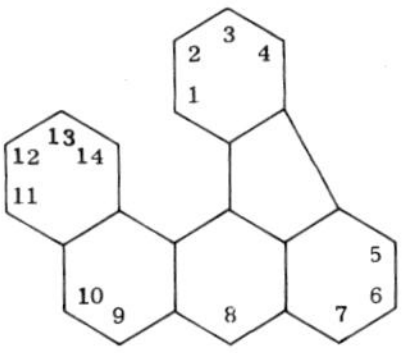

(*Continued*)

Appendix 1 (*Continued*)

Name	Molecular formula (weight)	Structure
Dibenz[*e,j*]aceanthrylene	$C_{24}H_{14}$ (302)	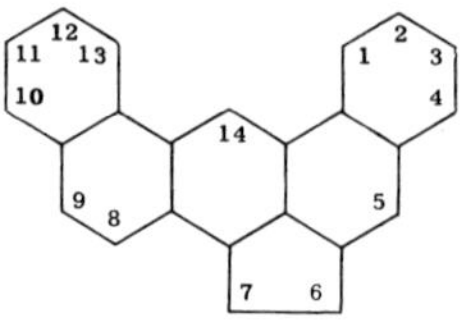
Dibenz[*e,l*]aceanthrylene	$C_{24}H_{14}$ (302)	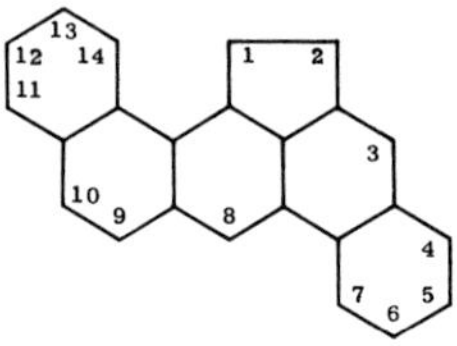
Dibenz[*a,k*]acephenanthrylene	$C_{24}H_{14}$ (302)	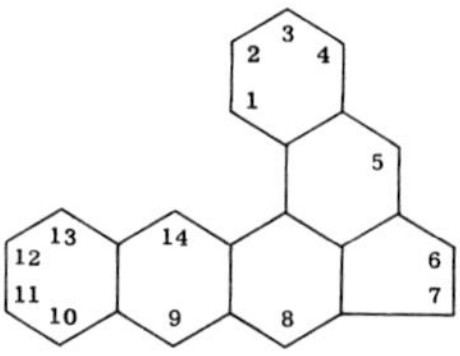
Dibenz[*e,k*]acephenanthrylene	$C_{24}H_{14}$ (302)	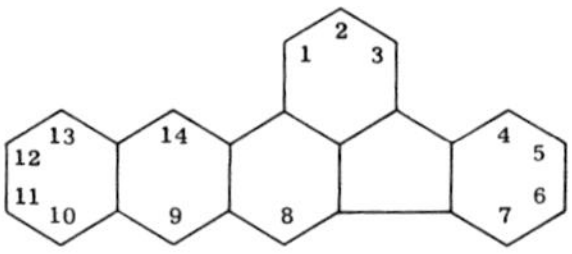
Dibenzo[*b,e*]fluoranthene	$C_{24}H_{14}$ (302)	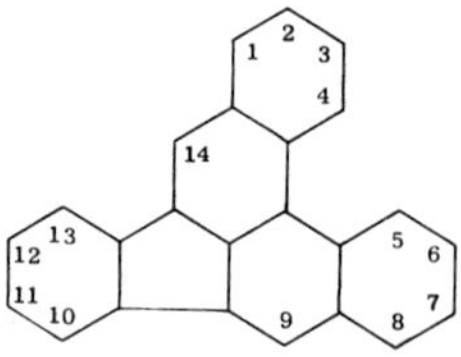
Dibenzo[*b,k*]fluoranthene	$C_{24}H_{14}$ (302)	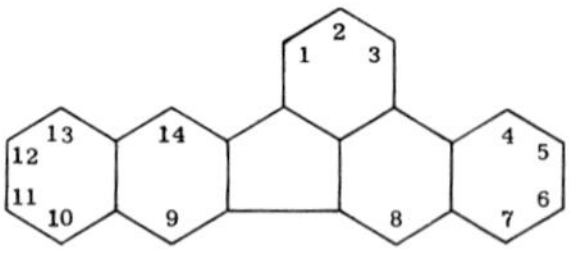
Dibenzo[*j,l*]fluoranthene	$C_{24}H_{14}$ (302)	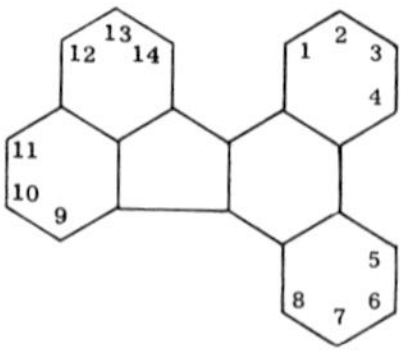

Appendix 1 (*Continued*)

Name	Molecular formula (weight)	Structure
Indeno[1,2,3-*hi*]chrysene	$C_{24}H_{14}$ (302)	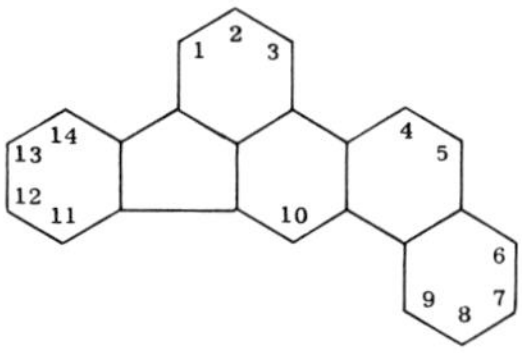
Indeno[1,7-*ab*]chrysene	$C_{24}H_{14}$ (302)	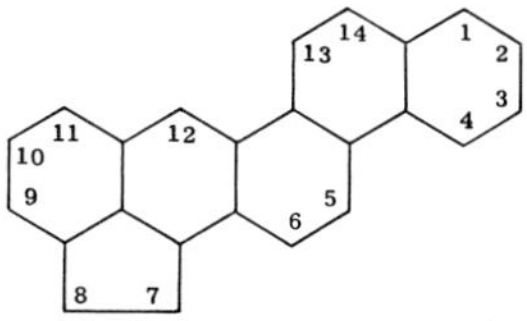
Cyclopenta[*de*]picene	$C_{24}H_{14}$ (302)	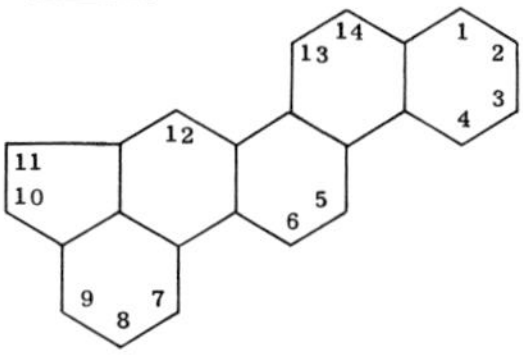
Indeno[7,1-*bc*]chrysene	$C_{24}H_{14}$ (302)	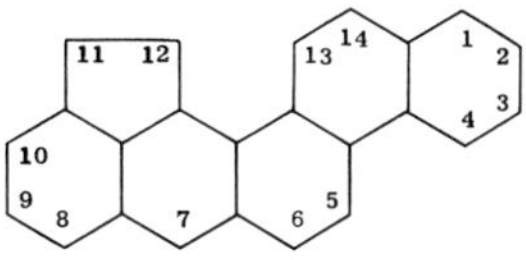
Indeno[1,2,3-*fg*]naphthacene	$C_{24}H_{14}$ (302)	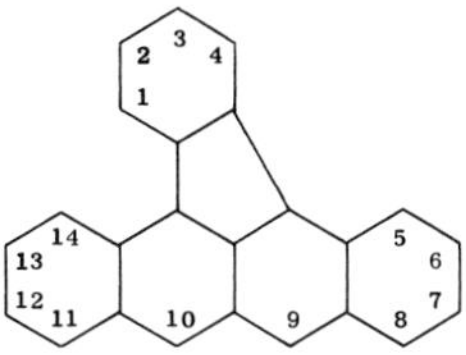
Indeno[1,2,3-*de*]naphthacene	$C_{24}H_{14}$ (302)	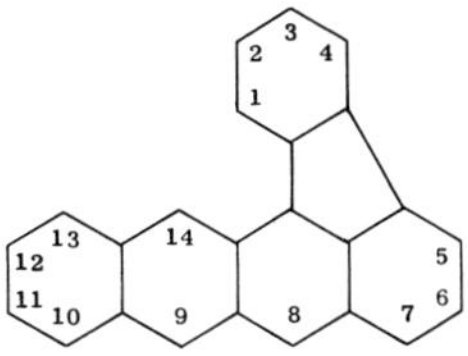
Indeno[7,1-*ab*]naphthacene	$C_{24}H_{14}$ (302)	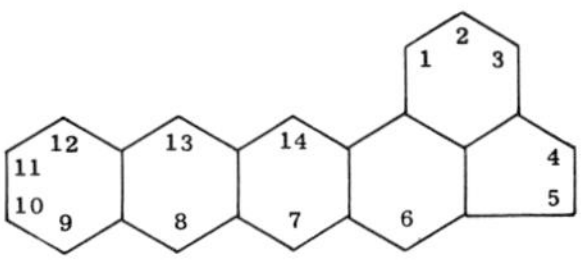

(*Continued*)

Appendix 1 (*Continued*)

Name	Molecular formula (weight)	Structure
Indeno[1,7-*ab*]triphenylene	C$_{24}$H$_{14}$ (302)	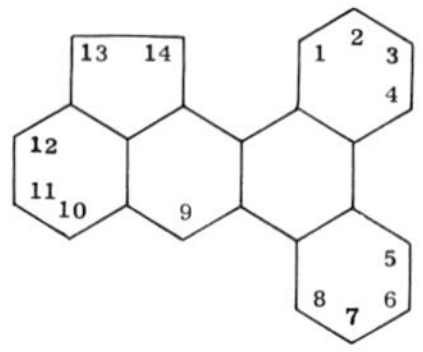
Indeno[7,1-*ab*]triphenylene	C$_{24}$H$_{14}$ (302)	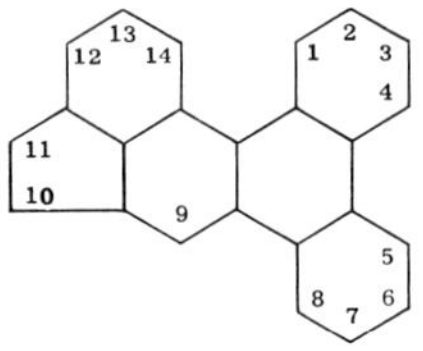
Naphth[1,2-*a*]aceanthrylene	C$_{24}$H$_{14}$ (302)	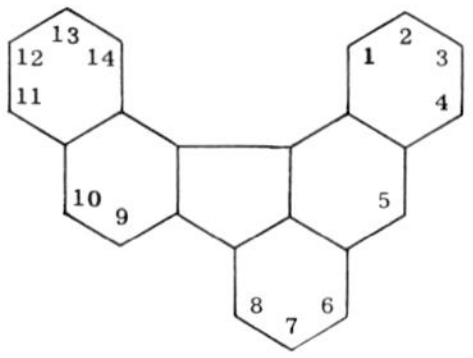
Naphth[1,2-*j*]aceanthrylene	C$_{24}$H$_{14}$ (302)	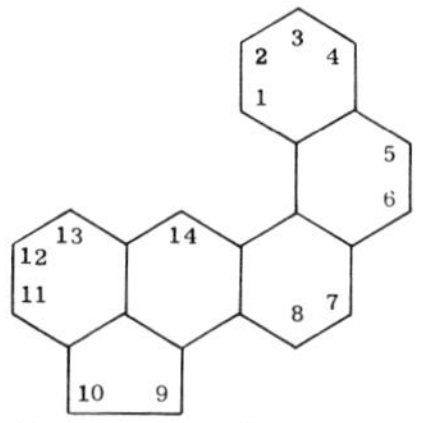
Naphth[2,1-*a*]aceanthrylene	C$_{24}$H$_{14}$ (302)	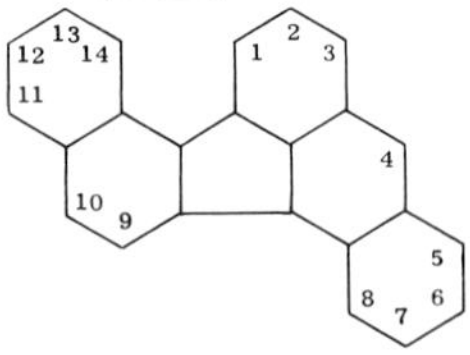
Naphth[2,1-*e*]aceanthrylene	C$_{24}$H$_{14}$ (302)	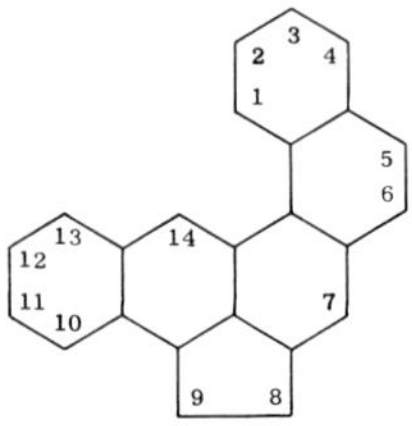

(*Continued*)

Appendix 1 (*Continued*)

Name	Molecular formula (weight)	Structure
Naphth[2,1-*l*]aceanthrylene	C$_{24}$H$_{14}$ (302)	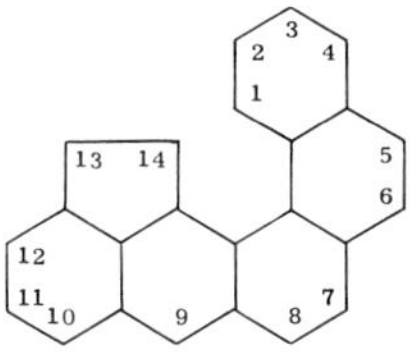
Naphth[2,3-*a*]aceanthrylene	C$_{24}$H$_{14}$ (302)	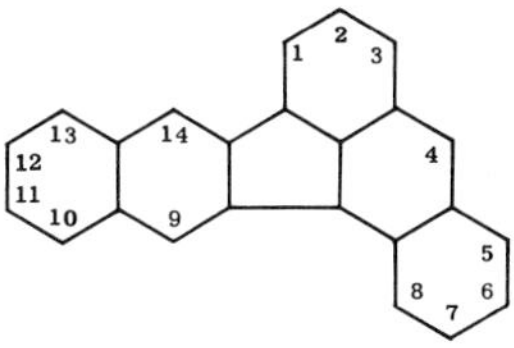
Naphth[1,2-*a*]acephenanthrylene	C$_{24}$H$_{14}$ (302)	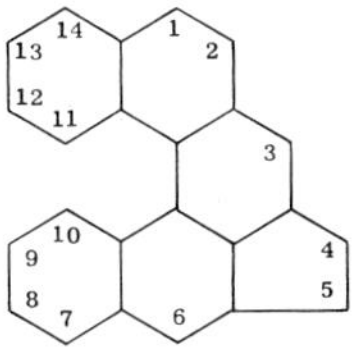
Naphth[1,2-*e*]acephenanthrylene	C$_{24}$H$_{14}$ (302)	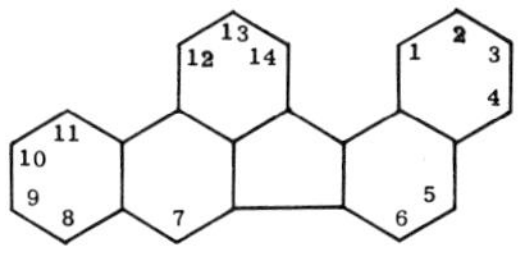
Naphth[1,2-*k*]fluoranthene	C$_{24}$H$_{14}$ (302)	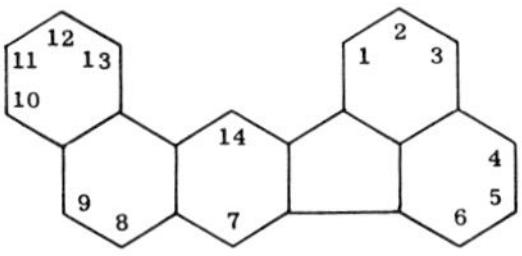
Naphth[2,1-*e*]acephenanthrylene	C$_{24}$H$_{14}$ (302)	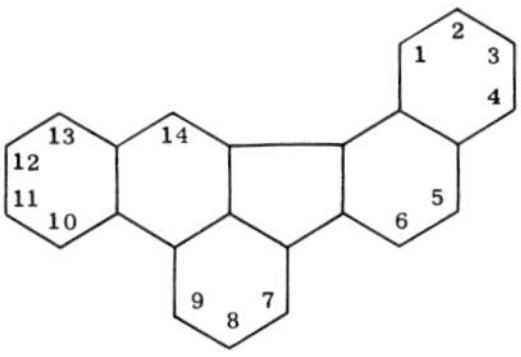
Naphth[2,1-*k*]acephenanthrylene	C$_{24}$H$_{14}$ (302)	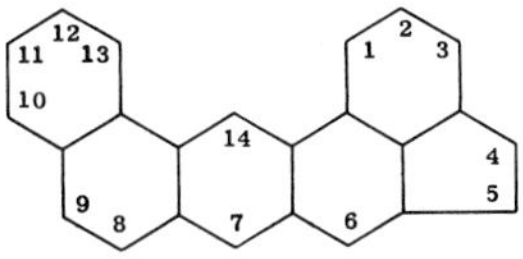

(*Continued*)

Appendix 1 (*Continued*)

Name	Molecular formula (weight)	Structure
Naphth[2,1-*l*]acephenanthrylene	$C_{24}H_{14}$ (302)	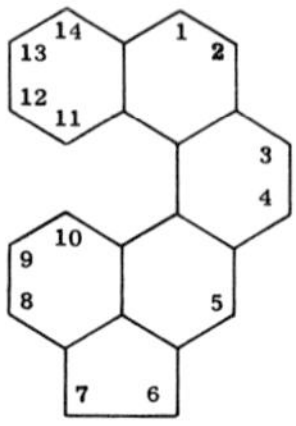
Naphth[2,3-*l*]acephenanthrylene	$C_{24}H_{14}$ (302)	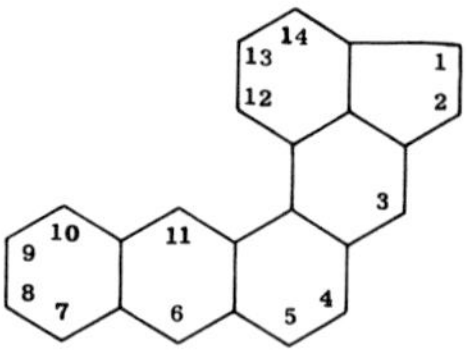
Naphtho[2,1-*b*]fluoranthene	$C_{24}H_{14}$ (302)	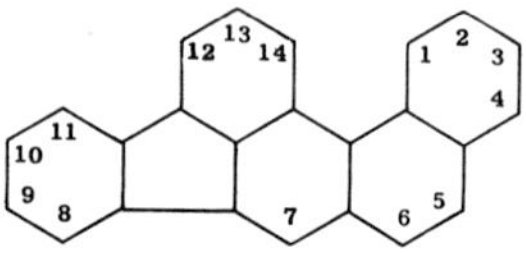
Naphtho[2,1-*j*]fluoranthene	$C_{24}H_{14}$ (302)	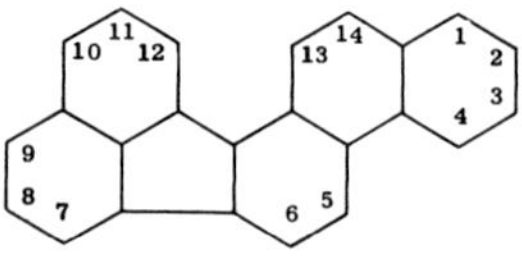
Naphtho[1,2-*k*]fluoranthene	$C_{24}H_{14}$ (302)	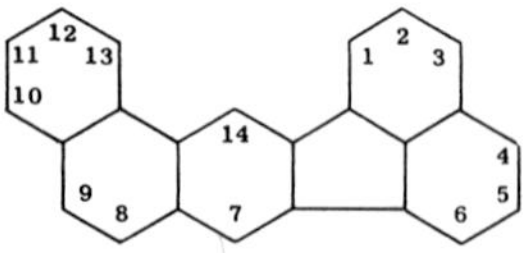
Naphtho[1,2-*j*]fluoranthene	$C_{24}H_{14}$ (302)	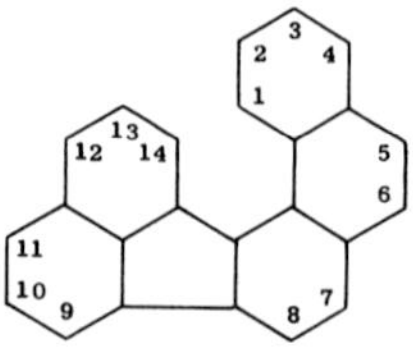

2. Polycyclic Aromatic Heterocycles Containing One Nitrogen Atom: Names, Formulas, Structures, and Numbering

Name	Molecular formula (weight)	Structure
Indole	C_8H_7N (117)	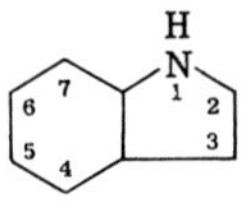
Isoquinoline (Benzo[c]pyridine)	C_9H_7N (129)	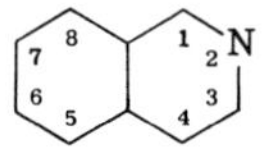
Quinoline (Benzo[b]pyridine)	C_9H_7N (129)	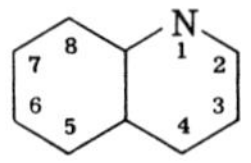
Carbazole	$C_{12}H_9N$ (167)	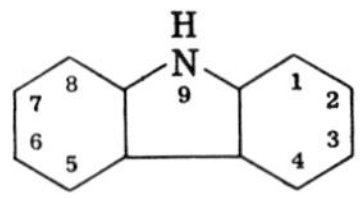
Acridine	$C_{13}H_9N$ (179)	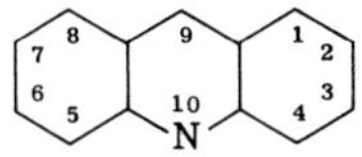
Benz[f]isoquinoline	$C_{13}H_9N$ (179)	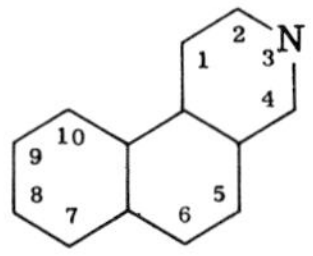
Benzo[f]quinoline	$C_{13}H_9N$ (179)	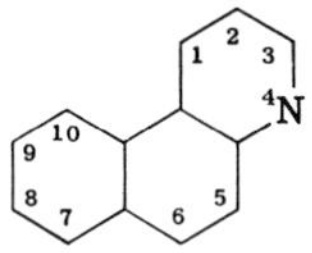
Benz[g]isoquinoline	$C_{13}H_9N$ (179)	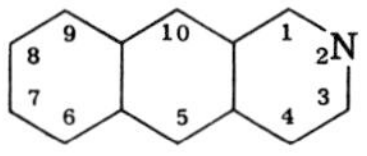
Benzo[g]quinoline	$C_{13}H_9N$ (179)	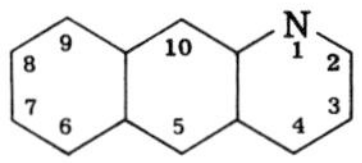
Benz[h]isoquinoline	$C_{13}H_9N$ (179)	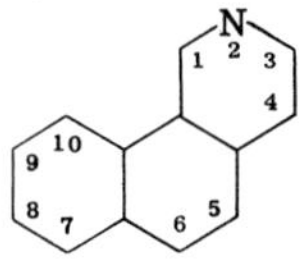

(*Continued*)

Appendix 2 (*Continued*)

Name	Molecular formula (weight)	Structure
Benzo[*h*]quinoline	$C_{13}H_9N$ (179)	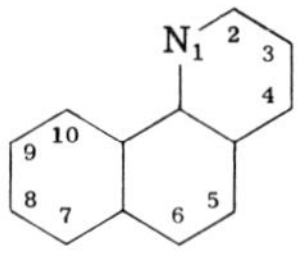
Phenanthridine	$C_{13}H_9N$ (179)	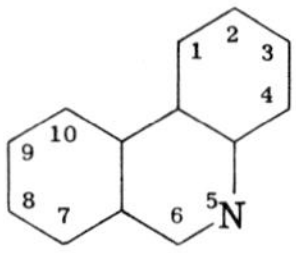
4*H*-Benzo[*def*]carbazole	$C_{14}H_9N$ (191)	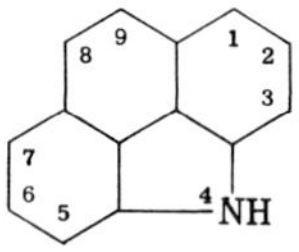
Naphth[2,1,8-*def*]isoquinoline	$C_{15}H_9N$ (203)	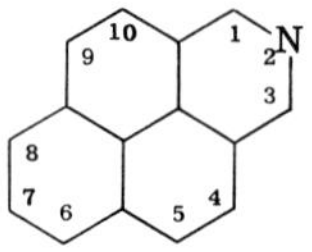
Naphtho[2,1,8-*def*]quinoline	$C_{15}H_9N$ (203)	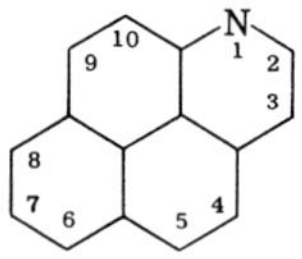
Thebenidine	$C_{15}H_9N$ (203)	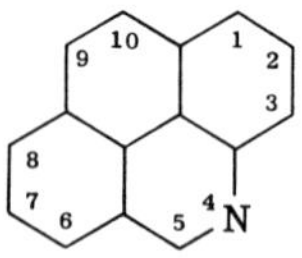
Indeno[1,2,3-*ij*]isoquinoline	$C_{15}H_9N$ (203)	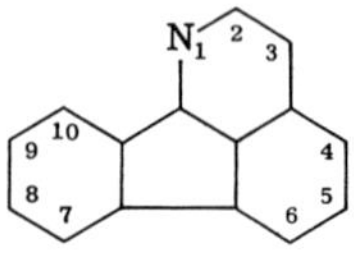
Indeno[1,2,3-*de*]isoquinoline	$C_{15}H_9N$ (203)	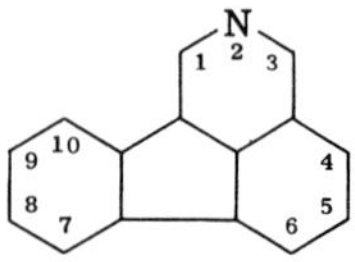

(*Continued*)

Appendix 2 (*Continued*)

Name	Molecular formula (weight)	Structure
Indeno[1,2,3-*de*]quinoline	$C_{15}H_9N$ (203)	
Acenaphtho[1,2-*b*]pyridine	$C_{15}H_9N$ (203)	
Acenaphtho[1,2-*c*]pyridine	$C_{15}H_9N$ (203)	
7*H*-Benzo[*c*]carbazole	$C_{16}H_{11}N$ (217)	
11*H*-Benzo[*a*]carbazole	$C_{16}H_{11}N$ (217)	
5*H*-Benzo[*b*]carbazole	$C_{16}H_{11}N$ (217)	
Benz[*b*]acridine	$C_{17}H_{11}N$ (229)	
Benz[*c*]acridine	$C_{17}H_{11}N$ (229)	

(*Continued*)

Appendix 2 (*Continued*)

Name	Molecular formula (weight)	Structure
Benzo[*a*]phenanthridine	$C_{17}H_{11}N$ (229)	
Benzo[*b*]phenanthridine	$C_{17}H_{11}N$ (229)	
Benzo[*c*]phenanthridine	$C_{17}H_{11}N$ (229)	
Benzo[*i*]phenanthridine	$C_{17}H_{11}N$ (229)	
Benzo[*j*]phenanthridine	$C_{17}H_{11}N$ (229)	
Benzo[*k*]phenanthridine	$C_{17}H_{11}N$ (229)	
Dibenz[*f*,*h*]isoquinoline	$C_{17}H_{11}N$ (229)	

(*Continued*)

Appendix 2 (*Continued*)

Name	Molecular formula (weight)	Structure
Dibenzo[*f*,*h*]quinoline	$C_{17}H_{11}N$ (229)	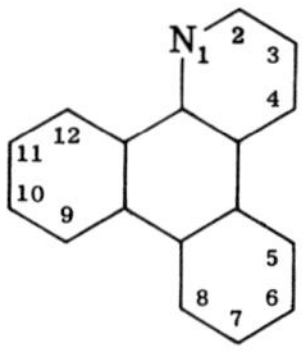
Naphth[1,2-*f*]isoquinoline	$C_{17}H_{11}N$ (229)	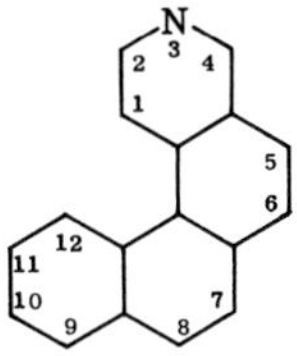
Naphtho[1,2-*f*]quinoline	$C_{17}H_{11}N$ (229)	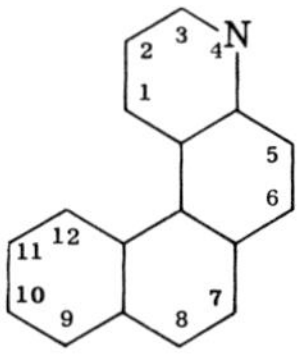
Naphth[1,2-*g*]isoquinoline	$C_{17}H_{11}N$ (229)	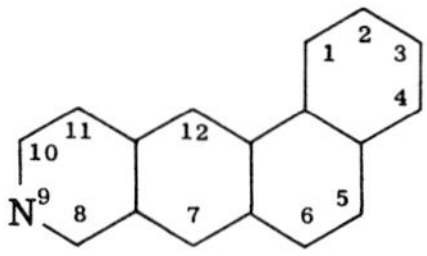
Naphtho[1,2-*g*]quinoline	$C_{17}H_{11}N$ (229)	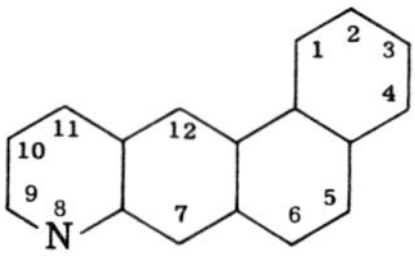
Naphth[1,2-*h*]isoquinoline	$C_{17}H_{11}N$ (229)	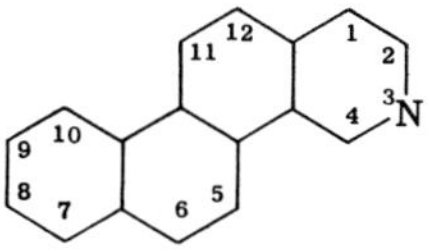
Naphtho[1,2-*h*]quinoline	$C_{17}H_{11}N$ (229)	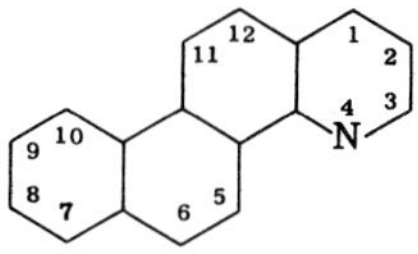

(*Continued*)

Appendix 2 (*Continued*)

Name	Molecular formula (weight)	Structure
Naphth[2,1-*f*]isoquinoline	$C_{17}H_{11}N$ (229)	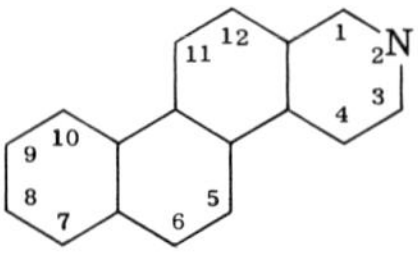
Naphtho[2,1-*f*]quinoline	$C_{17}H_{11}N$ (229)	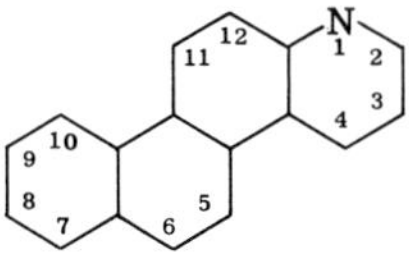
Naphth[2,1-*g*]isoquinoline	$C_{17}H_{11}N$ (229)	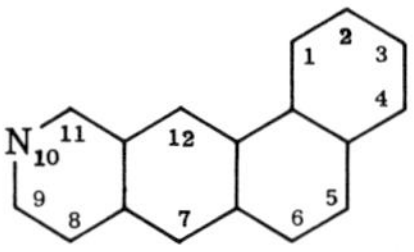
Naphtho[2,1-*g*]quinoline	$C_{17}H_{11}N$ (229)	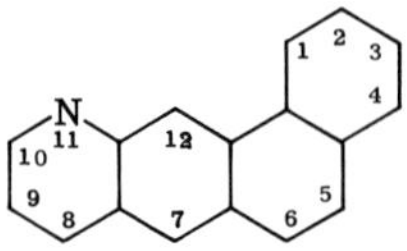
Naphth[2,1-*h*]isoquinoline	$C_{17}H_{11}N$ (229)	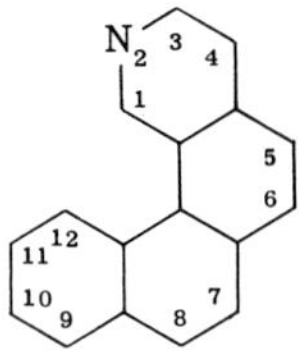
Naphtho[2,1-*h*]quinoline	$C_{17}H_{11}N$ (229)	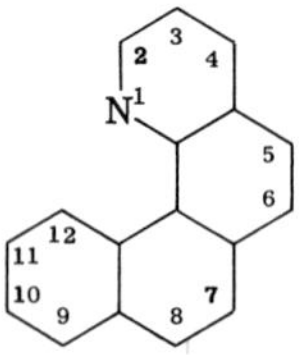
Naphth[2,3-*f*]isoquinoline	$C_{17}H_{11}N$ (229)	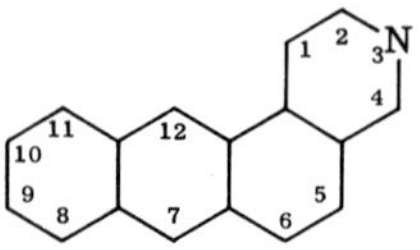
Naphtho[2,3-*f*]quinoline	$C_{17}H_{11}N$ (229)	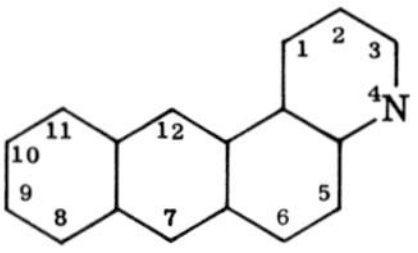

(*Continued*)

Appendix 2 (*Continued*)

Name	Molecular formula (weight)	Structure
Naphth[2,3-*g*]isoquinoline	$C_{17}H_{11}N$ (229)	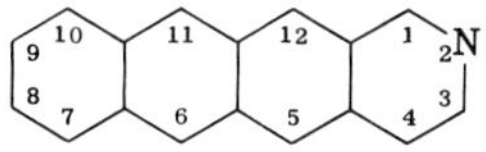
Naphtho[2,3-*g*]quinoline	$C_{17}H_{11}N$ (229)	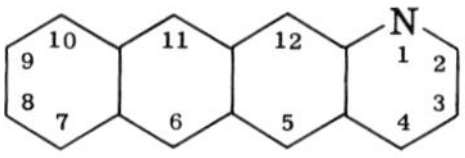
Benz[*a*]acridine	$C_{17}H_{11}N$ (229)	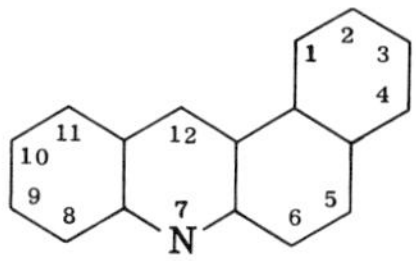
Naphth[2,3-*h*]isoquinoline	$C_{17}H_{11}N$ (229)	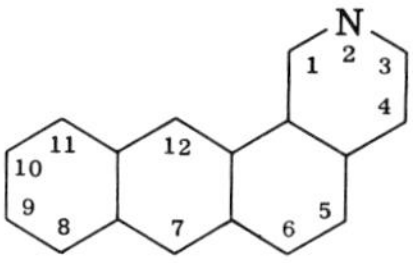
Naphtho[2,3-*h*]quinoline	$C_{17}H_{11}N$ (229)	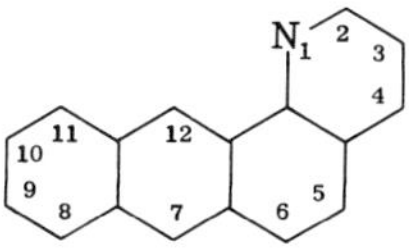
11*H*-Dibenzo[*a,def*]carbazole	$C_{18}H_{11}N$ (241)	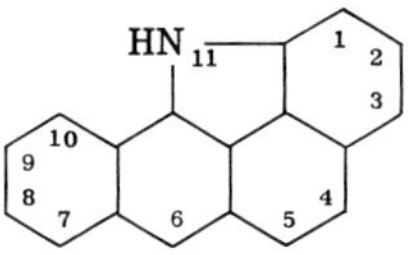
4*H*-Dibenzo[*b,def*]carbazole	$C_{18}H_{11}N$ (241)	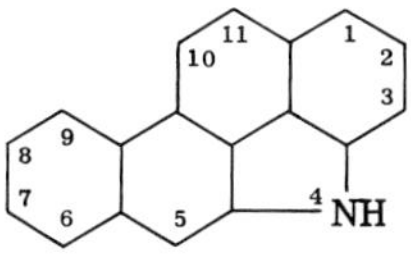
4*H*-Naphtho[1,2,3,4-*def*]-carbazole	$C_{18}H_{11}N$ (241)	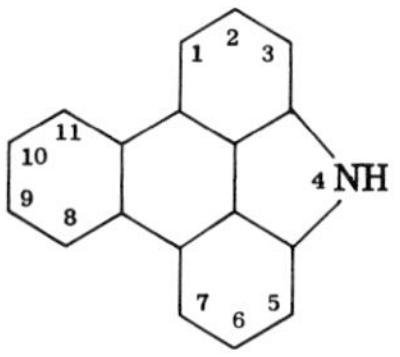

(*Continued*)

Appendix 2 (*Continued*)

Name	Molecular formula (weight)	Structure
Phenanthro[1,10,9-*hij*]iso-quinoline	$C_{19}H_{11}N$ (253)	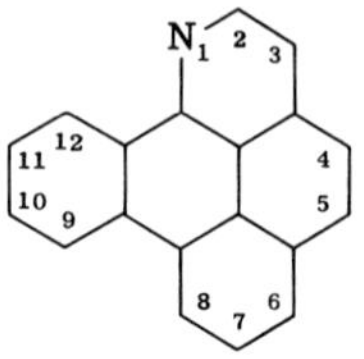
Phenanthro[4,5-*fgh*]isoquinoline	$C_{19}H_{11}N$ (253)	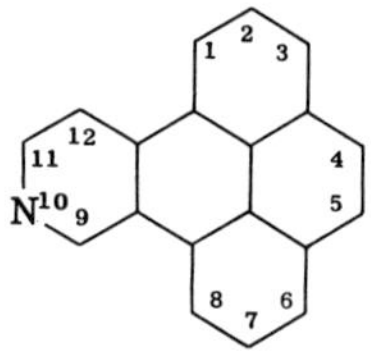
Phenanthro[4,5-*fgh*]quinoline	$C_{19}H_{11}N$ (253)	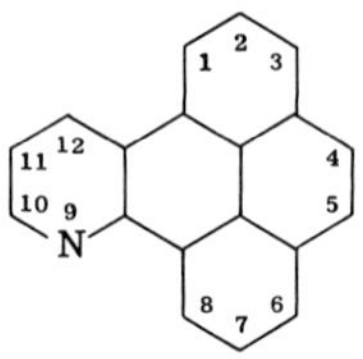
Phenanthro[9,10,1-*def*]iso-quinoline	$C_{19}H_{11}N$ (253)	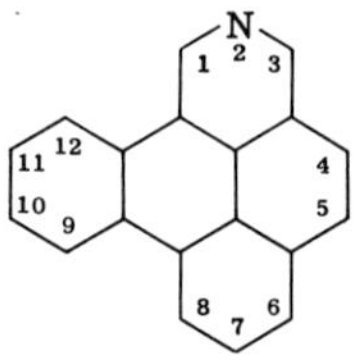
Phenanthro[9,10,1-*def*]quinoline	$C_{19}H_{11}N$ (253)	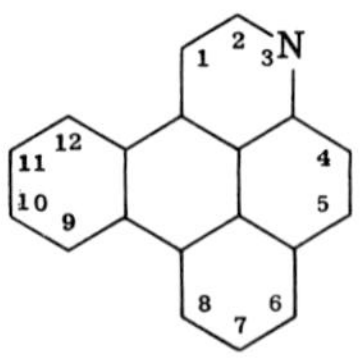
Anthra[9,1,2-*cde*]quinoline	$C_{19}H_{11}N$ (253)	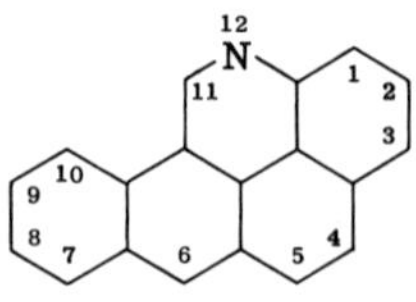

(*Continued*)

Appendix 2 (*Continued*)

Name	Molecular formula (weight)	Structure
Anthra[9,1,2-*hij*]isoquinoline	$C_{19}H_{11}N$ (253)	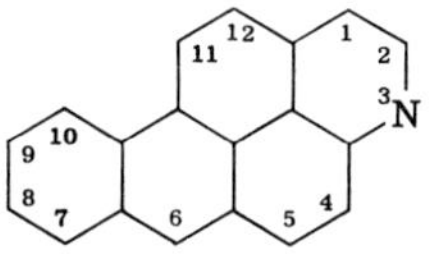
Dibenzo[*b,lmn*]phenanthridine	$C_{19}H_{11}N$ (253)	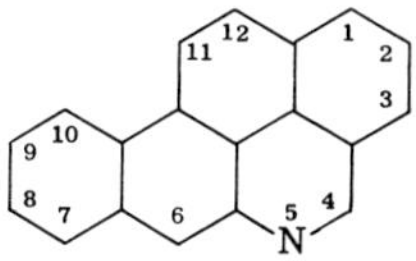
Dibenzo[*j,lmn*]phenanthridine	$C_{19}H_{11}N$ (253)	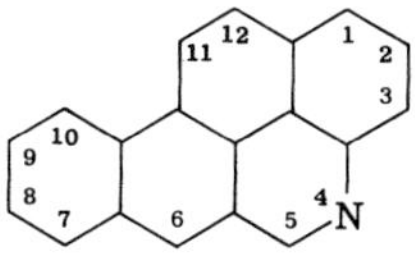
Naphtho[1,2,3,4-*lmn*]phenan-thridine	$C_{19}H_{11}N$ (253)	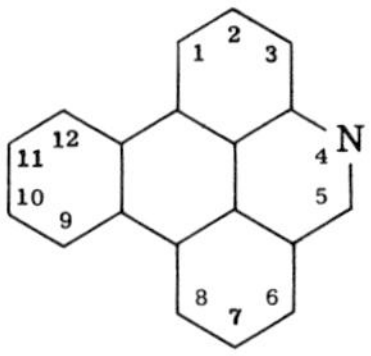
Phenaleno[1,2,3-*de*]isoquinoline	$C_{19}H_{11}N$ (253)	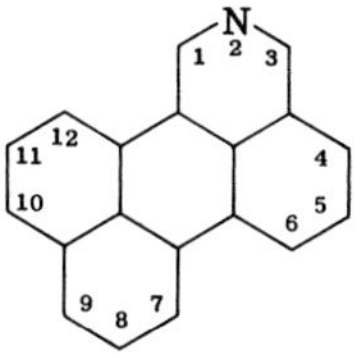
Phenaleno[1,2,3-*de*]quinoline	$C_{19}H_{11}N$ (253)	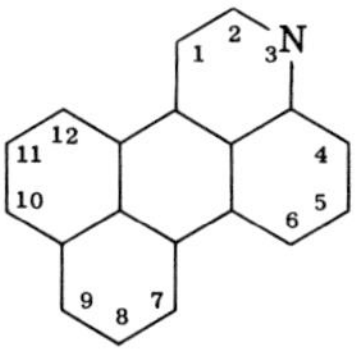
Phenaleno[1,2,3-*ij*]isoquinoline	$C_{19}H_{11}N$ (253)	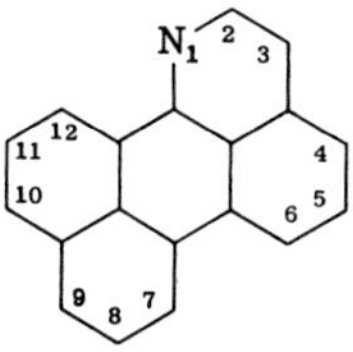

(*Continued*)

Appendix 2 (*Continued*)

Name	Molecular formula (weight)	Structure
Phenaleno[1,9-*fg*]quinoline	$C_{19}H_{11}N$ (253)	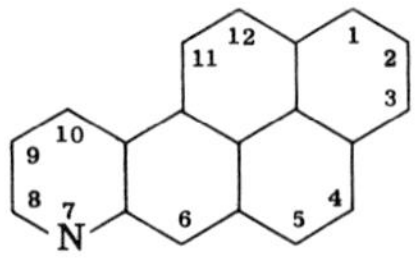
Phenaleno[1,9-*fg*]isoquinoline	$C_{19}H_{11}N$ (253)	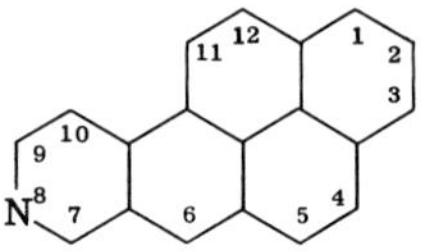
Phenaleno[1,9-*gh*]isoquinoline	$C_{19}H_{11}N$ (253)	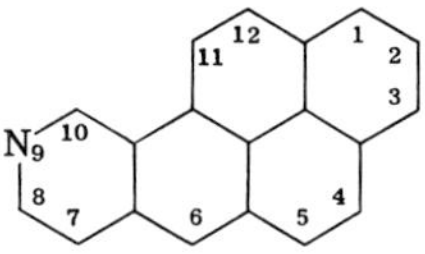
Phenaleno[1,9-*gh*]quinoline	$C_{19}H_{11}N$ (253)	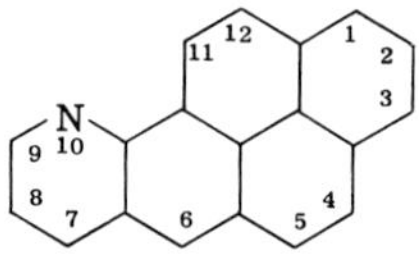
Naphth[2,1,8-*mna*]acridine	$C_{19}H_{11}N$ (253)	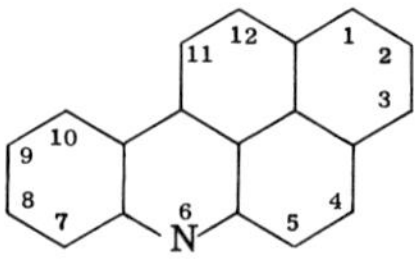
Anthra[2,1,9-*def*]isoquinoline	$C_{19}H_{11}N$ (253)	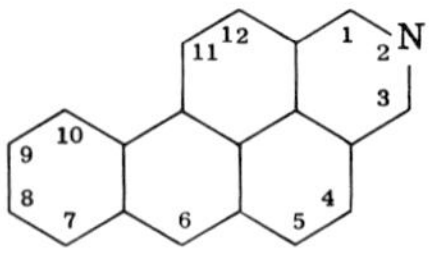
Anthra[2,1,9-*def*]quinoline	$C_{19}H_{11}N$ (253)	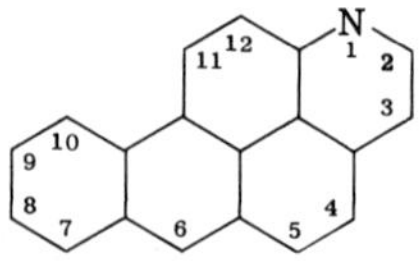
Anthra[9,1,2-*cde*]isoquinoline	$C_{19}H_{11}N$ (253)	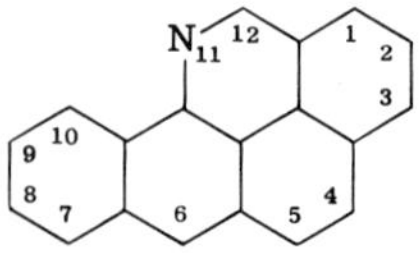

(Continued)

Appendix 2 (*Continued*)

Name	Molecular formula (weight)	Structure
10*H*-Phenanthro[2,3,4,5-*defgh*]-carbazole	$C_{20}H_{11}N$ (265)	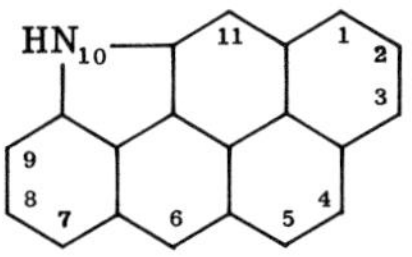
1*H*-Phenanthro[1,10,9,8-*cdefg*]carbazole	$C_{20}H_{11}N$ (265)	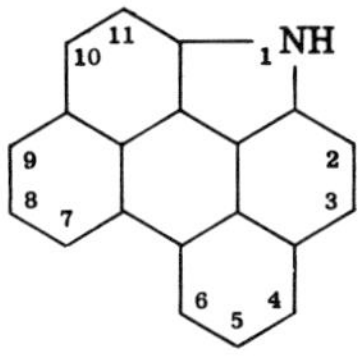
13*H*-Dibenzo[*a,i*]carbazole	$C_{20}H_{13}N$ (267)	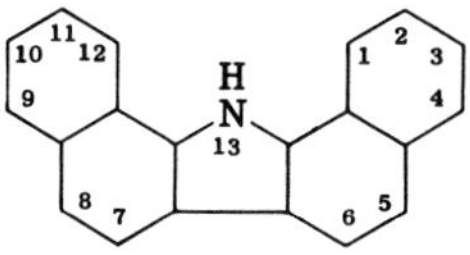
13*H*-Dibenzo[*a,h*]carbazole	$C_{20}H_{13}N$ (267)	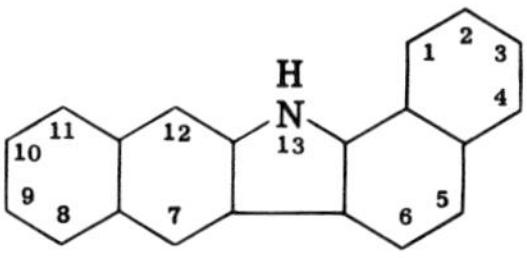
7*H*-Dibenzo[*c,g*]carbazole	$C_{20}H_{13}N$ (267)	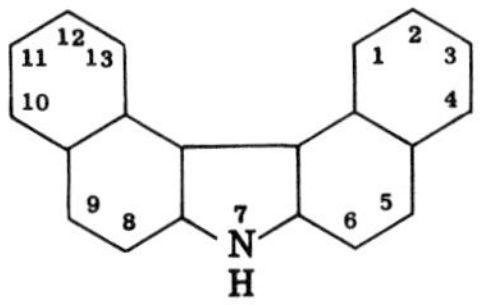
7*H*-Dibenzo[*b,g*]carbazole	$C_{20}H_{13}N$ (267)	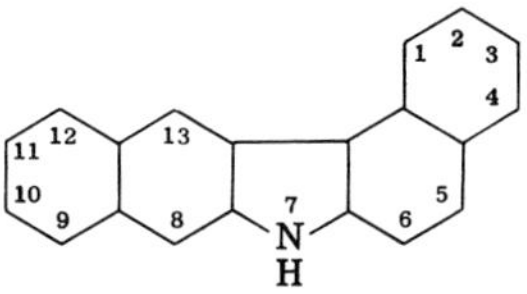
6*H*-Dibenzo[*b,h*]carbazole	$C_{20}H_{13}N$ (267)	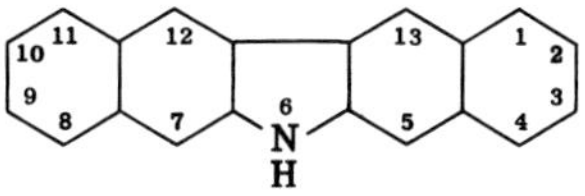

(*Continued*)

Appendix 2 (*Continued*)

Name	Molecular formula (weight)	Structure
12*H*-Naphtho[2,3-*a*]carbazole	$C_{20}H_{13}N$ (267)	
5*H*-Naphtho[2,3-*c*]carbazole	$C_{20}H_{13}N$ (267)	
11*H*-Naphtho[2,1-*a*]carbazole	$C_{20}H_{13}N$ (267)	
12*H*-Naphtho[1,2-*b*]carbazole	$C_{20}H_{13}N$ (267)	
8*H*-Naphtho[2,1-*b*]carbazole	$C_{20}H_{13}N$ (267)	
13*H*-Naphtho[1,2-*a*]carbazole	$C_{20}H_{13}N$ (267)	
9*H*-Naphtho[2,1-*c*]carbazole	$C_{20}H_{13}N$ (267)	

(*Continued*)

Appendix 2 (*Continued*)

Name	Molecular formula (weight)	Structure
9*H*-Dibenzo[*a,c*]carbazole	$C_{20}H_{13}N$ (267)	
7*H*-Dibenzo[*a,g*]carbazole	$C_{20}H_{13}N$ (267)	
5*H*-Naphtho[2,3-*b*]carbazole	$C_{20}H_{13}N$ (267)	
7*H*-Naphtho[1,2-*c*]carbazole	$C_{20}H_{13}N$ (267)	
Phenaleno[1,2,3,4-*lmna*]-phenanthridine	$C_{21}H_{11}N$ (277)	
Phenaleno[4,3,2,1-*klmn*]-phenanthridine	$C_{21}H_{11}N$ (277)	

(*Continued*)

Appendix 2 (*Continued*)

Name	Molecular formula (weight)	Structure
Phenanthro[8,9,10,1-*klmna*]-phenanthridine	$C_{21}H_{11}N$ (277)	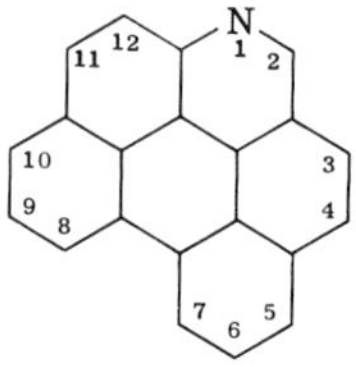
Pyreno[3,4,5-*hij*]isoquinoline	$C_{21}H_{11}N$ (277)	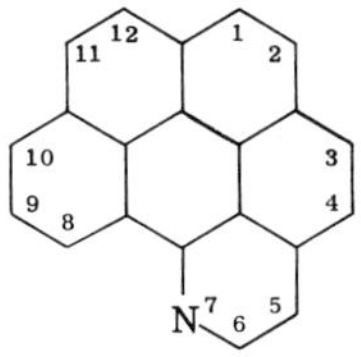
Phenanthro[4,3-*h*]quinoline	$C_{21}H_{13}N$ (279)	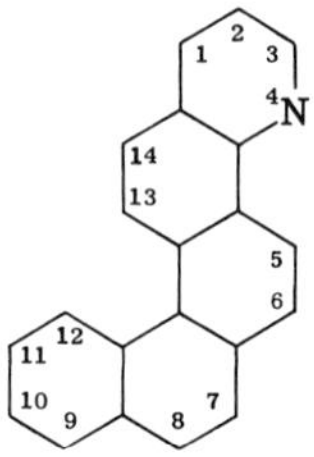
Phenanthro[9,10-*g*]isoquinoline	$C_{21}H_{13}N$ (279)	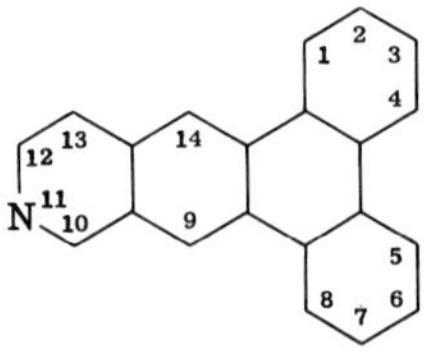
Pyreno[5,4,3-*def*]isoquinoline	$C_{21}H_{11}N$ (277)	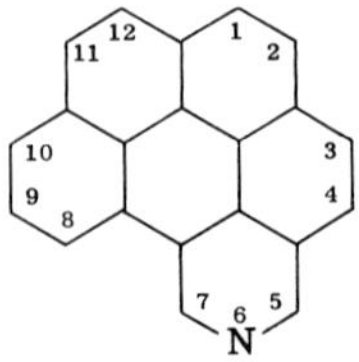
Pyreno[5,4,3-*def*]quinoline	$C_{21}H_{11}N$ (277)	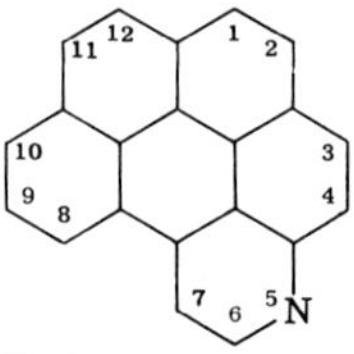

Appendix 2 (*Continued*)

Name	Molecular formula (weight)	Structure
Phenanthro[9,10-*g*]quinoline	$C_{21}H_{13}N$ (279)	
Anthra[2,1,9,8-*klmna*]acridine	$C_{21}H_{11}N$ (277)	
Pyreno[2,3,4-*def*]isoquinoline	$C_{21}H_{11}N$ (277)	
Pyreno[2,3,4-*def*]quinoline	$C_{21}H_{11}N$ (277)	
Pyreno[4,3,2-*cde*]isoquinoline	$C_{21}H_{11}N$ (277)	
Pyreno[4,3,2-*cde*]quinoline	$C_{21}H_{11}N$ (277)	
Pyreno[4,3,2-*hij*]isoquinoline	$C_{21}H_{11}N$ (277)	
Phenanthro[9,10-*h*]quinoline	$C_{21}H_{13}N$ (279)	

Appendix 2 (*Continued*)

Name	Molecular formula (weight)	Structure
Phenanthro[9,10-*f*]isoquinoline	$C_{21}H_{13}N$ (279)	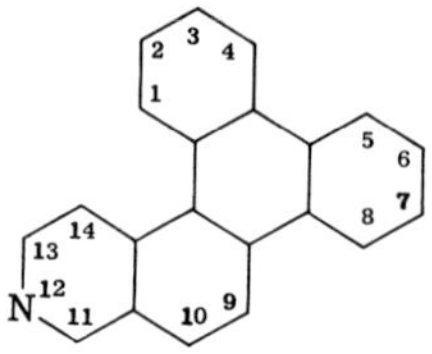
Phenanthro[9,10-*f*]quinoline	$C_{21}H_{13}N$ (279)	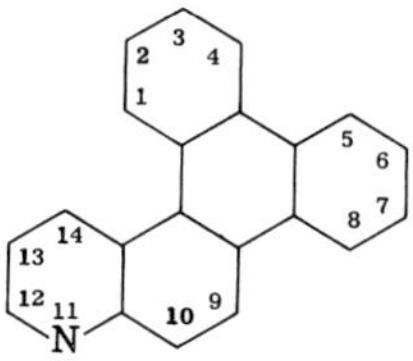
Naphtho[2,3-*c*]phenanthridine	$C_{21}H_{13}N$ (279)	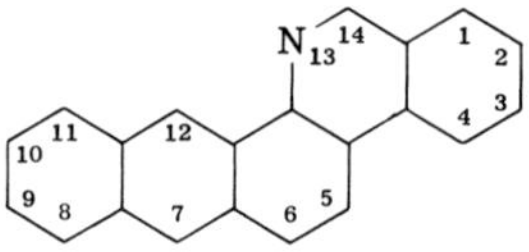
Naphtho[2,3-*i*]phenanthridine	$C_{21}H_{13}N$ (279)	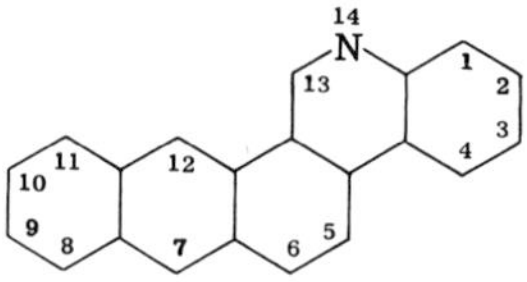
Anthra[1,2-*f*]isoquinoline	$C_{21}H_{13}N$ (279)	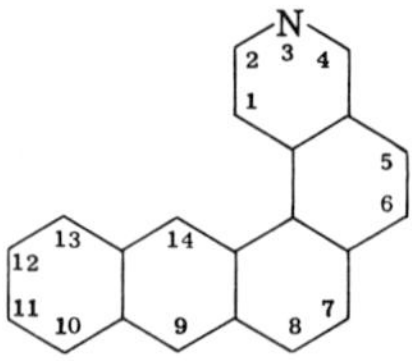
Anthra[1,2-*f*]quinoline	$C_{21}H_{13}N$ (279)	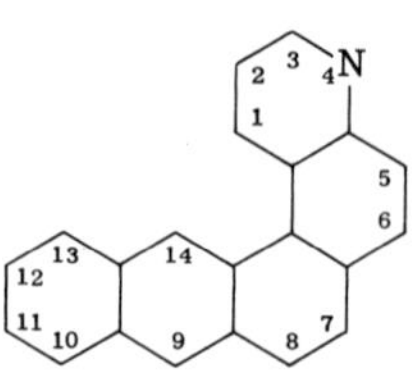

(*Continued*)

Appendix 2 (*Continued*)

Name	Molecular formula (weight)	Structure
Anthra[1,2-*g*]isoquinoline	$C_{21}H_{13}N$ (279)	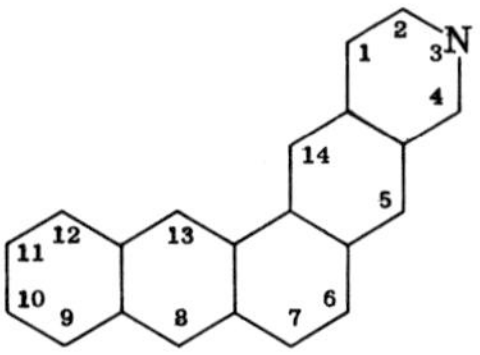
Anthra[1,2-*g*]quinoline	$C_{21}H_{13}N$ (279)	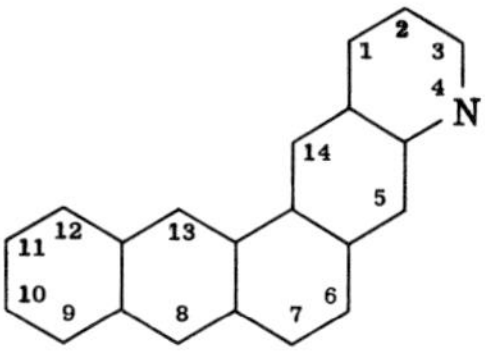
Anthra[1,2-*h*]isoquinoline	$C_{21}H_{13}N$ (279)	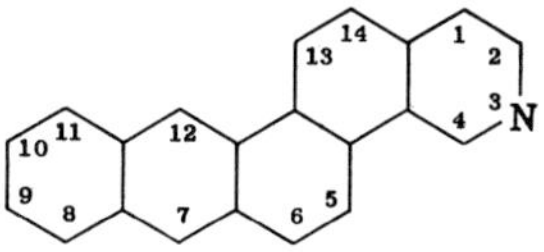
Anthra[1,2-*h*]quinoline	$C_{21}H_{13}N$ (279)	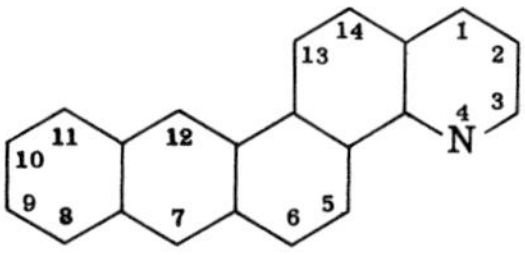
Anthra[2,1-*f*]isoquinoline	$C_{21}H_{13}N$ (279)	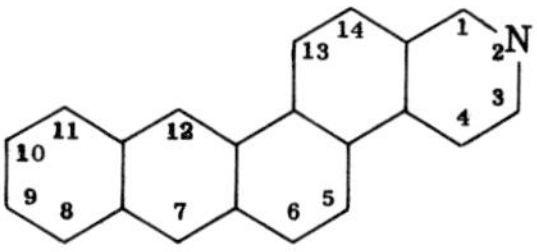
Anthra[2,1-*f*]quinoline	$C_{21}H_{13}N$ (279)	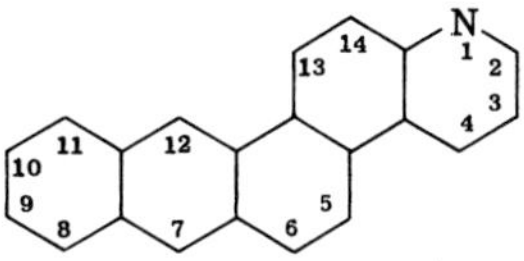
Anthra[2,1-*g*]isoquinoline	$C_{21}H_{13}N$ (279)	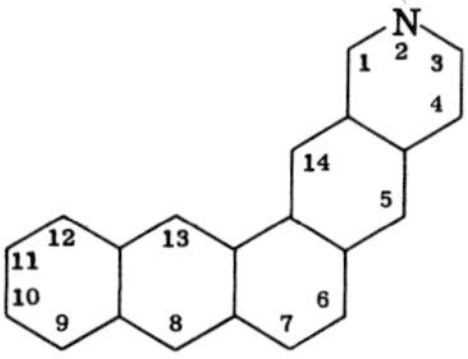

(*Continued*)

Appendix 2 (*Continued*)

Name	Molecular formula (weight)	Structure
Anthra[2,1-*g*]quinoline	$C_{21}H_{13}N$ (279)	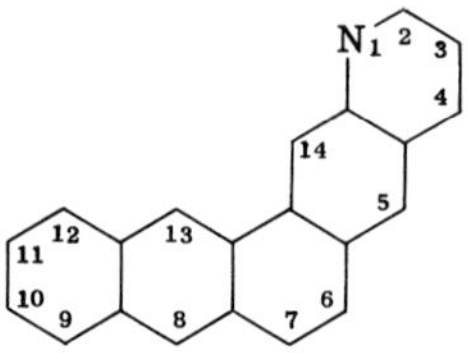
Anthra[2,1-*h*]isoquinoline	$C_{21}H_{13}N$ (279)	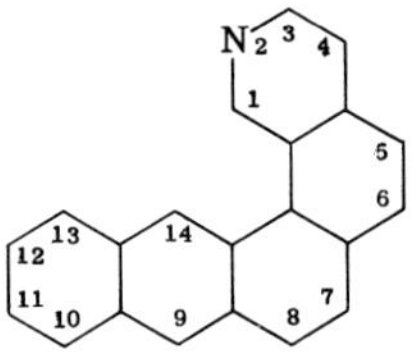
Anthra[2,1-*h*]quinoline	$C_{21}H_{13}N$ (279)	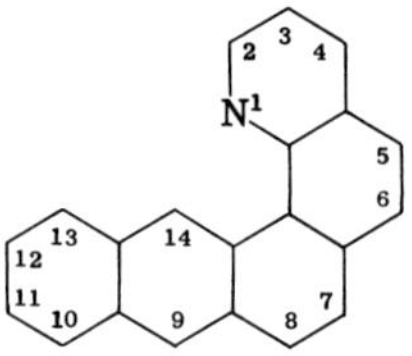
Anthra[2,3-*f*]isoquinoline	$C_{21}H_{13}N$ (279)	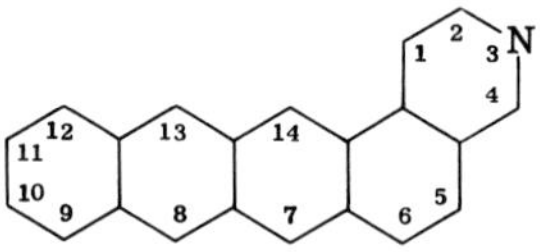
Anthra[2,3-*f*]quinoline	$C_{21}H_{13}N$ (279)	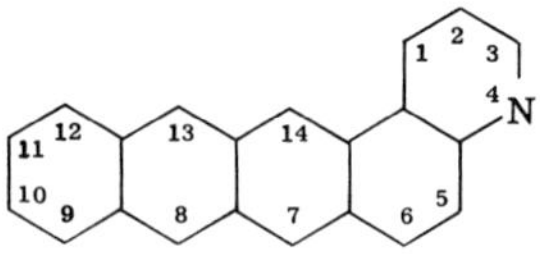
Anthra[2,3-*g*]isoquinoline	$C_{21}H_{13}N$ (279)	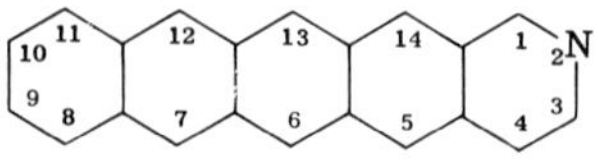
Anthra[2,3-*h*]isoquinoline	$C_{21}H_{13}N$ (279)	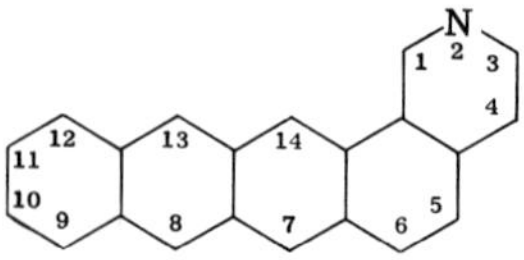
Anthra[2,3-*g*]quinoline	$C_{21}H_{13}N$ (279)	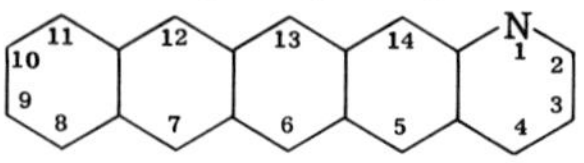

(*Continued*)

Appendix 2 (*Continued*)

Name	Molecular formula (weight)	Structure
Anthra[2,3-*h*]quinoline	$C_{21}H_{13}N$ (279)	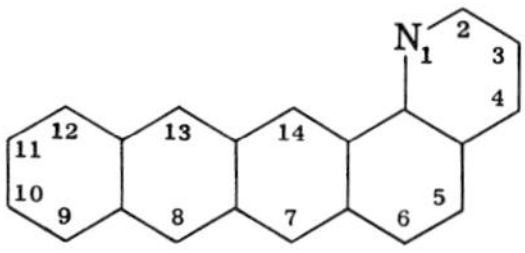
Benzo[*f*]naphth[1,2-*h*]-isoquinoline	$C_{21}H_{13}N$ (279)	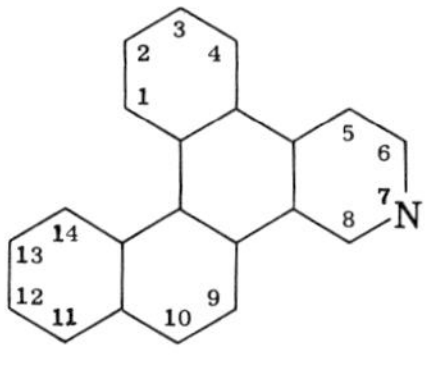
Benzo[*f*]naphtho[1,2-*h*]quinoline	$C_{21}H_{13}N$ (279)	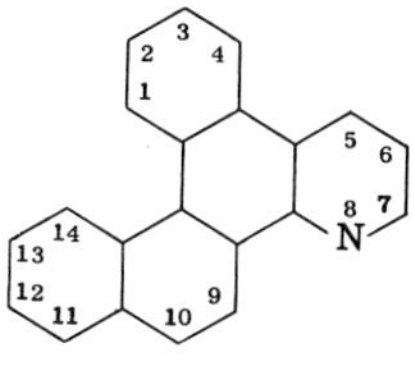
Benzo[*f*]naphth[2,1-*h*]-isoquinoline	$C_{21}H_{13}N$ (279)	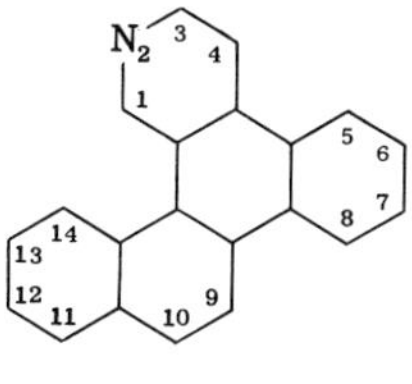
Benzo[*f*]naphtho[2,1-*h*]quinoline	$C_{21}H_{13}N$ (279)	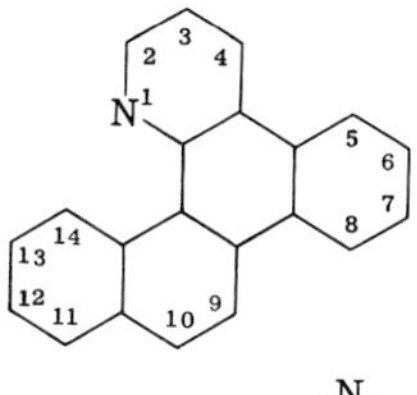
Benzo[*f*]naphth[2,3-*h*]-isoquinoline	$C_{21}H_{13}N$ (279)	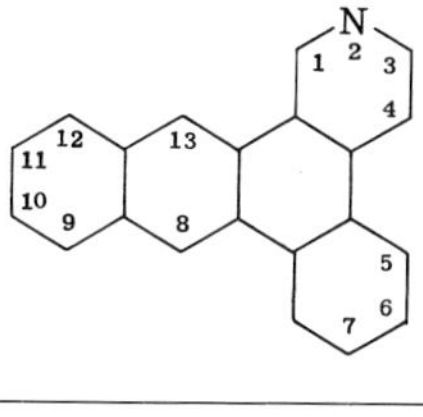

(*Continued*)

Appendix 2 (*Continued*)

Name	Molecular formula (weight)	Structure
Phenanthro[3,2-*f*]quinoline	$C_{21}H_{13}N$ (279)	
Phenanthro[3,2-*g*]isoquinoline	$C_{21}H_{13}N$ (279)	
Phenanthro[3,2-*g*]quinoline	$C_{21}H_{13}N$ (279)	
Phenanthro[3,2-*h*]isoquinoline	$C_{21}H_{13}N$ (279)	
Phenanthro[3,2-*h*]quinoline	$C_{21}H_{13}N$ (279)	
Naphtho[1,2-*a*]phenanthridine	$C_{21}H_{13}N$ (279)	
Phenanthro[2,1-*f*]quinoline	$C_{21}H_{13}N$ (279)	

Appendix 2 (*Continued*)

Name	Molecular formula (weight)	Structure
Phenanthro[2,1-*g*]isoquinoline	$C_{21}H_{13}N$ (279)	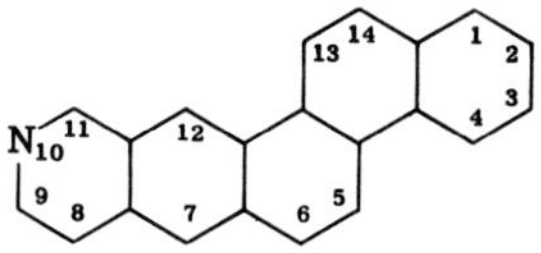
Phenanthro[2,1-*g*]quinoline	$C_{21}H_{13}N$ (279)	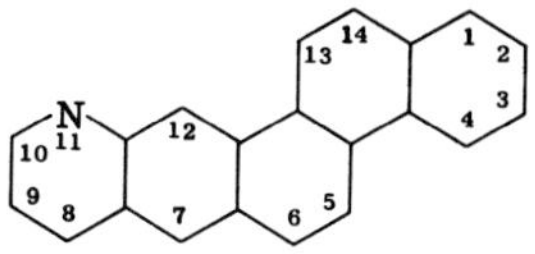
Phenanthro[2,1-*h*]isoquinoline	$C_{21}H_{13}N$ (279)	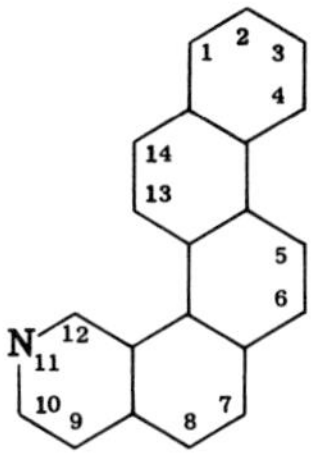
Phenanthro[2,1-*h*]quinoline	$C_{21}H_{13}N$ (279)	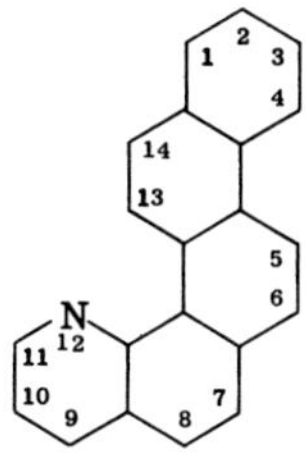
Phenanthro[2,3-*f*]isoquinoline	$C_{21}H_{13}N$ (279)	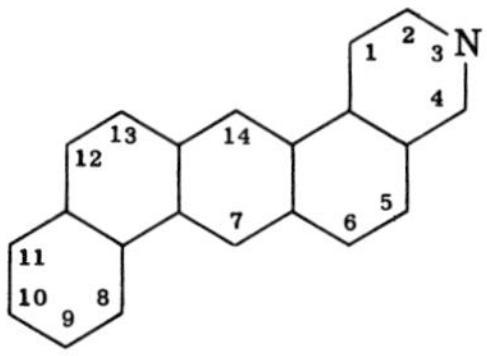
Phenanthro[2,3-*f*]quinoline	$C_{21}H_{13}N$ (279)	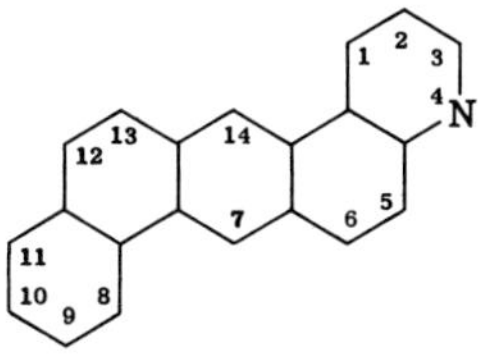

(*Continued*)

Appendix 2 (*Continued*)

Name	Molecular formula (weight)	Structure
Phenanthro[2,3-*g*]isoquinoline	$C_{21}H_{13}N$ (279)	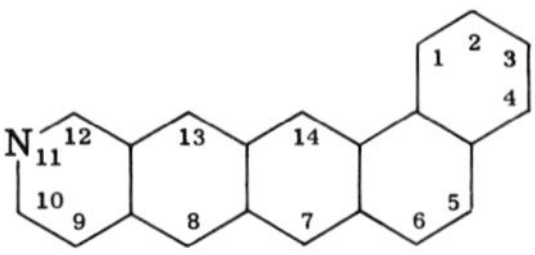
Phenanthro[2,3-*g*]quinoline	$C_{21}H_{13}N$ (279)	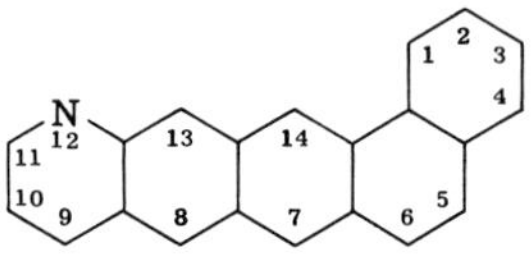
Phenanthro[2,3-*h*]isoquinoline	$C_{21}H_{13}N$ (279)	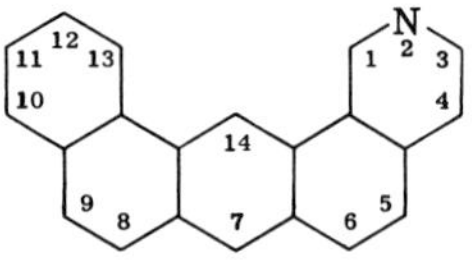
Phenanthro[2,3-*h*]quinoline	$C_{21}H_{13}N$ (279)	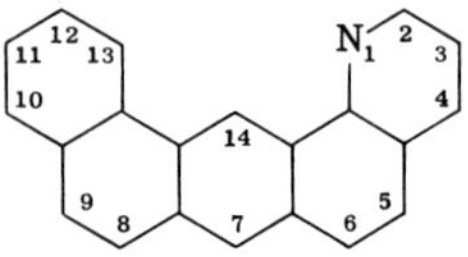
Phenanthro[3,2-*f*]isoquinoline	$C_{21}H_{13}N$ (279)	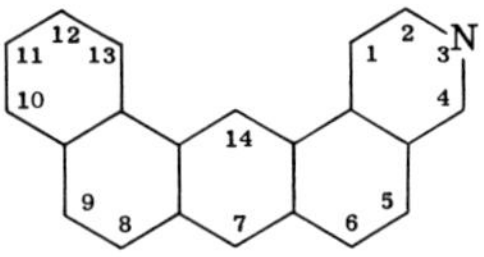
Benzo[*f*]naphtho[2,3-*h*]quinoline	$C_{21}H_{13}N$ (279)	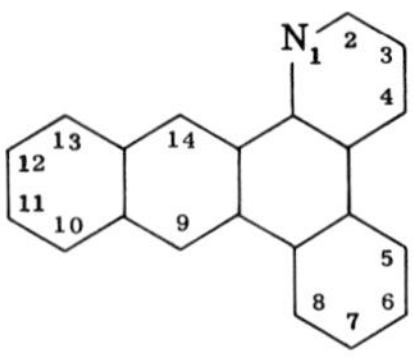
Phenanthro[9,10-*h*]isoquinoline	$C_{21}H_{13}N$ (279)	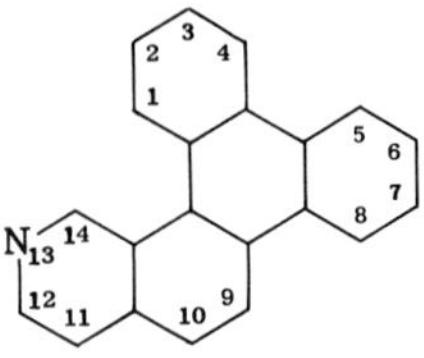

(*Continued*)

Appendix 2 (*Continued*)

Name	Molecular formula (weight)	Structure
Dibenzo[*c,i*]phenanthridine	$C_{21}H_{13}N$ (279)	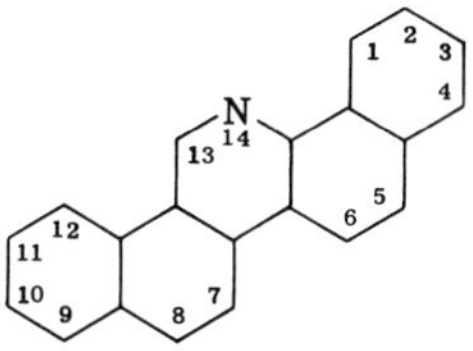
Benzo[*h*]naphth[1,2-*f*]isoquinoline	$C_{21}H_{13}N$ (279)	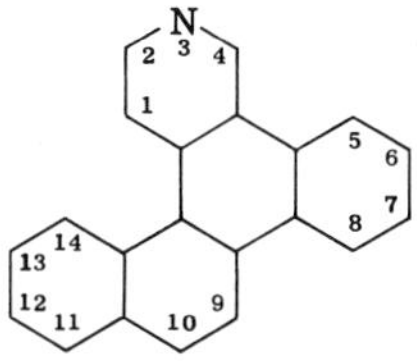
Benzo[*h*]naphtho[1,2-*f*]quinoline	$C_{21}H_{13}N$ (279)	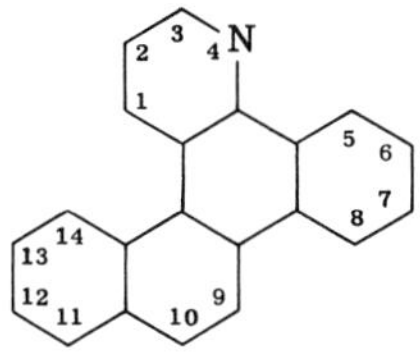
Benzo[*h*]naphth[2,1-*f*]isoquinoline	$C_{21}H_{13}N$ (279)	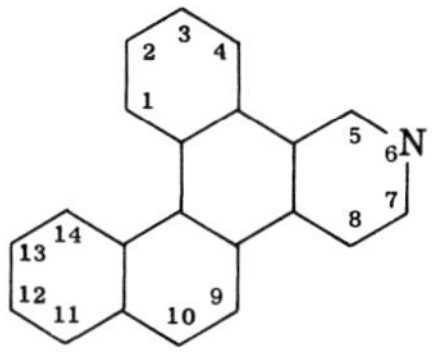
Benzo[*h*]naphtho[2,1-*f*]quinoline	$C_{21}H_{13}N$ (279)	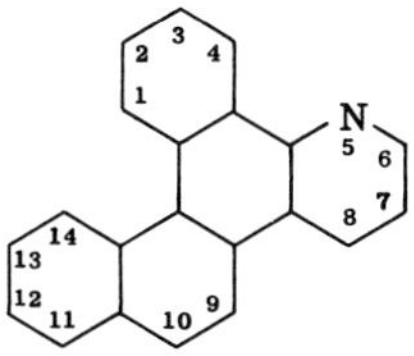
Benzo[*h*]naphth[2,3-*f*]isoquinoline	$C_{21}H_{13}N$ (279)	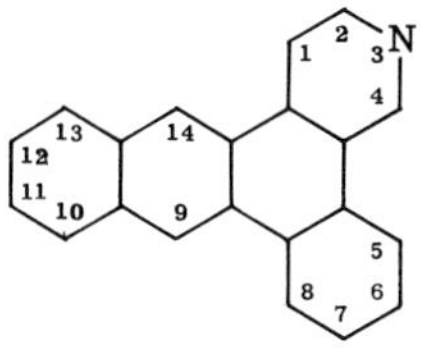

(*Continued*)

Appendix 2 (*Continued*)

Name	Molecular formula (weight)	Structure
Benzo[*h*]naphtho[2,3-*f*]quinoline	C$_{21}$H$_{13}$N (279)	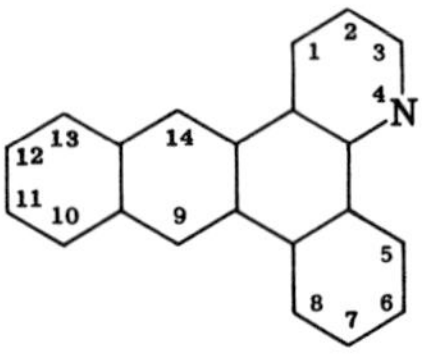
Dibenz[*a,c*]acridine	C$_{21}$H$_{13}$N (279)	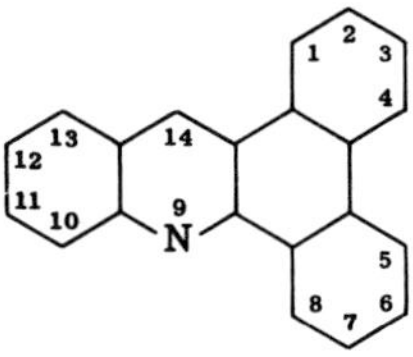
Dibenz[*a,h*]acridine	C$_{21}$H$_{13}$N (279)	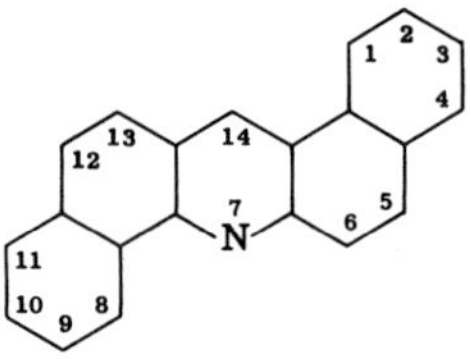
Dibenz[*a,i*]acridine	C$_{21}$H$_{13}$N (279)	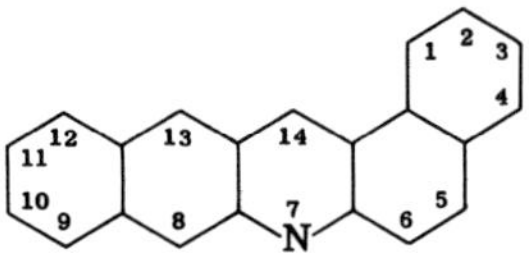
Dibenz[*a,j*]acridine	C$_{21}$H$_{13}$N (279)	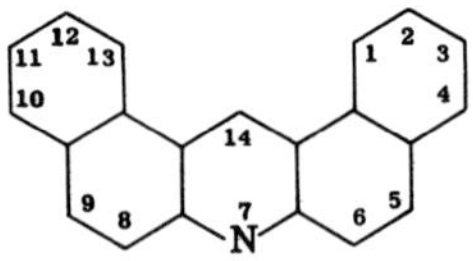
Dibenz[*b,h*]acridine	C$_{21}$H$_{13}$N (279)	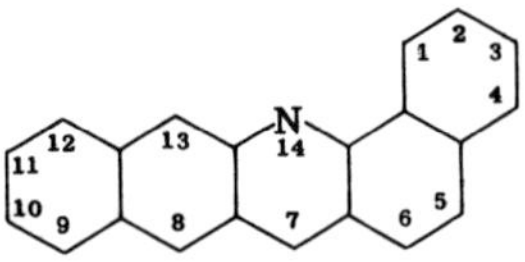
Dibenz[*b,i*]acridine	C$_{21}$H$_{13}$N (279)	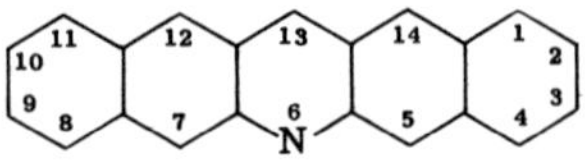

(*Continued*)

Appendix 2 (*Continued*)

Name	Molecular formula (weight)	Structure
Dibenz[c,h]acridine	$C_{21}H_{13}N$ (279)	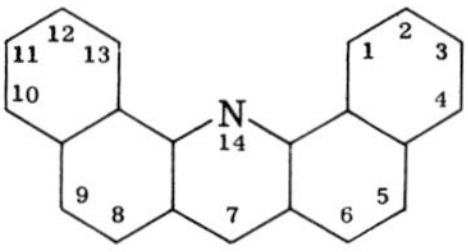
Dibenzo[a,c]phenanthridine	$C_{21}H_{13}N$ (279)	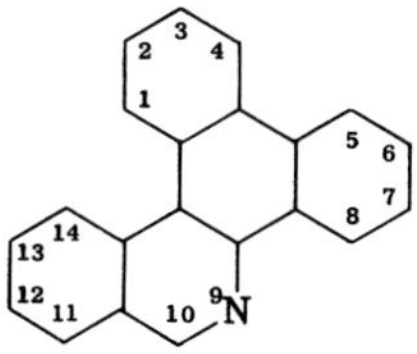
Dibenzo[a,i]phenanthridine	$C_{21}H_{13}N$ (279)	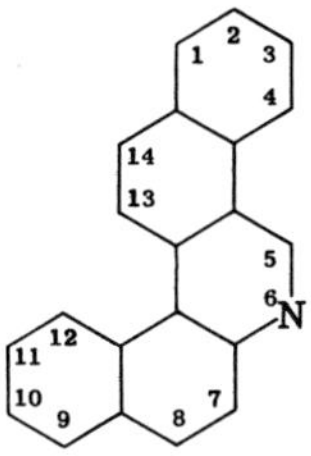
Dibenzo[a,j]phenanthridine	$C_{21}H_{13}N$ (279)	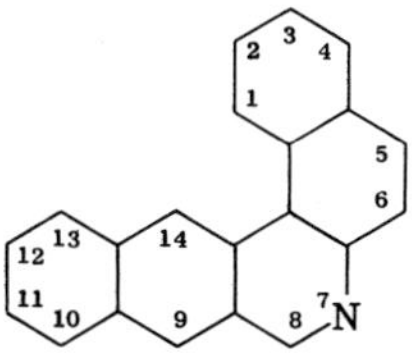
Dibenzo[a,k]phenanthridine	$C_{21}H_{13}N$ (279)	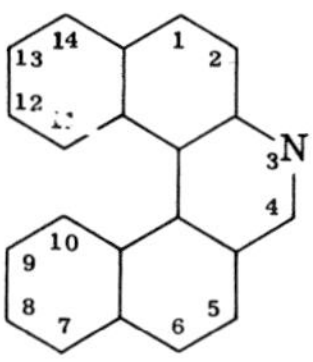
Dibenzo[b,i]phenanthridine	$C_{21}H_{13}N$ (279)	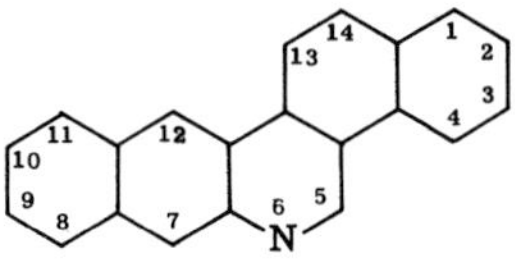

(*Continued*)

Appendix 2 (*Continued*)

Name	Molecular formula (weight)	Structure
Dibenzo[*b,j*]phenanthridine	C$_{21}$H$_{13}$N (279)	
Dibenzo[*b,k*]phenanthridine	C$_{21}$H$_{13}$N (279)	
Dibenzo[*c,j*]phenanthridine	C$_{21}$H$_{13}$N (279)	
Dibenzo[*c,k*]phenanthridine	C$_{21}$H$_{13}$N (279)	
Dibenzo[*i,k*]phenanthridine	C$_{21}$H$_{13}$N (279)	
Naphth[1,2-*a*]acridine	C$_{21}$H$_{13}$N (279)	

Appendix 2 (*Continued*)

Name	Molecular formula (weight)	Structure
Naphth[1,2-*b*]acridine	$C_{21}H_{13}N$ (279)	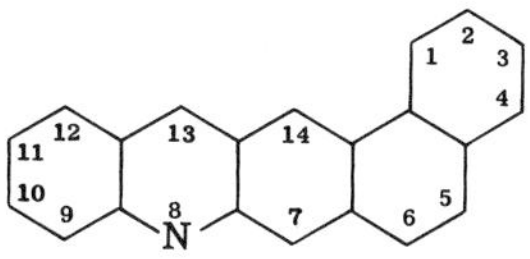
Naphth[1,2-*c*]acridine	$C_{21}H_{13}N$ (279)	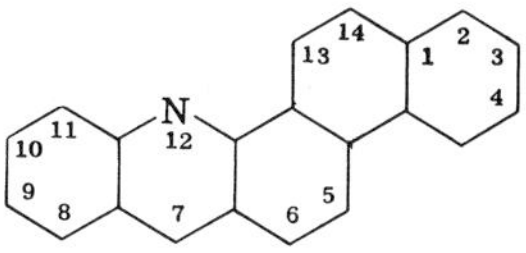
Naphth[2,1-*a*]acridine	$C_{21}H_{13}N$ (279)	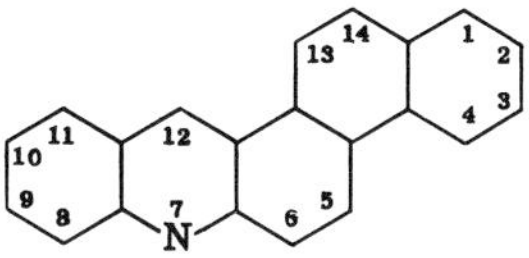
Naphth[2,1-*b*]acridine	$C_{21}H_{13}N$ (279)	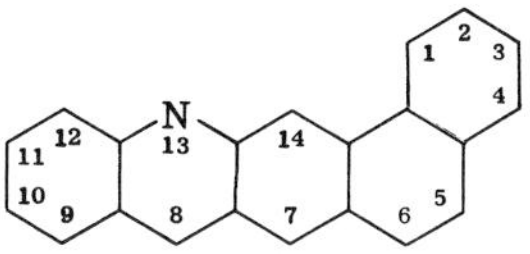
Naphth[2,1-*c*]acridine	$C_{21}H_{13}N$ (279)	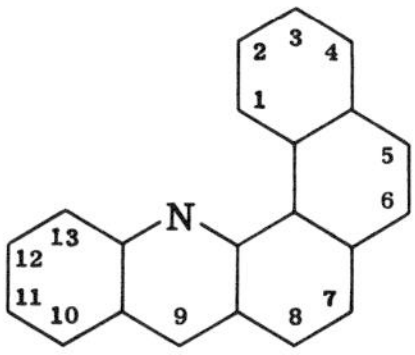
Naphth[2,3-*a*]acridine	$C_{21}H_{13}N$ (279)	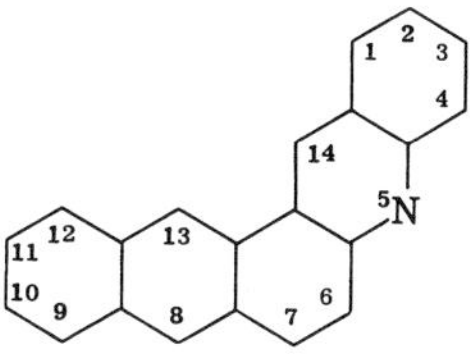
Naphth[2,3-*b*]acridine	$C_{21}H_{13}N$ (279)	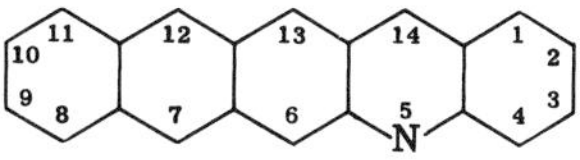

(*Continued*)

Appendix 2 (*Continued*)

Name	Molecular formula (weight)	Structure
Naphth[2,3-*c*]acridine	$C_{21}H_{13}N$ (279)	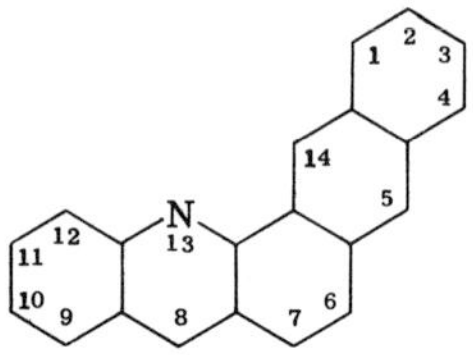
Naphtho[1,2-*b*]phenanthridine	$C_{21}H_{13}N$ (279)	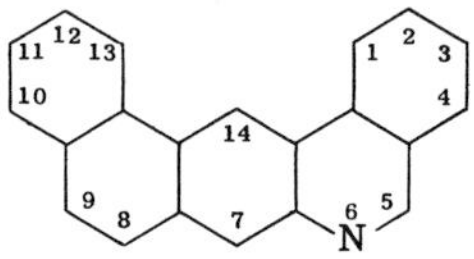
Naphtho[1,2-*c*]phenanthridine	$C_{21}H_{13}N$ (279)	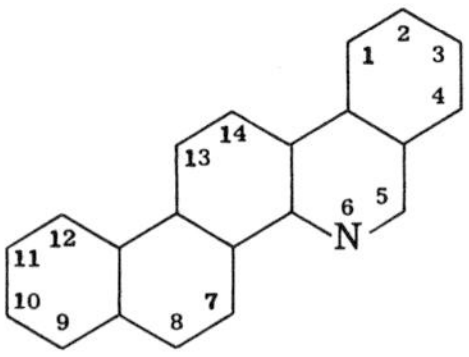
Naphtho[1,2-*j*]phenanthridine	$C_{21}H_{13}N$ (279)	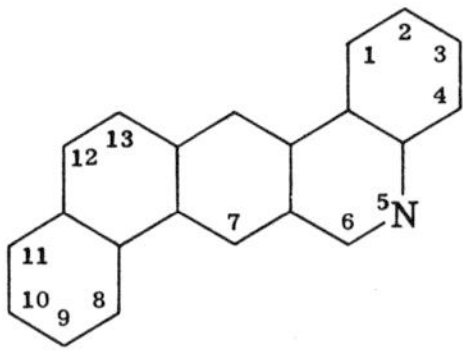
Naphtho[2,1-*b*]phenanthridine	$C_{21}H_{13}N$ (279)	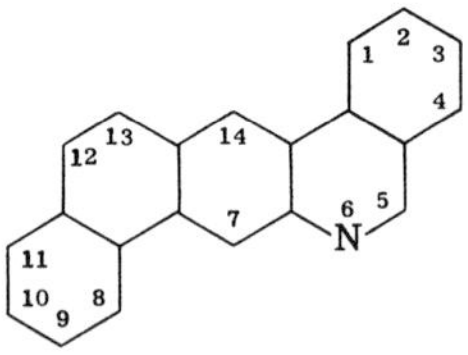
Naphtho[2,1-*i*]phenanthridine	$C_{21}H_{13}N$ (279)	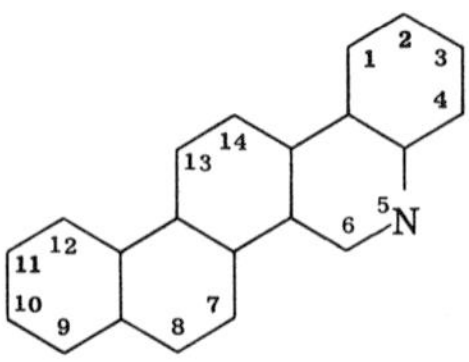

(*Continued*)

Appendix 2 (*Continued*)

Name	Molecular formula (weight)	Structure
Naphtho[2,1-*j*]phenanthridine	C$_{21}$H$_{13}$N (279)	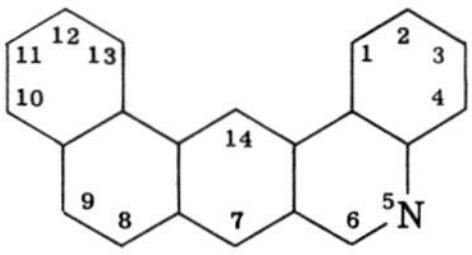
Naphtho[2,3-*a*]phenanthridine	C$_{21}$H$_{13}$N (279)	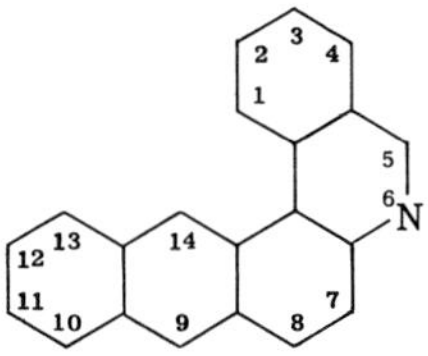
Naphtho[2,3-*b*]phenanthridine	C$_{21}$H$_{13}$N (279)	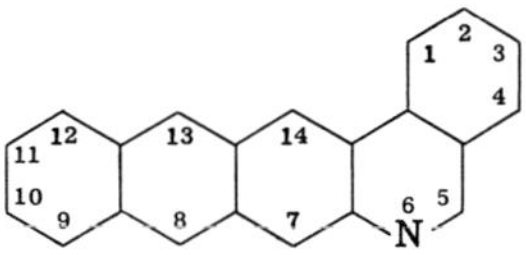
Naphtho[2,3-*j*]phenanthridine	C$_{21}$H$_{13}$N (279)	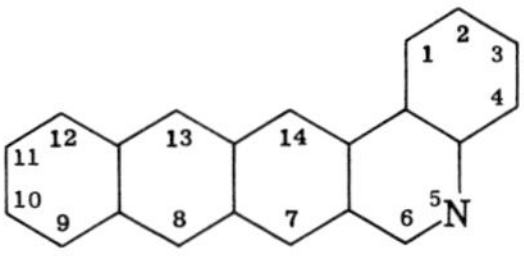
Naphtho[2,3-*k*]phenanthridine	C$_{21}$H$_{13}$N (279)	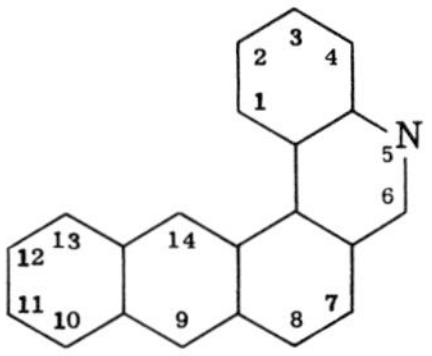
Phenanthro[1,2-*f*]isoquinoline	C$_{21}$H$_{13}$N (279)	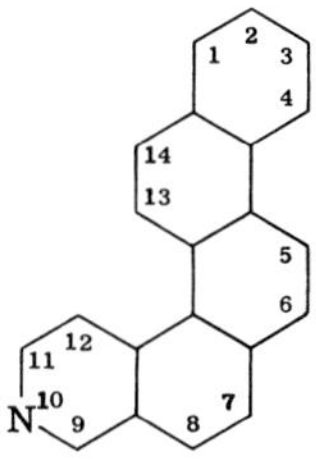

Appendix 2 (*Continued*)

Name	Molecular formula (weight)	Structure
Phenanthro[3,4-*g*]isoquinoline	C$_{21}$H$_{13}$N (279)	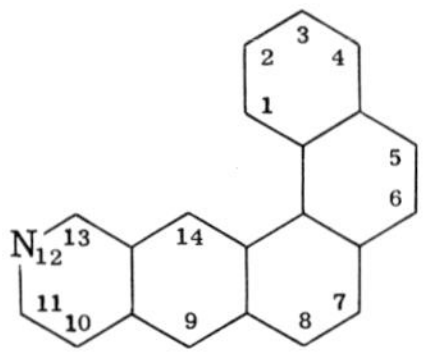
Phenanthro[3,4-*g*]quinoline	C$_{21}$H$_{13}$N (279)	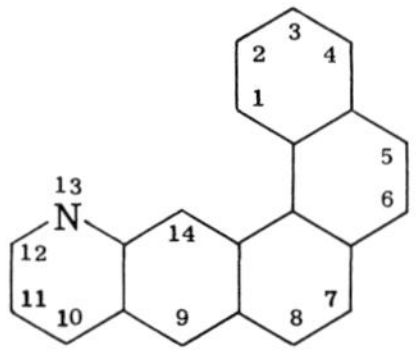
Phenanthro[3,4-*h*]isoquinoline	C$_{21}$H$_{13}$N (279)	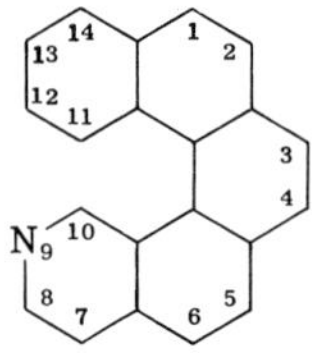
Phenanthro[3,4-*h*]quinoline	C$_{21}$H$_{13}$N (279)	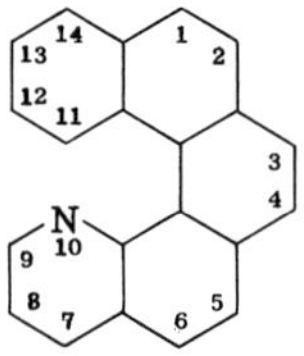
Naphtho[2,1-*c*]phenanthridine	C$_{21}$H$_{13}$N (279)	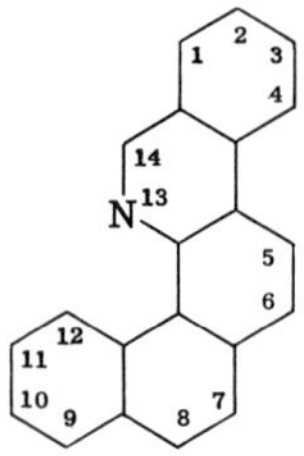

(*Continued*)

Appendix 2 (*Continued*)

Name	Molecular formula (weight)	Structure
Naphtho[1,2-*i*]phenanthridine	$C_{21}H_{13}N$ (279)	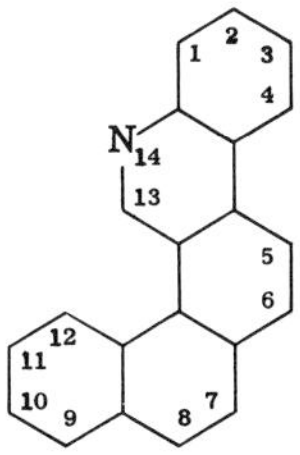
Phenanthro[4,3-*f*]quinoline	$C_{21}H_{13}N$ (279)	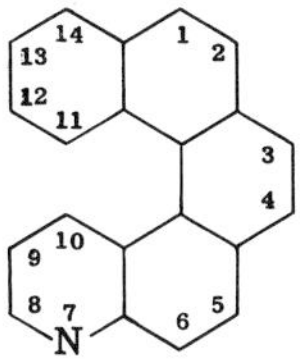
Phenanthro[4,3-*f*]isoquinoline	$C_{21}H_{13}N$ (279)	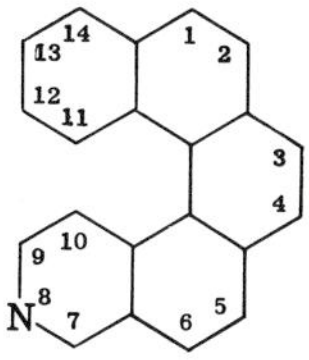
Phenanthro[4,3-*g*]isoquinoline	$C_{21}H_{13}N$ (279)	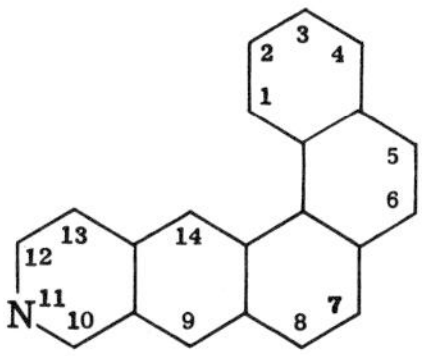
Phenanthro[4,3-*g*]quinoline	$C_{21}H_{13}N$ (279)	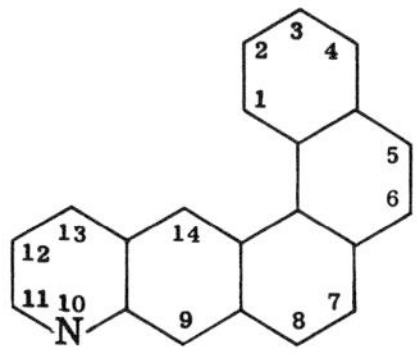

(*Continued*)

Appendix 2 (*Continued*)

Name	Molecular formula (weight)	Structure
Phenanthro[1,2-*f*]quinoline	$C_{21}H_{13}N$ (279)	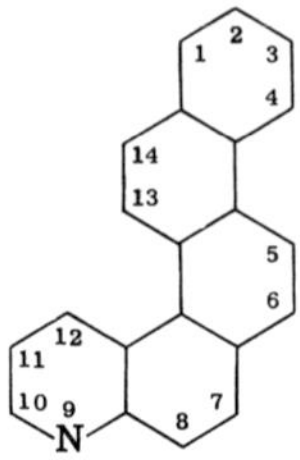
Phenanthro[1,2-*g*]isoquinoline	$C_{21}H_{13}N$ (279)	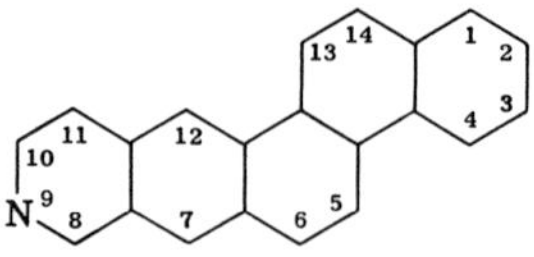
Phenanthro[1,2-*g*]quinoline	$C_{21}H_{13}N$ (279)	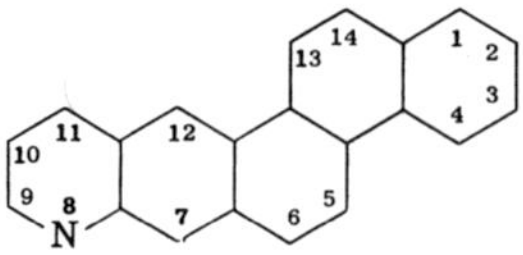
Phenanthro[1,2-*h*]isoquinoline	$C_{21}H_{13}N$ (279)	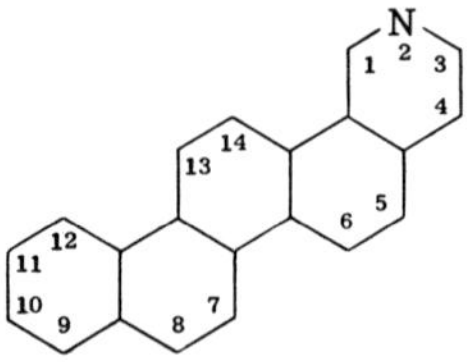
Phenanthro[1,2-*h*]quinoline	$C_{21}H_{13}N$ (279)	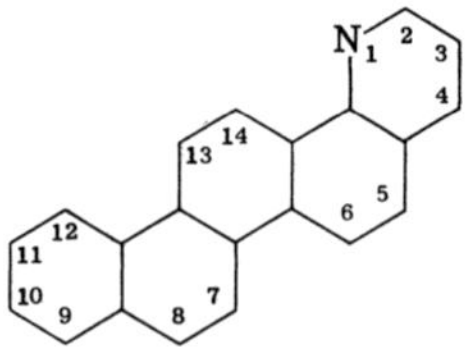
Naphtho[2,1-*a*]phenanthridine	$C_{21}H_{13}N$ (279)	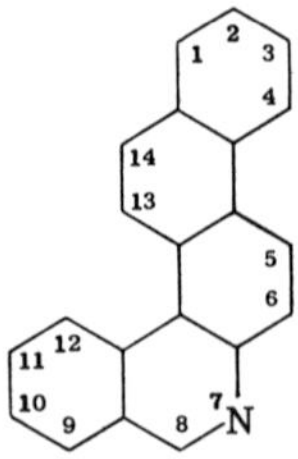

(*Continued*)

Appendix 2 (*Continued*)

Name	Molecular formula (weight)	Structure
Naphtho[1,2-*k*]phenanthridine	$C_{21}H_{13}N$ (279)	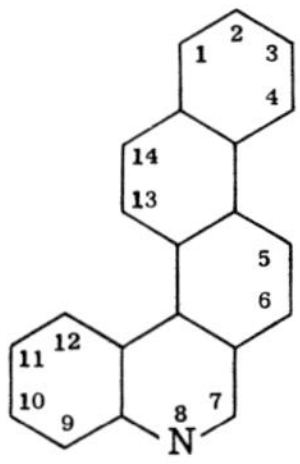
Phenanthro[2,1-*f*]isoquinoline	$C_{21}H_{13}N$ (279)	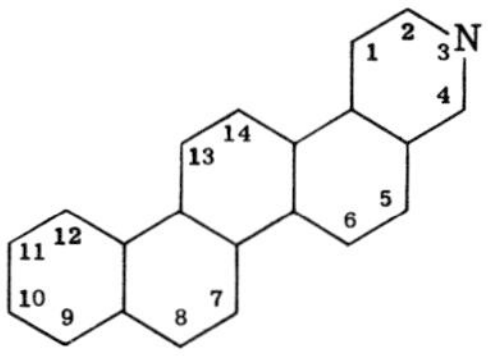
Naphtho[2,1-*k*]phenanthridine	$C_{21}H_{13}N$ (279)	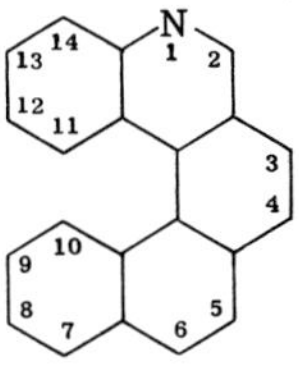
Phenanthro[3,4-*f*]isoquinoline	$C_{21}H_{13}N$ (279)	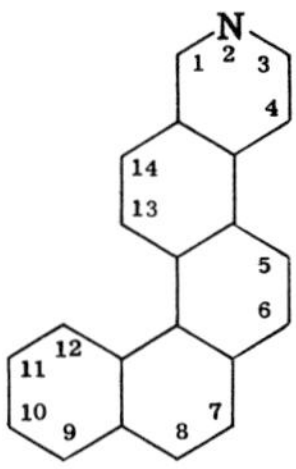
Phenanthro[3,4-*f*]quinoline	$C_{21}H_{13}N$ (279)	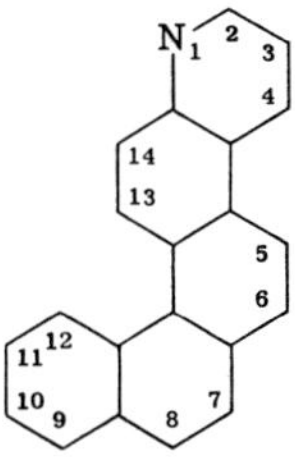

(*Continued*)

Appendix 2 (*Continued*)

Name	Molecular formula (weight)	Structure
Phenanthro[4,3-*h*]isoquinoline	$C_{21}H_{13}N$ (279)	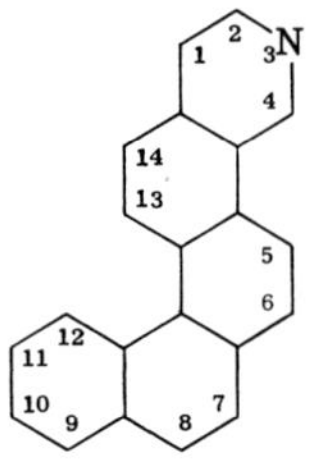

3. Polycyclic Aromatic Heterocycles Containing One Sulfur Atom: Names, Formulas, Structures, and Numbering

Name	Molecular formula (weight)	Structure
Benzo[b]thiophene	C_8H_6S (134)	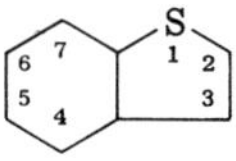
Benzo[c]thiophene	C_8H_6S (134)	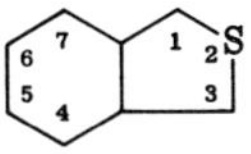
Dibenzo[b,d]thiophene	$C_{12}H_8S$ (184)	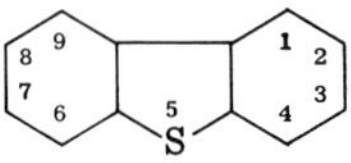
Naphtho[1,2-b]thiophene	$C_{12}H_8S$ (184)	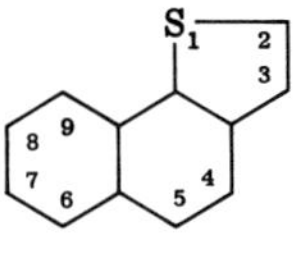
Naphtho[1,2-c]thiophene	$C_{12}H_8S$ (184)	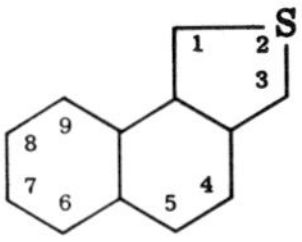
Naphtho[2,1-b]thiophene	$C_{12}H_8S$ (184)	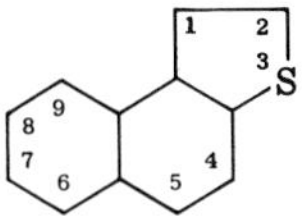
Naphtho[2,3-b]thiophene	$C_{12}H_8S$ (184)	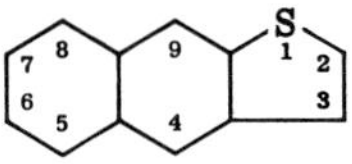
Naphtho[2,3-c]thiophene	$C_{12}H_8S$ (184)	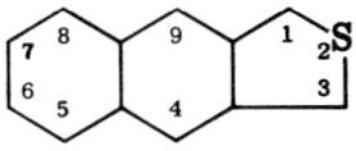
Phenanthro[4,5-bcd]thiophene	$C_{14}H_8S$ (208)	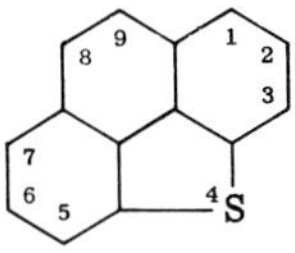

(*Continued*)

Appendix 3 (*Continued*)

Name	Molecular formula (weight)	Structure
Anthra[1,2-*b*]thiophene	$C_{16}H_{10}S$ (234)	
Anthra[1,2-*c*]thiophene	$C_{16}H_{10}S$ (234)	
Anthra[2,1-*b*]thiophene	$C_{16}H_{10}S$ (234)	
Anthra[2,3-*b*]thiophene	$C_{16}H_{10}S$ (234)	
Anthra[2,3-*c*]thiophene	$C_{16}H_{10}S$ (234)	
Benzo[*b*]naphtho[1,2-*d*]thiophene	$C_{16}H_{10}S$ (234)	
Benzo[*b*]naphtho[2,1-*d*]thiophene	$C_{16}H_{10}S$ (234)	
Benzo[*b*]naphtho[2,3-*d*]thiophene	$C_{16}H_{10}S$ (234)	
Phenanthro[1,2-*b*]thiophene	$C_{16}H_{10}S$ (234)	

Appendix 3 (*Continued*)

Name	Molecular formula (weight)	Structure
Phenanthro[1,2-*c*]thiophene	$C_{16}H_8S$ (234)	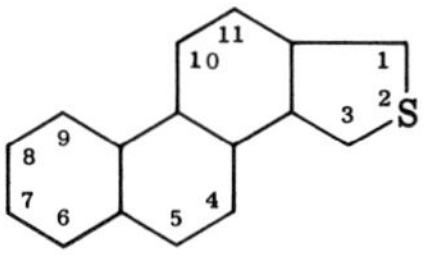
Phenanthro[2,1-*b*]thiophene	$C_{16}H_{10}S$ (234)	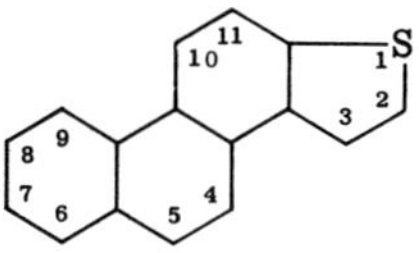
Phenanthro[2,3-*b*]thiophene	$C_{16}H_{10}S$ (234)	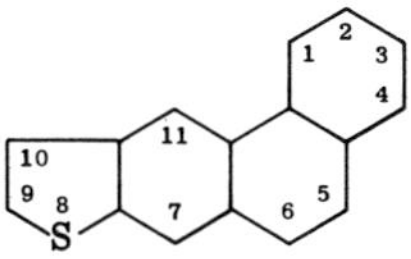
Phenanthro[2,3-*c*]thiophene	$C_{16}H_{10}S$ (234)	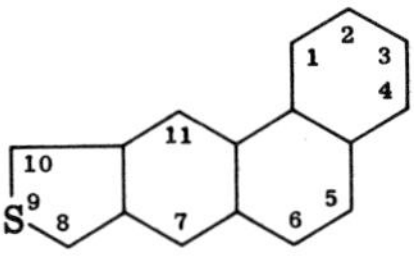
Phenanthro[3,2-*b*]thiophene	$C_{16}H_{10}S$ (234)	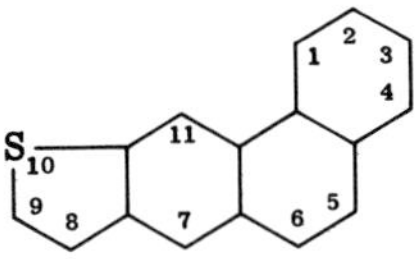
Phenanthro[3,4-*b*]thiophene	$C_{16}H_{10}S$ (234)	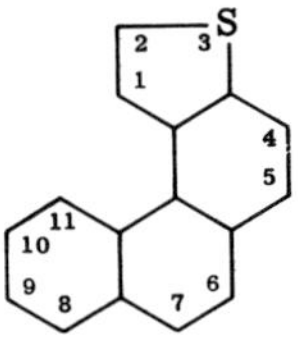
Phenanthro[3,4-*c*]thiophene	$C_{16}H_{10}S$ (234)	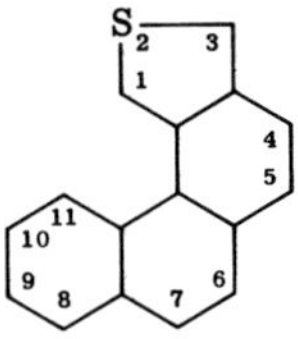

(*Continued*)

Name	Molecular formula (weight)	Structure
Phenanthro[4,3-*b*]thiophene	$C_{16}H_{10}S$ (234)	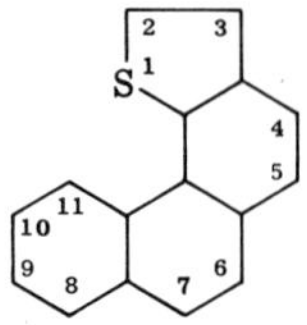
Phenanthro[9,10-*b*]thiophene	$C_{16}H_{10}S$ (234)	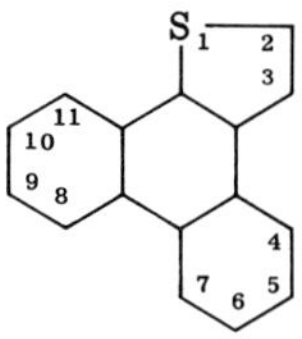
Phenanthro[9,10-*c*]thiophene	$C_{16}H_{10}S$ (234)	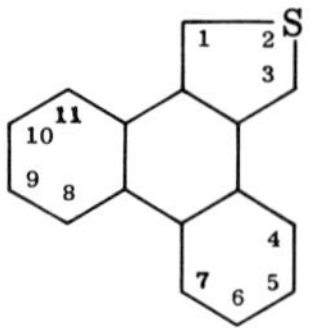
Benz[5,10]anthra[1,9-*bc*]thiophene	$C_{16}H_{10}S$ (258)	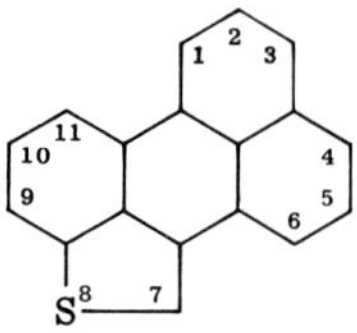
Benz[5,10]anthra[9,1-*bc*]thiophene	$C_{18}H_{10}S$ (258)	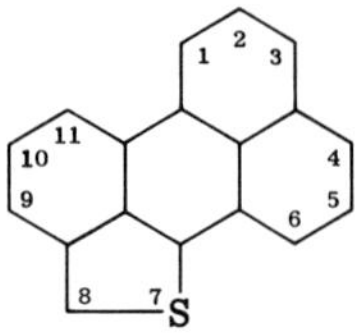
Benzo[1,2]phenaleno[3,4-*bc*]-thiophene	$C_{18}H_{10}S$ (258)	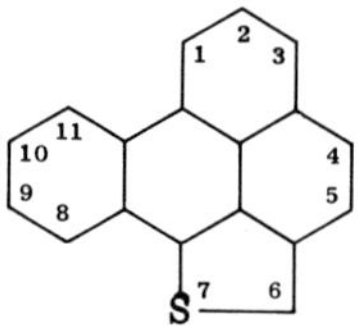

(*Continued*)

Appendix 3 (*Continued*)

Name	Molecular formula (weight)	Structure
Benzo[1,2]phenaleno[4,3-*bc*]-thiophene	$C_{18}H_{10}S$ (258)	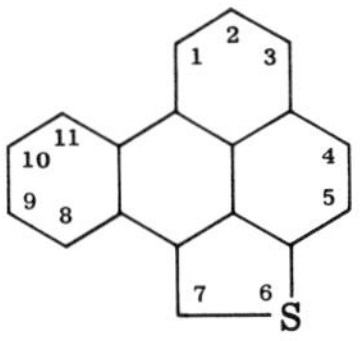
Benzo[4,5]phenaleno[1,9-*bc*]-thiophene	$C_{18}H_{10}S$ (258)	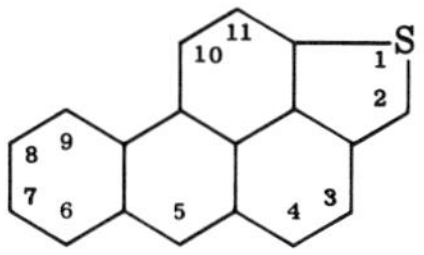
Benzo[4,5]phenaleno[9,1-*bc*]-thiophene	$C_{18}H_{10}S$ (258)	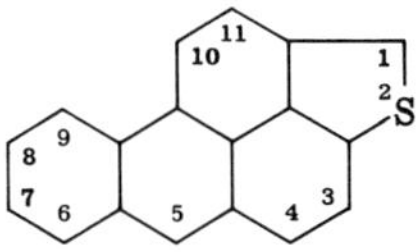
Benzo[2,3]phenanthro[4,5-*bcd*]-thiophene	$C_{18}H_{10}S$ (258)	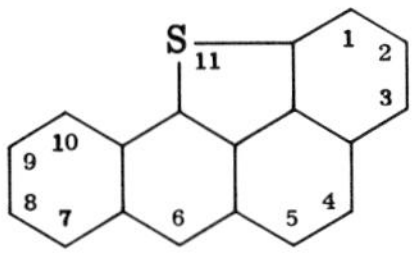
Chryseno[4,5-*bcd*]thiophene	$C_{18}H_{10}S$ (258)	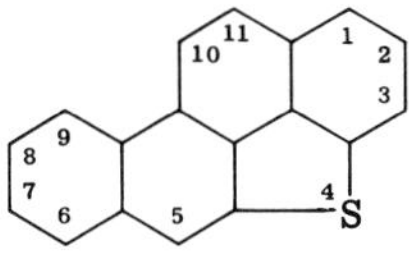
Pyreno[1,2-*b*]thiophene	$C_{18}H_{10}S$ (258)	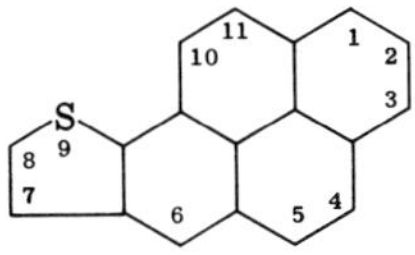
Pyreno[1,2-*c*]thiophene	$C_{18}H_{10}S$ (258)	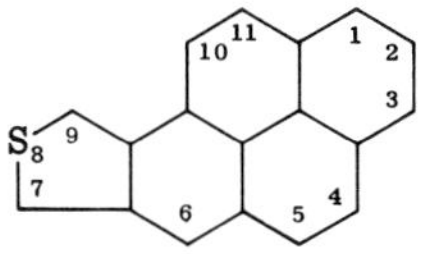
Pyreno[2,1-*b*]thiophene	$C_{18}H_{10}S$ (258)	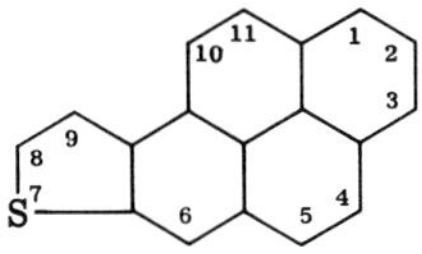

(*Continued*)

Appendix 3 (*Continued*)

Name	Molecular formula (weight)	Structure
Pyreno[4,5-*b*]thiophene	$C_{18}H_{10}S$ (258)	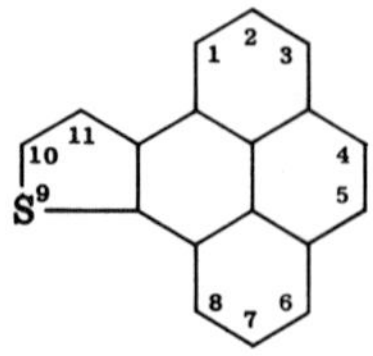
Pyreno[4,5-*c*]thiophene	$C_{18}H_{10}S$ (258)	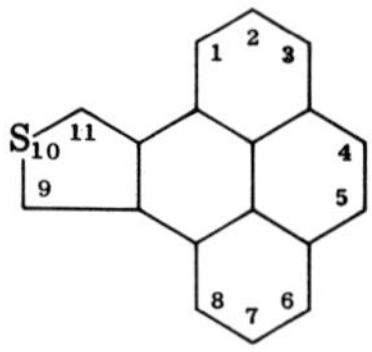
Triphenyleno[4,5-*bcd*]thiophene	$C_{18}H_{10}S$ (258)	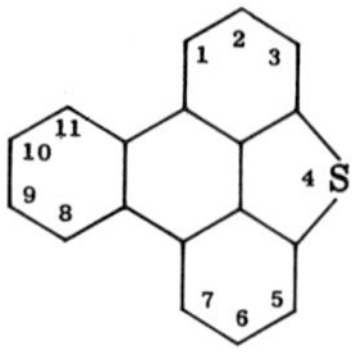
Anthra[1,2-*b*]benzo[*d*]thiophene	$C_{20}H_{12}S$ (284)	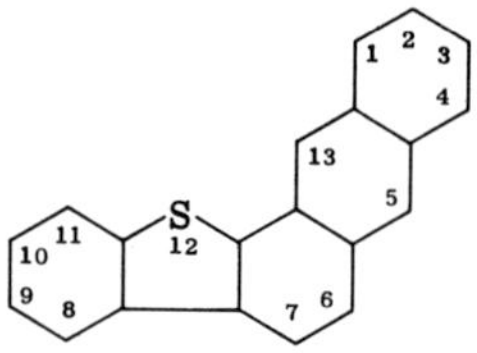
Anthra[2,1-*b*]benzo[*d*]thiophene	$C_{20}H_{12}S$ (284)	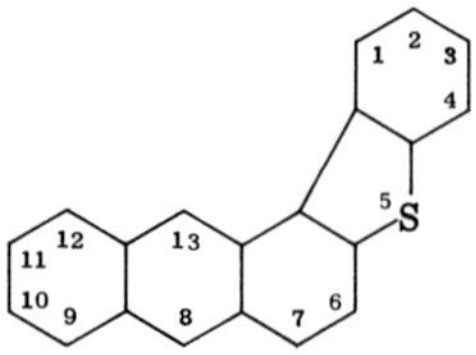
Anthra[2,3-*b*]benzo[*d*]thiophene	$C_{20}H_{12}S$ (284)	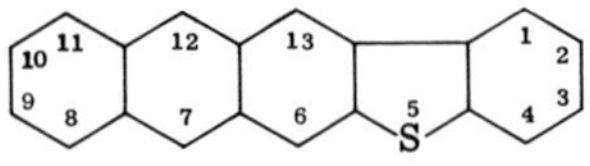

(*Continued*)

Appendix 3 (*Continued*)

Name	Molecular formula (weight)	Structure
Benz[3,4]anthra[1,2-*b*]thiophene	$C_{20}H_{12}S$ (284)	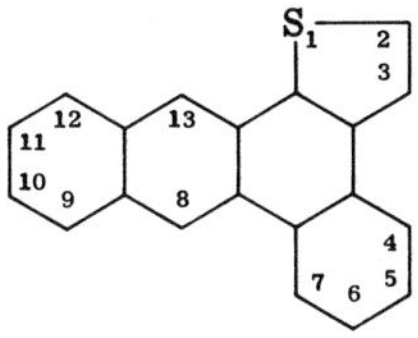
Benz[3,4]anthra[1,2-*c*]thiophene	$C_{20}H_{12}S$ (284)	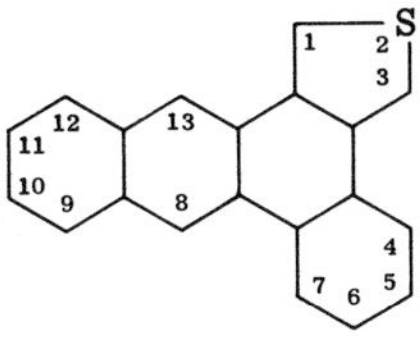
Benz[3,4]anthra[2,1-*b*]thiophene	$C_{20}H_{12}S$ (284)	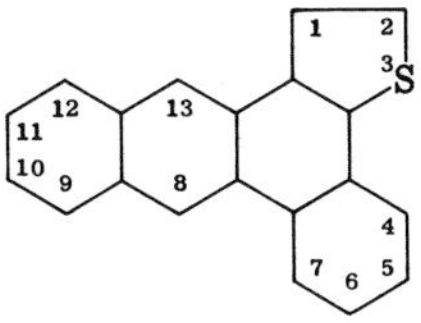
Benz[5,6]anthra[1,2-*b*]thiophene	$C_{20}H_{12}S$ (284)	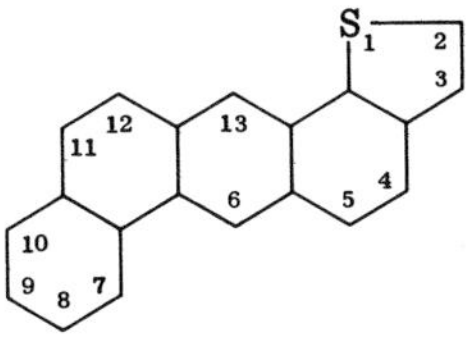
Benz[5,6]anthra[1,2-*c*]thiophene	$C_{20}H_{12}S$ (284)	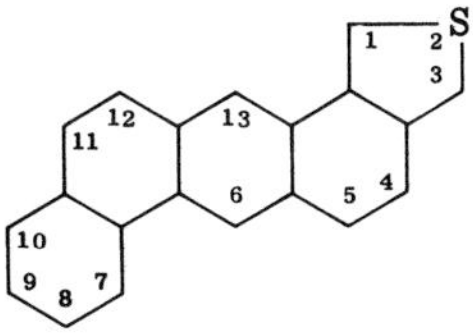
Benz[5,6]anthra[2,1-*b*]thiophene	$C_{20}H_{12}S$ (284)	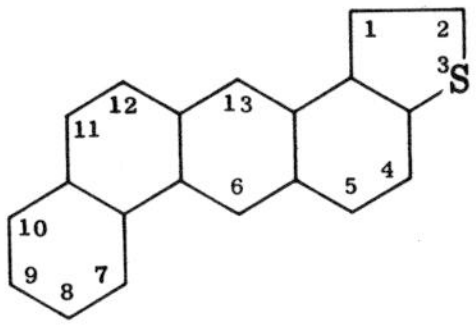

(*Continued*)

Appendix 3 (*Continued*)

Name	Molecular formula (weight)	Structure
Benz[5,6]anthra[2,3-*b*]thiophene	$C_{20}H_{12}S$ (284)	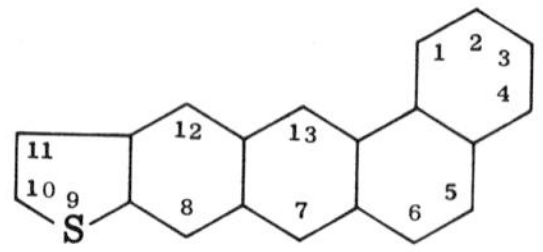
Benz[5,6]anthra[2,3-*c*]thiophene	$C_{20}H_{12}S$ (284)	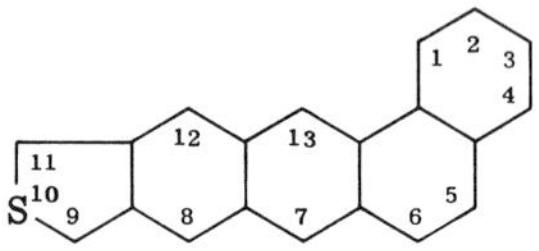
Benz[5,6]anthra[3,2-*b*]thiophene	$C_{20}H_{12}S$ (284)	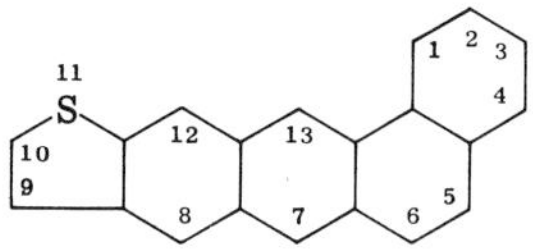
Benz[7.8]anthra[1,2-*b*]thiophene	$C_{20}H_{12}S$ (284)	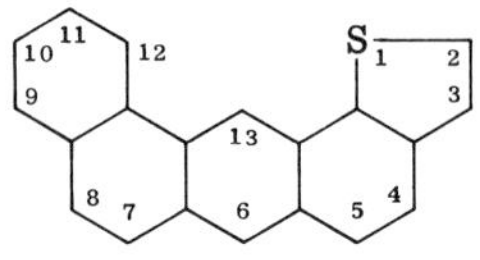
Benz[7,8]anthra[1,2-*c*]thiophene	$C_{20}H_{12}S$ (284)	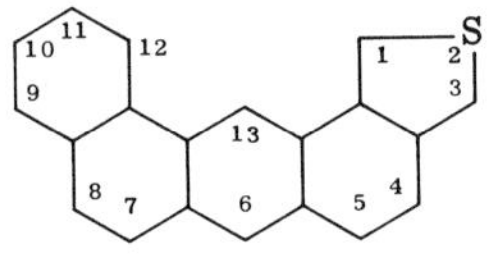
Benz[7,8]anthra[2,1-*b*]thiophene	$C_{20}H_{12}S$ (284)	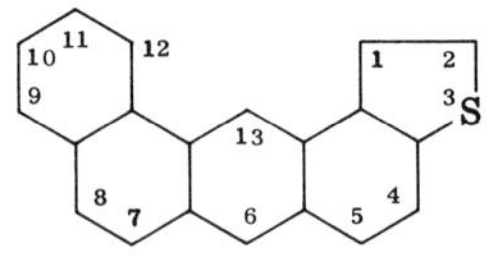
Benzo[*b*]phenanthro[1,2-*d*]-thiophene	$C_{20}H_{12}S$ (284)	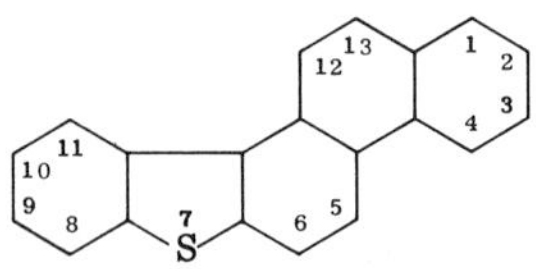
Benzo[*b*]phenanthro[2,1-*d*]-thiophene	$C_{20}H_{12}S$ (284)	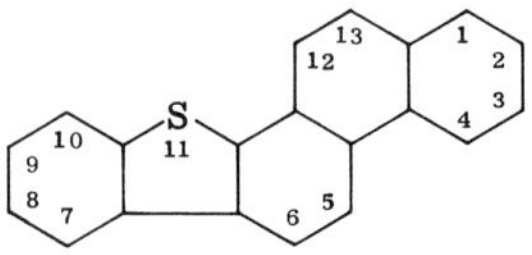

(*Continued*)

Appendix 3 (*Continued*)

Name	Molecular formula (weight)	Structure
Benzo[b]phenanthro[2,3-d]-thiophene	$C_{20}H_{12}S$ (284)	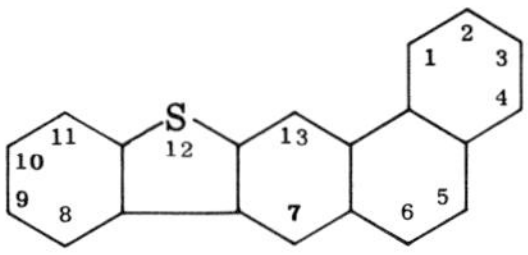
Benzo[b]phenanthro[3,2-d]-thiophene	$C_{20}H_{12}S$ (284)	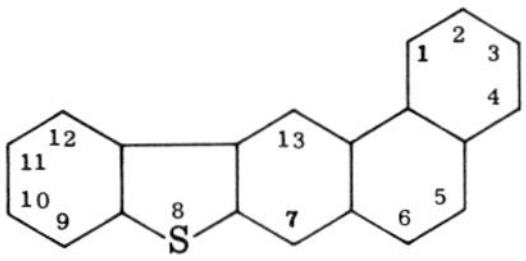
Benzo[b]phenanthro[3,4-d]-thiophene	$C_{20}H_{12}S$ (284)	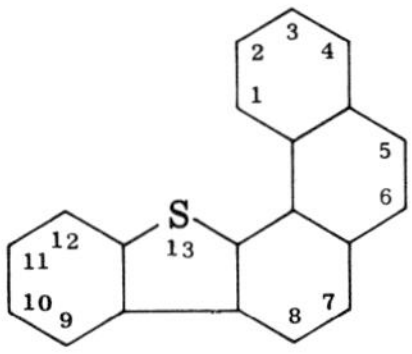
Benzo[b]phenanthro[4,3-d]-thiophene	$C_{20}H_{12}S$ (284)	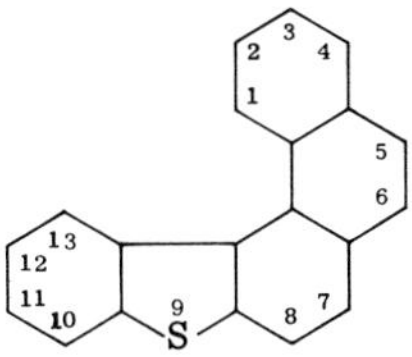
Benzo[b]phenanthro[9,10-d]-thiophene	$C_{20}H_{12}S$ (284)	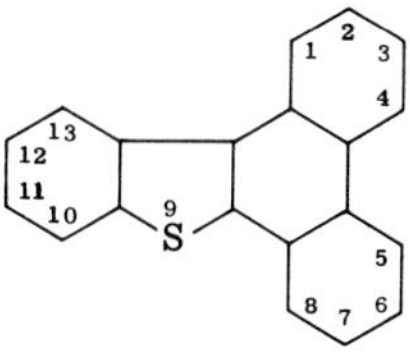
Benzo[2,3]phenanthro[5,6-b]-thiophene	$C_{20}H_{12}S$ (284)	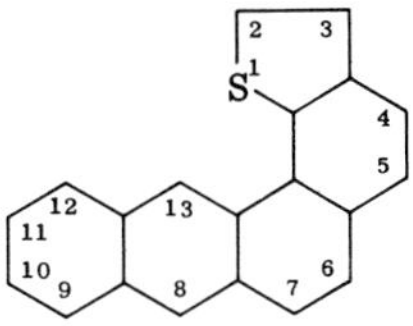

(*Continued*)

Appendix 3 (*Continued*)

Name	Molecular formula (weight)	Structure
Benzo[2,3]-phenanthro[5,6-c]-thiophene	$C_{20}H_{12}S$ (284)	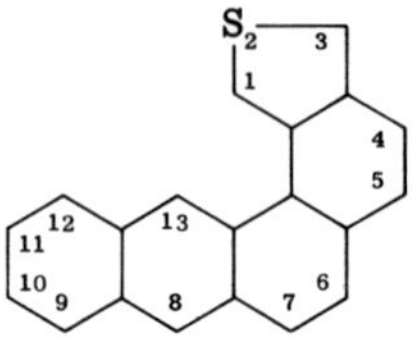
Benzo[2,3]phenanthro[6,5-b]-thiophene	$C_{20}H_{12}S$ (284)	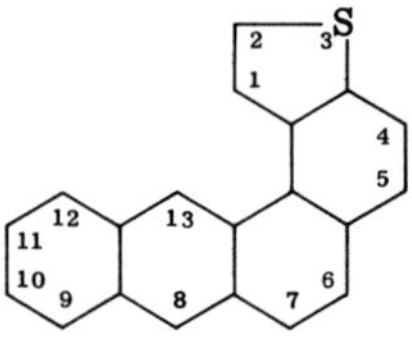
Benzo[3,4]phenanthro[1,2-b]-thiophene	$C_{20}H_{12}S$ (284)	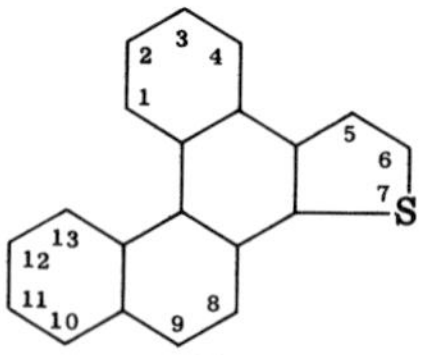
Benzo[3,4]phenanthro[1,2-c]-thiophene	$C_{20}H_{12}S$ (284)	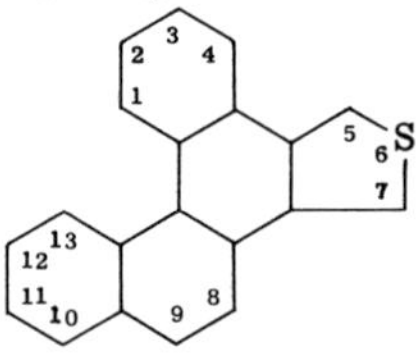
Benzo[3,4]phenanthro[2,1-b]-thiophene	$C_{20}H_{12}S$ (284)	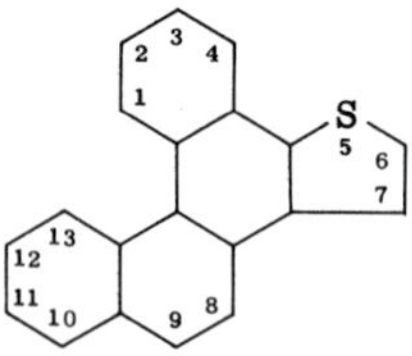
Benzo[3,4]phenanthro[5,6-b]-thiophene	$C_{20}H_{12}S$ (284)	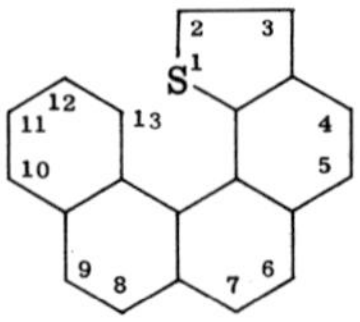

(*Continued*)

Appendix 3 (*Continued*)

Name	Molecular formula (weight)	Structure
Benzo[3,4]phenanthro[5,6-*c*]-thiophene	$C_{20}H_{12}S$ (284)	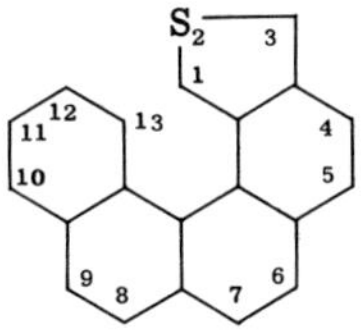
Benzo[3,4]phenanthro[6,5-*b*]-thiophene	$C_{20}H_{12}S$ (284)	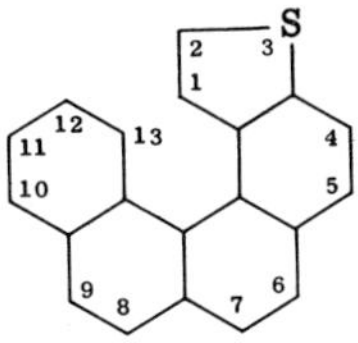
Benzo[5,6]phenanthro[1,2-*b*]-thiophene	$C_{20}H_{12}S$ (284)	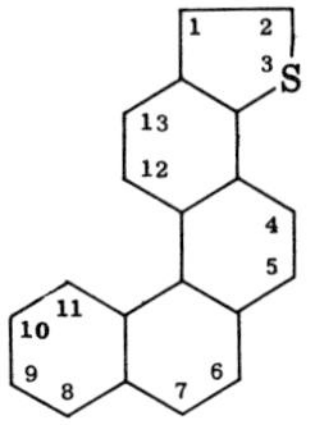
Benzo[5,6]phenanthro[1,2-*c*]-thiophene	$C_{20}H_{12}S$ (284)	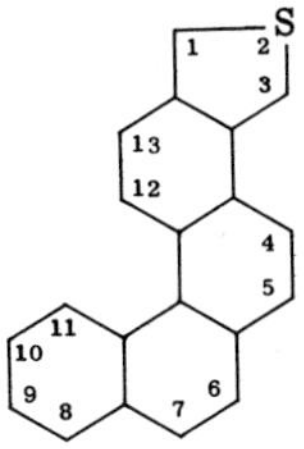
Benzo[5,6]phenanthro[2,1-*b*]-thiophene	$C_{20}H_{12}S$ (284)	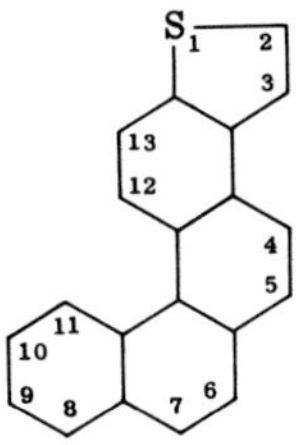

(*Continued*)

Appendix-3 (*Continued*)

Name	Molecular formula (weight)	Structure
Benzo[5,6]phenanthro[2,3-*b*]-thiophene	$C_{20}H_{12}S$ (284)	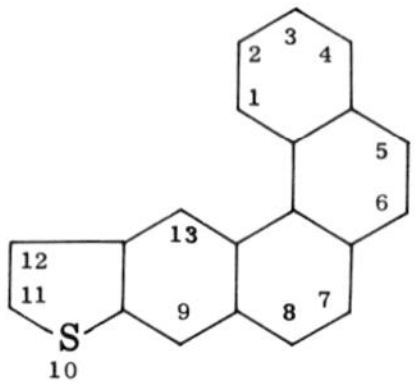
Benzo[5,6]phenanthro[2,3-*c*]-thiophene	$C_{20}H_{12}S$ (284)	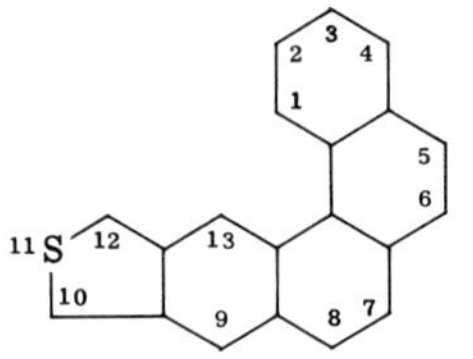
Benzo[5,6]phenanthro[3,2-*b*]-thiophene	$C_{20}H_{12}S$ (284)	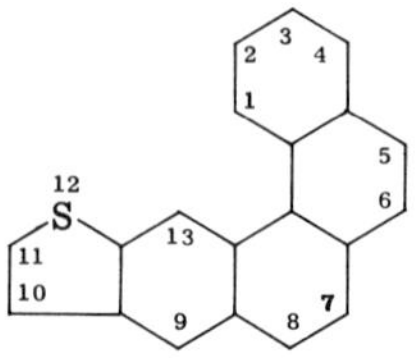
Benzo[6,7]phenanthro[1,2-*b*]-thiophene	$C_{20}H_{12}S$ (284)	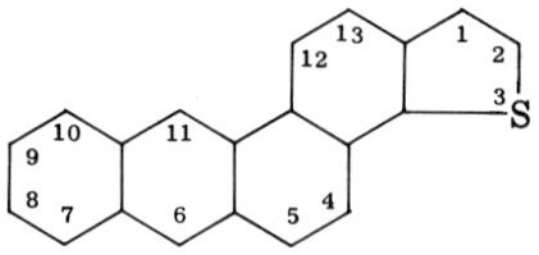
Benzo[6,7]phenanthro[1,2-*c*]-thiophene	$C_{20}H_{12}S$ (284)	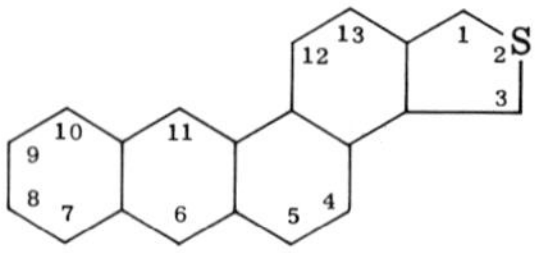
Benzo[6,7]phenanthro[2,1-*b*]-thiophene	$C_{20}H_{12}S$ (284)	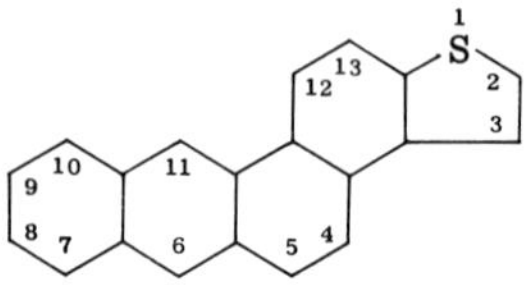

(*Continued*)

Appendix 3 (*Continued*)

Name	Molecular formula (weight)	Structure
Benzo[6,7]phenanthro[2,3-*b*]-thiophene	$C_{20}H_{12}S$ (284)	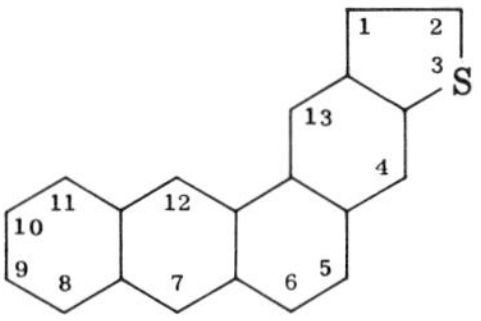
Benzo[6,7]phenanthro[2,3-*c*]-thiophene	$C_{20}H_{12}S$ (284)	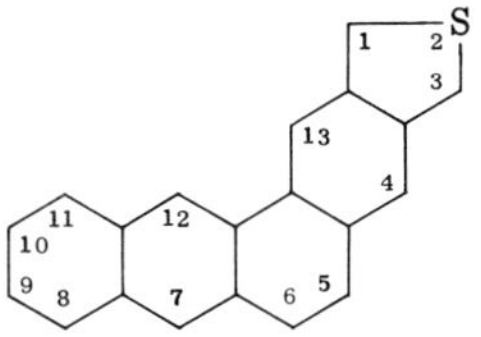
Benzo[6,7]phenanthro[3,2-*b*]-thiophene	$C_{20}H_{12}S$ (284)	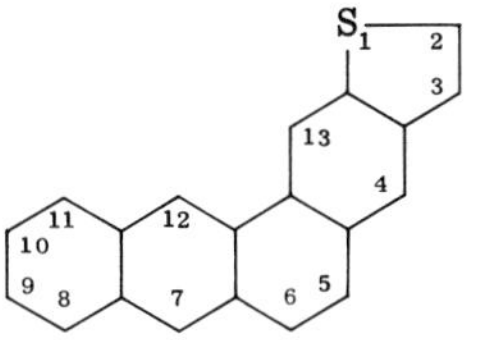
Chryseno[1,2-*b*]thiophene	$C_{20}H_{12}S$ (284)	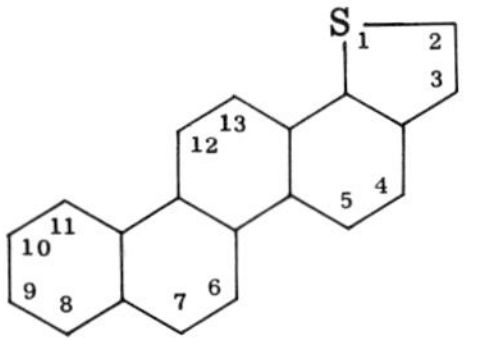
Chryseno[1,2-*c*]thiophene	$C_{20}H_{12}S$ (284)	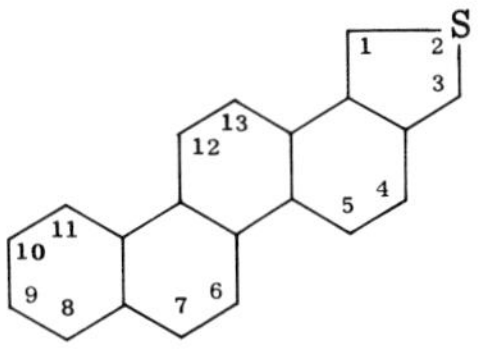
Chryseno[2,1-*b*]thiophene	$C_{20}H_{12}S$ (284)	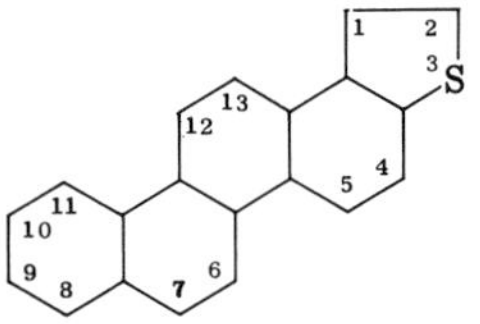

(*Continued*)

Appendix 3 (*Continued*)

Name	Molecular formula (weight)	Structure
Chryseno[2,3-*b*]thiophene	$C_{20}H_{12}S$ (284)	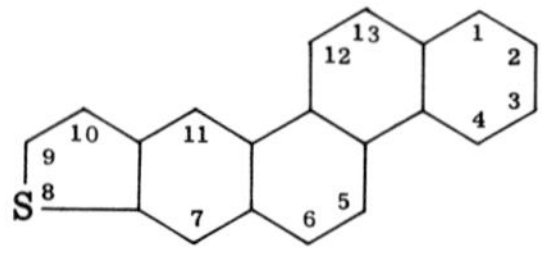
Chryseno[2,3-*c*]thiophene	$C_{20}H_{12}S$ (284)	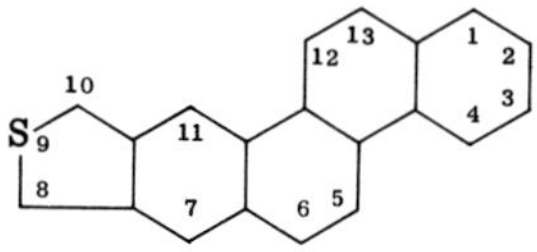
Chryseno[3,2-*b*]thiophene	$C_{20}H_{12}S$ (284)	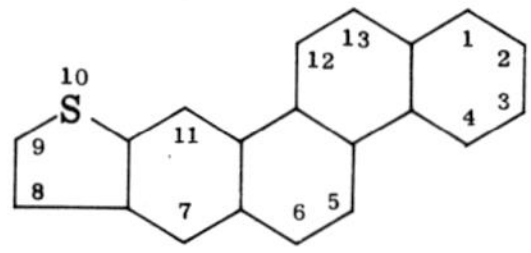
Chryseno[3,4-*b*]thiophene	$C_{20}H_{12}S$ (284)	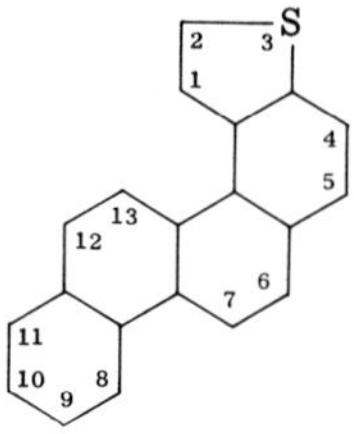
Chryseno[3,4-*c*]thiophene	$C_{20}H_{12}S$ (284)	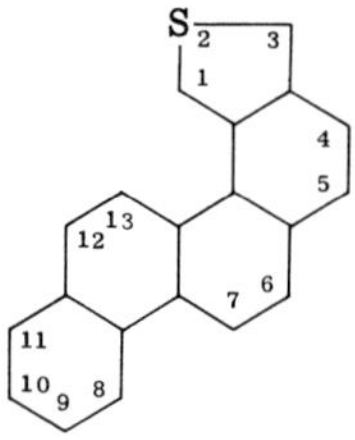
Chryseno[4,3-*b*]thiophene	$C_{20}H_{12}S$ (284)	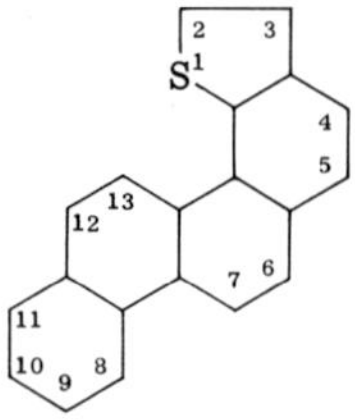

Appendix 3 (*Continued*)

Name	Molecular formula (weight)	Structure
Chryseno[5,6-*b*]thiophene	$C_{20}H_{12}S$ (284)	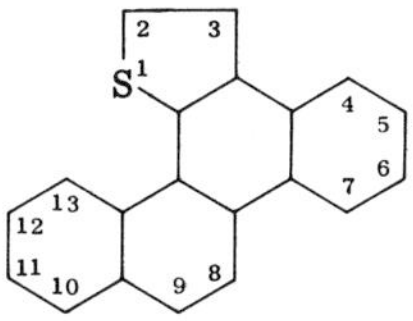
Chryseno[5,6-*c*]thiophene	$C_{20}H_{12}S$ (284)	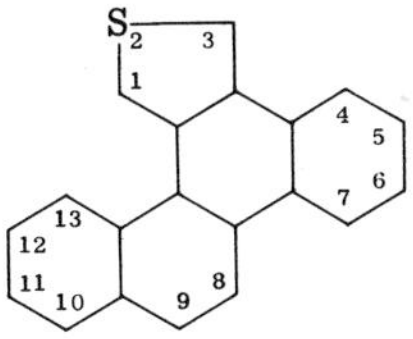
Chryseno[6,5-*b*]thiophene	$C_{20}H_{12}S$ (284)	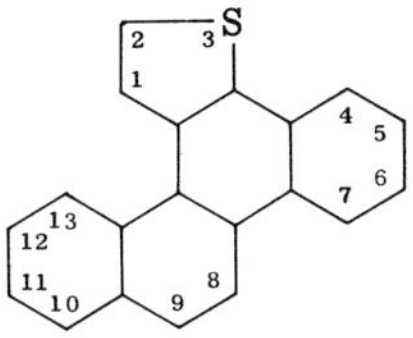
Dinaphtho[1,2-*b*:1′,2′-*d*]-thiophene	$C_{20}H_{12}S$ (284)	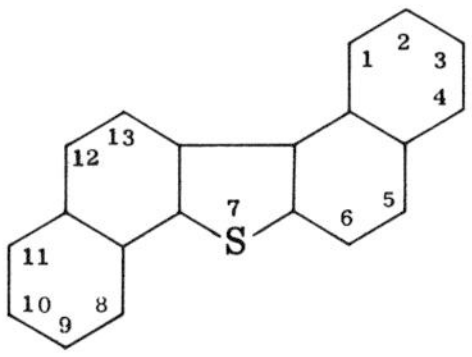
Dinaphtho[1,2-*b*:2′,1′-*d*]-thiophene	$C_{20}H_{12}S$ (284)	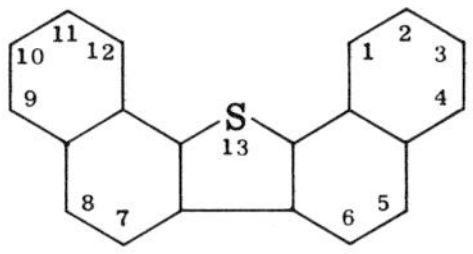
Dinaphtho[1,2-*b*:2′,3′-*d*]-thiophene	$C_{20}H_{12}S$ (284)	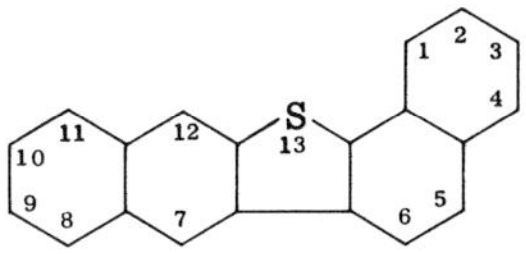
Dinaphtho[2,1-*b*:1′,2′-*d*]-thiophene	$C_{20}H_{12}S$ (284)	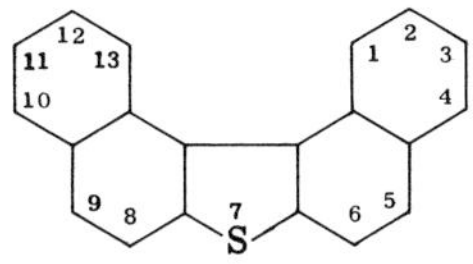

(*Continued*)

Appendix 3 (*Continued*)

Name	Molecular formula (weight)	Structure
Dinaphtho[2,1-*b*:2′,3′-*d*]-thiophene	$C_{20}H_{12}S$ (284)	
Dinaphtho[2,3-*b*:2′,3′-*d*]-thiophene	$C_{20}H_{12}S$ (284)	
Naphthaceno[1,2-*b*]thiophene	$C_{20}H_{12}S$ (284)	
Naphthaceno[1,2-*c*]thiophene	$C_{20}H_{12}S$ (284)	
Naphthaceno[2,3-*b*]thiophene	$C_{20}H_{12}S$ (284)	
Naphthaceno[2,3-*c*]thiophene	$C_{20}H_{12}S$ (284)	
Triphenyleno[1,2-*b*]thiophene	$C_{20}H_{12}S$ (284)	
Triphenyleno[1,2-*c*]thiophene	$C_{20}H_{12}S$ (284)	

Appendix 3 (*Continued*)

Name	Molecular formula (weight)	Structure
Triphenyleno[2,1-*b*]thiophene	C$_{20}$H$_{12}$S (284)	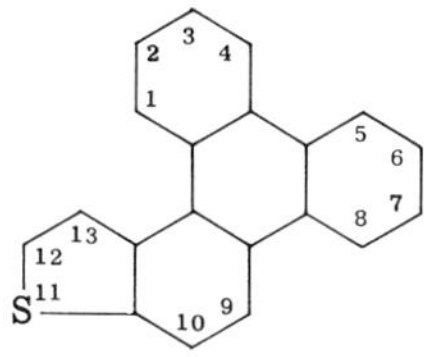
Triphenyleno[2,3-*b*]thiophene	C$_{20}$H$_{12}$S (284)	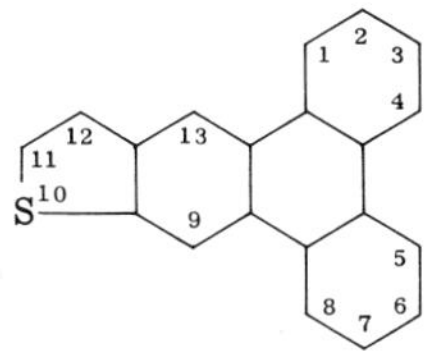
Triphenyleno[2,3-*c*]thiophene	C$_{20}$H$_{12}$S (284)	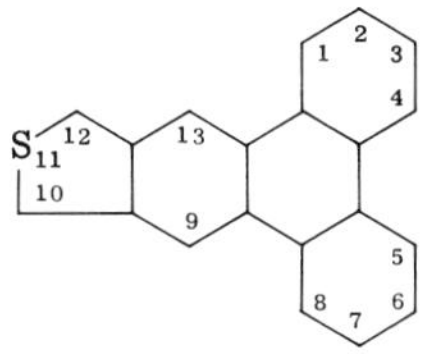

4. Polycyclic Aromatic Heterocycles Containing One Oxygen Atom: Names, Formulas, Structures, and Numbering

Name	Molecular formula (weight)	Structure
Benzofuran	C$_8$H$_6$O (118)	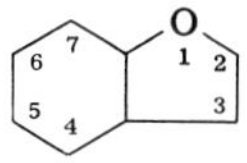
Dibenzofuran	C$_{12}$H$_8$O (168)	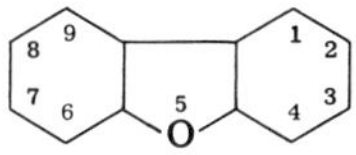
Phenanthro[4,5-*bcd*]furan	C$_{14}$H$_8$O (192)	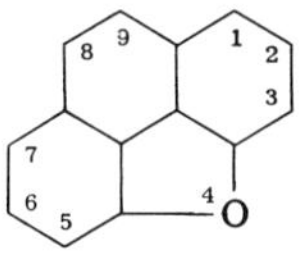

(*Continued*)

Appendix 4 (*Continued*)

Name	Molecular formula (weight)	Structure
Benzo[*b*]naphtho[1,2-*d*]furan	$C_{16}H_{10}O$ (208)	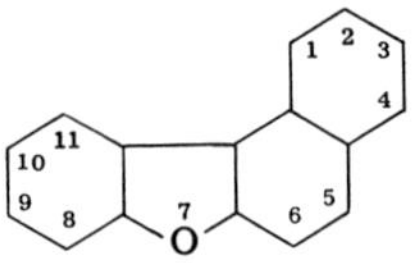
Benzo[*b*]naphtho[2,1-*d*]furan	$C_{16}H_{10}O$ (208)	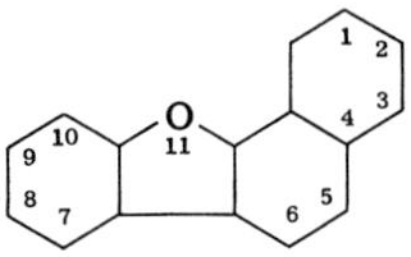
Benzo[*b*]naphtho[2,3-*d*]furan	$C_{16}H_{10}O$ (208)	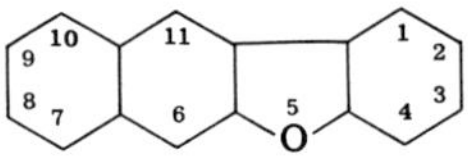
Anthra[9,1,2-*bcd*]benzofuran	$C_{18}H_{10}O$ (242)	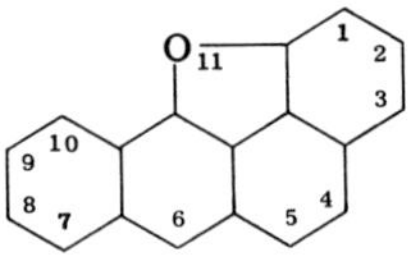
Chryseno[4,5-*bcd*]furan	$C_{18}H_{10}O$ (242)	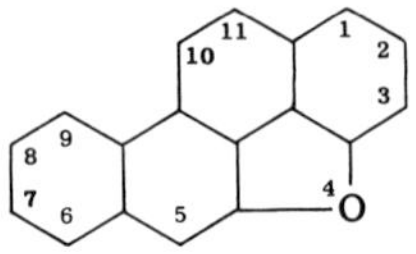
Triphenyleno[4,5-*bcd*]furan	$C_{18}H_{10}O$ (242)	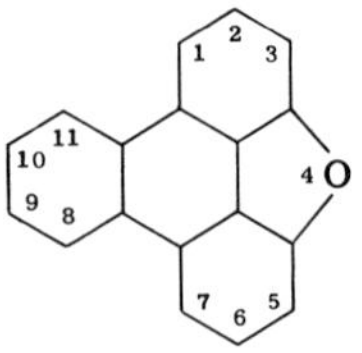
Anthra[1,2-*b*]benzofuran	$C_{20}H_{12}O$ (268)	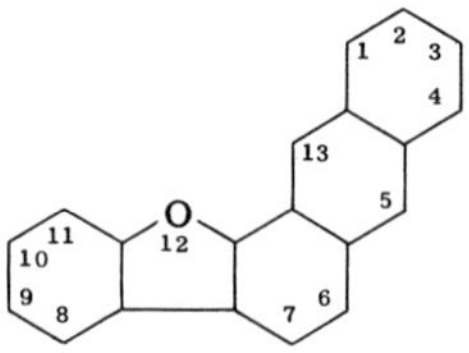
Anthra[2,1-*b*]benzofuran	$C_{20}H_{12}O$ (268)	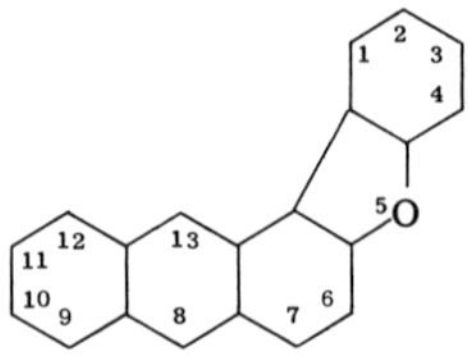

(*Continued*)

Appendix 4 (*Continued*)

Name	Molecular formula (weight)	Structure
Anthra[2,3-*b*]benzofuran	$C_{20}H_{12}O$ (268)	
Benzo[*b*]phenanthro[1,2-*d*]furan	$C_{20}H_{12}O$ (268)	
Benzo[*b*]phenanthro[2,1-*d*]furan	$C_{20}H_{12}O$ (268)	
Benzo[*b*]phenanthro[2,3-*d*]furan	$C_{20}H_{12}O$ (268)	
Benzo[*b*]phenanthro[3,2-*d*]furan	$C_{20}H_{12}O$ (268)	
Benzo[*b*]phenanthro[3,4-*d*]furan	$C_{20}H_{12}O$ (268)	
Benzo[*b*]phenanthro[4,3-*d*]furan	$C_{20}H_{12}O$ (268)	
Benzo[*b*]phenanthro[9,10-*d*]furan	$C_{20}H_{12}O$ (268)	

(*Continued*)

Appendix 4 (*Continued*)

Name	Molecular formula (weight)	Structure
Dinaphtho[1,2-*b*:1′,2′-*d*]furan	$C_{20}H_{12}O$ (268)	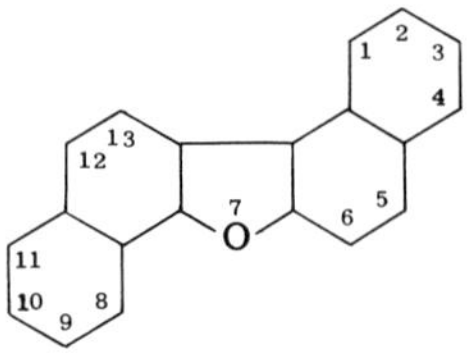
Dinaphtho[1,2-*b*:2′,1′-*d*]furan	$C_{20}H_{12}O$ (268)	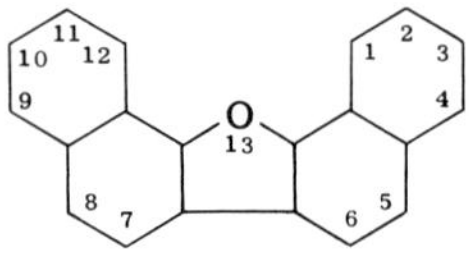
Dinaphtho[1,2-*b*:2′,3′-*d*]furan	$C_{20}H_{12}O$ (268)	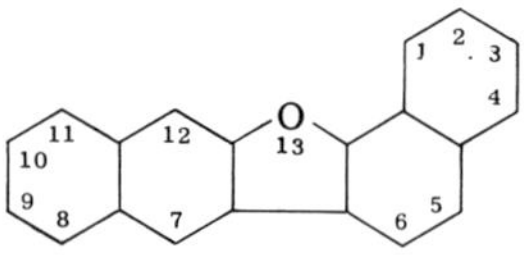
Dinaphtho[2,1-*b*:1′,2′-*d*]furan	$C_{20}H_{12}O$ (268)	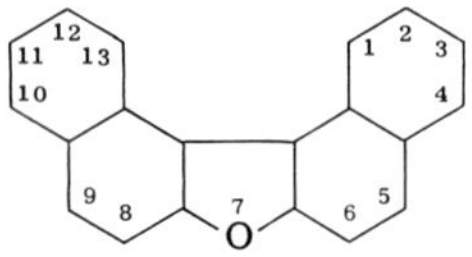
Dinaphtho[2,1-*b*:2′,3′-*d*]furan	$C_{20}H_{12}O$ (268)	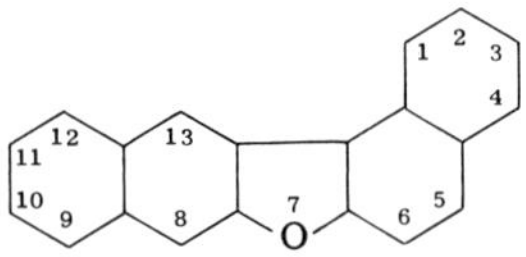
Dinaphtho[2,3-*b*:2′,3′-*d*]furan	$C_{20}H_{12}O$ (268)	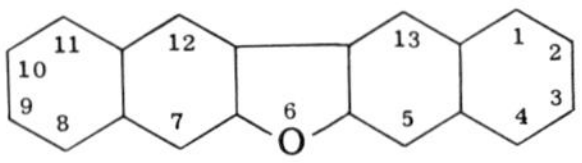
Pyreno[4,3,2-*bcd*]benzofuran	$C_{20}H_{10}O$ (266)	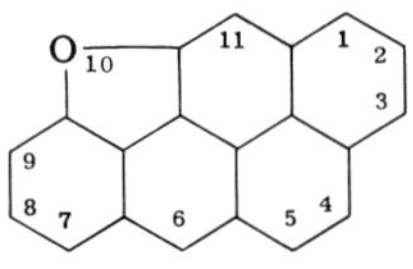
Perylo[1,12-*bcd*]furan	$C_{20}H_{10}O$ (266)	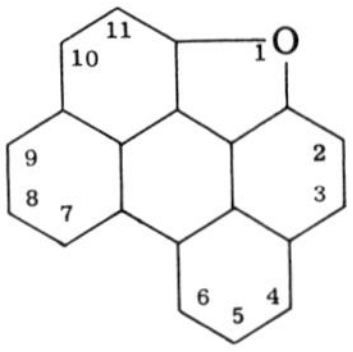

5. Polycyclic Aromatic Compounds that Have Been Tested for Carcinogenic Activity[a,b]

Compounds	Carcinogenic activity
Polycyclic aromatic hydrocarbons	
Naphthalene	0
2-Methylnaphthalene	0
Acenaphthylene	0
Acenaphthene	0
Fluorene	0
Anthracene	0
2-Methylanthracene	0
9-Methylanthracene	0
1,2-Dimethylanthracene	0
1,3-Dimethylanthracene	0
1,4-Dimethylanthracene	0
2,3-Dimethylanthracene	0
9,10-Dimethylanthracene	0/+
9-Ethylanthracene	0
2,9,10-Trimethylanthracene	0
2,3,6,7-Tetramethylanthracene	0
Phenanthrene	0
1,9-Dimethylphenanthrene	0
1,2,4-Trimethylphenanthrene	+
1,2,3,4-Tetramethylphenanthrene	0/+
1-Methyl-3-isopropylphenanthrene	0/+
1-Methyl-7-isopropylphenanthrene	0
Fluoranthene	0
2-Methylfluoranthene	+
3-Methylfluoranthene	+
Pyrene	0
1-Methylpyrene	0
2-Methylpyrene	0
Aceanthrylene	0
Benzo[a]fluorene	0
Benzo[c]fluorene	0
Benzo[ghi]fluoranthene	0
Cyclopenta[cd]pyrene	+
Benz[a]anthracene	+
1-Methylbenz[a]anthracene	0/+
2-Methylbenz[a]anthracene	0
3-Methylbenz[a]anthracene	0/+
4-Methylbenz[a]anthracene	0/+
5-Methylbenz[a]anthracene	0/+
6-Methylbenz[a]anthracene	+ +
7-Methylbenz[a]anthracene	+ +

(Continued)

Compounds	Carcinogenic activity
8-Methylbenz[a]anthracene	+ +
9-Methylbenz[a]anthracene	+ +
10-Methylbenz[a]anthracene	0/ +
11-Methylbenz[a]anthracene	0/ +
12-Methylbenz[a]anthracene	+ +
1,7-Dimethylbenz[a]anthracene	0
1,12-Dimethylbenz[a]anthracene	0
2,9-Dimethylbenz[a]anthracene	0
2,10-Dimethylbenz[a]anthracene	0
3,9-Dimethylbenz[a]anthracene	0
3,10-Dimethylbenz[a]anthracene	0
4,5-Dimethylbenz[a]anthracene	+ +
4,7-Dimethylbenz[a]anthracene	0
4,12-Dimethylbenz[a]anthracene	0
5,12-Dimethylbenz[a]anthracene	0
6,7-Dimethylbenz[a]anthracene	+ +
6,8-Dimethylbenz[a]anthracene	+ +
6,12-Dimethylbenz[a]anthracene	+ +
7,8-Dimethylbenz[a]anthracene	+ +
7,11-Dimethylbenz[a]anthracene	+ +
7,12-Dimethylbenz[a]anthracene	+ +
8,9-Dimethylbenz[a]anthracene	+ +
8,11-Dimethylbenz[a]anthracene	0
8,12-Dimethylbenz[a]anthracene	+ +
9,10-Dimethylbenz[a]anthracene	+
9,11-Dimethylbenz[a]anthracene	+
7-Ethylbenz[a]anthracene	+
8-Ethylbenz[a]anthracene	+
11-Ethylbenz[a]anthracene	0
12-Ethylbenz[a]anthracene	+
2,3,7-Trimethylbenz[a]anthracene	0
2,7,12-Trimethylbenz[a]anthracene	0
3,7,12-Trimethylbenz[a]anthracene	0
4,5,10-Trimethylbenz[a]anthracene	+ +
4,7,12-Trimethylbenz[a]anthracene	+ +
5,7,12-Trimethylbenz[a]anthracene	+
6,7,8-Trimethylbenz[a]anthracene	+ +
6,7,12-Trimethylbenz[a]anthracene	+ +
6,8,12-Trimethylbenz[a]anthracene	+ +
7,8,12-Trimethylbenz[a]anthracene	+ +
7,9,12-Trimethylbenz[a]anthracene	+ +
7,10,12-Trimethylbenz[a]anthracene	+ +
7-*n*-Propylbenz[a]anthracene	0
8-*n*-Propylbenz[a]anthracene	+

(*Continued*)

Appendix 5 (*Continued*)

Compounds	Carcinogenic activity
5-Isopropylbenz[*a*]anthracene	0
7-Isopropylbenz[*a*]anthracene	0
8-Isopropylbenz[*a*]anthracene	+
9-Isopropylbenz[*a*]anthracene	+
10-Isopropylbenz[*a*]anthracene	0
11-Isopropylbenz[*a*]anthracene	0
7-Ethyl-12-methylbenz[*a*]anthracene	+ +
8-Ethyl-7-methylbenz[*a*]anthracene	0
12-Ethyl-7-methylbenz[*a*]anthracene	+
7,8,9,12-Tetramethylbenz[*a*]anthracene	+
7,9,10,12-Tetramethylbenz[*a*]anthracene	+
7-*n*-Butylbenz[*a*]anthracene	0
8-*n*-Butylbenz[*a*]anthracene	0/+
7-Methyl-8-*n*-propylbenz[*a*]anthracene	+
12-Methyl-7-*n*-propylbenz[*a*]anthracene	+
6,8-Diethylbenz[*a*]anthracene	+ +
7,8-Diethylbenz[*a*]anthracene	0
7,9-Diethylbenz[*a*]anthracene	0
7,12-Diethylbenz[*a*]anthracene	0/+
8,12-Diethylbenz[*a*]anthracene	+
7,12-Dimethyl-8-ethylbenz[*a*]anthracene	0
7-*n*-Pentylbenz[*a*]anthracene	0
8-*n*-Pentylbenz[*a*]anthracene	0/+
7-*n*-Butyl-12-methylbenz[*a*]anthracene	0
7,12-Dimethyl-8-*n*-propylbenz[*a*]anthracene	+
8-*n*-Hexylbenz[*a*]anthracene	0/+
7,12-Di-*n*-propylbenz[*a*]anthracene	0
8-*n*-Heptylbenz[*a*]anthracene	0/+
Benzo[*c*]phenanthrene	+
2-Methylbenzo[*c*]phenanthrene	+
3-Methylbenzo[*c*]phenanthrene	0/+
4-Methylbenzo[*c*]phenanthrene	0/+
5-Methylbenzo[*c*]phenanthrene	+ +
6-Methylbenzo[*c*]phenanthrene	+ +
2,3-Dimethylbenzo[*c*]phenanthrene	0
5,8-Dimethylbenzo[*c*]phenanthrene	0
5-Ethylbenzo[*c*]phenanthrene	+
5-*n*-Propylbenzo[*c*]phenanthrene	+ +
5-Isopropylbenzo[*c*]phenanthrene	+ +
6-Isopropylbenzo[*c*]phenanthrene	0
5,8-Diethylbenzo[*c*]phenanthrene	0
Chrysene	+
1-Methylchrysene	0/+
2-Methylchrysene	+

(*Continued*)

Appendix 5 (*Continued*)

Compounds	Carcinogenic activity
3-Methylchrysene	+
4-Methylchrysene	+
5-Methylchrysene	+ +
6-Methylchrysene	+
2,3-Dimethylchrysene	0
4,5-Dimethylchrysene	+
5,6-Dimethylchrysene	+
2-Isopropylchrysene	0
Naphthacene	0
2-Isopropylnaphthacene	0
Triphenylene	0
1-Methyltriphenylene	0
1,4-Dimethyltriphenylene	0
11H-Benz[*bc*]aceanthrylene	0
6-Methyl-11H-benz[*bc*]aceanthrylene	+
4H-Cyclopenta[*def*]chrysene	+
Benzo[*b*]fluoranthene	+ +
Benzo[*j*]fluoranthene	+ +
Benzo[*k*]fluoranthene	+ +
Benzo[*a*]pyrene	+ +
2-Methylbenzo[*a*]pyrene	+ +
3-Methylbenzo[*a*]pyrene	+ +
4-Methylbenzo[*a*]pyrene	+ +
5-Methylbenzo[*a*]pyrene	+
6-Methylbenzo[*a*]pyrene	+ +
7-Methylbenzo[*a*]pyrene	+
8-Methylbenzo[*a*]pyrene	0
11-Methylbenzo[*a*]pyrene	+ +
12-Methylbenzo[*a*]pyrene	+ +
1,2-Dimethylbenzo[*a*]pyrene	+ +
1,3-Dimethylbenzo[*a*]pyrene	+ +
1,4-Dimethylbenzo[*a*]pyrene	+ +
1,6-Dimethylbenzo[*a*]pyrene	+ +
2,3-Dimethylbenzo[*a*]pyrene	+ +
3,6-Dimethylbenzo[*a*]pyrene	+ +
3,12-Dimethylbenzo[*a*]pyrene	+ +
4,5-Dimethylbenzo[*a*]pyrene	+ +
1,3,6-Trimethylbenzo[*a*]pyrene	+
Benzo[*e*]pyrene	0/+
Perylene	0
3-Methylperylene	0
Cholanthrene (3-Methylbenz[*j*]aceanthrylene)	+ +
3-Methylcholanthrene	+ +
4-Methylcholanthrene	+ +

(*Continued*)

Appendix 5 (*Continued*)

Compounds	Carcinogenic activity
5-Methylcholanthrene	+ +
1,3-Dimethylcholanthrene	+
2,3-Dimethylcholanthrene	+ +
3-Ethylcholanthrene	+ +
3-Isopropylcholanthrene	+
Dibenzo[*a,c*]fluorene	0/ +
Dibenzo[*a,g*]fluorene	0/ +
3-Methyldibenzo[*a,g*]fluorene	0
Dibenzo[*a,i*]fluorene	0/ +
Dibenzo[*c,g*]fluorene	0
13*H*-Indeno[1,2-*b*]anthracene	0
12*H*-Indeno[1,2-*b*]phenanthrene	0
12-Methyl-13*H*-indeno[1,2-*c*]phenanthrene	0
Benzo[*ghi*]perylene	+
Dibenzo[*def,mno*]chrysene	0
6-Methyldibenzo[*def,mno*]chrysene	+
6,12-Dimethyldibenzo[*def,mno*]chrysene	+
Indeno[1,2,3-*cd*]pyrene	+
Benzo[*b*]chrysene	0
7,12-Dimethylbenzo[*b*]chrysene	0
Benzo[*c*]chrysene	+
Benzo[*g*]chrysene	+
9-Methylbenzo[*g*]chrysene	0
10-Methylbenzo[*g*]chrysene	0
Benzo[*a*]naphthacene	0
Dibenz[*a,c*]anthracene	+
10-Methyldibenz[*a,c*]anthracene	+
9,14-Dimethyldibenz[*a,c*]anthracene	0
Dibenz[*a,h*]anthracene	+
2-Methyldibenz[*a,h*]anthracene	+
3-Methyldibenz[*a,h*]anthracene	+
6-Methyldibenz[*a,h*]anthracene	+
7-Methyldibenz[*a,h*]anthracene	+ +
7,14-Dimethyldibenz[*a,h*]anthracene	+
7,14-Di-*n*-butyldibenz[*a,h*]anthracene	+
Dibenz[*a,j*]anthracene	+
7,14-Dimethyldibenz[*a,j*]anthracene	+
Dibenzo[*b,g*]phenanthrene	0
Dibenzo[*c,g*]phenanthrene	0
Pentacene	0
Pentaphene	0
2,11-Dimethylpentaphene	0
Picene	0
2,9-Dimethylpicene	0

(*Continued*)

Appendix 5 (*Continued*)

Compounds	Carcinogenic activity
Coronene	0/+
Benzo[*rst*]pentaphene	+ +
5-Methylbenzo[*rst*]pentaphene	+
5,8-Dimethylbenzo[*rst*]pentaphene	0
7-Methylbenzo[*pqr*]picene	0/+
Dibenzo[*b,def*]chrysene	+ +
7-Methyldibenzo[*b,def*]chrysene	+
7,14-Dimethyldibenzo[*b,def*]chrysene	0
Dibenzo[*def,p*]chrysene	+ +
10-Methyldibenzo[*def,p*]chrysene	+ +
Dibenzo[*a,e*]fluoranthene	+
Dibenzo[*b,k*]fluoranthene	0
Dibenzo[*de,qr*]naphthacene	0
Dibenzo[*fg,op*]naphthacene	0
Naphtho[1,2,3,4-*def*]chrysene	+ +
5-Methylnaphtho[1,2,3,4-*def*]chrysene	+ +
6-Methylnaphtho[1,2,3,4-*def*]chrysene	+
Naphtho[2,1,8-*qra*]naphthacene	+
Dibenzo[*cd,lm*]perylene	+
Anthra[1,2-*a*]anthracene	0
Benzo[*b*]pentaphene	0
Benzo[*c*]pentaphene	0
Dibenzo[*b,k*]chrysene	0
Dibenzo[*a,c*]naphthacene	+
Dibenzo[*a,j*]naphthacene	0
Naphtho[1,2-*b*]triphenylene	0
Benzo[*a*]coronene	0
Benzo[*a*]naphtho[8,1,2-*cde*]naphthacene	+
Benzo[*e*]naphtho[8,1,2-*cde*]naphthacene	0
Dibenzo[*h,rst*]pentaphene	+
Tribenzo[*a,e,h*]pyrene	+
Pyranthrene	0
Benzo[*b*]naphtho[1,2-*k*]chrysene	0
Dibenzo[*a,l*]pentacene	0
Phenanthro[9,10-*b*]triphenylene	0
Tribenzo[*a,c,j*]naphthacene	+
Diindeno[1,2,3-*cd*:1′,2′,3′-*lm*]perylene	0
Dinaphtho[1,2-*b*:1,2-*k*]chrysene	0
Dinaphtho[1,2-*b*:2,1-*n*]perylene	0
Benzo[*a*]pyreno[2,3-*c*]fluorene	0
Nitrogen heterocycles	
Indole	0/+
3-Methylindole	0
Isoquinoline	0

(*Continued*)

Appendix 5 (*Continued*)

Compounds	Carcinogenic activity
Quinoline	+
2-Methylquinoline	+
3-Methylquinoline	+
4-Methylquinoline	+
5-Methylquinoline	+
6-Methylquinoline	+
7-Methylquinoline	+
8-Methylquinoline	+
Carbazole	0
Acridine	0
9-Methylacridine	0
2,7,9-Trimethylacridine	0
Benzo[*h*]quinoline	0
2,3,4-Trimethylbenzo[*h*]quinoline	0
Benzo[*f*]quinoline	0
Phenanthridine	+
11*H*-Benzo[*a*]carbazole	+
11-Methylbenzo[*a*]carbazole	0
7,10-Dimethylbenzo[*a*]carbazole	0
10-Methyl-7*H*-Benzo[*c*]carbazole	0/+
Benz[*a*]acridine	+
9-Methylbenz[*a*]acridine	0
10-Methylbenz[*a*]acridine	0
12-Methylbenz[*a*]acridine	0
8,9-Dimethylbenz[*a*]acridine	0
8,12-Dimethylbenz[*a*]acridine	+
9,12-Dimethylbenz[*a*]acridine	+
2,9,12-Trimethylbenz[*a*]acridine	0
3,8,12-Trimethylbenz[*a*]acridine	+
8,9,12-Trimethylbenz[*a*]acridine	0
8,10,12-Trimethylbenz[*a*]acridine	+
8,11,12-Trimethylbenz[*a*]acridine	0
9,10,12-Trimethylbenz[*a*]acridine	+
9,11,12-Trimethylbenz[*a*]acridine	0
2,8,10,12-Tetramethylbenz[*a*]acridine	0
3,8,10,12-Tetramethylbenz[*a*]acridine	0
10-Methyl-12-isopropylbenz[*a*]acridine	0
2,9-Dimethyl-1-ethylbenz[*a*]acridine	0
2,3,8,10,11-Pentamethylbenz[*a*]acridine	0
2,3,9,10,12-Pentamethylbenz[*a*]acridine	0
8,9-Dimethyl-12-isopropylbenz[*a*]acridine	0
10,11-Dimethyl-12-isopropylbenz[*a*]acridine	0
2,8,12-Trimethyl-11-ethylbenz[*a*]acridine	0
2,3,8,10,11,12-Hexamethylbenz[*a*]acridine	0

(*Continued*)

Appendix 5 (*Continued*)

Compounds	Carcinogenic activity
Benz[c]acridine	+
7-Methylbenz[c]acridine	+ +
8-Methylbenz[c]acridine	0
9-Methylbenz[c]acridine	0
10-Methylbenz[c]acridine	0
5,7-Dimethylbenz[c]acridine	0
7,9-Dimethylbenz[c]acridine	+
7,10-Dimethylbenz[c]acridine	+
7,11-Dimethylbenz[c]acridine	+
10,11-Dimethylbenz[c]acridine	0
5,7,11-Trimethylbenz[c]acridine	0/+
7,8,11-Trimethylbenz[c]acridine	0/+
7,9,10-Trimethylbenz[c]acridine	0/+
7,9,11-Trimethylbenz[c]acridine	+
7-*n*-Propylbenz[c]acridine	0
7-Methyl-1-ethylbenz[c]acridine	0/+
7-Ethyl-9-methylbenz[c]acridine	+
7-Ethyl-11-methylbenz[c]acridine	0
7,8,9,11-Tetramethylbenz[c]acridine	+ +
Naphtho[1,2-*f*]quinoline	0
Naphtho[1,2-*g*]quinoline	0
Naphtho[2,1-*f*]quinoline	0
Naphtho[2,3-*f*]quinoline	0/+
Fluoreno[9,9a,1-*gh*]quinoline	+
Phenaleno[1,9-*gh*]quinoline	+
7*H*-Dibenzo[a,g]carbazole	+
9-Methyl-7*H*-dibenzo[a,g]carbazole	0
8,11-Dimethyl-7*H*-dibenzo[a,g]carbazole	0
9,10-Dimethyl-7*H*-dibenzo[a,g]carbazole	0
13*H*-Dibenzo[a,i]carbazole	+
2,3-Dimethyl-13*H*-dibenzo[a,i]carbazole	0
7*H*-Dibenzo[c,g]carbazole	+ +
7-Methyldibenzo[c,g]carbazole	+ +
7-Ethyldibenzo[c,g]carbazole	+ +
Benzo[h]naphtho[1,2-*f*]quinoline	+
Dibenz[a,h]acridine	+
3-Methyldibenz[a,h]acridine	0
Dibenz[a,j]acridine	+
14-Methyldibenz[a,j]acridine	+ +
Dibenz[c,h]acridine	+
Sulfur heterocycles	
Benzo[b]naphtho[1,2-*d*]thiophene	0
6,11-Dimethylbenzo[b]naphtho[2,3-*d*]thiophene	+
Dinaphtho[2,1-*b*:1′,2′-*d*]thiophene	0

(*Continued*)

Appendix 5 (*Continued*)

Compounds	Carcinogenic activity
7,13-Dimethylbenzo[*b*]phenanthro[2,3-*d*]thiophene	+ +
7,13-Dimethylbenzo[*b*]phenanthro[3,2-*d*]thiophene	+ +
Oxygen heterocycles	
Dibenzofuran	0
Dinaphtho[2,1-*b*:2′,3′-*d*]furan	0/+
Dinaphtho[2,1-*b*:1′,2′-*d*]furan	0

[a] Attempts at standardizing carcinogenicity indices among the polycyclic aromatic hydrocarbons have been largely ineffective. Several numerical scales of carcinogenicity (*1–3*) have been proposed and are based on the statistics of cancer induced in small animals, as detected by pathological assay. It has been found that these indices are only useful when standard procedures have been applied to large numbers of standard animals (usually specific strains of mice) because of the large number of variables (including animal species, age, strain, sex, hair cycle, mitotic cycle, diet, general toxicity, and site of administration) which all affect the response. Nevertheless, it is useful to have a comprehensive list of polycyclic aromatic compounds that have been tested for carcinogenic activity. Therefore, in this table, a listing is presented of these compounds with their activities based on the percentage of treated animals which developed tumors, i.e., none, noncarcinogenic (0); up to 33%, weakly carcinogenic (+); and above 33%, strongly carcinogenic (+ +). The compounds listed have not necessarily been tested under strictly comparable conditions, and the relative activities given must be regarded as crude approximations. The information in the table was compiled from a number of sources (*4–9*) and contains data only on parent and alkyl-substituted polycyclic aromatic compounds. Data on various other derivatives can be obtained by consulting the references previously mentioned.

[b] References: *1*. J. Iball, *Am. J. Cancer* **35,** 188 (1939). *2*. I. Berenblum, *Cancer Res.* **5,** 561 (1945). *3*. G. M. Badger, *Br. J. Cancer* **2,** 309 (1948). *4*. J. L. Hartwell, "Survey of Compounds Which Have Been Tested for Carcinogenic Activity," U. S. Public Health Service, Publ. 149, Washington, D. C., 1951. *5*. P. Shubik and J. L. Hartwell, "Survey of Compounds Which Have Been Tested for Carcinogenic Activity," Suppl. 1, U. S. Public Health Service, Publ. 149, Washington, D. C., 1957. *6*. P. Shubik and J. L. Hartwell, "Survey of Compounds Which Have Been Tested for Carcinogenic Activity," Suppl. 2, U. S. Public Health Service, Publ. 149, Washington, D. C., 1969. *7*. J. I. Thompson and Company, "Survey of Compounds Which Have Been Tested for Carcinogenic Activity," 1968–1969 Volume, U. S. Public Health Service, Publ. 149, Washington, D. C., 1968–1969. *8*. D. W. Jones and R. S. Matthews, *Progr. Med. Chem.* **10,** 159 (1974). *9*. A. Dipple, *in* "Chemical Carcinogens" (C. E. Searle, ed.), p. 245. ACS Monogr. 173, Am. Chem. Soc., Washington, D.C., 1976.

Index

A

Abietic acid, conversion of to retene, 18–19

Acenaphthene, electron-impact mass spectrum of, 248

Acetonitrile, fluorescence spectra in, 310

Acetylated paper, in paper chromatography of PAH, 134

Acid–base interactions, in HPLC, 162–163

Acridine, mass spectra for, 246, 249

Adsorption gel chromatography, PAH elution in, 144; *see also* Column adsorption chromatography; Gel permeation chromatography

Air, PAC distribution in, 27–32

Air and combustion effluents, in PAC collection and extraction, 79–87

Airborne *N*-heterocycles, 31–32

Airborne PAH, 28–30, *see also* Polycyclic aromatic hydrocarbons

Airborne *S*-heterocycles, 33

Air particulates

capillary gas chromatogram of, 205

PAH recovery from, 88–89

Alkylated PAH isomers, capillary gas chromatogram of, 213

Alkylbenzenes

in HPLC, 160

in reversed-phase chromatography, 171

Alkyl homolog plots, 263–265

Alkyl group positions, chromatographic mobility of, 166

Alkyl side-chain carbon number, for alkylbenzenes and naphthalenes, 171

Allium cepa, 60

Alumina

in column adsorption chromatography, 124–125

in HPLC, 159–166

retention of isomeric alkylated aromatics on, 167

Alumina chromatography, standard fractions of, 111; *see also* Gas chromatography

Alumina-packed microcapillary section, 184

Amberlite XAD-2, in PAH concentration, 84, 90

Aminco-Bowman spectrofluorimeter, 232

Aminosilane-bonded silica, chromatogram of PAH on, 174

Amyrins, 23

Analytical chemistry, physical and chemical properties, 1–14

Annelation, defined, 5–6

Anthracene

addition and redox reactions of, 7

argon–methane chemical ionization mass spectrum of, 171

fluorescence spectra of, 234

ion kinetic energy spectrum of, 280–281

methane chemical ionization mass spectrum of, 269

negative-ion spectrum of, 276

photooxidation products of, 11, 13

selective separation of, 198

structural configuration of, 131–132

Apiezons, in liquid stationary phases, 191

Aromatic hydrocarbons, *see* Polycyclic aromatic hydrocarbons

Aromatic molecules, as UV light absorbers, 177

Arylhydrocarbon hydrolase activity, toxic effects and, 69

Atmosphere, PAH distribution in, 27–32

2-Azaanthracene, 3

Azaarenes

retention of on nematic crystal column, 199

separation of on silica, 1–2, 164

Azaaromatics, elution order of, 164

451

O

Octadecylsilane reversed-phase packing materials, 172
2-n-Octyltriphenylene, 148
OID, see Optoelectronic image detectors
Oils, heavy gas, 256, see also Petroleum (adj.)
Oils and fats, PAH extraction for, 91
Oil types, fluorescence emission spectra for, 303–304
Optoelectronic image detectors, 333
Oxyacetylene flame, PAC in, 22
Oxygen, as reagent in chemical ionization mass spectrometry, 272

P

PAC, see Polycyclic aromatic compounds
PAH, see Polycyclic aromatic hydrocarbons
PAH retention indices, 214–218
Paper chromatography, 133–135
Parallel optoelectronic image detectors, in luminescence analysis, 333
Partition coefficients, for PAH between solvent pairs, 94–95, 98
Partition liquid chromatography, limited miscibility in, 175
Pellicular packings, in HPLC, 156
Pentasil, 193
Perylene
changes in, 11
for erythroaphin pigments, 19
formation of, 22
Perylo[1,12-bcd]thiophene ion, 246
Petroleum base fractions, separation of, 113–114; see also Oil types
Petroleum heterocycles, integrated separation scheme for analysis of, 112–113
Petroleum hydrocarbons, structural range of, 43
Petroleum mixture, high-resolution low-voltage spectrum of, 261
Petroleum sulfur heterocycles, structural range of, 43
Phenanthrene
addition and redox reactions of, 7
argon-methane chemical ionization mass spectrum for, 271
electron-impact mass spectrum for, 245

K region of, 56
methane chemical ionization mass spectrum for, 269
selective separation of, 198
structural configuration of, 131–132
Phenanthrene–anthracene
capillary gas chromatography for, 204
gas chromatography resolution of, 193
Phenanthridine
electron-impact mass spectrum of, 249
mass spectrum of, 246
Phenylacetylene, in oxyacetylene flame, 22
Phosphorescence spectroscopy, 304–308
at room temperature, 307–308, 330
Photoionization techniques, in mass spectrometry, 285–286
Phthalimidopropylsilane-bonded packing, in HPLC, 174
Phytosterols, 23
π-antibonding orbitals, in PAC, 4
π-electron delocalization, in carcinogen molecules, 56
Picric acid, in PAH separation, 103
PI techniques, see Photoionization techniques
PMD, see Programmed multiple development
Polar columns, in high-performance liquid chromatography, 173–175
Polluted air, PAC in, 28–30
Polluted water, PAC in, 352–353
Polyaminoundecanoic acid, in thin-layer chromatography, 139
Polycarborane–siloxane polymers, 194
Polycyclic aromatic compounds, see also Polycyclic aromatic hydrocarbons
absorption of in particulate matter, 79
in air or atmosphere, 27–33
analytical schemes for, 105–110
"bay-region" theory in, 57
binding of to cellular and subcellular units, 60–61
biological testing of, 53–55
carcinogenic activity of, 44, 51–53, 62, 441–449
chemical properties of, 11–14
collection and extraction of, 79–93
cryostat head design for matrix isolation of, 318
distribution of, 26–44
as donors in charge-transfer molecular π-complexes, 7